Fuzzy Mathematics

Fuzzy Mathematics

Special Issue Editors

Etienne E. Kerre

John Mordeson

MDPI • Basel • Beijing • Wuhan • Barcelona • Belgrade

Special Issue Editors
Etienne E. Kerre
Ghent University
Belgium

John Mordeson
Creighton University
USA

Editorial Office
MDPI
St. Alban-Anlage 66
Basel, Switzerland

This is a reprint of articles from the Special Issue published online in the open access journal *Mathematics* (ISSN 2227-7390) from 2017 to 2018 (available at: https://www.mdpi.com/journal/mathematics/special_issues/Fuzzy_Mathematics)

For citation purposes, cite each article independently as indicated on the article page online and as indicated below:

LastName, A.A.; LastName, B.B.; LastName, C.C. Article Title. *Journal Name* **Year**, *Article Number*, Page Range.

ISBN 978-3-03897-322-5 (Pbk)
ISBN 978-3-03897-323-2 (PDF)

Contents

About the Special Issue Editors

Etienne E. Kerre was born in Zele, Belgium on 8 May 1945. He obtained his M. Sc. degree in Mathematics in 1967 and his Ph.D. in Mathematics in 1970 from Ghent University. He has been a lector since 1984, and has been a full professor at Ghent University since 1991. In 2010, he became a retired professor. He is a referee for more than 80 international scientific journals, and also a member of the editorial boards of many international journals and conferences on fuzzy set theory. He has been an honorary chairman at various international conferences. In 1976, he founded the Fuzziness and Uncertainty Modeling Research Unit (FUM). Since then, his research has been focused on the modeling of fuzziness and uncertainty, and has resulted in a great number of contributions in fuzzy set theory and its various generalizations. Especially the theories of fuzzy relational calculus and of fuzzy mathematical structures owe a very great deal to him. Over the years he has also been a promotor of 30 Ph.D.s on fuzzy set theory. His current research interests include fuzzy and intuitionistic fuzzy relations, fuzzy topology, and fuzzy image processing. He has authored or co-authored 25 books and more than 500 papers appearing in international refereed journals and proceedings.

John Mordeson is Professor Emeritus of Mathematics at Creighton University. He received his B.S., M.S., and Ph. D. from Iowa State University. He is a Member of Phi Kappa Phi. He is President of the Society for Mathematics of Uncertainty. He has published 15 books and over two hundred journal articles. He is on the editorial boards of numerous journals. He has served as an external examiner of Ph.D. candidates from India, South Africa, Bulgaria, and Pakistan. He has refereed for numerous journals and granting agencies. He is particularly interested in applying mathematics of uncertainty to combat the problems of human trafficking, modern slavery, and illegal immigration.

Preface to "Fuzzy Mathematics"

This Special Issue on fuzzy mathematics is dedicated to Lotfi A. Zadeh. In his 1965 seminal paper entitled "Fuzzy Sets", Zadeh extended Cantor's binary set theory to a gradual model by introducing degrees of belonging and relationship. Very soon, the extension was applied to almost all domains of contemporary mathematics, giving birth to new disciplines such as fuzzy topology, fuzzy arithmetic, fuzzy algebraic structures, fuzzy differential calculus, fuzzy geometry, fuzzy relational calculus, fuzzy databases, and fuzzy decision making. In the beginning, mostly direct fuzztfications of the classical mathematical domains were launched by simply changing Cantor's set-theoretic operations by Zadeh's max-min extensions. The 1980s were characterized by an extension of the possible fuzzifications due to the discovery of triangular norms and conorms. Starting in the 1990s, more profound analysis was performed by studying the axiomatization of fuzzy structures and searching for links between the different models to represent imprecise and uncertain information. It was our aim to have this Special Issue comprise a healthy mix of excellent state-of-the-art papers as well as brand-new material that can serve as a starting point for newcomers in the field to further develop this wonderful domain of fuzzy mathematics.

This Special Issue starts with a corner-stone paper that should be read by all working in the field of fuzzy mathematics. Using lattice isomorphisms, it shows that the results of many of the variations and extensions of fuzzy set theory can be obtained immediately from the results of set theory itself. The paper is extremely valuable to reviewers in the field. This paper is followed by one which gives the definition of the most extended modal operator of the first type over interval-valued sets, and presents some of its basic properties. A new function called a negative-valued function is presented and applied to various structures. Results concerning N-hyper sets and hypergraphs in m-polar fuzzy setting are also presented.

In the next paper, it is shown that the lower approximation of a subset of an LA-semigroup need not be an LA-subsemigroup/ideal of an LA-semigroup under a set valued homomorphism. A generalization of a bipolar fuzzy subsemigroup is given, and any regular semigroup is characterized in terms of generalized BF-semigroups. The T1-spaces induced by a fuzzy semi-metric space endowed with the special kind of triangle inequality are investigated. The limits in fuzzy semi-metric spaces are also studied. The consistency of limit concepts in the induced topologies is shown. Nilpotent fuzzy subgroups and neutrosophic triplet G-modules are also studied. Next we have a series of papers on BCK/BCI algebras. The notions of hyper fuzzy sets in BCK/BCI-algebras are introduced, and characterizations of hyper fuzzy ideals are established. The length-fuzzy set is introduced and applied to BCK/BCI algebras. Neutrosophic permeable values and energetic subsets with applications to BCK/BCI algebras are presented.

The following three papers concern decision making issue. An information measure for measuring the degree of fuzziness in an intuitionistic fuzzy set is introduced. An illustrative example related to a linguistic variable is given to illustrate it. The notion of occurring probabilistic values into hesitant fuzzy linguistic elements is introduced and studied.

The Special Issue ends with some applications of fuzzy mathematics. Several portfolio choice models are studied. A possibilistic model in which the return of the risk is a fuzzy number, and four models in which the background risk appears in addition to the investment risk are presented. The interwoven historical developments of the two mathematical theories by Zadeh and Rosenblatt which opened up into pattern classification and fuzzy clustering are presented. Credibility for intuitionistic fuzzy sets is presented. Expected values, entropy, and general formulae for the central moments are introduced and studied. Algorithms for modeling uncertain data using fuzzy tensor product surfaces are presented. In particular, fuzzification and defuzzification processes are applied to obtain crisp Bezier curves and surfaces from fuzzy data. Three numerical methods for solving linear systems are presented, namely Jacobi, Gauss–Seidel, and successive over-relaxation.

Etienne E. Kerre, John Mordeson
Special Issue Editors

mathematics MDPI

Article

L-Fuzzy Sets and Isomorphic Lattices: Are All the "New" Results Really New? †

Erich Peter Klement [1,*] **and Radko Mesiar** [2]

[1] Department of Knowledge-Based Mathematical Systems, Johannes Kepler University, 4040 Linz, Austria
[2] Department of Mathematics and Descriptive Geometry, Faculty of Civil Engineering, Slovak University of Technology, 810 05 Bratislava, Slovakia; radko.mesiar@stuba.sk
* Correspondence: ep.klement@jku.at; Tel.: +43-650-2468290
† These authors contributed equally to this work.

Received: 23 June 2018; Accepted: 20 August 2018; Published: 23 August 2018

Abstract: We review several generalizations of the concept of fuzzy sets with two- or three-dimensional lattices of truth values and study their relationship. It turns out that, in the two-dimensional case, several of the lattices of truth values considered here are pairwise isomorphic, and so are the corresponding families of fuzzy sets. Therefore, each result for one of these types of fuzzy sets can be directly rewritten for each (isomorphic) type of fuzzy set. Finally we also discuss some questionable notations, in particular, those of "intuitionistic" and "Pythagorean" fuzzy sets.

Keywords: fuzzy set; interval-valued fuzzy set; "intuitionistic" fuzzy set; "pythagorean" fuzzy set; isomorphic lattices; truth values

1. Introduction

In the paper "Fuzzy sets" [1] L. A. Zadeh suggested the unit interval $[0,1]$ (which we shall denote by $\mathbb{I}$ throughout the paper) as set of truth values for fuzzy sets, in a generalization of Boolean logic and Cantorian set theory where the two-element Boolean algebra $\{0,1\}$ is used.

Soon after a further generalization was proposed in J. Goguen [2]: to replace the unit interval $\mathbb{I}$ by an abstract set L (in most cases a lattice), noticing that the key feature of the unit interval in this context is its lattice structure. In yet another generalization L. A. Zadeh [3,4] introduced fuzzy sets of type 2 where the value of the membership function is a fuzzy subset of $\mathbb{I}$.

Since then, many more variants and generalizations of the original concept in [1] were presented, most of them being either L-fuzzy sets, type-n fuzzy sets or both. In a recent and extensive "historical account", H. Bustince et al. ([5], Table 1) list a total of 21 variants of fuzzy sets and study their relationships.

In this paper, we will deal with the concepts of (generalizations of) fuzzy sets where the set of truth values is either one-dimensional (the unit interval $\mathbb{I}$), two-dimensional (e.g., a suitable subset of the unit square $\mathbb{I} \times \mathbb{I}$) or three-dimensional (a subset of the unit cube $\mathbb{I}^3$).

The one-dimensional case (where the set of truth values equals $\mathbb{I}$) is exactly the case of fuzzy sets in the sense of [1].

Concerning the two-dimensional case, we mainly consider the following subsets of the unit square $\mathbb{I} \times \mathbb{I}$:

$$\begin{aligned} \mathbb{L}^* &= \{(x_1,x_2) \in \mathbb{I} \times \mathbb{I} \mid x_1 + x_2 \leq 1\}, \\ L_2(\mathbb{I}) &= \{(x_1,x_2) \in \mathbb{I} \times \mathbb{I} \mid 0 \leq x_1 \leq x_2 \leq 1\}, \\ P^* &= \{(x_1,x_2) \in \mathbb{I} \times \mathbb{I} \mid x_1^2 + x_2^2 \leq 1\}, \end{aligned}$$

and the related set of all closed subintervals of the unit interval $\mathbb{I}$:

$$\mathfrak{I}(\mathbb{I}) = \{[x_1, x_2] \subseteq \mathbb{I} \mid 0 \leq x_1 \leq x_2 \leq 1\}.$$

Equipped with suitable orders, these lattices of truth values give rise to several generalizations of fuzzy sets known from the literature: $\mathbb{L}^*$-fuzzy sets, "intuitionistic" fuzzy sets [6,7], grey sets [8,9], vague sets [10], 2-valued sets [11], interval-valued fuzzy sets [4,12–14], and "Pythagorean" fuzzy sets [15].

In the three-dimensional case, the following subsets of the unit cube $\mathbb{I}^3$ will play a major role:

$$\mathbb{D}^* = \{(x_1, x_2, x_3) \in \mathbb{I}^3 \mid x_1 + x_2 + x_3 \leq 1\},$$
$$L_3(\mathbb{I}) = \{(x_1, x_2, x_3) \in \mathbb{I}^3 \mid 0 \leq x_1 \leq x_2 \leq x_3 \leq 1\}.$$

Equipped with suitable orders, these lattices of truth values lead to the concepts of 3-valued sets [11] and picture fuzzy sets [16].

While it is not surprising that lattices of truth values of higher dimension correspond to more complex types of fuzzy sets, it is remarkable that in the two-dimensional case the lattices with the carriers $\mathbb{L}^*$, $L_2(\mathbb{I})$, P^*, and $\mathfrak{I}(\mathbb{I})$ are mutually isomorphic, i.e., the families of fuzzy sets with these truth values have the same lattice-based properties. This implies that mathematical results for one type of fuzzy sets can be carried over in a straightforward way to the other (isomorphic) types. This also suggests that, in a mathematical sense, often only one of these lattices of truth values (and only one of the corresponding types of fuzzy sets) is really needed.

Note that if some algebraic structures are isomorphic, then it is meaningful to consider all of them only if they have different meanings and interpretations.

This is, e.g., the case for the arithmetic mean (on $[-\infty, \infty[$) and for the geometric mean (on $[0, \infty[$). On the other hand, concerning results dealing with such isomorphic structures, it is enough to prove them once and then to transfer them to the other isomorphic structures simply using the appropriate isomorphisms. For example, in the case of the arithmetic and geometric means mentioned here, the additivity of the arithmetic mean is equivalent to the multiplicativity of the geometric mean.

Another example are pairs (a, b) of real numbers which can be interpreted as points in the real plane, as (planar) vectors, as complex numbers, and (if $a \leq b$) as closed sub-intervals of the real line. Most algebraic operations for these objects are defined for the representing pairs of real numbers; in the case of the addition, the exact same formula is used.

We only mention that in the case of three-dimensional sets of truth values, the corresponding lattices (and the families of fuzzy sets based on them) are not isomorphic, which means that they have substantially different properties.

The paper is organized as follows. In Section 2, we discuss the sets of truth values for Cantorian (or crisp) sets and for fuzzy sets and present the essential notions of abstract lattice theory, including the crucial concept of isomorphic lattices. In Section 3, we review the two- and three-dimensional sets of truth values mentioned above and study the isomorphisms between them and between the corresponding families of fuzzy sets. Finally, in Section 4, we discuss some further consequences of lattice isomorphisms as well as some questionable notations appearing in the literature, in particular "intuitionistic" fuzzy sets and "Pythagorean" fuzzy sets.

2. Preliminaries

Let us start with collecting some of the basic and important prerequisites from set theory, fuzzy set theory, and some generalizations thereof.

2.1. Truth Values and Bounded Lattices

The set of truth values in Cantorian set theory [17,18] (and in the underlying Boolean logic [19,20]) is the Boolean algebra $\{0,1\}$, which we will denote by **2** in this paper. Given a universe of discourse, i.e., a non-empty set X, each *Cantorian* (or *crisp*) *subset A of X* can be identified with its *indicator function* $\mathbf{1}_A\colon X \to \mathbf{2}$, defined by $\mathbf{1}_A(x) = 1$ if and only if $x \in A$.

In L. A. Zadeh's seminal paper on fuzzy sets [1] (compare also the work of K. Menger [21–23] and D. Klaua [24,25]), the unit interval $[0,1]$ was proposed as set of *truth values*, thus providing a natural extension of the Boolean case. As usual, a *fuzzy subset* A of the universe of discourse X is described by its *membership function* $\mu_A\colon X \to \mathbb{I}$, and $\mu_A(x)$ is interpreted as the *degree of membership* of the object x in the fuzzy set A. The standard order reversing involution (or double negation) $N_{\mathbb{I}}\colon \mathbb{I} \to \mathbb{I}$ is given by $N_{\mathbb{I}}(x) = 1 - x$.

For the rest of this paper, we will reserve the shortcut $\mathbb{I}$ for the unit interval $[0,1]$ of the real line $\mathbb{R}$. On each subset of the real line, the order $\leq$ will denote the standard linear order inherited from $\mathbb{R}$.

In a further generalization, J. Goguen [2] suggested to use the elements of an abstract set L as truth values and to describe an *L-fuzzy subset* A of X by means of its membership function $\mu_A\colon X \to L$, where $\mu_A(x)$ stands for the *degree of membership* of the object x in the L-fuzzy set A.

Several important examples for L were discussed in [2], such as *complete lattices* or *complete lattice-ordered semigroups*. There is an extensive literature on L-fuzzy sets dealing with various aspects of algebra, analysis, category theory, topology, and stochastics (see, e.g., [26–44]). For a more recent overview of these and other types and generalizations of fuzzy sets see [5].

In most of these papers the authors work with a *lattice* $(L, \leq_L)$, i.e., a non-empty, partially ordered set $(L, \leq_L)$ such that each finite subset of L has a *meet* (or *greatest lower bound*) and a *join* (or *least upper bound*) in L. If each arbitrary subset of L has a meet and a join then the lattice is called *complete*, and if there exist a *bottom* (or *smallest*) *element* $\mathbf{0}_L$ and a *top* (or *greatest*) *element* $\mathbf{1}_L$ in L, then the lattice is called *bounded*.

For notions and results in the theory of general lattices we refer to the book of G. Birkhoff [45]. There is an equivalent, purely algebraic approach to lattices without referring to a partial order: if $\wedge_L\colon L \times L \to L$ and $\vee_L\colon L \times L \to L$ are two commutative, associative operations on a set L that satisfy the two *absorption laws*, i.e., for all $x, y \in L$ we have $x \wedge_L (x \vee_L y) = x$ and $x \vee_L (x \wedge_L y) = x$, and if we define the binary relation $\leq_L$ on L by $x \leq_L y$ if and only if $x \wedge_L y = x$ (which is equivalent to saying that $x \leq_L y$ if and only if $x \vee_L y = y$), then $\leq_L$ is a partial order on L and $(L, \leq_L)$ is a lattice such that, for each set $\{x, y\} \subseteq L$, the elements $x \wedge_L y$ and $x \vee_L y$ coincide with the meet and the join, respectively, of the set $\{x, y\}$ with respect to the order $\leq_L$.

Clearly, the lattices $(\mathbf{2}, \leq)$ and $(\mathbb{I}, \leq)$ already mentioned are examples of complete bounded lattices: **2**-fuzzy sets are exactly crisp sets, $\mathbb{I}$-fuzzy sets are the fuzzy sets in the sense of [1].

If $(L_1, \leq_{L_1}), (L_2, \leq_{L_2}), \ldots, (L_n, \leq_{L_n})$ are lattices and $\prod\limits_{i=1}^{n} L_i = L_1 \times L_2 \times \cdots \times L_n$ is the Cartesian product of the underlying sets, then also

$$\left(\prod_{i=1}^{n} L_i, \leq_{\text{comp}}\right), \tag{1}$$

is a lattice, the so-called *product lattice* of $(L_1, \leq_{L_1})$, $(L_2, \leq_{L_2})$, ..., $(L_n, \leq_{L_n})$, where $\leq_{\text{comp}}$ is the *componentwise* partial order on the Cartesian product $\prod L_i$ given by

$$\begin{aligned}(x_1, x_2, \ldots, x_n) &\leq_{\text{comp}} (y_1, y_2, \ldots, y_n) \\ &\iff x_1 \leq_{L_1} y_1 \text{ AND } x_2 \leq_{L_2} y_2 \text{ AND } \ldots \text{ AND } x_n \leq_{L_n} y_n.\end{aligned} \tag{2}$$

The componentwise partial order is not the only partial order that can be defined on $\prod L_i$. An alternative is, for example, the *lexicographical* partial order $\leq_{\text{lexi}}$ given by $(x_1, x_2, \ldots, x_n) \leq_{\text{lexi}} (y_1, y_2, \ldots, y_n)$ if and only if $((x_1, x_2, \ldots, x_n) = (y_1, y_2, \ldots, y_n)$ or $(x_1, x_2, \ldots, x_n) <_{\text{lexi}} (y_1, y_2, \ldots, y_n))$, where the strict inequality $(x_1, x_2, \ldots, x_n) <_{\text{lexi}} (y_1, y_2, \ldots, y_n)$ holds if and only if there is an $i_0 \in \{1, 2, \ldots, n\}$ such that $x_i = y_i$ for each $i \in \{1, 2, \ldots, i_0 - 1\}$ and $x_{i_0} <_{L_{i_0}} y_{i_0}$.

Obviously, whenever $(L_1, \leq_{L_1}), (L_2, \leq_{L_2}), \ldots, (L_n, \leq_{L_n})$ are lattices then also

$$\left(\prod_{i=1}^{n} L_i, \leq_{\text{lexi}}\right)$$

is a lattice. Moreover, if each of the partial orders $\leq_{L_1}, \leq_{L_2}, \ldots, \leq_{L_n}$ is linear, then $\leq_{\text{lexi}}$ is also a linear order. Note that this is not the case for $\leq_{\text{comp}}$ whenever $n > 1$ and at least two of the sets $L_1, L_2, \ldots, L_n$ contain two or more elements. To take the simplest example: the lattice $(\mathbf{2} \times \mathbf{2}, \leq_{\text{lexi}})$ is a chain, i.e., $(0,0) <_{\text{lexi}} (0,1) <_{\text{lexi}} (1,0) <_{\text{lexi}} (1,0)$, but in the product lattice $(\mathbf{2} \times \mathbf{2}, \leq_{\text{comp}})$ the elements $(0,1)$ and $(1,0)$ are incomparable with respect to $\leq_{\text{comp}}$.

We only mention that also the product of infinitely many lattices may be a lattice. As an example, if $(L, \leq_L)$ is a lattice and X a non-empty set, then the set L^X of all functions from X to L, equipped with the componentwise partial order $\leq_{\text{comp}}$, is again a lattice. Recall that, for functions $f, g\colon X \to L$, the componentwise partial order $\leq_{\text{comp}}$ is defined by $f \leq_{\text{comp}} g$ if and only if $f(x) \leq_L g(x)$ for all $x \in X$. If no confusion is possible, we simply shall write $f \leq_L g$ rather than $f \leq_{\text{comp}} g$.

2.2. Isomorphic Lattices: Some General Consequences

For two partially ordered sets $(L_1, \leq_{L_1})$ and $(L_2, \leq_{L_2})$, a function $\varphi\colon L_1 \to L_2$ is called an *order homomorphism* if it preserves the monotonicity, i.e., if $x \leq_{L_1} y$ implies $\varphi(x) \leq_{L_2} \varphi(y)$.

If $(L_1, \leq_{L_1})$ and $(L_2, \leq_{L_2})$ are two lattices then a function $\varphi\colon L_1 \to L_2$ is called a *lattice homomorphism* if it preserves finite meets and joins, i.e., if for all $x, y \in L_1$

$$\varphi(x \wedge_{L_1} y) = \varphi(x) \wedge_{L_2} \varphi(y) \qquad \text{and} \qquad \varphi(x \vee_{L_1} y) = \varphi(x) \vee_{L_2} \varphi(y). \tag{3}$$

Each lattice homomorphism is an order homomorphism, but the converse is not true in general. A lattice homomorphism $\varphi\colon L_1 \to L_2$ is called an *embedding* if it is injective, an *epimorphism* if it is surjective, and an *isomorphism* if it is bijective, i.e., if it is both an embedding and an epimorphism.

If a function $\varphi\colon L_1 \to L_2$ is an embedding from a lattice $(L_1, \leq_{L_1})$ into a lattice $(L_2, \leq_{L_2})$ then the set $\{\varphi(x) \mid x \in L_1\}$ (equipped with the partial order inherited from $(L_2, \leq_{L_2})$) forms a sublattice of $(L_2, \leq_{L_2})$ which is isomorphic to $(L_1, \leq_{L_1})$. If $(L_1, \leq_{L_1})$ is bounded or complete, so is this sublattice of $(L_2, \leq_{L_2})$. Conversely, if $(L_1, \leq_{L_1})$ is a sublattice of $(L_2, \leq_{L_2})$ then $(L_1, \leq_{L_1})$ trivially can be embedded into $(L_2, \leq_{L_2})$ (the identity function $\text{id}_{L_1}\colon L_1 \to L_2$ provides an embedding).

The word "isomorphic" is derived from the composition of the two Greek words "isōs" (meaning *similar, equal, corresponding*) and "morphē" (meaning *shape, structure*), so it means having *the same shape* or *the same structure*.

If two lattices $(L_1, \leq_{L_1})$ and $(L_2, \leq_{L_2})$ are isomorphic this means that they have the same mathematical structure in the sense that there is a bijective function $\varphi\colon L_1 \to L_2$ that preserves the order as well as finite meets and joins, compare (3).

However, being isomorphic does not necessarily mean to be identical, for example (not in the lattice framework), consider the *arithmetic mean* on $[-\infty, \infty[$ and the *geometric mean* on $[0, \infty[$ [46,47] which are isomorphic aggregation functions on $\mathbb{R}^n$, but they have some different properties and they are used for different purposes.

If $(L_1, \leq_{L_1})$ and $(L_2, \leq_{L_2})$ are isomorphic and if $(L_1, \leq_{L_1})$ has additional order theoretical properties, these properties automatically carry over to the lattice $(L_2, \leq_{L_2})$.

For instance, if the lattice $(L_1, \leq_{L_1})$ is complete so is $(L_2, \leq_{L_2})$. Or, if the lattice $(L_1, \leq_{L_1})$ is bounded (with bottom element $\mathbf{0}_{L_1}$ and top element $\mathbf{1}_{L_1}$) then also $(L_2, \leq_{L_2})$ is bounded, and the bottom and top elements of $(L_2, \leq_{L_2})$ are obtained via $\mathbf{0}_{L_2} = \varphi(\mathbf{0}_{L_1})$ and $\mathbf{1}_{L_2} = \varphi(\mathbf{1}_{L_1})$.

Moreover, it is well-known that corresponding constructs over isomorphic structures are again isomorphic. Here are some particularly interesting cases:

Remark 1. *Suppose that $(L_1, \leq_{L_1})$ and $(L_2, \leq_{L_2})$ are isomorphic lattices and that $\varphi\colon L_1 \to L_2$ is a lattice isomophism between $(L_1, \leq_{L_1})$ and $(L_2, \leq_{L_2})$.*

(i) *If $f\colon L_1 \to L_1$ is a function then the composite function $\varphi \circ f \circ \varphi^{-1}\colon L_2 \to L_2$ has the same order theoretical properties as f.*

(ii) *If $F\colon L_1 \times L_1 \to L_1$ is a binary operation on L_1 and if we define $(\varphi^{-1}, \varphi^{-1})\colon L_2 \times L_2 \to L_2 \times L_2$ by $(\varphi^{-1}, \varphi^{-1})((x,y)) = (\varphi^{-1}(x), \varphi^{-1}(y))$, then the function $\varphi \circ F \circ (\varphi^{-1}, \varphi^{-1})\colon L_2 \times L_2 \to L_2$ is a binary operation on L_2 with the same order theoretical properties as F.*

(iii) *If $\mathbf{A}_1\colon (L_1)^n \to L_1$ is an n-ary operation on L_1 then, as a straightforward generalization, the composite function $\varphi \circ \mathbf{A}_1 \circ (\varphi^{-1}, \varphi^{-1}, \ldots, \varphi^{-1})\colon (L_2)^n \to L_2$ given by*

$$\varphi \circ \mathbf{A}_1 \circ (\varphi^{-1}, \varphi^{-1}, \ldots, \varphi^{-1})(x_1, x_2, \ldots, x_n) = \varphi\Big(\mathbf{A}_1(\varphi^{-1}(x_1), \varphi^{-1}(x_2), \ldots, \varphi^{-1}(x_n))\Big),$$

is an n-ary operation on L_2 with the same order theoretical properties as $\mathbf{A}_1$.

As a consequence of Remark 1, many structures used in fuzzy set theory can be carried over to any isomorphic lattice, for example, order reversing involutions or residua [45], which are used in BL-logics [48–62]. The same is true for many connectives (mostly on the unit interval $\mathbb{I}$ but also on more general and more abstract structures (see, e.g., [63,64])) for many-valued logics such as triangular norms and conorms (t-norms and t-conorms for short), going back to K. Menger [65] and B. Schweizer and A. Sklar [66–68] (see also [69–73]), uninorms [74], and nullnorms [75]. Another example are aggregation functions which have been extensively studied on the unit interval $\mathbb{I}$ in, e.g., [46,47,76–78].

Example 1. *Let $(L_1, \leq_{L_1})$ and $(L_2, \leq_{L_2})$ be isomorphic bounded lattices, suppose that $\varphi\colon L_1 \to L_2$ is a lattice isomorphism between $(L_1, \leq_{L_1})$ and $(L_2, \leq_{L_2})$, and denote the bottom and top elements of $(L_1, \leq_{L_1})$ by $\mathbf{0}_{L_1}$ and $\mathbf{1}_{L_1}$, respectively.*

(i) *Let $N_{L_1}\colon L_1 \to L_1$ be an order reversing involution (or double negation) on L_1, i.e., $x \leq_{L_1} y$ implies $N_{L_1}(y) \leq_{L_1} N_{L_1}(x)$, and $N_{L_1} \circ N_{L_1} = id_{L_1}$. Then the function $\varphi \circ \mathrm{N}_{L_1} \circ \varphi^{-1}$ is an order reversing involution on L_2, and the complemented lattice $(L_2, \leq_2, \varphi \circ \mathrm{N}_{L_1} \circ \varphi^{-1})$ is isomorphic to $(L_1, \leq_1, N_{L_1})$.*

(ii) *Let $(L_1, \leq_{L_1}, *_1, e_1, \rightarrow_1, \leftarrow_1)$ be a residuated lattice, i.e., $(L_1, *_1)$ is a (not necessarily commutative) monoid with neutral element e_1, and for the residua $\rightarrow_1, \leftarrow_1\colon L_1 \times L_1 \to L_1$ we have that for all $x, y, z \in L_1$ the assertion $(x *_1 y) \leq_{L_1} z$ is equivalent to both $y \leq_{L_1} (x \rightarrow_1 z)$ and $x \leq_{L_1} (z \leftarrow_1 y)$. Then*

$$\Big(L_2, \leq_{L_2}, \varphi \circ *_1 \circ (\varphi^{-1}, \varphi^{-1}), \varphi(e_1), \varphi \circ \rightarrow_1 \circ (\varphi^{-1}, \varphi^{-1}), \varphi \circ \leftarrow_1 \circ (\varphi^{-1}, \varphi^{-1})\Big)$$

is an isomorphic residuated lattice.

(iii) *Let $T_1\colon L_1 \times L_1 \to L_1$ be a triangular norm on L_1, i.e., T_1 is an associative, commutative order homomorphism with neutral element $\mathbf{1}_{L_1}$. Then the function $\varphi \circ T_1 \circ (\varphi^{-1}, \varphi^{-1})$ is a triangular norm on L_2.*

(iv) *Let $S_1\colon L_1 \times L_1 \to L_1$ be a triangular conorm on L_1, i.e., S_1 is an associative, commutative order homomorphism with neutral element $\mathbf{0}_{L_1}$. Then the function $\varphi \circ S_1 \circ (\varphi^{-1}, \varphi^{-1})$ is a triangular conorm on L_2.*

(v) *Let $U_1\colon L_1 \times L_1 \to L_1$ be a uninorm on L_1, i.e., U_1 is an associative, commutative order homomorphism with neutral element $e \in L_1$ such that $\mathbf{0}_{L_1} <_{L_1} e <_{L_1} \mathbf{1}_{L_1}$. Then the function $\varphi \circ U_1 \circ (\varphi^{-1}, \varphi^{-1})$ is a uninorm on L_2 with neutral element $\varphi(e)$.*

(vi) *Let $V_1\colon L_1 \times L_1 \to L_1$ be a nullnorm on L_1, i.e., V_1 is an associative, commutative order homomorphism such that there is an $a \in L_1$ with $\mathbf{0}_{L_1} <_{L_1} a <_{L_1} \mathbf{1}_{L_1}$ such that for all $x \leq_{L_1} a$ we have $V_1((x, \mathbf{0}_{L_1})) = x$, and for all $x \geq_{L_1} a$ we have $V_1((x, \mathbf{1}_{L_1})) = x$. Then the function $\varphi \circ V_1 \circ (\varphi^{-1}, \varphi^{-1})$ is a nullnorm on L_2.*

(vii) *Let* $\mathbf{A}_1\colon (L_1)^n \to L_1$ *be an n-ary aggregation function on* L_1, *i.e.,* $\mathbf{A}_1$ *is an order homomorphism which satisfies* $\mathbf{A}_1(\mathbf{0}_{L_1}, \mathbf{0}_{L_1}, \ldots, \mathbf{0}_{L_1}) = \mathbf{0}_{L_1}$ *and* $\mathbf{A}_1(\mathbf{1}_{L_1}, \mathbf{1}_{L_1}, \ldots, \mathbf{1}_{L_1}) = \mathbf{1}_{L_1}$. *Then the function* $\varphi \circ \mathbf{A}_1 \circ (\varphi^{-1}, \varphi^{-1}, \ldots, \varphi^{-1})$ *is an n-ary aggregation function on* L_2.

3. Some Generalizations of Truth Values and Fuzzy Sets

In this section we first review the lattices of truth values for crisp sets and for fuzzy sets as introduced in [1], followed by a detailed description of various generalizations thereof by means of sets of truth values of dimension two and higher.

3.1. The Classical Cases: Crisp and Fuzzy Sets

Now we shall consider different lattices of types of truth values and, for a fixed non-empty universe of discourse X, the corresponding classes of (fuzzy) subsets of X.

Recall that if the set of truth values is the classical Boolean algebra $\{0,1\}$ (denoted in this paper simply by $\mathbf{2}$), then the corresponding set of all *crisp* (or *Cantorian*) *subsets of* X will be denoted by $\mathscr{P}(X)$ (called the *power set of* X). Each crisp subset A of X can be identified with its *characteristic function* $\mathbf{1}_A\colon X \to \mathbf{2}$, which is defined by $\mathbf{1}_A(x) = 1$ if and only if $x \in A$. There are exactly two constant characteristic functions: $\mathbf{1}_\varnothing\colon X \to \mathbf{2}$ maps every $x \in X$ to 0, and $\mathbf{1}_X\colon X \to \mathbf{2}$ maps every $x \in X$ to 1.

Obviously, we have $A \subseteq B$ if and only if $\mathbf{1}_A \leq \mathbf{1}_B$, i.e., $\mathbf{1}_A(x) \leq \mathbf{1}_B(x)$ for all $x \in X$, and $(\mathscr{P}(X), \subseteq)$ is a complete bounded lattice with bottom element $\varnothing$ and top element X, i.e., $(\mathscr{P}(X), \subseteq)$ is isomorphic to the product lattice $(\mathbf{2}^X, \leq)$, where $\mathbf{2}^X$ is the set of all functions from X to $\mathbf{2}$, and $\leq$ is the componentwise standard order.

Switching to the unit interval (denoted by $\mathbb{I}$) as set of truth values in the sense of [1], the set of all *fuzzy subsets of* X will be denoted by $\mathscr{F}(X)$. As usual, each fuzzy subset $A \in \mathscr{F}(X)$ is characterized by its *membership function* $\mu_A\colon X \to \mathbb{I}$, where $\mu_A(x) \in \mathbb{I}$ describes the *degree of membership* of the object $x \in X$ in the fuzzy set A.

For fuzzy sets $A, B \in \mathscr{F}(X)$ we have $A \subseteq B$ if and only if $\mu_A \leq \mu_B$, i.e., $\mu_A(x) \leq \mu_B(x)$ for all $x \in X$. Therefore, $(\mathscr{F}(X), \subseteq)$ is a complete bounded lattice with bottom element $\varnothing$ and top element X, i.e., $(\mathscr{F}(X), \subseteq)$ is isomorphic to $(\mathbb{I}^X, \leq)$, where $\mathbb{I}^X$ is the set of all functions from X to $\mathbb{I}$.

Only for the sake of completeness we mention that the bottom and top elements in $\mathscr{F}(X)$ are also denoted by $\varnothing$ and X, and they correspond to the membership functions $\mu_\varnothing = \mathbf{1}_\varnothing$ and $\mu_X = \mathbf{1}_X$, respectively.

The lattice $(\mathscr{P}(X), \subseteq)$ of crisp subsets of X can be embedded into the lattice $(\mathscr{F}(X), \subseteq)$ of fuzzy sets of X: the function $\mathrm{emb}_{\mathscr{P}(X)}\colon \mathscr{P}(X) \to \mathscr{F}(X)$ given by $\mu_{\mathrm{emb}_{\mathscr{P}(X)}(A)} = \mathbf{1}_A$, i.e., the membership function of $\mathrm{emb}_{\mathscr{P}(X)}(A)$ is just the characteristic function of A, provides a natural embedding.

The membership function $\mu_{A^\mathsf{C}}\colon X \to \mathbb{I}$ of the *complement* A^C of a fuzzy set $A \in \mathscr{F}(X)$ is given by $\mu_{A^\mathsf{C}}(x) = N_\mathbb{I}(\mu_A(x)) = 1 - \mu_A(x)$.

For a fuzzy set $A \in \mathscr{F}(X)$ and $\alpha \in \mathbb{I}$, the *α-cut* (or *α-level set*) of A is defined as the crisp set $[A]_\alpha \in \mathscr{P}(X)$ given by

$$[A]_\alpha = \{x \in X \mid \mu_A(x) \geq \alpha\}.$$

The 1-cut $[A]_1 = \{x \in X \mid \mu_A(x) = 1\}$ of a fuzzy set $A \in \mathscr{F}(X)$ is often called the *kernel* of A, and the crisp set $\{x \in X \mid \mu_A(x) > 0\}$ usually is called the *support* of the fuzzy set A.

The family $([A]_\alpha)_{\alpha \in \mathbb{I}}$ of α-cuts of a fuzzy subset A of X carries the same information as the membership function $\mu_A\colon X \to \mathbb{I}$ in the sense that it is possible to reconstruct the membership function μ_A from the family of α-cuts of A: for all $x \in X$ we have [27,79]

$$\mu_A(x) = \sup\{\min(\alpha, \mathbf{1}_{[A]_\alpha}(x)) \mid \alpha \in \mathbb{I}\}.$$

We only mention that this is no more possible if the unit interval $\mathbb{I}$ is replaced by some lattice L which is not a chain.

3.2. Generalizations: The Two-Dimensional Case

A simple example of a two-dimensional lattice is $(\mathbb{I} \times \mathbb{I}, \leq_{\text{comp}})$ as defined by (1) and (2), i.e., the unit square of the real plane $\mathbb{R}^2$. In [63], triangular norms on this lattice (and on other product lattices) were studied. The standard order reversing involution $N_{\mathbb{I}\times\mathbb{I}} \colon \mathbb{I} \times \mathbb{I} \to \mathbb{I} \times \mathbb{I}$ in $(\mathbb{I} \times \mathbb{I}, \leq_{\text{comp}})$ is given by

$$N_{\mathbb{I}\times\mathbb{I}}((x,y)) = (1-y, 1-x). \tag{4}$$

This product lattice was considered in several expert systems [80–82]. There, the first coordinate was interpreted as a degree of positive information (*measure of belief*), and the second coordinate as a degree of negative information (*measure of disbelief*). Note that though several operations for this structure were considered in the literature (for a nice overview see [83]), a deeper algebraic investigation is still missing in this case.

To the best of our knowledge, K. T. Atanassov [6,7,84] (compare [85,86]) was the first to consider both the *degree of membership* and the *degree of non-membership* when using and studying the bounded lattice $(\mathbb{L}^*, \leq_{\mathbb{L}^*})$ of truth values given by (5) and (6). Unfortunately, he called the corresponding $\mathbb{L}^*$-fuzzy sets *"intuitionistic" fuzzy sets* because of the lack of the *law of excluded middle* (for a critical discussion of this terminology see Section 4.2):

$$\mathbb{L}^* = \{(x_1, x_2) \in \mathbb{I} \times \mathbb{I} \mid x_1 + x_2 \leq 1\}, \tag{5}$$

$$(x_1, x_2) \leq_{\mathbb{L}^*} (y_1, y_2) \iff x_1 \leq y_1 \text{ AND } x_2 \geq y_2. \tag{6}$$

Obviously, $(\mathbb{L}^*, \leq_{\mathbb{L}^*})$ is a complete bounded lattice: $\mathbf{0}_{\mathbb{L}^*} = (0,1)$ and $\mathbf{1}_{\mathbb{L}^*} = (1,0)$ are the bottom and top elements of $(\mathbb{L}^*, \leq_{\mathbb{L}^*})$, respectively, and the meet $\wedge_{\mathbb{L}^*}$ and the join $\vee_{\mathbb{L}^*}$ in $(\mathbb{L}^*, \leq_{\mathbb{L}^*})$ are given by

$$(x_1, x_2) \wedge_{\mathbb{L}^*} (y_1, y_2) = (\min(x_1, y_1), \max(x_2, y_2)),$$
$$(x_1, x_2) \vee_{\mathbb{L}^*} (y_1, y_2) = (\max(x_1, y_1), \min(x_2, y_2)).$$

Moreover, $(\mathbb{I}, \leq)$ can be embedded in a natural way into $(\mathbb{L}^*, \leq_{\mathbb{L}^*})$: the function $\text{emb}_{\mathbb{I}} \colon \mathbb{I} \to \mathbb{L}^*$ given by $\text{emb}_{\mathbb{I}}(x) = (x, 1-x)$ is an embedding. Observe that there are also other embeddings of $(\mathbb{I}, \leq)$ into $(\mathbb{L}^*, \leq_{\mathbb{L}^*})$, e.g., $\varphi \colon \mathbb{I} \to \mathbb{L}^*$ given by $\varphi(x) = (x, 0)$.

Note that the order $\leq_{\mathbb{L}^*}$ is not linear. However, it is possible to construct refinements of $\leq_{\mathbb{L}^*}$ which are linear [87].

Mirroring the set $\mathbb{L}^*$ about the axis passing through the points $(0, 0.5)$ and $(1, 0.5)$ of the unit square $\mathbb{I} \times \mathbb{I}$ one immediately sees that there is some other lattice which is isomorphic to $(\mathbb{L}^*, \leq_{\mathbb{L}^*})$. Both lattices are visualized in Figure 1.

Proposition 1. *The complete bounded lattice $(\mathbb{L}^*, \leq_{\mathbb{L}^*})$ is isomorphic to the upper left triangle $L_2(\mathbb{I})$ in $\mathbb{I} \times \mathbb{I}$ (with vertexes $(0,0)$, $(0,1)$ and $(1,1)$), i.e.,*

$$L_2(\mathbb{I}) = \{(x_1, x_2) \in \mathbb{I} \times \mathbb{I} \mid 0 \leq x_1 \leq x_2 \leq 1\}, \tag{7}$$

equipped with the componentwise partial order $\leq_{\text{comp}}$, whose bottom and top elements are $\mathbf{0}_{L_2(\mathbb{I})} = (0,0)$ and $\mathbf{1}_{L_2(\mathbb{I})} = (1,1)$, respectively. A canonical isomorphism between the lattices $(\mathbb{L}^*, \leq_{\mathbb{L}^*})$ and $(L_2(\mathbb{I}), \leq_{\text{comp}})$ is provided by the function $\varphi_{\mathbb{L}^*}^{L_2(\mathbb{I})} \colon \mathbb{L}^* \to L_2(\mathbb{I})$ defined by $\varphi_{\mathbb{L}^*}^{L_2(\mathbb{I})}((x_1, x_2)) = (x_1, 1 - x_2)$.

It is readily seen that $(L_2(\mathbb{I}), \leq_{\text{comp}})$ is a sublattice of the product lattice $(\mathbb{I} \times \mathbb{I}, \leq_{\text{comp}})$, and the standard order reversing involution $N_{L_2(\mathbb{I})} \colon L_2(\mathbb{I}) \to L_2(\mathbb{I})$ is given by

$$N_{L_2(\mathbb{I})}((x,y)) = (1-y, 1-x) \tag{8}$$

(compare (4)). On the other hand, the lattice $(\mathbb{L}^*, \le_{\mathbb{L}^*})$ is not a sublattice of $(\mathbb{I} \times \mathbb{I}, \le_{\text{comp}})$, but it can be embedded into $(\mathbb{I} \times \mathbb{I}, \le_{\text{comp}})$ using, e.g., the lattice monomorphism (as visualized in Figure 2)

$$\text{id}_{L_2(\mathbb{I})} \circ \varphi_{\mathbb{L}^*}^{L_2(\mathbb{I})} \colon \mathbb{L}^* \longrightarrow L_2(\mathbb{I}).$$

Several other lattices "look" different when compared with $(\mathbb{L}^*, \le_{\mathbb{L}^*})$ or seem to address a different context, but in fact they carry the same structural information as $(\mathbb{L}^*, \le_{\mathbb{L}^*})$.

Well-known examples of this phenomenon are the lattices $(\mathfrak{I}(\mathbb{I}), \le_{\mathfrak{I}(\mathbb{I})})$, providing the basis of interval-valued (or grey) fuzzy sets [4,8,9,12–14], and $(P^*, \le_{\mathbb{L}^*})$, giving rise to the so-called "Pythagorean" fuzzy sets [15,88,89], both turning out to be isomorphic to the lattice $(\mathbb{L}^*, \le_{\mathbb{L}^*})$. The following statements can be verified by simply checking the required properties.

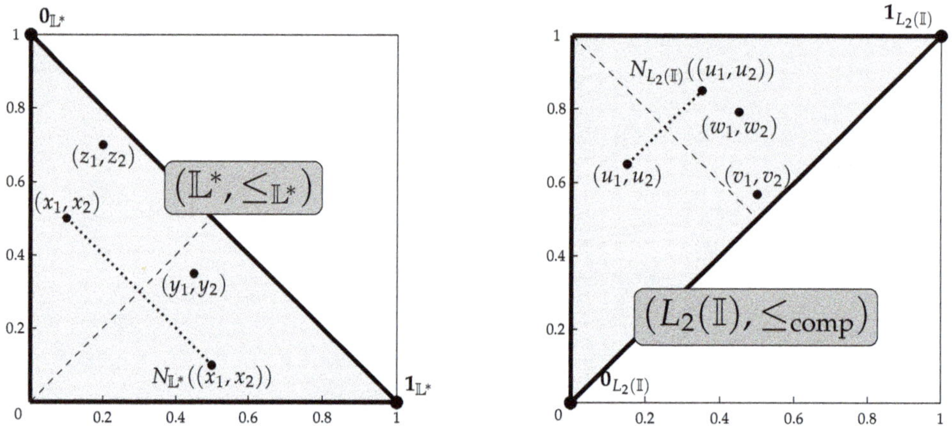

Figure 1. The lattices $(\mathbb{L}^*, \le_{\mathbb{L}^*})$ (**left**) and $(L_2(\mathbb{I}), \le_{\text{comp}})$ (**right**). Note the difference of two orders: we have $(x_1, x_2) \le_{\mathbb{L}^*} (y_1, y_2)$, $(z_1, z_2) \le_{\mathbb{L}^*} (y_1, y_2)$, and $(u_1, u_2) \le_{\text{comp}} (w_1, w_2)$, but (x_1, x_2) and (z_1, z_2) are not comparable in $(\mathbb{L}^*, \le_{\mathbb{L}^*})$, and (v_1, v_2) is neither comparable to (u_1, u_2) nor to (w_1, w_2) with respect to $\le_{\text{comp}}$. Also, a hint for the constructions of the order reversing involutions $N_{\mathbb{L}^*}$ and $N_{L_2(\mathbb{I})}$ as reflections through the appropriate diagonal (dashed line) of $\mathbb{I} \times \mathbb{I}$ is given.

Proposition 2. *The complete bounded lattice $(\mathbb{L}^*, \le_{\mathbb{L}^*})$ is isomorphic to the following two lattices:*

(i) *to the lattice $(\mathfrak{I}(\mathbb{I}), \le_{\mathfrak{I}(\mathbb{I})})$ of all closed subintervals of the unit interval $\mathbb{I}$, given by*

$$\mathfrak{I}(\mathbb{I}) = \{[x_1, x_2] \subseteq \mathbb{I} \mid 0 \le x_1 \le x_2 \le 1\}, \tag{9}$$

$$[x_1, x_2] \le_{\mathfrak{I}(\mathbb{I})} [y_1, y_2] \iff x_1 \le y_1 \text{ AND } x_2 \le y_2, \tag{10}$$

with bottom and top elements $\mathbf{0}_{\mathfrak{I}(\mathbb{I})} = [0, 0]$ and $\mathbf{1}_{\mathfrak{I}(\mathbb{I})} = [1, 1]$, respectively; a canonical example of an isomorphism between $(\mathbb{L}^*, \le_{\mathbb{L}^*})$ and $(\mathfrak{I}(\mathbb{I}), \le_{\mathfrak{I}(\mathbb{I})})$ is provided by the function $\varphi_{\mathbb{L}^*}^{\mathfrak{I}(\mathbb{I})} \colon \mathbb{L}^* \to \mathfrak{I}(\mathbb{I})$ defined by $\varphi_{\mathbb{L}^*}^{\mathfrak{I}(\mathbb{I})}((x_1, x_2)) = [x_1, 1 - x_2]$;

(ii) *to the lattice $(P^*, \le_{\mathbb{L}^*})$ of all points in the intersection of the unit square $\mathbb{I} \times \mathbb{I}$ and the unit disk with center $(0, 0)$, i.e.,*

$$P^* = \{(x_1, x_2) \in \mathbb{I} \times \mathbb{I} \mid x_1^2 + x_2^2 \le 1\}; \tag{11}$$

a canonical example of a lattice isomorphism from $(P^*, \le_{\mathbb{L}^*})$ to $(\mathbb{L}^*, \le_{\mathbb{L}^*})$ is provided by the function $\varphi_{P^*}^{\mathbb{L}^*} \colon P^* \to \mathbb{L}^*$ defined by $\varphi_{P^*}^{\mathbb{L}^*}((x_1, x_2)) = (x_1^2, x_2^2)$.

Example 2. *Let us start with the standard order reversing involution* $N_{L_2(\mathbb{I})}$ *on* $(L_2(\mathbb{I}), \leq_{\text{comp}})$ *given by* (8). *The fact that* $(L_2(\mathbb{I}), \leq_{\text{comp}})$ *is isomorphic to each of the lattices* $(\mathbb{L}^*, \leq_{\mathbb{L}^*})$, $(P^*, \leq_{\mathbb{L}^*})$, *and* $(\mathfrak{I}(\mathbb{I}), \leq_{\mathfrak{I}(\mathbb{I})})$ *(see Propositions 1 and 2) and Example 1(i) allow us to construct the order reversing involutions* $N_{\mathbb{L}^*}: \mathbb{L}^* \to \mathbb{L}^*$, $N_{P^*}: P^* \to P^*$, *and* $N_{\mathfrak{I}(\mathbb{I})}: \mathfrak{I}(\mathbb{I}) \to \mathfrak{I}(\mathbb{I})$ *on the lattices* $(\mathbb{L}^*, \leq_{\mathbb{L}^*})$, $(P^*, \leq_{\mathbb{L}^*})$, *and* $(\mathfrak{I}(\mathbb{I}), \leq_{\mathfrak{I}(\mathbb{I})})$ *are given by*

$$N_{\mathbb{L}^*}((x_1, x_2)) = N_{P^*}((x_1, x_2)) = (x_2, x_1), \qquad N_{\mathfrak{I}(\mathbb{I})}([x_1, x_2]) = [1 - x_2, 1 - x_1].$$

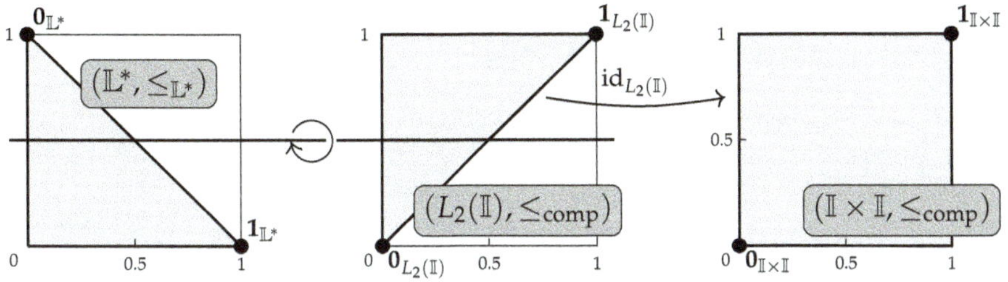

Figure 2. The lattices $(\mathbb{L}^*, \leq_{\mathbb{L}^*})$ **(left)**, $(L_2(\mathbb{I}), \leq_{\text{comp}})$ **(center)**, and $(\mathbb{I} \times \mathbb{I}, \leq_{\text{comp}})$ **(right)**. The mirror symmetry between $\mathbb{L}^*$ and $L_2(\mathbb{I})$ shows that $(\mathbb{L}^*, \leq_{\mathbb{L}^*})$ and $(L_2(\mathbb{I}), \leq_{\text{comp}})$ are isomorphic, and $(L_2(\mathbb{I}), \leq_{\text{comp}})$ is a sublattice of $(\mathbb{I} \times \mathbb{I}, \leq_{\text{comp}})$.

Given a universe of discourse X, i.e., a non-empty set X, and fixing a bounded lattice $(L, \leq_L)$, we obtain a special type of *L-fuzzy subsets* of X in the sense of [2] and, on the other hand, a particular case of *type-2 fuzzy sets* (also proposed by L. A. Zadeh [3,4]; see [90,91] for some algebraic aspects of truth values for type-2 fuzzy sets).

An $\mathbb{L}^*$*-fuzzy subset* A of X is characterized by its membership function $\mu_A^{\mathbb{L}^*}: X \to \mathbb{L}^*$, where the bounded lattice $(\mathbb{L}^*, \leq_{\mathbb{L}^*})$ is given by (5) and (6). The bottom and top elements of $(\mathbb{L}^*, \leq_{\mathbb{L}^*})$ are $\mathbf{0}_{\mathbb{L}^*} = (0, 1)$ and $\mathbf{1}_{\mathbb{L}^*} = (1, 0)$, respectively.

Over the years, different names for fuzzy sets based on the lattices that are isomorphic to $(\mathbb{L}^*, \leq_{\mathbb{L}^*})$ according to Propositions 1 and 2 were used in the literature: in the mid-seventies $\mathfrak{I}(\mathbb{I})$-fuzzy sets were called *interval-valued* in [4,12–14], in the eighties first the name *"intuitionistic"* fuzzy sets was used for $\mathbb{L}^*$-fuzzy sets in [6,7] (compare also [84–86]) and then *grey* sets in [8,9]), and even later *vague* sets in [10] (see also [10,92]). More recently, the name "Pythagorean" fuzzy sets was introduced for P^*-fuzzy sets in [15,88,89].

As a function $\mu_A^{\mathbb{L}^*}: X \to \mathbb{L}^* \subseteq \mathbb{I} \times \mathbb{I}$, the membership function $\mu_A^{\mathbb{L}^*}$ has two components $\mu_A, \nu_A: X \to \mathbb{I}$ such that for each $x \in X$ we have $\mu_A^{\mathbb{L}^*}(x) = (\mu_A(x), \nu_A(x))$ and $\mu_A(x) + \nu_A(x) \leq 1$.

Both $\mu_A: X \to \mathbb{I}$ and $\nu_A: X \to \mathbb{I}$ can be seen as membership functions of fuzzy subsets of X, say $A^+, A^- \in \mathscr{F}(X)$, respectively, i.e., for each $x \in X$ we have

$$\mu_{A^+}(x) = \mu_A(x), \quad \mu_{A^-}(x) = \nu_A(x), \quad \text{and} \quad \mu_{A^+}(x) + \mu_{A^-}(x) \leq 1. \tag{12}$$

The value $\mu_{A^+}(x)$ is usually called the *degree of membership* of the object x in the $\mathbb{L}^*$-fuzzy set A, while $\mu_{A^-}(x)$ is said to be the *degree of non-membership* of the object x in the $\mathbb{L}^*$-fuzzy set A.

Denoting the set of all $\mathbb{L}^*$*-fuzzy subsets of* X by $\mathscr{F}_{\mathbb{L}^*}(X)$ and keeping the notations from (12), for each $A \in \mathscr{F}_{\mathbb{L}^*}(X)$ and its membership function $\mu_A^{\mathbb{L}^*}: X \to \mathbb{L}^* \subseteq \mathbb{I} \times \mathbb{I}$ we may write

$$\mu_A^{\mathbb{L}^*} = (\mu_A, \nu_A) = (\mu_{A^+}, \mu_{A^-}).$$

As a consequence of (12), for the fuzzy sets A^+ and A^- we have $A^+ \subseteq (A^-)^{\complement}$. In other words, we can identify each $\mathbb{L}^*$-fuzzy subset $A \in \mathscr{F}_{\mathbb{L}^*}(X)$ with a pair of fuzzy sets (A^+, A^-) with $A^+ \subseteq (A^-)^{\complement}$, i.e.,

$$\mathscr{F}_{\mathbb{L}^*}(X) = \{(A^+, A^-) \in \mathscr{F}(X) \times \mathscr{F}(X) \mid A^+ \subseteq (A^-)^{\complement}\},$$

and for two $\mathbb{L}^*$-fuzzy subsets $A = (A^+, A^-)$ and $B = (B^+, B^-)$ of X the assertion $A \subseteq_{\mathbb{L}^*} B$ is equivalent to $A^+ \subseteq B^+$ and $B^- \subseteq A^-$. The complement of an $\mathbb{L}^*$-fuzzy subset $A = (A^+, A^-)$ is the $\mathbb{L}^*$-fuzzy set $A^{\complement} = (A^-, A^+)$.

Then $(\mathscr{F}_{\mathbb{L}^*}(X), \subseteq_{\mathbb{L}^*})$ is a complete bounded lattice with bottom element $\varnothing = (\varnothing, X)$ and top element $X = (X, \varnothing)$, and the lattice $(\mathscr{F}_{\mathbb{L}^*}(X), \subseteq_{\mathbb{L}^*})$ of L^*-fuzzy sets is isomorphic to $(\mathbb{L}^{*X}, \leq_{\mathbb{L}^*})$. Clearly, $(\mathscr{F}(X), \subseteq)$ can be embedded into $(\mathscr{F}_{\mathbb{L}^*}(X), \subseteq_{\mathbb{L}^*})$: a natural embedding is provided by the function $\mathrm{emb}_{\mathscr{F}(X)} \colon \mathscr{F}(X) \to \mathscr{F}_{\mathbb{L}^*}(X)$ defined by $\mathrm{emb}_{\mathscr{F}(X)}(A) = (A, A^{\complement})$.

An *interval-valued fuzzy subset* A of the universe X (introduced independently in [4,12–14], some authors called them *grey* sets [8,9]) is characterized by its membership function $\mu_A \colon X \to \mathfrak{I}(\mathbb{I})$, where $(\mathfrak{I}(\mathbb{I}), \leq_{\mathfrak{I}(\mathbb{I})})$ is the bounded lattice of all closed subintervals of the unit interval $\mathbb{I}$ given by (9) and (10). The bottom and top elements of $(\mathfrak{I}(\mathbb{I}), \leq_{\mathfrak{I}(\mathbb{I})})$ are then$\mathbf{0}_{\mathfrak{I}(\mathbb{I})} = [0,0]$ and $\mathbf{1}_{\mathfrak{I}(\mathbb{I})} = [1,1]$, respectively.

A *"Pythagorean" fuzzy subset* A of the universe X (first considered in [15,88,89]) is characterized by its membership function $\mu_A \colon X \to P^*$, where the bounded lattice $(P^*, \leq_{\mathbb{L}^*})$ is given by (11) and (6). The bottom and top elements of $(P^*, \leq_{\mathbb{L}^*})$ are the same as in $(\mathbb{L}^*, \leq_{\mathbb{L}^*})$, i.e., we have $\mathbf{0}_{P^*} = (0,1)$ and $\mathbf{1}_{P^*} = (1,0)$.

From Propositions 1 and 2 we know that the four bounded lattices $(\mathbb{L}^*, \leq_{\mathbb{L}^*})$, $(\mathfrak{I}(\mathbb{I}), \leq_{\mathfrak{I}(\mathbb{I})})$, $(P^*, \leq_{\mathbb{L}^*})$, and $(L_2(\mathbb{I}), \leq_{\mathrm{comp}})$ are isomorphic to each other. As an immediate consequence we obtain the following result.

Proposition 3. *Let X be a universe of discourse. Then we have:*

(i) *The product lattices $((\mathbb{L}^*)^X, \leq_{\mathbb{L}^*})$, $((\mathfrak{I}(\mathbb{I}))^X, \leq_{\mathfrak{I}(\mathbb{I})})$, $((P^*)^X, \leq_{\mathbb{L}^*})$, and $((L_2(\mathbb{I}))^X, \leq_{\mathrm{comp}})$ are isomorphic to each other.*

(ii) *The lattices of all $\mathbb{L}^*$-fuzzy subsets of X, of all "intuitionistic" fuzzy subsets of X, of all interval-valued fuzzy subsets of X, of all "Pythagorean" fuzzy subsets of X, and of all $L_2(\mathbb{I})$-fuzzy subsets of X are isomorphic to each other.*

This means that, mathematically speaking, all the function spaces mentioned in Proposition 3(i) and all the "different" classes of fuzzy subsets of X referred to in Proposition 3(ii) share an identical (lattice) structure. Any differences between them only come from the names used for individual objects, and from the interpretation or meaning of these objects. In other words, since any mathematical result for one of these lattices immediately can be carried over to all isomorphic lattices, in most cases there is no need to use different names for them.

3.3. Generalizations to Higher Dimensions

As a straightforward generalization of the product lattice $(\mathbb{I} \times \mathbb{I}, \leq_{\mathrm{comp}})$, for each $n \in \mathbb{N}$ the *n-dimensional unit cube* $(\mathbb{I}^n, \leq_{\mathrm{comp}})$, i.e., the n-dimensional product of the lattice $(\mathbb{I}, \leq)$, can be defined by means of (1) and (2).

The so-called *"neutrosophic"* sets introduced by F. Smarandache [93] (see also [94–97] are based on the bounded lattices $(\mathbb{I}^3, \leq_{\mathbb{I}^3})$ and $(\mathbb{I}^3, \leq^{\mathbb{I}^3})$, where the orders $\leq_{\mathbb{I}^3}$ and $\leq^{\mathbb{I}^3}$ on the unit cube $\mathbb{I}^3$ are defined by

$$(x_1, x_2, x_3) \leq_{\mathbb{I}^3} (y_1, y_2, y_3) \iff x_1 \leq y_1 \text{ AND } x_2 \leq y_2 \text{ AND } x_3 \geq y_3, \tag{13}$$

$$(x_1, x_2, x_3) \leq^{\mathbb{I}^3} (y_1, y_2, y_3) \iff x_1 \leq y_1 \text{ AND } x_2 \geq y_2 \text{ AND } x_3 \geq y_3. \tag{14}$$

Observe that $\leq_{\mathbb{I}^3}$ is a variant of the order $\leq_{\mathrm{comp}}$: it is defined componentwise, but in the third component the order is reversed. The top element of $(\mathbb{I}^3, \leq_{\mathbb{I}^3})$ is $(1,1,0)$, and $(0,0,1)$ is its bottom element. Analogous assertions are true for the lattice $(\mathbb{I}^3, \leq^{\mathbb{I}^3})$.

Clearly, the three lattices $(\mathbb{I}^3, \leq_{\mathrm{comp}})$, $(\mathbb{I}^3, \leq_{\mathbb{I}^3})$, and $(\mathbb{I}^3, \leq^{\mathbb{I}^3})$ are mutually isomorphic: the functions $\varphi, \psi \colon \mathbb{I}^3 \to \mathbb{I}^3$ given by $\varphi((x_1, x_2, x_3)) = (x_1, x_2, 1 - x_3)$ and $\psi((x_1, x_2, x_3)) = (x_1, 1 - x_2, x_3)$ are canonical isomorphisms between $(\mathbb{I}^3, \leq_{\text{comp}})$ and $(\mathbb{I}^3, \leq_{\mathbb{I}^3})$, on the one hand, and between $(\mathbb{I}^3, \leq_{\mathbb{I}^3})$ and $(\mathbb{I}^3, \leq^{\mathbb{I}^3})$, on the other hand.

For each $n \in \mathbb{N}$ the *n-fuzzy sets* introduced by B. Bedregal et al. in [11] (see also [98,99]) are based on the bounded lattice $(L_n(\mathbb{I}), \leq_{\text{comp}})$, where the set $L_n(\mathbb{I})$ is a straightforward generalization of $L_2(\mathbb{I})$ defined in (7):

$$L_n(\mathbb{I}) = \{(x_1, x_2, \ldots, x_n) \in \mathbb{I}^n \mid x_1 \leq x_2 \leq \cdots \leq x_n\}. \tag{15}$$

The order $\leq_{\text{comp}}$ on $L_n(\mathbb{I})$ coincides with the restriction of the componentwise order $\leq_{\text{comp}}$ on $\mathbb{I}^n$ to $L_n(\mathbb{I})$, implying that $(L_n(\mathbb{I}), \leq_{\text{comp}})$ is a sublattice of the product lattice $(\mathbb{I}^n, \leq_{\text{comp}})$. As a consequence, we also have the standard order reversing involution $N_{L_n(\mathbb{I})} \colon L_n(\mathbb{I}) \to L_n(\mathbb{I})$ which is defined coordinatewise, i.e., $N_{L_n(\mathbb{I})}((x_1, x_2, \ldots, x_n)) = (1 - x_n, \ldots, 1 - x_2, 1 - x_1)$ (compare (8)). Considering, for $n > 3$, lattices which are isomorphic to $(L_n(\mathbb{I}), \leq_{\text{comp}})$, further generalizations of "neuthrosophic" sets can be introduced.

B. C. Cuong and V. Kreinovich [16] proposed the concept of so-called *picture fuzzy sets* which are based on the set $\mathbb{D}^* \subseteq \mathbb{I}^3$ of truth values given by

$$\mathbb{D}^* = \{(x_1, x_2, x_3) \in \mathbb{I}^3 \mid x_1 + x_2 + x_3 \leq 1\}. \tag{16}$$

The motivation for the set $\mathbb{D}^*$ came from a simple voting scenario where each voter can act in one of the four following ways: to vote for the nominated candidate (the proportion of these voters being equal to x_1), to vote against the candidate (described by x_2), to have no preference and to abstain so this vote will not be counted (described by x_3), or to be absent (described by $1 - x_1 - x_2 - x_3$).

In the original proposal [16] the set $\mathbb{D}^*$ was equipped with the partial order $\leq_{\mathbb{I}^3}$ given by (13), as inherited from the lattice $(\mathbb{I}^3, \leq_{\mathbb{I}^3})$. As $\{(x_1, x_3) \in \mathbb{I} \times \mathbb{I} \mid (x_1, 0, x_3) \in \mathbb{D}^*\} = \mathbb{L}^*$ and (6), we may also write $(x_1, x_2, x_3) \leq_{\mathbb{I}^3} (y_1, y_2, y_3)$ if and only if $(x_1, x_3) \leq_{\mathbb{L}^*} (y_1, y_3)$ and $x_2 \leq y_2$. However, $(\mathbb{D}^*, \leq_{\mathbb{I}^3})$ is not a lattice, but only a meet-semilattice with bottom element $\mathbf{0}_{\mathbb{D}^*} = (0, 0, 1)$; indeed, the set $\{(1, 0, 0), (0, 1, 0)\}$ has no join in $\mathbb{D}^*$ with respect to $\leq_{\mathrm{P}}$ (to be more precise, the semi-lattice $(\mathbb{D}^*, \leq_{\mathrm{P}})$ has infinitely many pairwise incomparable maximal elements of the form $(a, 1 - a, 0)$ with $a \in \mathbb{I}$).

Therefore, (without modifications) it is impossible [100] to introduce logical operations such as t-norms or t-conorms [69] and, in general, aggregation functions [46] on $(\mathbb{D}^*, \leq_{\mathbb{I}^3})$.

As a consequence, the order $\leq_{\mathbb{I}^3}$ on $\mathbb{D}^*$ was replaced by the following partial order $\leq_{\mathbb{D}^*}$ on $\mathbb{D}^*$ (compare [16,100–102]) which is a refinement of $\leq_{\mathbb{I}^3}$:

$$\begin{aligned} &(x_1, x_2, x_3) \leq_{\mathbb{D}^*} (y_1, y_2, y_3) \\ &\qquad \Longleftrightarrow \quad (x_1, x_3) <_{\mathbb{L}^*} (y_1, y_3) \text{ OR } \Big((x_1, x_3) = (y_1, y_3) \text{ AND } x_2 \leq y_2\Big). \end{aligned} \tag{17}$$

Note that the order $\leq_{\mathbb{D}^*}$ can be seen as a kind of lexicographical order related to two orders: to the order $\leq_{\mathbb{L}^*}$ on $\mathbb{L}^*$ and to the standard order $\leq$ on $\mathbb{I}$.

It is easy to see that $(\mathbb{D}^*, \leq_{\mathbb{D}^*})$ is a bounded lattice with bottom element $\mathbf{0}_{\mathbb{D}^*} = (0, 0, 1)$ and top element $\mathbf{1}_{\mathbb{D}^*} = (1, 0, 0)$. This allows aggregation functions (as studied on the unit interval $\mathbb{I}$ in, e.g., [46,47,76–78]) to be introduced on $(\mathbb{D}^*, \leq_{\mathbb{D}^*})$. Observe also that the lattice $(\mathbb{D}^*, \leq_{\mathbb{D}^*})$ was considered in recent applications of picture fuzzy sets [103,104].

We only recall [105] that the join $\vee_{\leq_{\mathbb{D}^*}}$ and the meet $\wedge_{\leq_{\mathbb{D}^*}}$ in the lattice $(\mathbb{D}^*, \leq_{\mathbb{D}^*})$ are given by

$$(x_1,x_2,x_3) \vee_{\leq_{\mathbb{D}^*}} (y_1,y_2,y_3) = \begin{cases} (x_1,x_2,x_3) & \text{if } (x_1,x_2,x_3) \geq_{\mathbb{D}^*} (y_1,y_2,y_3), \\ (y_1,y_2,y_3) & \text{if } (x_1,x_2,x_3) \leq_{\mathbb{D}^*} (y_1,y_2,y_3), \\ (\max(x_1,y_1), 0, \min(x_3,y_3)) & \text{otherwise,} \end{cases}$$

$$(x_1,x_2,x_3) \wedge_{\leq_{\mathbb{D}^*}} (y_1,y_2,y_3) = \begin{cases} (x_1,x_2,x_3) & \text{if } (x_1,x_2,x_3) \leq_{\mathbb{D}^*} (y_1,y_2,y_3), \\ (y_1,y_2,y_3) & \text{if } (x_1,x_2,x_3) \geq_{\mathbb{D}^*} (y_1,y_2,y_3), \\ (\min(x_1,y_1), 1-\min(x_1,y_1) & \\ \quad -\max(x_3,y_3), \max(x_3,y_3)) & \text{otherwise,} \end{cases}$$

and the standard order reversing involution $N_{\mathbb{D}^*} \colon \mathbb{D}^* \to \mathbb{D}^*$ by $N_{\mathbb{D}^*}((x_1,x_2,x_3)) = (x_3,x_2,x_1)$.

From the definition of $\leq_{\mathbb{D}^*}$ in (17) it is obvious that $(\mathbb{L}^*, \leq_{\mathbb{L}^*})$ can be embedded in a natural way into $(\mathbb{D}^*, \leq_{\mathbb{D}^*})$; an example of an embedding is given by

$$\begin{aligned} \mathrm{emb}_{\mathbb{L}^*} \colon \quad \mathbb{L}^* &\longrightarrow \mathbb{D}^* \\ (x_1,x_2) &\longmapsto (x_1,0,x_2). \end{aligned} \tag{18}$$

Let us now have a look at the relationship between the lattice $(\mathbb{D}^*, \leq_{\mathbb{D}^*})$ and the lattice $(L_3(\mathbb{I}), \leq_{\mathrm{comp}})$ given by (2) and (15). It is not difficult to see that the function $\psi \colon \mathbb{D}^* \to L_3(\mathbb{I})$ given by $\psi((x_1,x_2,x_3)) = (x_1, x_1+x_2, 1-x_3)$ is a bijection, its inverse $\psi^{-1} \colon L_3(\mathbb{I}) \to \mathbb{D}^*$ being given by $\psi^{-1}((x_1,x_2,x_3)) = (x_1, x_2-x_1, 1-x_3)$.

Observe that that the bijection ψ is not order preserving: we have $(0.2,0.5,0) \leq_{\mathbb{D}^*} (0.3,0,0)$, but $\psi((0.2,0.5,0)) = (0.2,0.7,1)$ and $\psi((0.2,0.5,0)) = (0.3,0.3,1)$ are incomparable with respect to $\leq_{\mathrm{comp}}$.

From ([105], Propositions 1 and 2) we have the following result:

Proposition 4. *The lattices $(L_3(\mathbb{I}), \leq_{\mathrm{comp}})$ and $(\mathbb{D}^*, \leq_{\mathbb{D}^*})$ are not isomorphic. However, we have*

(i) *The lattice $(L_3(\mathbb{I}), \leq_{\mathrm{comp}})$ is isomorphic to the lattice $(\mathbb{D}^*, \leq_{\mathbb{D}_3^*})$ with top element $(1,0,0)$ and bottom element $(0,0,1)$, where the order $\leq_{\mathbb{D}_3^*}$ is given by*

$$(x_1,x_2,x_3) \leq_{\mathbb{D}_3^*} (y_1,y_2,y_3) \quad \Longleftrightarrow \quad x_1 \leq y_1 \text{ AND } x_1+x_2 \leq y_1+y_2 \text{ AND } x_3 \geq y_3.$$

(ii) *The lattice $(\mathbb{D}^*, \leq_{\mathbb{D}^*})$ is isomorphic to the lattice $(L_3(\mathbb{I}), \leq_{L_3(\mathbb{I})})$ with top element $(1,1,1)$ and bottom element $(0,0,0)$, where the order $\leq_{L_3(\mathbb{I})}$ is given by*

$$\begin{aligned} &(x_1,x_2,x_3) \leq_{L_3(\mathbb{I})} (y_1,y_2,y_3) \\ &\qquad \Longleftrightarrow \quad (x_1,x_3) <_{\mathrm{comp}} (y_1,y_3) \text{ OR } ((x_1,x_3) = (y_1,y_3) \text{ AND } x_2-x_1 \leq y_2-y_1). \end{aligned}$$

In summary, if a universe of discourse X is fixed, then a *picture* fuzzy subset A of X is based on the bounded lattice $(\mathbb{D}^*, \leq_{\mathbb{D}^*})$ defined in (16) and (17). It is characterized by its membership function $\mu_A^{\mathbb{D}^*} \colon X \to \mathbb{D}^*$ [16,100,106–109] where $\mu_A^{\mathbb{D}^*}(x) = (\mu_{A_1}(x), \mu_{A_2}(x), \mu_{A_3}(x)) \in \mathbb{D}^*$ for some functions $\mu_{A_1}, \mu_{A_2}, \mu_{A_3} \colon X \to \mathbb{I}$.

Clearly, the function $\mu_{A_1}: X \to \mathbb{I}$ can be interpreted as the membership function of the fuzzy set $A_1 \in \mathscr{F}(X)$ and, analogously, $\mu_{A_2}: X \to \mathbb{I}$ and $\mu_{A_3}: X \to \mathbb{I}$ as membership functions of the fuzzy sets A_2 and A_3, respectively. In other words, for each picture fuzzy set A we may write $A = (A_1, A_2, A_3)$.

In this context, the values $\mu_{A_1}(x)$, $\mu_{A_2}(x)$ and $\mu_{A_3}(x)$ are called the *degree of positive membership*, the *degree of neutral membership*, and the *degree of negative membership* of the object x in the picture fuzzy set A, respectively. The value $1 - (\mu_{A_1}(x) + \mu_{A_2}(x) + \mu_{A_3}(x)) \in \mathbb{I}$ is called the *degree of refusal membership* of the object x in A.

If X is a fixed universe of discourse, then we denote the set of all picture fuzzy subsets of X by $\mathscr{F}_{\mathbb{D}^*}(X)$. Obviously, for two picture fuzzy sets $A, B \in \mathscr{F}_{\mathbb{D}^*}(X)$ the assertion $A \subseteq_{\mathbb{D}^*} B$ is equivalent to $(\mu_{A_1}, \mu_{A_2}, \mu_{A_3}) \leq_{\mathbb{D}^*} (\mu_{B_1}, \mu_{B_2}, \mu_{B_3})$, i.e., $(\alpha_A(x), \beta_A(x), \gamma_A(x)) \leq_{\mathbb{D}^*} (\alpha_B(x), \beta_B(x), \gamma_B(x))$ for all $x \in X$, and the membership function of the complement $A^{\complement}$ of a picture fuzzy set $A \in \mathscr{F}_{\mathbb{D}^*}(X)$ with membership function $\mu_A^{\mathbb{D}^*} = (\mu_{A_1}, \mu_{A_2}, \mu_{A_3})$ is given by $\mu_{A^{\complement}}^{\mathbb{D}^*} = (\mu_{A_3}, \mu_{A_2}, \mu_{A_1})$.

This means that $(\mathscr{F}_{\mathbb{D}^*}(X)), \subseteq_{\mathbb{D}^*})$ is a bounded lattice with bottom element $\emptyset = (\emptyset, \emptyset, X)$ and top element $X = (X, \emptyset, \emptyset)$, and it is isomorphic to the product lattice $((\mathbb{D}^*)^X, \leq_{\text{comp}})$ of all functions from X to $\mathbb{D}^*$ (clearly, $\mu_A^{\mathbb{D}^*} \leq_{\text{comp}} \mu_B^{\mathbb{D}^*}$ means here $\mu_A^{\mathbb{D}^*}(x) \leq_{\mathbb{D}^*} \mu_B^{\mathbb{D}^*}(x)$ for all $x \in X$).

As a consequence, the lattice $(\mathscr{F}_{\mathbb{L}^*}(X)), \subseteq_{\mathbb{L}^*})$ of all $\mathbb{L}^*$-fuzzy subsets of X can be embedded into the lattice $(\mathscr{F}_{\mathbb{D}^*}(X)), \subseteq_{\mathbb{D}^*})$ of all picture fuzzy subsets of X via

$$\begin{aligned} \text{emb}_{\mathscr{F}_{\mathbb{L}^*}(X)}\colon \quad \mathscr{F}_{\mathbb{L}^*}(X) &\longrightarrow \mathscr{F}_{\mathbb{D}^*}(X) \\ (A^+, A^-) &\longmapsto (A^+, \emptyset, A^-), \end{aligned}$$

and, using the embedding $\text{emb}_{\mathbb{L}^*}: \mathbb{L}^* \to \mathbb{D}^*$ defined in (18), the product lattice $((\mathbb{L}^*)^X, \leq_{\text{comp}})$ can be embedded into the product lattice $((\mathbb{D}^*)^X, \leq_{\text{comp}})$.

We recognize a chain of subsets of X of increasing generality and complexity: crisp sets $\mathscr{P}(X)$, fuzzy sets $\mathscr{F}(X)$, $\mathbb{L}^*$-fuzzy sets $\mathscr{F}_{\mathbb{L}^*}(X)$, and picture fuzzy sets $\mathscr{F}_{\mathbb{D}^*}(X)$. This corresponds to the increasing complexity and dimensionality of the lattices of truth values $(\mathbf{2}, \leq)$, $(\mathbb{I}, \leq)$, $(\mathbb{L}^*, \leq_{\mathbb{L}^*})$, and $(\mathbb{D}^*, \leq_{\mathbb{D}^*})$. The commutative diagram in Figure 3 visualizes the relationship between these types of (fuzzy) sets and their respective membership functions, and also of the corresponding lattices of truth values.

The content of this subsection also makes clear that the situation in the case of three-dimensional sets of truth values is much more complex than for the two-dimensional truth values considered before.

In Proposition 3, we have seen that several classes of fuzzy sets with two-dimensional sets of truth values are isomorphic to each other, while, in the case of three-dimensional truth values, we have given a number of lattices of truth values that are not isomorphic to each other.

Obviously, continuing in the series of generalizations from $\mathbb{I}$ over $\mathbb{L}^*$ to $\mathbb{D}^*$, for any arity $n \in \mathbb{N}$ one can define a carrier

$$\mathbb{D}_n^* = \left\{ (x_1, ..., x_n) \in \mathbb{I}^n \;\middle|\; \sum_{i=1}^{n} x_i \leq 1 \right\}$$

and equip it with some order $\preceq$ such that $(\mathbb{D}_n^*, \preceq)$ is a bounded lattice with top element $(1, 0, ..., 0)$ and bottom element $(0, ..., 0, 1)$. The problematic question is whether such a generalization is meaningful and can be used to model some real problem.

If the arrow $\hookrightarrow$ indicates an embedding, $\twoheadrightarrow$ an epimorphism, and $\hookrightarrow\!\!\!\!\twoheadrightarrow$ an isomorphism, and if the homomorphisms are defined by

$$
\begin{aligned}
&\mathrm{emb}_{\mathbb{I}}(x) = (x, 1-x), && \mathrm{con}_{\mathbb{L}^*}((\alpha_1, \alpha_2)) = (\alpha_1 \cdot \mathbf{1}_X, \alpha_2 \cdot \mathbf{1}_X),\\
&\mathrm{emb}_{\mathbb{L}^*}((x_1, x_2)) = (x_1, 0, x_2), && \mathrm{con}_{\mathbb{D}^*}((\alpha_1, \alpha_2, \alpha_3)) = (\alpha_1 \cdot \mathbf{1}_X, \alpha_2 \cdot \mathbf{1}_X, \alpha_3 \cdot \mathbf{1}_X),\\
&\mathrm{emb}_{\mathbb{I}^X}(f) = (f, 1-f), && \mathrm{emb}_{(\mathbb{L}^*)^X}((\mu_{A^+}, \mu_{A^-})) = (\mu_{A^+}, \mathbf{1}_{\emptyset}, \mu_{A^-}),\\
&\mathrm{emb}_{\mathscr{F}(X)}(A) = (A, \complement A), && \mathrm{emb}_{\mathscr{F}_{\mathbb{L}^*}(X)}((A^+, A^-)) = (A^+, \emptyset, A^-),\\
&\pi_\alpha(f) = f(\alpha), && \mathrm{mem}_{\mathbb{I}}(A) = \mu_A,\\
&\mathrm{ind}(A) = \mathbf{1}_A, && \mathrm{mem}_{\mathbb{L}^*}(A) = (\mathrm{mem}_{\mathbb{I}}(A^+), \mathrm{mem}_{\mathbb{I}}(A^-)),\\
&\mathrm{con}(\alpha) = \alpha \cdot \mathbf{1}_X, && \mathrm{mem}_{\mathbb{D}^*}(A) = (\mathrm{mem}_{\mathbb{I}}(A^+), \mathrm{mem}_{\mathbb{I}}(A^{(n)}), \mathrm{mem}_{\mathbb{I}}(A^-)),
\end{aligned}
$$

then we obtain the following commutative diagram:

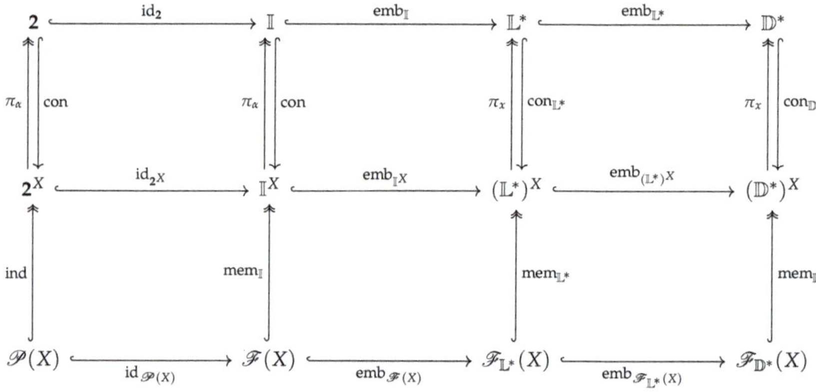

Figure 3. Crisp sets, fuzzy sets, $\mathbb{L}^*$-fuzzy sets, and picture fuzzy sets, and the corresponding sets of truth values.

4. Discussion: Isomorphisms and Questionable Notations

In this section, we first mention some further consequences of isomorphic lattices for the construction of logical and other connectives, and then we argue why, in our opinion, notations like "intuitionistic" fuzzy sets and "Pythagorean" fuzzy sets are questionable and why it would be better to avoid them.

4.1. Isomorphic Lattices: More Consequences

From Propositions 1 and 2, we know that the bounded lattice $(L_2(\mathbb{I}), \leq_{\mathrm{comp}})$ is isomorphic to each of the lattices $(\mathbb{L}^*, \leq_{\mathbb{L}^*})$, $(\mathfrak{I}(\mathbb{I}), \leq_{\mathfrak{I}(\mathbb{I})})$, and $(P^*, \leq_{\mathbb{L}^*})$.

Many results for and constructions of operations on the lattices $(\mathbb{L}^*, \leq_{\mathbb{L}^*})$, $(\mathfrak{I}(\mathbb{I}), \leq_{\mathfrak{I}(\mathbb{I})})$, and $(P^*, \leq_{\mathbb{L}^*})$, and, subsequently, for $\mathbb{L}^*$-fuzzy sets ("intuitionistic" fuzzy sets), interval-valued fuzzy sets, and "Pythagorean" fuzzy sets are a consequence of a rather general result for operations on the lattice $(L_2(\mathbb{I}), \leq_{\mathrm{comp}})$ and, because of the isomorphisms given in Propositions 1 and 2, they automatically can be carried over to the isomorphic lattices $(\mathbb{L}^*, \leq_{\mathbb{L}^*})$, $(\mathfrak{I}(\mathbb{I}), \leq_{\mathfrak{I}(\mathbb{I})})$, and $(P^*, \leq_{\mathbb{L}^*})$.

The following result makes use of the fact that $(L_2(\mathbb{I}), \leq_{\mathrm{comp}})$ is a sublattice of the product lattice $(\mathbb{I} \times \mathbb{I}, \leq_{\mathrm{comp}})$ and is based on [63]. It can be verified in a straightforward way by checking the required properties:

Proposition 5. *Let* $F_1, F_2\colon \mathbb{I} \times \mathbb{I} \to \mathbb{I}$ *be two functions such that* $F_1 \leq F_2$*, i.e., for each* $(x_1, x_2) \in \mathbb{I} \times \mathbb{I}$ *we have* $F_1((x_1, x_2)) \leq F_2((x_1, x_2))$*, and consider the function* $F\colon L_2(\mathbb{I}) \times L_2(\mathbb{I}) \to L_2(\mathbb{I})$ *given by*

$$F((x_1, x_2), (y_1, y_2)) = (F_1((x_1, y_1)), F_2((x_2, y_2))).$$

Then we have:

(i) *if* F_1 *and* F_2 *are two binary aggregation functions then the function* F *is a binary aggregation function on* $L_2(\mathbb{I})$*;*
(ii) *if* F_1 *and* F_2 *are two triangular norms then the function* F *is a triangular norm on* $L_2(\mathbb{I})$*;*
(iii) *if* F_1 *and* F_2 *are two triangular conorms then the function* F *is a triangular conorm on* $L_2(\mathbb{I})$*;*
(iv) *if* F_1 *and* F_2 *are two uninorms then the function* F *is a uninorm on* $L_2(\mathbb{I})$*;*
(v) *if* F_1 *and* F_2 *are two nullnorms then the function* F *is a nullnorm on* $L_2(\mathbb{I})$*.*

Not all t-(co)norms, uninorms and nullnorms on the lattice $(L_2(\mathbb{I}), \leq_{\text{comp}}$ can be obtained by means of Proposition 5, as the following example shows (see [110], Theorem 5):

Example 3. *Let* $T\colon \mathbb{I} \times \mathbb{I} \to \mathbb{I}$ *be a t-norm on the unit interval* $\mathbb{I}$*. Then, for each* $\alpha \in \mathbb{I} \setminus \{1\}$*, the function* $T_\alpha\colon L_2(\mathbb{I}) \times L_2(\mathbb{I}) \to L_2(\mathbb{I})$ *defined by*

$$T_\alpha((x_1, x_2), (y_1, y_2)) = (T((x_1, x_2)), \max(T(\alpha, T((y_1, y_2))), T((x_1, y_2)), T((y_1, x_2))))$$

is a t-norm on $L_2(\mathbb{I})$ *which cannot be obtained applying Proposition 5.*

The characterization of those connectives on $L_2(\mathbb{I}), \leq_{\text{comp}})$ is an interesting problem that has been investigated in several papers (e.g., in [110–120]). Again, each of these results is automatically valid for connectives on the isomorphic lattices $(\mathbb{L}^*, \leq_{\mathbb{L}^*})$, $(\mathfrak{I}(\mathbb{I}), \leq_{\mathfrak{I}(\mathbb{I})})$, and $(P^*, \leq_{\mathbb{L}^*})$.

The result of Proposition 5(i) can be carried over to the n-dimensional case in a straightforward way:

Corollary 1. *Let* $\mathbf{A}_1, \mathbf{A}_2\colon \mathbb{I}^n \to \mathbb{I}$ *be two n-ary aggregation functions such that* $\mathbf{A}_1 \leq \mathbf{A}_2$*, i.e., for each* $(x_1, x_2, \ldots, x_n) \in \mathbb{I}^n$ *we have* $\mathbf{A}_1((x_1, x_2, \ldots, x_n)) \leq \mathbf{A}_2((x_1, x_2, \ldots, x_n))$*. Then also the function* $\mathbf{A}\colon (L_2(\mathbb{I}))^n \to L_2(\mathbb{I})$ *given by*

$$\mathbf{A}((x_1, y_1), (x_2, y_2), \ldots, (x_n, y_n)) = (\mathbf{A}_1(x_1, x_2, \ldots, x_n), \mathbf{A}_2(y_1, y_2, \ldots, y_n))$$

is an n-ary aggregation function on $L_2(\mathbb{I})$*.*

4.2. The Case of "Intuitionistic" Fuzzy Sets

As already mentioned, $\mathbb{L}^*$-fuzzy sets have been called "intuitionistic" fuzzy sets in [6,7,84] and in a number of other papers (e.g., in [86,92,116,117,121–147]). In ([7], p. 87) K. T. Atanassov points out

> [...] the logical law of the excluded middle is not valid, similarly to the case in intuitionistic mathematics. Herein emerges the name of that set. [...]

Looking at Zadeh's first paper on fuzzy sets [1] one readily sees that the elements of $\mathscr{F}(X)$ also violate the law of the excluded middle if the unit interval $\mathbb{I}$ is equipped with the standard order reversing involution and if the t-norm min and the t-conorm max are used to model intersection and union of elements of $\mathscr{F}(X)$, respectively. In other words, the violation of the law of the excluded middle is no specific feature of the $\mathbb{L}^*$-fuzzy sets.

A short look at the history of mathematics and logic at the beginning of the 20th century shows that the philosophy of *intuitionism* goes back to the work of the Dutch mathematician L. E. J. Brouwer who suggested and discussed (for the first time 1912 in his inaugural address at the University of Amsterdam [148]) a foundation of mathematics independent of the law of excluded middle (see also [149–157]), a proposal eventually leading to a major controversy with the German mathematician D. Hilbert [158–160] (compare also [161]).

There are only a few papers (most remarkably, those by G. Takeuti and S. Titani [162,163]) where the original concept of intuitionistic logic was properly extended to the fuzzy case (see also [164–168])—here the use of the term "intuitionistic" fuzzy set is fully justified (see [169]).

As a consequence, the use of the name "intuitionistic" fuzzy sets in [6,7,84] and in a number of other papers in the same spirit has been criticized (mainly in [169–172]—compare Atanassov's reply [173] where he defended his original naming) because of its lack of relationship with the original concept of intuitionism and intuitionistic logic.

Here are the main arguments against using the term "intuitionistic" fuzzy sets in the context of $\mathbb{L}^*$-fuzzy sets, as given in [169]:

- the mere fact that the law of the excluded middle is violated in the case of $\mathbb{L}^*$-fuzzy sets does not justify to call them "intuitionistic" (also the fuzzy sets in the sense of [1] do not satisfy the law of the excluded middle, in general); moreover (see [53,170,174,175]), the use of an order reversing involution for $\mathbb{L}^*$-fuzzy sets contradicts intuitionistic logic [176]:

  > [...] the connectives of IFS theory violate properties of intuitionistic logic by validating the double negation (involution) axiom [...], which is not valid in intuitionistic logic. (Recall that axioms of intuitionistic logic extended by the axiom of double negation imply classical logic, and thus imply excluded middle [...]

- intuitionistic logic has a close relationship to constructivism:

  > [...] the philosophical ideas behind intuitionism in general, and intuitionistic mathematics and intuitionistic logic in particular have a strong tendency toward constructivist points of view. There are no relationship between these ideas and the basic intuitive ideas of IFS theory [...]

The redundancy of the names "intuitionistic" fuzzy sets, "$\mathbb{L}^*$-fuzzy sets" and "interval-valued fuzzy sets" is also mentioned by J. Gutiérrez García and S. E. Rodabaugh in the abstract of [172]:

> ...(1) the term "intuitionistic" in these contexts is historically inappropriate given the standard mathematical usage of "intuitionistic"; and (2), at *every* level of existence—powerset level, topological fibre level, categorical level—interval-valued sets, [...], and "intuitionistic" fuzzy sets [...] are *redundant* ...

Also in a more recent paper by H. Bustince et al. ([5], p. 189) one can find an extensive discussion of the "terminological problem with the name intuitionistic", and the correctness of the notion chosen in [162,163] is explicitly acknowledged.

To summarize, the name "intuitionistic" in the context of $\mathbb{L}^*$-fuzzy sets is not compatible with the meaning of this term in the history of mathematics, and it would be better to avoid it.

Instead, because of the isomorphism between the lattice $(\mathbb{L}^*, \leq_{\mathbb{L}^*})$ and the lattice $(\mathfrak{I}(\mathbb{I}), \leq_{\mathfrak{I}(\mathbb{I})})$ of all closed subintervals of the unit interval $\mathbb{I}$, it is only a matter of personal taste and of the meaning given to the corresponding fuzzy sets to use one of the terms "$\mathbb{L}^*$-fuzzy sets" or "interval-valued fuzzy sets".

4.3. The Case of "Pythagorean" Fuzzy Sets

From Propositions 1 and 2 we know that the lattice $(P^*, \leq_{\mathbb{L}^*})$ given by (6) and (11) is isomorphic to each of the lattices $(\mathbb{L}^*, \leq_{\mathbb{L}^*})$, $(L_2(\mathbb{I}), \leq_{\mathrm{comp}})$, and $(\mathfrak{I}(\mathbb{I}), \leq_{\mathfrak{I}(\mathbb{I})})$.

Recently, in [15,88,89] the term "Pythagorean" fuzzy set was coined and used, which turns out to be a special case of an L-fuzzy set in the sense of [2], to be precise, an L-fuzzy set with P^* as lattice of truth values.

No justification for the choice of the adjective "Pythagorean" in this context was offered. One may only guess that the fact that, in the definition of the set P^* in (11), a sum of two squares occurs, indicating some similarity with the famous formula $a^2 + b^2 = c^2$ for right triangles—usually attributed to the Greek philosopher and mathematician Pythagoras who lived in the sixth century B.C.

The mutual isomorphism between the lattices $(P^*, \leq_{\mathbb{L}^*})$, $(\mathbb{L}^*, \leq_{\mathbb{L}^*})$, $(L_2(\mathbb{I}), \leq_{\text{comp}})$, and $(\mathfrak{J}(\mathbb{I}), \leq_{\mathfrak{J}(\mathbb{I})})$ implies that the families of L-fuzzy sets based on these lattices of truth values as well as the families of their corresponding membership functions are also isomorphic, i.e., have the same mathematical structure, as pointed out in Proposition 3. The identity of "Pythagorean" and "intuitionistic" fuzzy sets was also noted in ([5], Corollary 8.1).

Therefore, each mathematical result for $\mathbb{L}^*$-fuzzy sets, interval-valued fuzzy sets, "intuitionistic" fuzzy sets, etc., can be immediately translated into a result for "Pythagorean" fuzzy sets, and vice versa.

In other words, the term "Pythagorean" fuzzy sets is not only a fantasy name with no meaning whatsoever, it is absolutely useless, superfluous and even misleading, because it gives the impression to investigate something new, while isomorphic concepts have been studied already for many years. Therefore, the name "Pythagorean" fuzzy sets should be completely avoided.

Instead, because of the pairwise isomorphism between the lattices $(P^*, \leq_{\mathbb{L}^*})$, $(\mathbb{L}^*, \leq_{\mathbb{L}^*})$ and the lattice $(\mathfrak{J}(\mathbb{I}), \leq_{\mathfrak{J}(\mathbb{I})})$ of all closed subintervals of the unit interval $\mathbb{I}$, it is only a matter of personal taste to use one of the synonymous terms "$\mathbb{L}^*$-fuzzy sets" or "interval-valued fuzzy sets"—in any case, this can be done without any problem.

5. Concluding Remarks

As already mentioned, in the case of isomorphic lattices, any result known for one lattice can be rewritten in a straightforward way for each isomorphic lattice.

As a typical situation, recall that $(\mathbb{L}^*, \leq_{\mathbb{L}^*})$ and $(L_2(\mathbb{I}), \leq_{\text{comp}})$ are isomorphic lattices. Then, for each aggregation function $\mathbf{A}\colon \mathbb{I}^n \to \mathbb{I}$, the function $\mathbf{A}_{(2)}\colon (L_2(\mathbb{I}))^n \to L_2(\mathbb{I})$ given by

$$\mathbf{A}_{(2)}((x_1, y_1), (x_2, y_2), \ldots, (x_n, y_n)) = (\mathbf{A}(x_1, x_2, \ldots, x_n), \mathbf{A}(y_1, y_2, \ldots, y_n))$$

is an aggregation function on $L_2(\mathbb{I})$ (called representable in [85,110–120]), and any properties of $\mathbf{A}$ are inherited by $A_{(2)}$. For example, if $\mathbf{A}$ is a t-norm or t-conorm, uninorm, nullnorm, so is $\mathbf{A}_{(2)}$. If $\mathbf{A}$ is an averaging (conjunctive, disjunctive) aggregation function [46] so is $\mathbf{A}_{(2)}$, etc.

Due to the isomorphism between the lattices $(\mathbb{L}^*, \leq_{\mathbb{L}^*})$ and $(L_2(\mathbb{I}), \leq_{\text{comp}})$ (see Proposition 1), one can easily, for each aggregation function $\mathbf{A}\colon \mathbb{I}^n \to \mathbb{I}$, define the corresponding aggregation function $\mathbf{A}^*\colon (\mathbb{L}^*)^n \to \mathbb{L}^*$ by

$$\mathbf{A}^*((x_1, y_1), \ldots, (x_n, y_n)) = (\mathbf{A}(x_1, \ldots, x_n), 1 - \mathbf{A}(1 - y_1, \ldots, 1 - y_n)).$$

In doing so, it is superfluous to give long and tedious proofs that, whenever $\mathbf{A}$ is a t-norm (t-conorm, uninorm, nullnorm) on $\mathbb{I}$, then $\mathbf{A}^*$ is a t-norm (t-conorm, uninorm, nullnorm) on $\mathbb{L}^*$. Similarly, considering any averaging aggregation function $\mathbf{A}$ [46] (e.g., a weighted quasi-arithmetic mean based on an additive generator of some continuous Archimedean t-norm, e.g., the Einstein t-norm [177]), then evidently also $\mathbf{A}^*$ is an averaging (thus idempotent) aggregation function on $\mathbb{L}^*$.

In the same way, one can easily re-define aggregation functions on the "Pythagorean" lattice $(P^*, \leq_{\mathbb{L}^*})$, and again there is no need of proving their properties (automatically inherited from the original aggregation function $\mathbf{A}$ acting on $\mathbb{I}$), as it was done in, e.g., [178].

Finally, let us stress that we are not against reasonable generalizations of fuzzy sets in the sense of [1], in particular if they proved to be useful in certain applications.

However, as one of the referees for this paper noted, "the crucial point is: not to introduce the same under different name" and "not to re-prove the same [...] facts". Therefore we have underlined that it is superfluous to (re-)prove "new" results for isomorphic lattices when the corresponding results are already known for at least one of the (already existing) isomorphic lattices. Also, we will continue to argue against "new" fantasy names for known mathematical objects and against the (ab)use of established (historical) mathematical notions in an improper context.

Author Contributions: These authors contributed equally to this work.

Acknowledgments: We gratefully acknowledge the support by the "Technologie-Transfer-Förderung" of the Upper Austrian Government (Wi-2014-200710/13-Kx/Kai). The second author was also supported by the Slovak grant APVV-14-0013. Finally, we would like to thank the four anonymous referees for their valuable comments.

Conflicts of Interest: The authors declare no conflict of interest.

References

1. Zadeh, L.A. Fuzzy sets. *Inf. Control* **1965**, *8*, 338–353. [CrossRef]
2. Goguen, J.A. *L*-fuzzy sets. *J. Math. Anal. Appl.* **1967**, *18*, 145–174. [CrossRef]
3. Zadeh, L.A. Quantitative fuzzy semantics. *Inf. Sci.* **1971**, *3*, 159–176. [CrossRef]
4. Zadeh, L.A. The concept of a linguistic variable and its applications to approximate reasoning. Part I. *Inform. Sci.* **1975**, *8*, 199–251. [CrossRef]
5. Bustince, H.; Barrenechea, E.; Pagola, M.; Fernandez, J.; Xu, Z.; Bedregal, B.; Montero, J.; Hagras, H.; Herrera, F.; De Baets, B. A historical account of types of fuzzy sets and their relationships. *IEEE Trans. Fuzzy Syst.* **2016**, *24*, 179–194. [CrossRef]
6. Atanassov, K. Intuitionistic fuzzy sets. In *VII ITKR's Session, Sofia, June* 1983; Sgurev, V., Ed.; Central Science and Technology Library, Bulgarian Academy of Sciences: Sofia, Bulgaria, 1984.
7. Atanassov, K.T. Intuitionistic fuzzy sets. *Fuzzy Sets Syst.* **1986**, *20*, 87–96. [CrossRef]
8. Deng, J.L. Introduction to grey system theory. *J. Grey Syst.* **1989**, *1*, 1–24.
9. Deng, J.L. Grey information space. *J. Grey Syst.* **1989**, *1*, 103–117.
10. Gau, W.-L.; Buehrer, D.J. Vague sets. *IEEE Trans. Syst. Man Cybern.* **1993**, *23*, 610–614. [CrossRef]
11. Bedregal, B.; Beliakov, G.; Bustince, H.; Calvo, T.; Mesiar, R.; Paternain, D. A class of fuzzy multisets with a fixed number of memberships. *Inf. Sci.* **2012**, *189*, 1–17. [CrossRef]
12. Jahn, K.-U. Intervall-wertige Mengen. *Math. Nachr.* **1975**, *68*, 115–132. [CrossRef]
13. Sambuc, R. Fonctions φ-Floues: Application à L'aide au Diagnostic en Pathologie Thyroidienne. Ph.D. Thesis, Université Aix-Marseille II, Faculté de Médecine, Marseille, France, 1975. (In French)
14. Grattan-Guinness, I. Fuzzy membership mapped onto intervals and many-valued quantities. *Zeitschrift für Mathematische Logik und Grundlagen der Mathematik* **1976**, *22*, 149–160. [CrossRef]
15. Yager, R.R.; Abbasov, A.M. Pythagorean membership grades, complex numbers, and decision making. *Int. J. Intell. Syst.* **2013**, *28*, 436–452. [CrossRef]
16. Cuong, B.C.; Kreinovich, V. Picture fuzzy sets—A new concept for computational intelligence problems. In Proceedings of the Third World Congress on Information and Communication Technologies (WICT 2013), Hanoi, Vietnam, 15–18 December 2013; pp. 1–6.
17. Cantor, G. Beiträge zur Begründung der transfiniten Mengenlehre. Art. I. *Math. Ann.* **1895**, *46*, 481–512. (In German) [CrossRef]
18. Hausdorff, F. *Grundzüge der Mengenlehre*; Veit und Comp.: Leipzig, Germany, 1914. (In German)
19. Boole, G. *The Mathematical Analysis of Logic, Being an Essay Towards a Calculus of Deductive Reasoning*; Macmillan, Barclay, & Macmillan: Cambridge, UK, 1847.
20. Boole, G. *An Investigation of the Laws of Thought, on Which Are Founded the Mathematical Theories of Logic and Probabilities*; Walton: London, UK, 1854.
21. Menger, K. Probabilistic theories of relations. *Proc. Natl. Acad. Sci. USA* **1951**, *37*, 178–180. [CrossRef] [PubMed]
22. Menger, K. Probabilistic geometry. *Proc. Natl. Acad. Sci. USA* **1951**, *37*, 226–229. [CrossRef] [PubMed]
23. Menger, K. Ensembles flous et fonctions aléatoires. *C. R. Acad. Sci. Paris Sér. A* **1951**, *232*, 2001–2003. (In French)
24. Klaua, D. Über einen Ansatz zur mehrwertigen Mengenlehre. *Monatsb. Deutsch. Akad. Wiss.* **1965**, *7*, 859–867. (In German)
25. Klaua, D. Einbettung der klassischen Mengenlehre in die mehrwertige. *Monatsb. Deutsch. Akad. Wiss.* **1967**, *9*, 258–272. (In German)
26. De Luca, A.; Termini, S. Entropy of *L*-fuzzy sets. *Inf. Control* **1974**, *24*, 55–73. [CrossRef]

27. Negoita, C.V.; Ralescu, D.A. *Applications of Fuzzy Sets to Systems Analysis*; John Wiley & Sons: New York, NY, USA, 1975.
28. Negoita, C.V.; Ralescu, D.A. L-fuzzy sets and L-flou sets. *Elektronische Informationsverarbeitung und Kybernetik* **1976**, *12*, 599–605.
29. Höhle, U. Representation theorems for *L*-fuzzy quantities. *Fuzzy Sets Syst.* **1981**, *5*, 83–107. [CrossRef]
30. Sarkar, M. On *L*-fuzzy topological spaces. *J. Math. Anal. Appl.* **1981**, *84*, 431–442. [CrossRef]
31. Höhle, U. Probabilistic topologies induced by *L*-fuzzy uniformities. *Manuscr. Math.* **1982**, *38*, 289–323. [CrossRef]
32. Rodabaugh, S.E. Connectivity and the *L*-fuzzy unit interval. *Rocky Mt. J. Math.* **1982**, *12*, 113–121. [CrossRef]
33. Rodabaugh, S.E. Fuzzy addition in the *L*-fuzzy real line. *Fuzzy Sets Syst.* **1982**, *8*, 39–51. [CrossRef]
34. Cerruti, U. Completion of *L*-fuzzy relations. *J. Math. Anal. Appl.* **1983**, *94*, 312–327. [CrossRef]
35. Sugeno, M.; Sasaki, M. *L*-fuzzy category. *Fuzzy Sets Syst.* **1983**, *11*, 43–64. [CrossRef]
36. Klein, A.J. Generalizing the *L*-fuzzy unit interval. *Fuzzy Sets Syst.* **1984**, *12*, 271–279. [CrossRef]
37. Kubiak, T. *L*-fuzzy normal spaces and Tietze Extension Theorem. *J. Math. Anal. Appl.* **1987**, *125*, 141–153. [CrossRef]
38. Flüshöh, W.; Höhle, U. *L*-fuzzy contiguity relations and *L*-fuzzy closure operators in the case of completely distributive, complete lattices *L*. *Math. Nachr.* **1990**, *145*, 119–134. [CrossRef]
39. Kudri, S.R.T. Compactness in *L*-fuzzy topological spaces. *Fuzzy Sets Syst.* **1994**, *67*, 329–336. [CrossRef]
40. Kudri, S.R.T.; Warner, M.W. *L*-fuzzy local compactness. *Fuzzy Sets Syst.* **1994**, *67*, 337–345. [CrossRef]
41. Kubiak, T. On *L*-Tychonoff spaces. *Fuzzy Sets Syst.* **1995**, *73*, 25–53. [CrossRef]
42. Ovchinnikov, S. On the image of an *L*-fuzzy group. *Fuzzy Sets Syst.* **1998**, *94*, 129–131. [CrossRef]
43. Kubiak, T.; Zhang, D. On the *L*-fuzzy Brouwer fixed point theorem. *Fuzzy Sets Syst.* **1999**, *105*, 287–292. [CrossRef]
44. Jäger, G. A category of *L*-fuzzy convergence spaces. *Quaest. Math.* **2001**, *24*, 501–517. [CrossRef]
45. Birkhoff, G. *Lattice Theory*; American Mathematical Society: Providence, RI, USA, 1973.
46. Grabisch, M.; Marichal, J.-L.; Mesiar, R.; Pap, E. *Aggregation Functions*; Cambridge University Press: Cambridge, UK, 2009.
47. Grabisch, M.; Marichal, J.-L.; Mesiar, R.; Pap, E. Aggregation functions: Means. *Inf. Sci.* **2011**, *181*, 1–22. [CrossRef]
48. Pavelka, J. On fuzzy logic. II. Enriched residuated lattices and semantics of propositional calculi. *Z. Math. Log. Grundl. Math.* **1979**, *25*, 119–134. [CrossRef]
49. Höhle, U. Commutative, residuated l-monoids. In *Non-Classical Logics and Their Applications to Fuzzy Subsets. A Handbook of the Mathematical Foundations of Fuzzy Set Theory*; Höhle, U., Klement, E.P., Eds.; Kluwer Academic Publishers: Dordrecht, The Netherlands, 1995; Chapter IV, pp. 53–106.
50. Hájek, P. Basic fuzzy logic and BL-algebras. *Soft Comput.* **1998**, *2*, 124–128. [CrossRef]
51. Hájek, P. *Metamathematics of Fuzzy Logic*; Kluwer Academic Publishers: Dordrecht, The Netherlands, 1998.
52. Turunen, E. BL-algebras of basic fuzzy logic. *Mathw. Soft Comput.* **1999**, *6*, 49–61.
53. Esteva, F.; Godo, L.; Hájek, P.; Navara, M. Residuated fuzzy logics with an involutive negation. *Arch. Math. Log.* **2000**, *39*, 103–124. [CrossRef]
54. Jenei, S. New family of triangular norms via contrapositive symmetrization of residuated implications. *Fuzzy Sets Syst.* **2000**, *110*, 157–174. [CrossRef]
55. Jenei, S.; Kerre, E.E. Convergence of residuated operators and connective stability of non-classical logics. *Fuzzy Sets Syst.* **2000**, *114*, 411–415. [CrossRef]
56. Esteva, F.; Godo, L. Monoidal t-norm based logic: Towards a logic for left-continuous t-norms. *Fuzzy Sets Syst.* **2001**, *124*, 271–288. [CrossRef]
57. Hájek, P. Observations on the monoidal t-norm logic. *Fuzzy Sets Syst.* **2002**, *132*, 107–112. [CrossRef]
58. Esteva, F.; Godo, L.; Montagna, F. Axiomatization of any residuated fuzzy logic defined by a continuous t-norm. In Proceedings of the Congress of the International Fuzzy Systems Association (IFSA), Istanbul, Turkey, 30 June–2 July 2003; pp. 172–179.
59. Mesiar, R.; Mesiarová, A. Residual implications and left-continuous t-norms which are ordinal sums of semigroups. *Fuzzy Sets Syst.* **2004**, *143*, 47–57. [CrossRef]
60. Montagna, F. On the predicate logics of continuous t-norm BL-algebras. *Arch. Math. Log.* **2005**, *44*, 97–114. [CrossRef]

61. Van Gasse, B.; Cornelis, C.; Deschrijver, G.; Kerre, E.E. A characterization of interval-valued residuated lattices. *Int. J. Approx. Reason.* **2008**, *49*, 478–487. [CrossRef]
62. Van Gasse, B.; Cornelis, C.; Deschrijver, G.; Kerre, E.E. Triangle algebras: A formal logic approach to interval-valued residuated lattices. *Fuzzy Sets Syst.* **2008**, *159*, 1042–1060. [CrossRef]
63. De Baets, B.; Mesiar, R. Triangular norms on product lattices. *Fuzzy Sets Syst.* **1999**, *104*, 61–75. [CrossRef]
64. Saminger-Platz, S.; Klement, E.P.; Mesiar, R. On extensions of triangular norms on bounded lattices. *Indag. Math.* **2008**, *19*, 135–150. [CrossRef]
65. Menger, K. Statistical metrics. *Proc. Natl. Acad. Sci. USA* **1942**, *8*, 535–537. [CrossRef]
66. Schweizer, B.; Sklar, A. Espaces métriques aléatoires. *C. R. Acad. Sci. Paris Sér. A* **1958**, *247*, 2092–2094. (In French)
67. Schweizer, B.; Sklar, A. Statistical metric spaces. *Pac. J. Math.* **1960**, *10*, 313–334. [CrossRef]
68. Schweizer, B.; Sklar, A. *Probabilistic Metric Spaces*; North-Holland: New York, NY, USA, 1983.
69. Klement, E.P.; Mesiar, R.; Pap, E. *Triangular Norms*; Kluwer Academic Publishers: Dordrecht, The Netherlands, 2000.
70. Klement, E.P.; Mesiar, R.; Pap, E. Triangular norms. Position paper I: Basic analytical and algebraic properties. *Fuzzy Sets Syst.* **2004**, *143*, 5–26. [CrossRef]
71. Klement, E.P.; Mesiar, R.; Pap, E. Triangular norms. Position paper II: General constructions and parameterized families. *Fuzzy Sets Syst.* **2004**, *145*, 411–438. [CrossRef]
72. Klement, E.P.; Mesiar, R.; Pap, E. Triangular norms. Position paper III: Continuous t-norms. *Fuzzy Sets Syst.* **2004**, *145*, 439–454. [CrossRef]
73. Alsina, C.; Frank, M.J.; Schweizer, B. *Associative Functions: Triangular Norms and Copulas*; World Scientific: Singapore, 2006.
74. Yager, R.R.; Rybalov, A. Uninorm aggregation operators. *Fuzzy Sets Syst.* **1996**, *80*, 111–120. [CrossRef]
75. Calvo, T.; De Baets, B.; Fodor, J. The functional equations of Frank and Alsina for uninorms and nullnorms. *Fuzzy Sets Syst.* **2001**, *120*, 385–394. [CrossRef]
76. Calvo, T.; Mayor, G.; Mesiar, R. (Eds.) *Aggregation Operators. New Trends and Applications*; Physica-Verlag: Heidelberg, Germany, 2002.
77. Beliakov, G.; Pradera, A.; Calvo, T. *Aggregation Functions: A Guide for Practitioners*; Springer: Berlin/Heidelberg, Germany, 2007.
78. Grabisch, M.; Marichal, J.-L.; Mesiar, R.; Pap, E. Aggregation functions: Construction methods, conjunctive, disjunctive and mixed classes. *Inf. Sci.* **2011**, *181*, 23–43. [CrossRef]
79. Negoita, C.V.; Ralescu, D.A. Representation theorems for fuzzy concepts. *Kybernetes* **1975**, *4*, 169–174. [CrossRef]
80. Shortliffe, E.H.; Buchanan, B.G. A model of inexact reasoning in medicine. *Math. Biosci.* **1975**, *23*, 351–379. [CrossRef]
81. Shortliffe, E.H. *Computer Based Medical Consultation—'MYCIN'*; Elsevier: New York, NY, USA, 1976.
82. Jackson, P. *Introduction to Expert Systems*; Addison-Wesley: Wokingham, UK, 1986.
83. Hájek, P.; Havránek, T.; Jiroušek, R. *Uncertain Information Processing in Expert Systems*; CRC Press: Boca Raton, FL, USA, 1992.
84. Atanassov, K.T. *Intuitionistic Fuzzy Sets*; Physica-Verlag: Heidelberg, Germany, 1999.
85. Deschrijver, G.; Kerre, E.E. On the relationship between some extensions of fuzzy set theory. *Fuzzy Sets Syst.* **2003**, *133*, 227–235. [CrossRef]
86. Wang, G.-J.; He, Y.-Y. Intuitionistic fuzzy sets and *L*-fuzzy sets. *Fuzzy Sets Syst.* **2000**, *110*, 271–274. [CrossRef]
87. De Miguel, L.; Bustince, H.; Fernandez, J.; Induráin, E.; Kolesárová, A.; Mesiar, R. Construction of admissible linear orders for interval-valued Atanassov intuitionistic fuzzy sets with an application to decision making. *Inf. Fusion* **2016**, *27*, 189–197. [CrossRef]
88. Dick, S.; Yager, R.R.; Yazdanbakhsh, O. On Pythagorean and complex fuzzy set operations. *IEEE Trans. Fuzzy Syst.* **2016**, *24*, 1009–1021. [CrossRef]
89. Yager, R.R. Pythagorean membership grades in multi-criteria decision making. *IEEE Trans. Fuzzy Syst.* **2014**, *22*, 958–965. [CrossRef]
90. Harding, J.; Walker, C.; Walker, E. The variety generated by the truth value algebra of type-2 fuzzy sets. *Fuzzy Sets Syst.* **2010**, *161*, 735–749. [CrossRef]
91. Walker, C.; Walker, E. The algebra of fuzzy truth values. *Fuzzy Sets Syst.* **2005**, *149*, 309–347. [CrossRef]
92. Bustince, H.; Burillo, P. Vague sets are intuitionistic fuzzy sets. *Fuzzy Sets Syst.* **1996**, *79*, 403–405. [CrossRef]

93. Smarandache, F. *Neutrosophy: Neutrosophic Probability, Set, and Logic: Analytic Synthesis & Synthetic Analysis*; American Research Press: Rehoboth, NM, USA, 1998.
94. Smarandache, F. A unifying field in logics: Neutrosophic logic. *Multiple-Valued Log.* **2002**, *8*, 385–482.
95. Smarandache, F. Definition of neutrosophic logic—A generalization of the intuitionistic fuzzy logic. In Proceedings of the 3rd Conference of the European Society for Fuzzy Logic and Technology, Zittau, Germany, 10–12 September 2003; pp. 141–146.
96. Smarandache, F. Neutrosophic set—A generalization of the intuitionistic fuzzy set. In Proceedings of the 2006 IEEE International Conference on Granular Computing, Atlanta, GA, USA, 12–12 May 2006; pp. 38–42.
97. Wang, H.; Smarandache, F.; Zhang, Y.; Sunderraman, R. Single valued neutrosophic sets. In *Multispace & Multistructure. Neutrosophic Transdisciplinarity (100 Collected Papers of Sciences)*; Smarandache, F., Ed.; North-European Scientific Publishers: Hanko, Finland, 2010; Volume IV, pp. 410–413.
98. Bedregal, B.; Mezzomo, I.; Reiser, R.H.S. *n*-Dimensional Fuzzy Negations. 2017. Available online: arXiv.org/pdf/1707.08617v1 (accessed on 26 July 2017).
99. Bedregal, B.; Beliakov, G.; Bustince, H.; Calvo, T.; Fernández, J.; Mesiar, R. A characterization theorem for t-representable *n*-dimensional triangular norms. In *Eurofuse 2011. Workshop on Fuzzy Methods for Knowledge-Based Systems*; Melo-Pinto, P., Couto, P., Serôdio, C., Fodor, J., De Baets, B., Eds.; Springer: Berlin/Heidelberg, Germany, 2012; pp. 103–112.
100. Cuong, B.C.; Kreinovich, V.; Ngan, R.T. A classification of representable t-norm operators for picture fuzzy sets. In Proceedings of the Eighth International Conference on Knowledge and Systems Engineering (KSE 2016), Hanoi, Vietnam, 6–8 October 2016; pp. 19–24.
101. Cuong, B.C.; Ngan, R.T.; Ngoc, L.C. Some algebraic properties of picture fuzzy t-norms and picture fuzzy t-conorms on standard neutrosophic sets. *arXiv* **2017**, arXiv:1701.0144
102. Son, L.H.; Viet, P.V.; Hai, P.V. Picture inference system: A new fuzzy inference system on picture fuzzy set. *Appl. Intell.* **2017**, *46*, 652–669. [CrossRef]
103. Bo, C.; Zhang, X. New operations of picture fuzzy relations and fuzzy comprehensive evaluation. *Symmetry* **2017**, *9*, 268. [CrossRef]
104. Wang, C.; Zhou, X.; Tu, H.; Tao, S. Some geometric aggregation operators based on picture fuzzy sets and their application in multiple attribute decision making. *Ital. J. Pure Appl. Math.* **2017**, *37*, 477–492.
105. Klement, E.P.; Mesiar, R.; Stupňanová, A. Picture fuzzy sets and 3-fuzzy sets. In Proceedings of the 2018 IEEE International Conference on Fuzzy Systems (FUZZ-IEEE), Rio de Janeiro, Brazil, 8–13 July 2018; pp. 476–482.
106. Cuong, B.C. Picture fuzzy sets. *J. Comput. Sci. Cybern.* **2014**, *30*, 409–420.
107. Cuong, B.C.; Hai, P.V. Some fuzzy logic operators for picture fuzzy sets. In Proceedings of the Seventh International Conference on Knowledge and Systems Engineering (KSE 2015), Ho Chi Minh City, Vietnam, 8–10 October 2015; pp. 132–137.
108. Cuong, B.C.; Ngan, R.T.; Hai, B.D. An involutive picture fuzzy negation on picture fuzzy sets and some De Morgan triples. In Proceedings of the Seventh International Conference on Knowledge and Systems Engineering (KSE 2015), Ho Chi Minh City, Vietnam, 8–10 October 2015; pp. 126–131.
109. Thong, P.H.; Son, L.H. Picture fuzzy clustering: A new computational intelligence method. *Soft Comput.* **2016**, *20*, 3549–3562. [CrossRef]
110. Deschrijver, G.; Kerre, E.E. Classes of intuitionistic fuzzy t-norms satisfying the residuation principle. *Int. J. Uncertain. Fuzziness Knowl.-Based Syst.* **2003**, *11*, 691–709. [CrossRef]
111. Deschrijver, G. The Archimedean property for t-norms in interval-valued fuzzy set theory. *Fuzzy Sets Syst.* **2006**, *157*, 2311–2327. [CrossRef]
112. Deschrijver, G. Arithmetic operators in interval-valued fuzzy set theory. *Inf. Sci.* **2007**, *177*, 2906–2924. [CrossRef]
113. Deschrijver, G. A representation of t-norms in interval-valued *L*-fuzzy set theory. *Fuzzy Sets Syst.* **2008**, *159*, 1597–1618. [CrossRef]
114. Deschrijver, G. Characterizations of (weakly) Archimedean t-norms in interval-valued fuzzy set theory. *Fuzzy Sets Syst.* **2009**, *160*, 778–801. [CrossRef]
115. Deschrijver, G.; Cornelis, C. Representability in interval-valued fuzzy set theory. *Int. J. Uncertain. Fuzziness Knowl.-Based Syst.* **2007**, 15, 345–361. [CrossRef]
116. Deschrijver, G.; Cornelis, C.; Kerre, E.E. On the representation of intuitionistic fuzzy t-norms and t-conorms. *IEEE Trans. Fuzzy Syst.* **2004**, *12*, 45–61. [CrossRef]

117. Deschrijver, G.; Kerre, E.E. On the composition of intuitionistic fuzzy relations. *Fuzzy Sets Syst.* **2003**, *136*, 333–361. [CrossRef]
118. Deschrijver, G.; Kerre, E.E. Uninorms in L^*-fuzzy set theory. *Fuzzy Sets Syst.* **2004**, *148*, 243–262. [CrossRef]
119. Deschrijver, G.; Kerre, E.E. Implicators based on binary aggregation operators in interval-valued fuzzy set theory. *Fuzzy Sets Syst.* **2005**, *153*, 229–248. [CrossRef]
120. Deschrijver, G.; Kerre, E.E. Triangular norms and related operators in L^*-fuzzy set theory. In *Logical, Algebraic, Analytic, and Probabilistic Aspects of Triangular Norms;* Klement, E.P., Mesiar, R., Eds.; Elsevier: Amsterdam, The Netherlands, 2005; Chapter 8, pp. 231–259.
121. Abbas, S.E. Intuitionistic supra fuzzy topological spaces. *Chaos Solitons Fractals* **2004**, *21*, 1205–1214. [CrossRef]
122. Atanassov, K.; Gargov, G. Interval valued intuitionistic fuzzy sets. *Fuzzy Sets Syst.* **1989**, *31*, 343–349. [CrossRef]
123. Atanassov, K.; Gargov, G. Elements of intuitionistic fuzzy logic. Part I. *Fuzzy Sets Syst.* **1998**, *95*, 39–52. [CrossRef]
124. Atanassov, K.T. More on intuitionistic fuzzy sets. *Fuzzy Sets Syst.* **1989**, *33*, 37–45. [CrossRef]
125. Atanassov, K.T. Remarks on the intuitionistic fuzzy sets. *Fuzzy Sets Syst.* **1992**, *51*, 117–118. [CrossRef]
126. Atanassov, K.T. New operations defined over the intuitionistic fuzzy sets. *Fuzzy Sets Syst.* **1994**, *61*, 137–142. [CrossRef]
127. Atanassov, K.T. Operators over interval valued intuitionistic fuzzy sets. *Fuzzy Sets Syst.* **1994**, *64*, 159–174. [CrossRef]
128. Atanassov, K.T. Remarks on the intuitionistic fuzzy sets—III. *Fuzzy Sets Syst.* **1995**, *75*, 401–402. [CrossRef]
129. Atanassov, K.T. An equality between intuitionistic fuzzy sets. *Fuzzy Sets Syst.* **1996**, *79*, 257–258. [CrossRef]
130. Atanassov, K.T. Remark on the intuitionistic fuzzy logics. *Fuzzy Sets Syst.* **1998**, *95*, 127–129. [CrossRef]
131. Atanassov, K.T. Two theorems for intuitionistic fuzzy sets. *Fuzzy Sets Syst.* **2000**, *110*, 267–269. [CrossRef]
132. Atanassova, L.C. Remark on the cardinality of the intuitionistic fuzzy sets. *Fuzzy Sets Syst.* **1995**, *75*, 399–400. [CrossRef]
133. Ban, A.I.; Gal, S.G. Decomposable measures and information measures for intuitionistic fuzzy sets. *Fuzzy Sets Syst.* **2001**, *123*, 103–117. [CrossRef]
134. Burillo, P.; Bustince, H. Construction theorems for intuitionistic fuzzy sets. *Fuzzy Sets Syst.* **1996**, *84*, 271–281. [CrossRef]
135. Burillo, P.; Bustince, H. Entropy on intuitionistic fuzzy sets and on interval-valued fuzzy sets. *Fuzzy Sets Syst.* **1996**, *78*, 305–316. [CrossRef]
136. Bustince, H. Construction of intuitionistic fuzzy relations with predetermined properties. *Fuzzy Sets Syst.* **2000**, *109*, 379–403. [CrossRef]
137. Bustince, H.; Burillo, P. Structures on intuitionistic fuzzy relations. *Fuzzy Sets Syst.* **1996**, *78*, 293–300. [CrossRef]
138. Çoker, D. An introduction to intuitionistic fuzzy topological spaces. *Fuzzy Sets Syst.* **1997**, *88*, 81–89. [CrossRef]
139. Çoker, D.; Demirci, M. An introduction to intuitionistic fuzzy topological spaces in šostak's sense. *Busefal* **1996**, *67*, 67–76.
140. De, S.K.; Biswas, R.; Roy, A.R. Some operations on intuitionistic fuzzy sets. *Fuzzy Sets Syst.* **2000**, *114*, 477–484. [CrossRef]
141. De, S.K.; Biswas, R.; Roy, A.R. An application of intuitionistic fuzzy sets in medical diagnosis. *Fuzzy Sets Syst.* **2001**, *117*, 209–213. [CrossRef]
142. Demirci, M. Axiomatic theory of intuitionistic fuzzy sets. *Fuzzy Sets Syst.* **2000**, *110*, 253–266. [CrossRef]
143. Lee, S.J.; Lee, E.P. The category of intuitionistic fuzzy topological spaces. *Bull. Korean Math. Soc.* **2000**, *37*, 63–76.
144. Mondal, T.K.; Samanta, S.K. On intuitionistic gradation of openness. *Fuzzy Sets Syst.* **2002**, *131*, 323–336. [CrossRef]
145. Samanta, S.K.; Mondal, T.K. Intuitionistic gradation of openness: Intuitionistic fuzzy topology. *Busefal* **1997**, *73*, 8–17.
146. Szmidt, E.; Kacprzyk, J. Distances between intuitionistic fuzzy sets. *Fuzzy Sets Syst.* **2000**, *114*, 505–518. [CrossRef]

147. Szmidt, E.; Kacprzyk, J. Entropy for intuitionistic fuzzy sets. *Fuzzy Sets Syst.* **2001**, *118*, 467–477. [CrossRef]
148. Brouwer, L.E.J. Intuitionism and formalism. *Bull. Am. Math. Soc.* **1913**, *20*, 81–96. [CrossRef]
149. Brouwer, L.E.J. Intuitionistische verzamelingsleer. *Amst. Ak. Versl.* **1921**, *29*, 797–802. (In Dutch)
150. Brouwer, L.E.J. Intuitionistische splitsing van mathematische grondbegrippen. *Amst. Ak. Versl.* **1923**, *32*, 877–880. (In Dutch)
151. Brouwer, L.E.J. Über die Bedeutung des Satzes vom ausgeschlossenen Dritten in der Mathematik, insbesondere in der Funktionentheorie. *J. Reine Angew. Math.* **1925**, *154*, 1–7. (In German)
152. Brouwer, L.E.J. Zur Begründung der intuitionistischen Mathematik. I. *Math. Ann.* **1925**, *93*, 244–257. (In German) [CrossRef]
153. Brouwer, L.E.J. Zur Begründung der intuitionistischen Mathematik. II. *Math. Ann.* **1926**, *95*, 453–472. (In German) [CrossRef]
154. Brouwer, L.E.J. Zur Begründung der intuitionistischen Mathematik. III. *Math. Ann.* **1927**, *96*, 451–488. (In German) [CrossRef]
155. Brouwer, L.E.J. Intuitionistische Betrachtungen über den Formalismus. *Sitz. Preuß. Akad. Wiss. Phys. Math. Kl.* **1928**, 48–52. (In German)
156. Brouwer, L.E.J. On the significance of the principle of excluded middle in mathematics, especially in function theory. With two Addenda and corrigenda. In *From Frege to Gödel. A Source Book in Mathematical Logic, 1879–1931*; van Heijenoort, J., Ed.; Harvard University Press: Cambridge, MA, USA, 1967; pp. 334–345.
157. Van Heijenoort, J. *From Frege to Gödel. A Source Book in Mathematical Logic*, 1879–1931; Harvard University Press: Cambridge, MA, USA, 1967.
158. Hilbert, D. Die Grundlagen der Mathematik. Vortrag, gehalten auf Einladung des Mathematischen Seminars im Juli 1927 in Hamburg. *Abh. Math. Semin. Univ. Hamb.* **1928**, *6*, 65–85. (In German) [CrossRef]
159. Hilbert, D.; Bernays, P. *Grundlagen der Mathematik. I*; Springer: Berlin/Heidelberg, Germany, 1934. (In German)
160. Hilbert, D. The foundations of mathematics. In *From Frege to Gödel. A Source Book in Mathematical Logic, 1879–1931*; van Heijenoort, J., Ed.; Harvard University Press: Cambridge, MA, USA, 1967; pp. 464–480.
161. Kolmogorov, A.N. On the principle of excluded middle. In *From Frege to Gödel. A Source Book in Mathematical Logic, 1879–1931*; van Heijenoort, J., Ed.; Harvard University Press: Cambridge, MA, USA, 1967; pp. 414–437.
162. Takeuti, G.; Titani, S. Intuitionistic fuzzy logic and intuitionistic fuzzy set theory. *J. Symb. Log.* **1984**, *49*, 851–866. [CrossRef]
163. Takeuti, G.; Titani, S. Globalization of intuitionistic set theory. *Ann. Pure Appl. Log.* **1987**, *33*, 195–211. [CrossRef]
164. Baaz, M.; Fermüller, C.G. Intuitionistic counterparts of finitely-valued logics. In Proceedings of the 26th International Symposium on Multiple-Valued Logic, Santiago de Compostela, Spain, 19–31 January 1996; pp. 136–141.
165. Ciabattoni, A. A proof-theoretical investigation of global intuitionistic (fuzzy) logic. *Arch. Math. Log.* **2005**, *44*, 435–457. [CrossRef]
166. Gottwald, S. Universes of fuzzy sets and axiomatizations of fuzzy set theory. I. Model-based and axiomatic approaches. *Stud. Log.* **2006**, *82*, 211–244. [CrossRef]
167. Gottwald, S. Universes of fuzzy sets and axiomatizations of fuzzy set theory. II. Category theoretic approaches. *Stud. Log.* **2006**, *84*, 23–50. [CrossRef]
168. Hájek, P.; Cintula, P. On theories and models in fuzzy predicate logics. *J. Symb. Log.* **2006**, *71*, 863–880. [CrossRef]
169. Dubois, D.; Gottwald, S.; Hajek, P.; Kacprzyk, J.; Prade, H. Terminological difficulties in fuzzy set theory—The case of "Intuitionistic Fuzzy Sets". *Fuzzy Sets Syst.* **2005**, *156*, 485–491. [CrossRef]
170. Cattaneo, G.; Ciucci, D. Generalized negations and intuitionistic fuzzy sets—A criticism to a widely used terminology. In Proceedings of the 3rd Conference of the European Society for Fuzzy Logic and Technology, Zittau, Germany, 10–12 September 2003; pp. 147–152.
171. Grzegorzewski, P.; Mrówka, E. Some notes on (Atanassov's) intuitionistic fuzzy sets. *Fuzzy Sets Syst.* **2005**, *156*, 492–495. [CrossRef]
172. Gutiérrez García, J.; Rodabaugh, S.E. Order-theoretic, topological, categorical redundancies of interval-valued sets, grey sets, vague sets, interval-valued "intuitionistic" sets, "intuitionistic" fuzzy sets and topologies. *Fuzzy Sets Syst.* **2005**, *156*, 445–484. [CrossRef]

173. Atanassov, K. Answer to D. Dubois, S. Gottwald, P. Hajek, J. Kacprzyk and H. Prade's paper "Terminological difficulties in fuzzy set theory—The case of "Intuitionistic Fuzzy Sets"". *Fuzzy Sets Syst.* **2005**, *156*, 496–499. [CrossRef]
174. Butnariu, D.; Klement, E.P.; Mesiar, R.; Navara, M. Sufficient triangular norms in many-valued logics with standard negation. *Arch. Math. Log.* **2005**, *44*, 829–849. [CrossRef]
175. Cintula, P.; Klement, E.P.; Mesiar, R.; Navara, M. Residuated logics based on strict triangular norms with an involutive negation. *Math. Log. Quart.* **2006**, *52*, 269–282. [CrossRef]
176. Kleene, S.C. *Introduction to Metamathematics*; North-Holland: Amsterdam, The Netherlands, 1952.
177. Xia, M.; Xu, Z.; Zhu, B. Some issues on intuitionistic fuzzy aggregation operators based on Archimedean t-conorm and t-norm. *Knowl.-Based Syst.* **2012**, *31*, 78–88. [CrossRef]
178. Rahman, K.; Abdullah, S.; Ahmed, R.; Ullah, M. Pythagorean fuzzy Einstein weighted geometric aggregation operator and their application to multiple attribute group decision making. *J. Intell. Fuzzy Syst.* **2017**, *33*, 635–647. [CrossRef]

mathematics

Article

On the Most Extended Modal Operator of First Type over Interval-Valued Intuitionistic Fuzzy Sets

Krassimir Atanassov [1,2]

[1] Department of Bioinformatics and Mathematical Modelling, Institute of Biophysics and Biomedical Engineering, Bulgarian Academy of Sciences, 105 Acad. G. Bonchev Str., 1113 Sofia, Bulgaria; krat@bas.bg
[2] Intelligent Systems Laboratory, Prof. Asen Zlatarov University, 8010 Bourgas, Bulgaria

Received: 30 May 2018; Accepted: 4 July 2018; Published: 13 July 2018

Abstract: The definition of the most extended modal operator of first type over interval-valued intuitionistic fuzzy sets is given, and some of its basic properties are studied.

Keywords: interval-valued intuitionistic fuzzy set; intuitionistic fuzzy set; modal operator

---

1. Introduction

Intuitionistic fuzzy sets (IFSs; see [1–5]) were introduced in 1983 as an extension of the fuzzy sets defined by Lotfi Zadeh (4.2.1921–6.9.2017) in [6]. In recent years, the IFSs have also been extended: intuitionistic L-fuzzy sets [7], IFSs of second [8] and nth [9–12] types, temporal IFSs [4,5,13], multidimensional IFSs [5,14], and others. Interval-valued intuitionistic fuzzy sets (IVIFSs) are the most detailed described extension of IFSs. They appeared in 1988, when Georgi Gargov (7.4.1947–9.11.1996) and the author read Gorzalczany's paper [15] on the interval-valued fuzzy set (IVFS). The idea of IVIFS was announced in [16,17] and extended in [4,18], where the proof that IFSs and IVIFSs are equipollent generalizations of the notion of the fuzzy set is given.

Over IVIFS, many (more than the ones over IFSs) relations, operations, and operators are defined. Here, similar to the IFS case, the standard modal operators $\Box$ and $\Diamond$ have analogues, but their extensions—the intuitionistic fuzzy extended modal operators of the first type—already have two different forms. In the IFS case, there is an operator that includes as a partial case all other extended modal operators. In the present paper, we construct a similar operator for the case of IVIFSs and study its properties.

2. Preliminaries

Let us have a fixed universe E and its subset A. The set

$$A = \{\langle x, M_A(x), N_A(x)\rangle \mid x \in E\},$$

where $M_A(x) \subset [0,1]$ and $N_A(x) \subset [0,1]$ are closed intervals and for all $x \in E$:

$$\sup M_A(x) + \sup N_A(x) \leq 1 \quad (1)$$

is called IVIFS, and functions $M_A : E \to \mathcal{P}([0,1])$ and $N_A : E \to \mathcal{P}([0,1])$ represent the *set of degrees of membership (validity, etc.)* and *the set of degrees of non-membership (non-validity, etc.)* of element $x \in E$ to a fixed set $A \subseteq E$, where $\mathcal{P}(Z) = \{Y | Y \subseteq Z\}$ for an arbitrary set Z.

Obviously, both intervals have the representation:

$$M_A(x) = [\inf M_A(x), \sup M_A(x)],$$

$$N_A(x) = [\inf N_A(x), \sup N_A(x)].$$

Therefore, when

$$\inf M_A(x) = \sup M_A(x) = \mu_A(x) \quad \text{and} \quad \inf N_A(x) = \sup N_A(x) = \nu_A(x),$$

the IVIFS A is transformed to an IFS.

We must mention that in [19,20] the second geometrical interpretation of the IFSs is given (see Figure 1).

IVIFSs have geometrical interpretations similar to, but more complex than, those of the IFSs. For example, the analogue of the geometrical interpretation from Figure 1 is shown in Figure 2.

Obviously, each IVFS A can be represented by an IVIFS as

$$A = \{\langle x, M_A(x), N_A(x)\rangle \mid x \in E\}$$
$$= \{\langle x, M_A(x), [1 - \sup M_A(x), 1 - \inf M_A(x)]\rangle \mid x \in E\}.$$

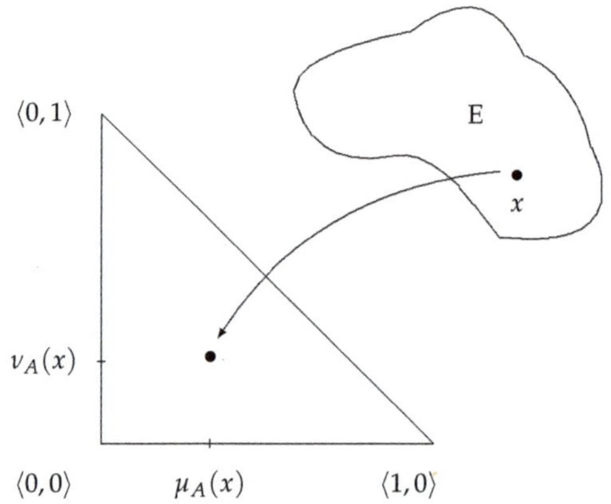

Figure 1. The second geometrical interpretation of an intuitionistic fuzzy set (IFS).

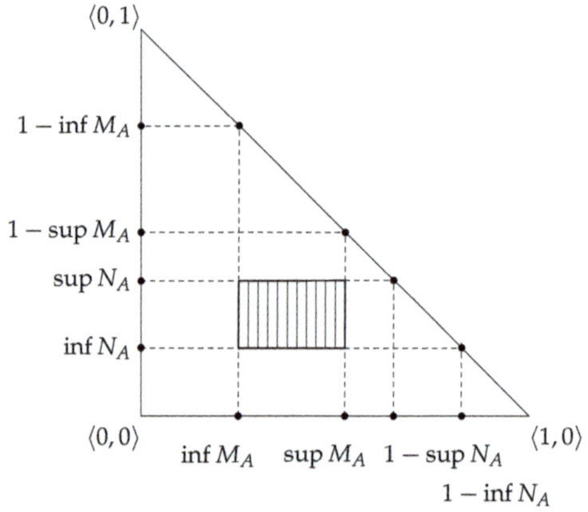

Figure 2. The second geometrical interpretation of an interval-valued intuitionistic fuzzy set (IVIFS).

The geometrical interpretation of the IVFS A is shown in Figure 3. It has the form of a section lying on the triangle's hypotenuse.

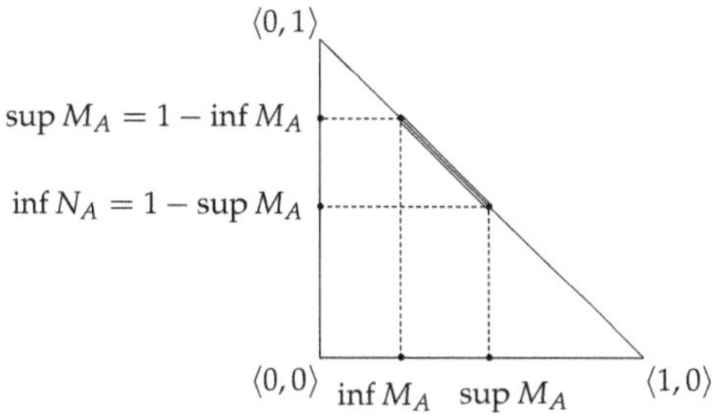

Figure 3. The second geometrical interpretation of an IVFS.

Modal-type operators are defined similarly to those defined for IFSs, but here they have two forms: shorter and longer. The shorter form is:

$$\begin{aligned}
\square A = &\ \{\langle x, M_A(x), [\inf N_A(x), 1 - \sup M_A(x)]\rangle \mid x \in E\},\\
\Diamond A = &\ \{\langle x, [\inf M_A(x), 1 - \sup N_A(x)], N_A(x)\rangle \mid x \in E\},\\
D_\alpha(A) = &\ \{\langle x, [\inf M_A(x), \sup M_A(x) + \alpha(1 - \sup M_A(x) - \sup N_A(x))],\\
&\ [\inf N_A(x), \sup N_A(x) + (1-\alpha)(1 - \sup M_A(x) - \sup N_A(x))]\rangle\\
&\ \mid x \in E\},\\
F_{\alpha,\beta}(A) = &\ \{\langle x, [\inf M_A(x), \sup M_A(x) + \alpha(1 - \sup M_A(x) - \sup N_A(x))],\\
&\ [\inf N_A(x), \sup N_A(x) + \beta(1 - \sup M_A(x) - \sup N_A(x))]\rangle\\
&\ \mid x \in E\}, \text{ for } \alpha + \beta \leq 1,\\
G_{\alpha,\beta}(A) = &\ \{\langle x, [\alpha \inf M_A(x), \alpha \sup M_A(x)], [\beta \inf N_A(x), \beta \sup N_A(x)]\rangle\\
&\ \mid x \in E\},\\
H_{\alpha,\beta}(A) = &\ \{\langle x, [\alpha \inf M(x), \alpha \sup M_A(x)], [\inf N_A(x), \sup N_A(x)\\
&\ +\beta(1 - \sup M_A(x) - \sup N_A(x))]\rangle \mid x \in E\},\\
H^*_{\alpha,\beta}(A) = &\ \{\langle x, [\alpha \inf M_A(x), \alpha \sup M_A(x)], [\inf N_A(x), \sup N_A(x)\\
&\ +\beta(1 - \alpha \sup M_A(x) - \sup N_A(x))]\rangle \mid x \in E\},\\
J_{\alpha,\beta}(A) = &\ \{\langle x, [\inf M_A(x), \sup M_A(x) + \alpha(1 - \sup M_A(x)\\
&\ - \sup N_A(x))], [\beta \inf N_A(x), \beta \sup N_A(x)]\rangle \mid x \in E\},\\
J^*_{\alpha,\beta}(A) = &\ \{\langle x, [\inf M_A(x), \sup M_A(x) + \alpha(1 - \sup M_A(x)\\
&\ -\beta \sup N_A(x))], [\beta \inf N_A(x), \beta . \sup N_A(x)]\rangle \mid x \in E\},
\end{aligned}$$

where $\alpha, \beta \in [0, 1]$.

Obviously, as in the case of IFSs, the operator D_α is an extension of the intuitionistic fuzzy forms of (standard) modal logic operators $\square$ and $\Diamond$, and it is a partial case of $F_{\alpha,\beta}$.

The longer form of these operators (operators $\Box$, $\Diamond$, and D do not have two forms—only the one above) is (see [4]):

$$\begin{aligned}
\overline{F}_{\begin{pmatrix} \alpha & \gamma \\ \beta & \delta \end{pmatrix}}(A) = \;& \{\langle x, [\inf M_A(x) + \alpha(1 - \sup M_A(x) - \sup N_A(x)), \\
& \sup M_A(x) + \beta(1 - \sup M_A(x) - \sup N_A(x))], \\
& [\inf N_A(x) + \gamma(1 - \sup M_A(x) - \sup N_A(x)), \\
& \sup N_A(x) + \delta(1 - \sup M_A(x) - \sup N_A(x))]\rangle \mid x \in E\} \\
& \text{where } \beta + \delta \leq 1, \\
\overline{G}_{\begin{pmatrix} \alpha & \gamma \\ \beta & \delta \end{pmatrix}}(A) = \;& \{\langle x, [\alpha \inf M_A(x), \beta \sup M_A(x)], \\
& [\gamma \inf N_A(x), \delta \sup N_A(x)]\rangle \mid x \in E\}, \\
\overline{H}_{\begin{pmatrix} \alpha & \gamma \\ \beta & \delta \end{pmatrix}}(A) = \;& \{\langle x, [\alpha \inf M_A(x), \beta \sup M_A(x)], \\
& [\inf N_A(x) + \gamma(1 - \sup M_A(x) - \sup N_A(x)), \\
& \sup N_A(x) + \delta(1 - \sup M_A(x) - \sup N_A(x))]\rangle \mid x \in E\}, \\
\overline{H}^*_{\begin{pmatrix} \alpha & \gamma \\ \beta & \delta \end{pmatrix}}(A) = \;& \{\langle x, [\alpha \inf M_A(x), \beta \sup M_A(x)], \\
& [\inf N_A(x) + \gamma(1 - \beta \sup M_A(x) - \sup N_A(x)), \\
& \sup N_A(x) + \delta(1 - \beta \sup M_A(x) - \sup N_A(x))]\rangle \mid x \in E\}, \\
\overline{J}_{\begin{pmatrix} \alpha & \gamma \\ \beta & \delta \end{pmatrix}} = \;& \{\langle x, [\inf M_A(x) + \alpha(1 - \sup M_A(x) - \sup N_A(x)), \\
& \sup M_A(x) + \beta(1 - \sup M_A(x) - \sup N_A(x))], \\
& [\gamma \inf N_A(x), \delta \sup N_A(x)]\rangle \mid x \in E\}, \\
\overline{J}^*_{\begin{pmatrix} \alpha & \gamma \\ \beta & \delta \end{pmatrix}}(A) = \;& \{\langle x, [\inf M_A(x) + \alpha(1 - \delta \sup M_A(x) - \sup N_A(x)), \\
& \sup M_A(x) + \beta(1 - \sup M_A(x) - \delta \sup N_A(x))], \\
& [\gamma . \inf N_A(x), \delta . \sup N_A(x)]\rangle \mid x \in E\},
\end{aligned}$$

where $\alpha, \beta, \gamma, \delta \in [0, 1]$ such that $\alpha \leq \beta$ and $\gamma \leq \delta$.

Figure 4 shows to which region of the triangle the element $x \in E$ (represented by the small rectangular region in the triangle) will be transformed by the operators $F, G, \ldots$, irrespective of whether they have two or four indices.

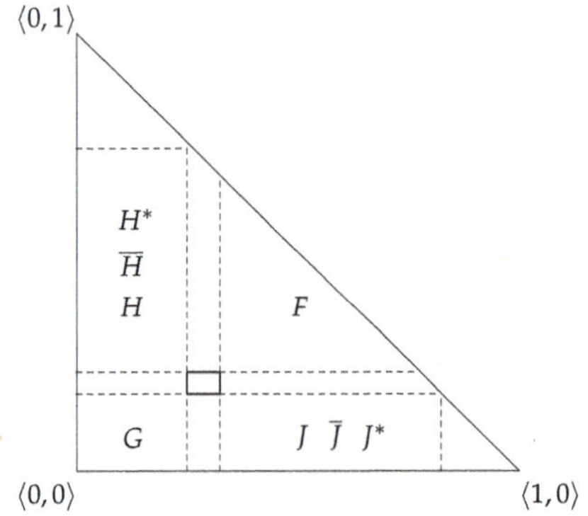

Figure 4. Region of transformation by the application of the operators.

3. Operator X

Now, we introduce the new operator

$$X_{\begin{pmatrix} a_1 & b_1 & c_1 & d_1 & e_1 & f_1 \\ a_2 & b_2 & c_2 & d_2 & e_2 & f_2 \end{pmatrix}}(A)$$

$$= \{\langle x, [a_1 \inf M_A(x) + b_1(1 - \inf M_A(x) - c_1 \inf N_A(x)),$$
$$a_2 \sup M_A(x) + b_2(1 - \sup M_A(x) - c_2 \sup N_A(x))],$$
$$[d_1 \inf N_A(x) + e_1(1 - f_1 \inf M_A(x) - \inf N_A(x)),$$
$$d_2 \sup N_A(x) + e_2(1 - f_2 \sup M_A(x) - \sup N_A(x))]\rangle | x \in E\},$$

where $a_1, b_1, c_1, d_1, e_1, f_1, a_2, b_2, c_2, d_2, e_2, f_2 \in [0,1]$, the following three conditions are valid for $i = 1, 2$:

$$a_i + e_i - e_i f_i \leq 1, \tag{2}$$

$$b_i + d_i - b_i c_i \leq 1, \tag{3}$$

$$b_i + e_i \leq 1, \tag{4}$$

and

$$a_1 \leq a_2, b_1 \leq b_2, c_1 \leq c_2, d_1 \leq d_2, e_1 \leq e_2, f_1 \leq f_2. \tag{5}$$

Theorem 1. *For every IVIFS A and for every $a_1, b_1, c_1, d_1, e_1, f_1, a_2, b_2, c_2, d_2, e_2, f_2 \in [0,1]$ that satisfy (2)–(5), $X_{\begin{pmatrix} a_1 & b_1 & c_1 & d_1 & e_1 & f_1 \\ a_2 & b_2 & c_2 & d_2 & e_2 & f_2 \end{pmatrix}}(A)$ is an IVIFS.*

Proof. Let $a_1, b_1, c_1, d_1, e_1, f_1, a_2, b_2, c_2, d_2, e_2, f_2 \in [0,1]$ satisfy (2)–(5) and let A be a fixed IVIFS. Then, from (5) it follows that

$$a_1 \inf M_A(x) + b_1(1 - \inf M_A(x) - c_1 \inf N_A(x))$$
$$\leq a_2 \sup M_A(x) + b_2(1 - \sup M_A(x) - c_2 \sup N_A(x))$$

and

$$d_1 \inf N_A(x) + e_1(1 - f_1 \inf M_A(x) - \inf N_A(x))$$
$$\leq d_2 \sup N_A(x) + e_2(1 - f_2 \sup M_A(x) - \sup N_A(x)).$$

Now, from (5) it is clear that it will be enough to check that

$$X = a_2 \sup M_A(x) + b_2(1 - \sup M_A(x) - c_2 \sup N_A(x))$$
$$+ d_2 \sup N_A(x) + e_2(1 - f_2 \sup M_A(x) - \sup N_A(x))$$
$$= (a_2 - b_2 - e_2 f_2) \sup M_A(x) + (d_2 - e_2 - b_2 c_2) \sup N_A(x) + b_2 + e_2 \leq 1.$$

In fact, from (2),

$$a_2 - b_2 - e_2 f_2 \leq 1 - b_2 - e_2$$

and from (3):

$$d_2 - e_2 - b_2 c_2 \leq 1 - b_2 - e_2.$$

Then, from (1),

$$X \leq (1 - b_2 - e_2)(\sup M_A(x) + \sup N_A(x)) + b_2 + e_2$$

$$\leq 1 - b_2 - e_2 + b_2 + e_2 = 1.$$

Finally, when $\sup M_A(x) = \inf N_A(x) = 0$ and from (4),

$$X = b_2(1 - 0 - 0) + e_2(1 - 0 - 0) = b_2 + e_2 \leq 1.$$

Therefore, the definition of the IVIFS is correct. □

All of the operators described above can be represented by the operator $X_{\left(\begin{smallmatrix} a_1 & b_1 & c_1 & d_1 & e_1 & f_1 \\ a_2 & b_2 & c_2 & d_2 & e_2 & f_2 \end{smallmatrix}\right)}$ at suitably chosen values of its parameters. These representations are the following:

$$\begin{aligned}
\square A &= X_{\left(\begin{smallmatrix} 1 & 0 & r_1 & 1 & 0 & s_1 \\ 1 & 0 & r_2 & 1 & 1 & 1 \end{smallmatrix}\right)}(A), \\
\lozenge A &= X_{\left(\begin{smallmatrix} 1 & 0 & r_1 & 1 & 0 & s_1 \\ 1 & 1 & 1 & 1 & 0 & s_2 \end{smallmatrix}\right)}(A), \\
D_\alpha(A) &= X_{\left(\begin{smallmatrix} 1 & 0 & r_1 & 1 & \alpha & 1 \\ 1 & 0 & r_2 & 1 & 1-\alpha & 1 \end{smallmatrix}\right)}(A), \\
F_{\alpha,\beta}(A) &= X_{\left(\begin{smallmatrix} 1 & 0 & r_1 & 1 & 0 & s_1 \\ 1 & \alpha & 1 & 1 & \beta & 1 \end{smallmatrix}\right)}(A), \\
G_{\alpha,\beta}(A) &= X_{\left(\begin{smallmatrix} \alpha & 0 & r_1 & \beta & 0 & s_1 \\ \alpha & 0 & r_2 & \beta & 0 & s_2 \end{smallmatrix}\right)}(A), \\
H_{\alpha,\beta}(A) &= X_{\left(\begin{smallmatrix} \alpha & 0 & r_1 & 1 & 0 & s_1 \\ \alpha & 0 & r_2 & 1 & \beta & 1 \end{smallmatrix}\right)}(A), \\
H^*_{\alpha,\beta}(A) &= X_{\left(\begin{smallmatrix} \alpha & 0 & r_1 & \alpha & 0 & s_1 \\ 1 & 0 & r_2 & 1 & \beta & \alpha \end{smallmatrix}\right)}(A), \\
J_{\alpha,\beta}(A) &= X_{\left(\begin{smallmatrix} 1 & 0 & r_1 & \beta & 0 & s_1 \\ 1 & \alpha & 1 & \beta & 0 & s_2 \end{smallmatrix}\right)}(A), \\
J^*_{\alpha,\beta}(A) &= X_{\left(\begin{smallmatrix} 1 & 0 & r_1 & \beta & 0 & s_1 \\ 1 & \alpha & \beta & \beta & 0 & s_2 \end{smallmatrix}\right)}(A), \\
\overline{F}_{\left(\begin{smallmatrix} \alpha & \gamma \\ \beta & \delta \end{smallmatrix}\right)}(A) &= X_{\left(\begin{smallmatrix} 1 & \alpha & 1 & 1 & \gamma & 1 \\ 1 & \beta & 1 & 1 & \delta & s_1 \end{smallmatrix}\right)}(A), \\
\overline{G}_{\left(\begin{smallmatrix} \alpha & \gamma \\ \beta & \delta \end{smallmatrix}\right)}(A) &= X_{\left(\begin{smallmatrix} \alpha & 0 & r_1 & \beta & 0 & s_1 \\ \gamma & 0 & r_2 & \delta & 0 & s_2 \end{smallmatrix}\right)}(A), \\
\overline{H}_{\left(\begin{smallmatrix} \alpha & \gamma \\ \beta & \delta \end{smallmatrix}\right)}(A) &= X_{\left(\begin{smallmatrix} \alpha & 0 & r_1 & 1 & \gamma & 1 \\ \beta & 0 & r_2 & 1 & \delta & 1 \end{smallmatrix}\right)}(A), \\
\overline{H}^*_{\left(\begin{smallmatrix} \alpha & \gamma \\ \beta & \delta \end{smallmatrix}\right)}(A) &= X_{\left(\begin{smallmatrix} \alpha & 0 & r_1 & 1 & \gamma & 1 \\ \beta & 0 & r_2 & 1 & \delta & \beta \end{smallmatrix}\right)}(A), \\
\overline{J}_{\left(\begin{smallmatrix} \alpha & \gamma \\ \beta & \delta \end{smallmatrix}\right)}(A) &= X_{\left(\begin{smallmatrix} 1 & \alpha & 1 & \gamma & 0 & s_1 \\ 1 & \beta & 1 & \delta & 0 & s_2 \end{smallmatrix}\right)}(A), \\
\overline{J}^*_{\left(\begin{smallmatrix} \alpha & \gamma \\ \beta & \delta \end{smallmatrix}\right)}(A) &= X_{\left(\begin{smallmatrix} 1 & \alpha & \delta & \gamma & 0 & s_1 \\ 1 & \beta & \delta & \delta & 0 & s_2 \end{smallmatrix}\right)}(A),
\end{aligned}$$

where r_1, r_2, s_1, s_2 are arbitrary real numbers in the interval $[0, 1]$.

Three of the operations, defined over two IVIFSs A and B, are the following:

$$\begin{aligned}
\neg A &= \{\langle x, N_A(x), M_A(x)\rangle \mid x \in E\}, \\
A \cap B &= \{\langle x, [\min(\inf M_A(x), \inf M_B(x)), \min(\sup M_A(x), \sup M_B(x))], \\
&\qquad [\max(\inf N_A(x), \inf N_B(x)), \max(\sup N_A(x), \sup N_B(x))]\rangle \mid x \in E\}, \\
A \cup B &= \{\langle x, [\max(\inf M_A(x), \inf M_B(x)), \max(\sup M_A(x), \sup M_B(x))], \\
&\qquad [\min(\inf N_A(x), \inf N_B(x)), \min(\sup N_A(x), \sup N_B(x))]\rangle \mid x \in E\}.
\end{aligned}$$

For any two IVIFSs A and B, the following relations hold:

$$\begin{aligned} A \subset B \quad &\text{iff} \quad \forall x \in E,\ \inf M_A(x) \leq \inf M_B(x),\ \ \inf N_A(x) \geq \inf N_B(x), \\ &\qquad \sup M_A(x) \leq \sup M_B(x) \ \text{ and } \ \sup N_A(x) \geq \sup N_B(x)), \\ A \supset B \quad &\text{iff} \quad B \subset A, \\ A = B \quad &\text{iff} \quad A \subset B \text{ and } B \subset A. \end{aligned}$$

Theorem 2. *For every two IVIFSs A and B and for every* $a_1, b_1, c_1, d_1, e_1, f_1, a_2, b_2, c_2, d_2, e_2, f_2 \in [0,1]$ *that satisfy (2)–(5),*

(a) $\neg X_{\left(\begin{smallmatrix} a_1 & b_1 & c_1 & d_1 & e_1 & f_1 \\ a_2 & b_2 & c_2 & d_2 & e_2 & f_2 \end{smallmatrix}\right)}(\neg A) = X_{d,e,f,a,b,c}(A)$,

(b) $X_{\left(\begin{smallmatrix} a_1 & b_1 & c_1 & d_1 & e_1 & f_1 \\ a_2 & b_2 & c_2 & d_2 & e_2 & f_2 \end{smallmatrix}\right)}(A \cap B)$

$$\subset X_{\left(\begin{smallmatrix} a_1 & b_1 & c_1 & d_1 & e_1 & f_1 \\ a_2 & b_2 & c_2 & d_2 & e_2 & f_2 \end{smallmatrix}\right)}(A) \cap X_{\left(\begin{smallmatrix} a_1 & b_1 & c_1 & d_1 & e_1 & f_1 \\ a_2 & b_2 & c_2 & d_2 & e_2 & f_2 \end{smallmatrix}\right)}(B),$$

(c) $X_{\left(\begin{smallmatrix} a_1 & b_1 & c_1 & d_1 & e_1 & f_1 \\ a_2 & b_2 & c_2 & d_2 & e_2 & f_2 \end{smallmatrix}\right)}(A \cup B)$

$$\supset X_{\left(\begin{smallmatrix} a_1 & b_1 & c_1 & d_1 & e_1 & f_1 \\ a_2 & b_2 & c_2 & d_2 & e_2 & f_2 \end{smallmatrix}\right)}(A) \cup X_{\left(\begin{smallmatrix} a_1 & b_1 & c_1 & d_1 & e_1 & f_1 \\ a_2 & b_2 & c_2 & d_2 & e_2 & f_2 \end{smallmatrix}\right)}(B).$$

Proof. (c) Let $a_1, b_1, c_1, d_1, e_1, f_1, a_2, b_2, c_2, d_2, e_2, f_2 \in [0,1]$ satisfy (2)–(5) , and let A and B be fixed IVIFSs. First, we obtain:

$$\begin{aligned} Y &= X_{\left(\begin{smallmatrix} a_1 & b_1 & c_1 & d_1 & e_1 & f_1 \\ a_2 & b_2 & c_2 & d_2 & e_2 & f_2 \end{smallmatrix}\right)}(A \cup B) \\ &= X_{\left(\begin{smallmatrix} a_1 & b_1 & c_1 & d_1 & e_1 & f_1 \\ a_2 & b_2 & c_2 & d_2 & e_2 & f_2 \end{smallmatrix}\right)}(\{\langle x, [\max(\inf M_A(x), \inf M_B(x)), \\ &\qquad \max(\sup M_A(x), \sup M_B(x))], \\ &\qquad [\min(\inf N_A(x), \inf N_B(x)), \min(\sup N_A(x), \sup N_B(x))]\rangle \mid x \in E\}) \\ &= \{\langle x, [a_1 \max(\inf M_A(x), \inf M_B(x)) + b_1(1 - \max(\inf M_A(x), \inf M_B(x)) \\ &\qquad -c_1 \min(\inf N_A(x), \inf N_B(x))), a_2 \max(\sup M_A(x) \sup M_B(x)) \\ &\qquad +b_2(1 - \max(\sup M_A(x) \sup M_B(x)) - c_2 \min(\sup N_A(x), \sup N_B(x)))], \\ &\qquad [d_1 \min(\inf N_A(x), \inf N_B(x)) + e_1(1 - f_1 \max(\inf M_A(x), \inf M_B(x)) \\ &\qquad - \min(\inf N_A(x), \inf N_B(x))), d_2 \min(\sup N_A(x), \sup N_B(x)) \\ &\qquad +e_2(1 - f_2 \max(\sup M_A(x) \sup M_B(x)) - \min(\sup N_A(x), \sup N_B(x)))]\rangle | x \in E\}. \end{aligned}$$

Second, we calculate:

$$\begin{aligned} Z &= X_{\left(\begin{smallmatrix} a_1 & b_1 & c_1 & d_1 & e_1 & f_1 \\ a_2 & b_2 & c_2 & d_2 & e_2 & f_2 \end{smallmatrix}\right)}(A) \cup X_{\left(\begin{smallmatrix} a_1 & b_1 & c_1 & d_1 & e_1 & f_1 \\ a_2 & b_2 & c_2 & d_2 & e_2 & f_2 \end{smallmatrix}\right)}(B) \\ &= \{\langle x, [a_1 \inf M_A(x) + b_1(1 - \inf M_A(x) - c_1 \inf N_A(x)), \\ &\qquad a_2 \sup M_A(x) + b_2(1 - \sup M_A(x) - c_2 \sup N_A(x))], \\ &\qquad [d_1 \inf N_A(x) + e_1(1 - f_1 \inf M_A(x) - \inf N_A(x)), \\ &\qquad d_2 \sup N_A(x) + e_2(1 - f_2 \sup M_A(x) - \sup N_A(x))]\rangle | x \in E\} \end{aligned}$$

$$\cup\{\langle x,[a_1 \inf M_B(x)+b_1(1-\inf M_B(x)-c_1 \inf N_B(x)),$$
$$a_2 \sup M_B(x)+b_2(1-\sup M_B(x)-c_2 \sup N_B(x))],$$
$$[d_1 \inf N_B(x)+e_1(1-f_1 \inf M_B(x)-\inf N_B(x)),$$
$$d_2 \sup N_B(x)+e_2(1-f_2 \sup M_B(x)-\sup N_B(x))]\rangle | x\in E\}$$

$$=\{\langle x,[\max(a_1 \inf M_A(x)+b_1(1-\inf M_A(x)-c_1 \inf N_A(x)),$$
$$a_1 \inf M_B(x)+b_1(1-\inf M_B(x)-c_1 \inf N_B(x))),$$
$$\max(a_2 \sup M_A(x)+b_2(1-\sup M_A(x)-c_2 \sup N_A(x)),$$
$$a_2 \sup M_B(x)+b_2(1-\sup M_B(x)-c_2 \sup N_B(x)))],$$
$$[\min(d_1 \inf N_A(x)+e_1(1-f_1 \inf M_A(x)-\inf N_A(x)),$$
$$d_1 \inf N_B(x)+e_1(1-f_1 \inf M_B(x)-\inf N_B(x))),$$
$$\min(d_2 \sup N_A(x)+e_2(1-f_2 \sup M_A(x)-\sup N_A(x)),$$
$$d_2 \sup N_B(x)+e_2(1-f_2 \sup M_B(x)-\sup N_B(x)))]\rangle \mid x\in E\}.$$

Let

$$P=a_1 \max(\inf M_A(x),\inf M_B(x))+b_1(1-\max(\inf M_A(x),\inf M_B(x))$$
$$-c_1 \min(\inf N_A(x),\inf N_B(x)))-\max(a_1 \inf M_A(x)+b_1(1-\inf M_A(x)-c_1 \inf N_A(x)),$$
$$a_1 \inf M_B(x)+b_1(1-\inf M_B(x)-c_1 \inf N_B(x)))$$

$$=a_1 \max(\inf M_A(x),\inf M_B(x))+b_1-b_1 \max(\inf M_A(x),\inf M_B(x))$$
$$-b_1c_1 \min(\inf N_A(x),\inf N_B(x)))-\max((a_1-b_1) \inf M_A(x)+b_1-b_1c_1 \inf N_A(x),$$
$$(a_1-b_1) \inf M_B(x)+b_1-b_1c_1 \inf N_B(x))$$

$$=a_1 \max(\inf M_A(x),\inf M_B(x))-b_1 \max(\inf M_A(x),\inf M_B(x))$$
$$-b_1c_1 \min(\inf N_A(x),\inf N_B(x))-\max((a_1-b_1) \inf M_A(x)-b_1c_1 \inf N_A(x),$$
$$(a_1-b_1) \inf M_B(x)-b_1c_1 \inf N_B(x)).$$

Let $\inf M_A(x)\geq \inf M_B(x)$. Then

$$P=(a_1-b_1) \inf M_A(x)-b_1c_1 \min(\inf N_A(x),\inf N_B(x))-\max((a_1-b_1) \inf M_A(x)$$
$$-b_1c_1 \inf N_A(x),(a_1-b_1) \inf M_B(x)-b_1c_1 \inf N_B(x)).$$

Let $(a_1-b_1) \inf M_A(x)-b_1c_1 \inf N_A(x)\geq (a_1-b_1) \inf M_B(x)-b_1c_1 \inf N_B(x)$. Then

$$P=(a_1-b_1) \inf M_A(x)-b_1c_1 \min(\inf N_A(x),\inf N_B(x))-(a_1-b_1) \inf M_A(x)$$
$$+b_1c_1 \inf N_A(x)$$
$$=b_1c_1 \inf N_A(x)-b_1c_1 \min(\inf N_A(x),\inf N_B(x))\geq 0.$$

If $(a_1-b_1) \inf M_A(x)-b_1c_1 \inf N_A(x)<(a_1-b_1) \inf M_B(x)-b_1c_1 \inf N_B(x)$. Then

$$P=(a_1-b_1) \inf M_A(x)-b_1c_1 \min(\inf N_A(x),\inf N_B(x))-(a_1-b_1) \inf M_B(x)$$

$$+b_1c_1 \inf N_B(x)).$$
$$= b_1c_1 \inf N_B(x) - b_1c_1 \min(\inf N_A(x), \inf N_B(x)) \geq 0.$$

Therefore, the $\inf M_A$-component of IVIFS Y is higher than or equal to the $\inf M_A$-component of IVIFS Z. In the same manner, it can be checked that the same inequality is valid for the $\sup M_A$-components of these IVIFSs. On the other hand, we can check that that the $\inf N_A$- and $\sup N_A$-components of IVIFS Y are, respectively, lower than or equal to the $\inf N_A$ and $\sup N_A$-components of IVIFS Z. Therefore, the inequality (c) is valid. □

4. Conclusions

In the near future, the author plans to study some other properties of the new operator $X_{\begin{pmatrix} a_1 & b_1 & c_1 & d_1 & e_1 & f_1 \\ a_2 & b_2 & c_2 & d_2 & e_2 & f_2 \end{pmatrix}}$.

In [21], it is shown that the IFSs are a suitable tool for the evaluation of data mining processes and objects. In the near future, we plan to discuss the possibilities of using IVIFSs as a similar tool.

Funding: This research was funded by the Bulgarian National Science Fund under Grant Ref. No. DN-02-10/2016.

Conflicts of Interest: The author declares no conflict of interest.

References

1. Atanassov, K. Intuitionistic Fuzzy Sets. 1983, VII ITKR Session. Deposed in Centr. Sci.-Techn. Library of the Bulg. Acad. of Sci., 1697/84. Available online: http://www.biomed.bas.bg/bioautomation/2016/vol_20.s1/files/20.s1_03.pdf (accessed on 13 July 2018)
2. Atanassov, K. Intuitionistic fuzzy sets. *Int. J. Bioautom.* **2016**, *20*, S1–S6.
3. Atanassov, K. Intuitionistic fuzzy sets. *Fuzzy Sets Syst.* **1986**, *20*, 87–96. [CrossRef]
4. Atanassov, K. *Intuitionistic Fuzzy Sets*; Springer: Heidelberg, Germany, 1999.
5. Atanassov, K. *On Intuitionistic Fuzzy Sets Theory*; Springer: Berlin, Germany, 2012.
6. Zadeh, L. Fuzzy sets. *Inf. Control* **1965**, *8*, 338–353. [CrossRef]
7. Atanassov, K.; Stoeva, S. Intuitionistic *L*-fuzzy sets. In *Cybernetics and Systems Research 2*; Trappl, R., Ed.; Elsevier Sci. Publ.: Amsterdam, The Nertherlands, 1984; pp. 539–540.
8. Atanassov, K. A second type of intuitionistic fuzzy sets. *BUSEFAL* **1993**, *56*, 66–70.
9. Atanassov, K.; Szmidt, E.; Kacprzyk, J.; Vassilev, P. On intuitionistic fuzzy pairs of n-th type. *Issues Intuit. Fuzzy Sets Gen. Nets* **2017**, *13*, 136–142.
10. Atanassov, K.T.; Vassilev, P. On the Intuitionistic Fuzzy Sets of n-th Type. In *Advances in Data Analysis with Computational Intelligence Methods*; Studies in Computational Intelligence; Gawęda, A., Kacprzyk, J., Rutkowski, L., Yen, G., Eds.; Springer: Cham, Switzerland, 2018; Volume 738, pp. 265–274.
11. Parvathi, R.; Vassilev, P.; Atanassov, K. A note on the bijective correspondence between intuitionistic fuzzy sets and intuitionistic fuzzy sets of p-th type. In *New Developments in Fuzzy Sets, Intuitionistic Fuzzy Sets, Generalized Nets and Related Topics*; SRI PAS IBS PAN: Warsaw, Poland, 2012; Volume 1, pp. 143–147.
12. Vassilev, P.; Parvathi, R.; Atanassov, K. Note on intuitionistic fuzzy sets of p-th type. *Issues Intuit. Fuzzy Sets Gen. Nets* **2008**, *6*, 43–50.
13. Atanassov, K. *Temporal Intuitionistic Fuzzy Sets*; Comptes Rendus de l'Academie bulgare des Sciences; Bulgarian Academy of Sciences: Sofia, Bulgaria, 1991; pp. 5–7.
14. Atanassov, K.; Szmidt, E.; Kacprzyk, J. On intuitionistic fuzzy multi-dimensional sets. *Issues Intuit. Fuzzy Sets Gen. Nets* **2008**, *7*, 1–6.
15. Gorzalczany, M. Interval-valued fuzzy fuzzy inference method—Some basic properties. *Fuzzy Sets Syst.* **1989**, *31*, 243–251. [CrossRef]
16. Atanassov, K. Review and New Results on Intuitionistic Fuzzy Sets. 1988, IM-MFAIS-1-88, Mathematical Foundations of Artificial Intelligence Seminar. Available online: http://www.biomed.bas.bg/bioautomation/2016/vol_20.s1/files/20.s1_03.pdf (accessed on 13 July 2018)
17. Atanassov, K. Review and new results on intuitionistic fuzzy sets. *Int. J. Bioautom.* **2016**, *20*, S7–S16.

18. Atanassov, K.; Gargov, G. Interval valued intuitionistic fuzzy sets. *Fuzzy Sets Syst.* **1989**, *31*, 343–349. [CrossRef]
19. Atanassov, K. Geometrical Interpretation of the Elements of the Intuitionistic Fuzzy Objects. 1989, IM-MFAIS-1-89, Mathematical Foundations of Artificial Intelligence Seminar. Available online: http://biomed.bas.bg/bioautomation/2016/vol_20.s1/files/20.s1_05.pdf (accessed on 13 July 2018)
20. Atanassov, K. Geometrical interpretation of the elements of the intuitionistic fuzzy objects. *Int. J. Bioautom.* **2016**, *20*, S27–S42.
21. Atanassov, K. Intuitionistic fuzzy logics as tools for evaluation of Data Mining processes. *Knowl.-Based Syst.* **2015**, *80*, 122–130. [CrossRef]

mathematics

Article

$\mathcal{N}$-Hyper Sets

Young Bae Jun [1], Seok-Zun Song [2,*] and Seon Jeong Kim [3]

1 Department of Mathematics Education, Gyeongsang National University, Jinju 52828, Korea; skywine@gmail.com
2 Department of Mathematics, Jeju National University, Jeju 63243, Korea
3 Department of Mathematics, Natural Science of College, Gyeongsang National University, Jinju 52828, Korea; skim@gnu.ac.kr
* Correspondence: szsong@jejunu.ac.kr

Received: 21 April 2018; Accepted: 21 May 2018; Published: 23 May 2018

Abstract: To deal with the uncertainties, fuzzy set theory can be considered as one of the mathematical tools by Zadeh. As a mathematical tool to deal with negative information, Jun et al. introduced a new function, which is called a negative-valued function, and constructed $\mathcal{N}$-structures in 2009. Since then, $\mathcal{N}$-structures are applied to algebraic structures and soft sets, etc. Using the $\mathcal{N}$-structures, the notions of (extended) $\mathcal{N}$-hyper sets, $\mathcal{N}$-substructures of type 1, 2, 3 and 4 are introduced, and several related properties are investigated in this research paper.

Keywords: $\mathcal{N}$-structure; (extended) $\mathcal{N}$-hyper set; $\mathcal{N}$-substructure of types $1,2,3,4$

JEL Classification: 06F35; 03G25; 08A72

---

1. Introduction

Most mathematical tools for computing, formal modeling and reasoning are crisp, deterministic and precise in many characters. However, several problems in economics, environment, engineering, social science, medical science, etc. do not always involve crisp data in real life. Consequently, we cannot successfully use the classical method because of various types of uncertainties presented in the problem. To deal with the uncertainties, fuzzy set theory can be considered as one of the mathematical tools (see [1]). A (crisp) set A in a universe X can be defined in the form of its characteristic function $\mu_A : X \rightarrow \{0,1\}$ yielding the value 1 for elements belonging to the set A and the value 0 for elements excluded from the set A. Thus far, most of the generalization of the crisp set has been conducted on the unit interval $[0,1]$ and they are consistent with the asymmetry observation. In other words, the generalization of the crisp set to fuzzy sets relied on spreading positive information that fit the crisp point $\{1\}$ into the interval $[0,1]$. Because no negative meaning of information is suggested, we now feel a need to deal with negative information. To do so, we also feel a need to supply mathematical tools. To attain such object, Jun et al. [2] introduced a new function, which is called negative-valued function, and constructed $\mathcal{N}$-structures. Since then, $\mathcal{N}$-structures are applied to rings (see [3]), BCH-algebras (see [4]), (ordered) semigroups (see [5–8]). The combination of soft sets and $\mathcal{N}$-structures is dealt with in [9,10] and [11]. The purpose of this paper is to introduce the notions of (extended) $\mathcal{N}$-hyper sets, $\mathcal{N}$-substructures of type 1, 2, 3 and 4, and to investigate several related properties. In our consecutive research in future, we will try to study several applications based on $\mathcal{N}$-structures, for example, another type of algebra, soft and rough set theory, decision-making problems, etc. In particular, we will study complex dynamics through $\mathcal{N}$-structures based on the paper [12].

2. Preliminaries

Denote by $\mathcal{F}(X,[-1,0])$ the collection of all functions from a set X to $[-1,0]$. We say that an element of $\mathcal{F}(X,[-1,0])$ is a *negative-valued function* from X to $[-1,0]$ (briefly, $\mathcal{N}$-*function* on X). By an $\mathcal{N}$-*structure*, we mean an ordered pair (X,ρ) of X and an $\mathcal{N}$-function ρ on X (see [2]).

For any family $\{a_i \mid i \in \Lambda\}$ of real numbers, we define

$$\bigvee\{a_i \mid i \in \Lambda\} := \begin{cases} \max\{a_i \mid i \in \Lambda\}, & \text{if } \Lambda \text{ is finite,} \\ \sup\{a_i \mid i \in \Lambda\}, & \text{otherwise.} \end{cases}$$

$$\bigwedge\{a_i \mid i \in \Lambda\} := \begin{cases} \min\{a_i \mid i \in \Lambda\} & \text{if } \Lambda \text{ is finite,} \\ \inf\{a_i \mid i \in \Lambda\} & \text{otherwise.} \end{cases}$$

Given a subset A of $[-1,0]$, we define

$$\ell(A) = \bigvee\{a \mid a \in A\} - \bigwedge\{a \mid a \in A\}.$$

3. (Extended) $\mathcal{N}$-Hyper Sets

Definition 1. *Let X be an initial universe set. By an $\mathcal{N}$-hyper set over X, we mean a mapping $\mu : X \to \mathcal{P}^*([-1,0])$, where $\mathcal{P}^*([-1,0])$ is the collection of all nonempty subsets of $[-1,0]$.*

In an $\mathcal{N}$-hyper set $\mu : X \to \mathcal{P}^*([-1,0])$ over X, we consider two $\mathcal{N}$-structures $(X,\mu_\wedge)$, $(X,\mu_\vee)$ and a fuzzy structure (X,μ_ℓ) in which

$$\mu_\wedge : X \to [-1,0],\ x \mapsto \bigwedge\{\mu(x)\}, \tag{1}$$

$$\mu_\vee : X \to [-1,0],\ x \mapsto \bigvee\{\mu(x)\}, \tag{2}$$

$$\mu_\ell : X \to [0,1],\ x \mapsto \ell(\mu(x)). \tag{3}$$

It is clear that $\mu_\ell(x) = \mu_\vee(x) - \mu_\wedge(x)$ for all $x \in X$.

Example 1. *Let $X = \{a,b,c,d\}$ and define an $\mathcal{N}$-hyper set $\mu : X \to \mathcal{P}^*([-1,0])$ over X by Table 1.*

Table 1. $\mathcal{N}$-hyper set.

X	*a*	*b*	*c*	*d*
μ	$[-0.5,0]$	$(-0.6,-0.3)$	$[-0.4,-0.2)$	$(-1,-0.8]$

Then, μ generates two $\mathcal{N}$-structures $(X,\mu_\wedge)$ and $(X,\mu_\vee)$, and a fuzzy structure (X,μ_ℓ) as Table 2.

Table 2. $\mathcal{N}$-structures $(X,\mu_\wedge)$, $(X,\mu_\vee)$ and (X,μ_ℓ).

X	*a*	*b*	*c*	*d*
$\mu_\wedge$	−0.5	−0.6	−0.4	−1
$\mu_\vee$	0	−0.3	−0.2	−0.8
μ_ℓ	0.5	0.3	0.2	0.2

Definition 2. *Given an $\mathcal{N}$-structure (X,φ) over X, define a map*

$$\varphi^e : \mathcal{P}^*(X) \to \mathcal{P}^*([-1,0]),\ A \mapsto \{\varphi(a) \mid a \in A\}, \tag{4}$$

where $\mathcal{P}^*(X)$ is the set of all nonempty subsets of X. We call φ^e the extended $\mathcal{N}$-hyper set over X.

Example 2. *Let $X = \{a, b, c, d\}$ be an initial universe set and let (X, φ) be an $\mathcal{N}$-structure over X given by Table 3.*

Table 3. $\mathcal{N}$-structure (X, φ).

X	a	b	c	d
φ	-0.5	-0.3	-0.4	-0.8

Then, the extended $\mathcal{N}$-hyper set φ^e over X is described as Table 4.

Table 4. The extended $\mathcal{N}$-hyper set φ^e over X.

$A \in \mathcal{P}^*(X)$	$\varphi^e(A)$	$A \in \mathcal{P}^*(X)$	$\varphi^e(A)$
$\{a\}$	$\{-0.5\}$	$\{b\}$	$\{-0.3\}$
$\{c\}$	$\{-0.4\}$	$\{d\}$	$\{-0.8\}$
$\{a,b\}$	$\{-0.5,-0.3\}$	$\{a,c\}$	$\{-0.5,-0.4\}$
$\{a,d\}$	$\{-0.5,-0.8\}$	$\{a,b,c\}$	$\{-0.5,-0.4,-0.3\}$
$\{a,b,d\}$	$\{-0.5,-0.3,-0.8\}$	$\{a,b,c,d\}$	$\{-0.5,-0.4,-0.3,-0.8\}$

Definition 3. *Let X be an initial universe set with a binary operation $*$. An $\mathcal{N}$-structure (X, φ) over X is called.*

- *an $\mathcal{N}$-substructure of $(X, *)$ with type 1 (briefly, $\mathcal{N}_1$-substructure of $(X, *)$) if it satisfies:*

$$(\forall x, y \in X)\left(\varphi(x * y) \leq \bigvee\{\varphi(x), \varphi(y)\}\right), \tag{5}$$

- *an $\mathcal{N}$-substructure of $(X, *)$ with type 2 (briefly, $\mathcal{N}_2$-substructure of $(X, *)$) if it satisfies:*

$$(\forall x, y \in X)\left(\varphi(x * y) \geq \bigwedge\{\varphi(x), \varphi(y)\}\right), \tag{6}$$

- *an $\mathcal{N}$-substructure of $(X, *)$ with type 3 (briefly, $\mathcal{N}_3$-substructure of $(X, *)$) if it satisfies:*

$$(\forall x, y \in X)\left(\varphi(x * y) \geq \bigvee\{\varphi(x), \varphi(y)\}\right), \tag{7}$$

- *an $\mathcal{N}$-substructure of $(X, *)$ with type 4 (briefly, $\mathcal{N}_4$-substructure of $(X, *)$) if it satisfies:*

$$(\forall x, y \in X)\left(\varphi(x * y) \leq \bigwedge\{\varphi(x), \varphi(y)\}\right). \tag{8}$$

It is clear that every $\mathcal{N}_4$-substructure of $(X, *)$ is an $\mathcal{N}_1$-substructure of $(X, *)$, and every $\mathcal{N}_3$-substructure of $(X, *)$ is an $\mathcal{N}_2$-substructure of $(X, *)$.

Example 3. *Let X be the set of all integers and let $*$ be a binary operation on X defined by*

$$(\forall x, y \in X)\,(x * y = -(|x| + |y|)).$$

(1) Define an $\mathcal{N}$-structure (X, φ) over X by

$$\varphi : X \to [-1, 0],\ \ x \mapsto -1 + \tfrac{1}{e^{|x|}}.$$

Then, $\varphi(0) = 0$, $\lim\limits_{|x| \to \infty} \varphi(x) = -1$ and

$$\varphi(x * y) = -1 + \tfrac{1}{e^{|x|+|y|}} \leq \bigwedge\left\{-1 + \tfrac{1}{e^{|x|}},\ -1 + \tfrac{1}{e^{|y|}}\right\} = \bigwedge\{\varphi(x), \varphi(y)\}$$

for all $x, y \in X$. *Therefore,* (X, φ) *is an* $\mathcal{N}_4$*-substructure of* $(X, *)$, *and hence it is also an* $\mathcal{N}_1$*-substructure of* $(X, *)$.

(2) Let (X, φ) *be an* $\mathcal{N}$*-structure over* X *in which* φ *is given by*

$$\varphi : X \to [-1, 0],\ \ x \mapsto \tfrac{-1}{1+|x|}.$$

Then,

$$\begin{aligned}\varphi(x * y) &= \varphi(-(|x| + |y|)) = \tfrac{-1}{1+|-(|x|+|y|)|} \\ &= \tfrac{-1}{1+|x|+|y|} \geq \bigvee\left\{\tfrac{-1}{1+|x|},\ \tfrac{-1}{1+|y|}\right\} \\ &= \bigvee\{\varphi(x),\ \varphi(y)\}\end{aligned}$$

for all $x, y \in X$. *Therefore,* (X, φ) *is an* $\mathcal{N}_3$*-substructure of* $(X, *)$, *and hence it is also an* $\mathcal{N}_2$*-substructure of* $(X, *)$.

For any initial universe set X with binary operations, let $\mathcal{H}(X)$ denote the set of all $(X, *)$ where $*$ is a binary operation on X, that is,

$$\mathcal{H}(X) := \{(X, *) \mid * \text{ is a binary operation on } X\}.$$

We consider the following subsets of $\mathcal{H}(X)$:

$$\begin{aligned}\mathcal{N}_1(\varphi) &:= \{(X, *) \in \mathcal{H}(X) \mid \varphi \text{ is an } \mathcal{N}_1\text{-substructure of } (X, *)\}, \\ \mathcal{N}_2(\varphi) &:= \{(X, *) \in \mathcal{H}(X) \mid \varphi \text{ is an } \mathcal{N}_2\text{-substructure of } (X, *)\}, \\ \mathcal{N}_3(\varphi) &:= \{(X, *) \in \mathcal{H}(X) \mid \varphi \text{ is an } \mathcal{N}_3\text{-substructure of } (X, *)\}, \\ \mathcal{N}_4(\varphi) &:= \{(X, *) \in \mathcal{H}(X) \mid \varphi \text{ is an } \mathcal{N}_4\text{-substructure of } (X, *)\}.\end{aligned}$$

Theorem 1. *Given an* $\mathcal{N}$*-structure* (X, φ) *over an initial universe set* X, *if* $(X, *) \in \mathcal{N}_1(\varphi)$, *then* $(\mathcal{P}^*(X), *) \in \mathcal{N}_1(\varphi^e_\wedge)$.

Proof. If $(X, *) \in \mathcal{N}_1(\varphi)$, then φ is an $\mathcal{N}_1$-substructure of $(X, *)$, that is, Equation (5) is valid. Let $A, B \in \mathcal{P}^*(X)$. Then,

$$\varphi^e_\wedge(A * B) = \bigwedge\{\varphi(a * b) \mid a \in A,\ b \in B\}. \tag{9}$$

Note that

$$(\forall \varepsilon > 0)(\exists a_0 \in X)\left(\varphi(a_0) < \bigwedge\{\varphi(a) \mid a \in A\} + \varepsilon\right)$$

and

$$(\forall \varepsilon > 0)(\exists b_0 \in X)\left(\varphi(b_0) < \bigwedge\{\varphi(b) \mid b \in B\} + \varepsilon\right).$$

It follows that

$$\begin{aligned}
&\bigwedge\{\varphi(a*b) \mid a \in A,\ b \in B\} \le \varphi(a_0 * b_0) \le \bigvee\{\varphi(a_0), \varphi(b_0)\} \\
&\le \bigvee\left\{\bigwedge\{\varphi(a) \mid a \in A\} + \varepsilon,\ \bigwedge\{\varphi(b) \mid b \in B\} + \varepsilon\right\} \\
&= \bigvee\{\varphi^e_\wedge(A) + \varepsilon,\ \varphi^e_\wedge(A) + \varepsilon\} \\
&= \bigvee\{\varphi^e_\wedge(A),\ \varphi^e_\wedge(A)\} + \varepsilon.
\end{aligned}$$

Since ε is arbitrary, it follows that

$$\varphi^e_\wedge(A * B) \le \bigvee\{\varphi^e_\wedge(A),\ \varphi^e_\wedge(A)\}.$$

Therefore, $(\mathcal{P}^*(X), *) \in \mathcal{N}_1(\varphi^e_\wedge)$. □

Theorem 2. *Given an $\mathcal{N}$-structure (X, φ) over an initial universe set X, if $(X, *) \in \mathcal{N}_2(\varphi)$, then $(\mathcal{P}^*(X), *) \in \mathcal{N}_2(\varphi^e_\vee)$.*

Proof. If $(X, *) \in \mathcal{N}_2(\varphi)$, then φ is an $\mathcal{N}_2$-substructure of $(X, *)$, that is, Equation (6) is valid. Let $A, B \in \mathcal{P}^*(X)$. Then,

$$\varphi^e_\vee(A * B) = \bigvee\{\varphi(a*b) \mid a \in A,\ b \in B\}. \tag{10}$$

Let ε be any positive number. Then, there exist $a_0, b_0 \in X$ such that

$$\begin{aligned}
&\left(\varphi(a_0) > \bigvee\{\varphi(a) \mid a \in A\} - \varepsilon\right), \\
&\left(\varphi(b_0) > \bigvee\{\varphi(b) \mid b \in B\} - \varepsilon\right),
\end{aligned}$$

respectively. It follows that

$$\begin{aligned}
&\bigvee\{\varphi(a*b) \mid a \in A,\ b \in B\} \ge \varphi(a_0 * b_0) \ge \bigwedge\{\varphi(a_0),\ \varphi(b_0)\} \\
&\ge \bigwedge\left\{\bigvee\{\varphi(a) \mid a \in A\} - \varepsilon,\ \bigvee\{\varphi(b) \mid b \in B\} - \varepsilon\right\} \\
&= \bigwedge\{\varphi^e_\vee(A) - \varepsilon,\ \varphi^e_\vee(B) - \varepsilon\} \\
&= \bigwedge\{\varphi^e_\vee(A),\ \varphi^e_\vee(B)\} - \varepsilon,
\end{aligned}$$

which shows that $\varphi^e_\vee(A * B) \ge \bigwedge\{\varphi^e_\vee(A),\ \varphi^e_\vee(B)\}$. Therefore, $(\mathcal{P}^*(X), *) \in \mathcal{N}_2(\varphi^e_\vee)$. □

Definition 4. *Given $\mathcal{N}$-hyper sets μ and λ over an initial universe set X, we define hyper-union ($\tilde{\cup}$), hyper-intersection ($\tilde{\cap}$), hyper complement ($'$) and hyper difference ($\setminus$) as follows:*

$$\begin{aligned}
&\mu \,\tilde{\cup}\, \lambda : X \to \mathcal{P}^*([-1,0]),\ x \mapsto \mu(x) \cup \lambda(x), \\
&\mu \,\tilde{\cap}\, \lambda : X \to \mathcal{P}^*([-1,0]),\ x \mapsto \mu(x) \cap \lambda(x), \\
&\mu \setminus \lambda : X \to \mathcal{P}^*([-1,0]),\ x \mapsto \mu(x) \setminus \lambda(x), \\
&\mu' : X \to \mathcal{P}^*([-1,0]),\ x \mapsto [-1,0] \setminus \{t \in [-1,0] \mid t \notin \mu(x)\}.
\end{aligned}$$

Proposition 1. *If μ and λ are $\mathcal{N}$-hyper sets over an initial universe set X, then*

$$(\forall x \in X)\left((\mu \,\tilde{\cup}\, \lambda)_\ell(x) \ge \bigvee\{\mu_\ell(x), \lambda_\ell(x)\}\right), \tag{11}$$

and

$$(\forall x \in X)\left((\mu \tilde{\cap} \lambda)_\ell(x) \leq \bigwedge\{\mu_\ell(x), \lambda_\ell(x)\}\right). \tag{12}$$

Proof. Let $x \in X$. Then,

$$(\mu \tilde{\cup} \lambda)_\vee(x) = \bigvee\{\mu(x) \cup \lambda(x)\} \geq \bigvee\{\mu(x)\} \text{ (and } \bigvee\{\lambda(x)\})$$

and

$$(\mu \tilde{\cup} \lambda)_\wedge(x) = \bigwedge\{\mu(x) \cup \lambda(x)\} \leq \bigwedge\{\mu(x)\} \text{ (and } \bigwedge\{\lambda(x)\}).$$

It follows that

$$(\mu \tilde{\cup} \lambda)_\vee(x) \geq \bigvee\left\{\bigvee\{\mu(x)\},\ \bigvee\{\lambda(x)\}\right\}$$

and

$$(\mu \tilde{\cup} \lambda)_\wedge(x) \leq \bigwedge\left\{\bigwedge\{\mu(x)\},\ \bigwedge\{\lambda(x)\}\right\}.$$

Note that $\bigvee\{a,b\} + \bigvee\{c,d\} \geq \bigvee\{a+c,\ b+d\}$ for all $a,b,c,d \in [-1,0]$. Hence,

$$\begin{aligned}
(\mu \tilde{\cup} \lambda)_\ell(x) &= (\mu \tilde{\cup} \lambda)_\vee(x) - (\mu \tilde{\cup} \lambda)_\wedge(x) \\
&\geq \bigvee\left\{\bigvee\{\mu(x)\},\ \bigvee\{\lambda(x)\}\right\} - \bigwedge\left\{\bigwedge\{\mu(x)\},\ \bigwedge\{\lambda(x)\}\right\} \\
&\geq \bigvee\left\{\bigvee\{\mu(x)\},\ \bigvee\{\lambda(x)\}\right\} + \bigvee\left\{-\bigwedge\{\mu(x)\},\ -\bigwedge\{\lambda(x)\}\right\} \\
&\geq \bigvee\left\{\bigvee\{\mu(x)\} - \bigwedge\{\mu(x)\},\ \bigvee\{\lambda(x)\} - \bigwedge\{\lambda(x)\}\right\} \\
&= \bigvee\left\{\mu_\ell(x),\ \lambda_\ell(x)\right\},
\end{aligned}$$

and so Equation (11) is valid. For any $x \in X$, we have

$$(\mu \tilde{\cap} \lambda)_\vee(x) = \bigvee\{\mu(x) \cap \lambda(x)\} \leq \bigvee\{\mu(x)\} \text{ (and } \bigvee\{\lambda(x)\})$$

and

$$(\mu \tilde{\cap} \lambda)_\wedge(x) = \bigwedge\{\mu(x) \cap \lambda(x)\} \geq \bigwedge\{\mu(x)\} \text{ (and } \bigwedge\{\lambda(x)\}),$$

which imply that

$$(\mu \tilde{\cap} \lambda)_\vee(x) \leq \bigwedge\left\{\bigvee\{\mu(x)\},\ \bigvee\{\lambda(x)\}\right\}$$

and

$$(\mu \tilde{\cap} \lambda)_\wedge(x) \geq \bigvee\left\{\bigwedge\{\mu(x)\},\ \bigwedge\{\lambda(x)\}\right\}.$$

Since $\bigwedge\{a,b\} + \bigwedge\{c,d\} \leq \bigwedge\{a+c,\ b+d\}$ for all $a,b,c,d \in [-1,0]$, we have

$$
\begin{aligned}
(\mu \tilde{\cap} \lambda)_\ell(x) &= (\mu \tilde{\cap} \lambda)_\vee(x) - (\mu \tilde{\cap} \lambda)_\wedge(x) \\
&\leq \bigwedge\left\{\bigvee\{\mu(x)\},\ \bigvee\{\lambda(x)\}\right\} - \bigvee\left\{\bigwedge\{\mu(x)\},\ \bigwedge\{\lambda(x)\}\right\} \\
&= \bigwedge\left\{\bigvee\{\mu(x)\},\ \bigvee\{\lambda(x)\}\right\} + \bigwedge\left\{-\bigwedge\{\mu(x)\},\ -\bigwedge\{\lambda(x)\}\right\} \\
&\leq \bigwedge\left\{\bigvee\{\mu(x)\} - \bigwedge\{\mu(x)\},\ \bigvee\{\lambda(x)\} - \bigwedge\{\lambda(x)\}\right\} \\
&= \bigwedge\{\mu_\ell(x), \lambda_\ell(x)\}.
\end{aligned}
$$

This completes the proof. □

Proposition 2. *If μ is an $\mathcal{N}$-hyper set over an initial universe set X, then*

$$(\forall x \in X)\left((\mu \tilde{\cup} \mu')_\ell(x) \geq \bigvee\{\mu_\ell(x),\ \mu'_\ell(x)\}\right). \tag{13}$$

Proof. Note that

$$(\mu \tilde{\cup} \mu')(x) = \mu(x) \cup \mu'(x) = \mu(x) \cup ([-1,0] \setminus \mu(x)) = [-1,0]$$

for all $x \in X$. It follows that

$$(\mu \tilde{\cup} \mu')_\ell(x) = (\mu \tilde{\cup} \mu')_\vee(x) - (\mu \tilde{\cup} \mu')_\wedge(x) = 1 \geq \bigvee\{\mu_\ell(x),\ \mu'_\ell(x)\}$$

for all $x \in X$. □

Proposition 3. *If μ and λ are $\mathcal{N}$-hyper sets over an initial universe set X, then*

$$(\forall x \in X)\left((\mu \setminus \lambda)_\ell(x) \leq \mu_\ell(x)\right). \tag{14}$$

Proof. Note that $(\mu \setminus \lambda)(x) = \mu(x) \setminus \lambda(x) \subseteq \mu(x)$ for all $x \in X$. Hence,

$$(\mu \setminus \lambda)_\vee(x) \leq \mu_\vee(x) \text{ and } (\mu \setminus \lambda)_\wedge(x) \geq \mu_\wedge(x).$$

It follows that

$$
\begin{aligned}
(\mu \setminus \lambda)_\ell(x) &= (\mu \setminus \lambda)_\vee(x) - (\mu \setminus \lambda)_\wedge(x) \\
&\leq \mu_\vee(x) - \mu_\wedge(x) = \mu_\ell(x),
\end{aligned}
$$

proving the proposition. □

Given $\mathcal{N}$-hyper sets μ and λ over an initial universe set X, we define

$$\mu_\xi : X \to [-1,0],\ x \mapsto \mu_\wedge(x) - \mu_\vee(x), \tag{15}$$

$$\mu \tilde{\vee} \lambda : X \to \mathcal{P}^*([-1,0]),\ x \mapsto \left\{\bigvee\{a,b\} \in [-1,0] \mid a \in \mu(x), b \in \lambda(x)\right\}, \tag{16}$$

$$\mu \tilde{\wedge} \lambda : X \to \mathcal{P}^*([-1,0]),\ x \mapsto \left\{\bigwedge\{a,b\} \in [-1,0] \mid a \in \mu(x), b \in \lambda(x)\right\}. \tag{17}$$

Example 4. *Let μ and λ be $\mathcal{N}$-hyper sets over $X = \{a,b,c,d\}$ defined by Table 5.*

Table 5. $\mathcal{N}$-hyper sets μ and λ.

X	a	b	c	d
μ	$[-0.5, 0]$	$\{-1, -0.6\}$	$[-0.4, -0.2)$	$(-1, -0.8]$
λ	$[-0.6, -0.3]$	$\{-1, -0.8, -0.5\}$	$[-0.5, -0.3)$	$(-0.9, -0.7]$

Then, μ_{ξ} is given as Table 6

Table 6. $\mathcal{N}$-function (X, μ_{ξ}).

X	a	b	c	d
$\mu_{\wedge}$	-0.5	-1	-0.4	-1
$\mu_{\vee}$	0	-0.6	-0.2	-0.8
μ_{ξ}	-0.5	-0.4	-0.2	-0.2

and

$$\begin{aligned}(\mu \tilde{\vee} \lambda)(b) &= \Big\{ \bigvee\{-1,-1\}, \bigvee\{-1,-0.8\}, \bigvee\{-1,-0.5\}, \\ &\qquad \bigvee\{-0.6,-1\}, \bigvee\{-0.6,-0.8\}, \bigvee\{-0.6,-0.5\}\Big\} \\ &= \{-1,-0.8,-0.5,-0.6\}, \\ (\mu \tilde{\wedge} \lambda)(b) &= \Big\{ \bigwedge\{-1,-1\}, \bigwedge\{-1,-0.8\}, \bigwedge\{-1,-0.5\}, \\ &\qquad \bigwedge\{-0.6,-1\}, \bigwedge\{-0.6,-0.8\}, \bigwedge\{-0.6,-0.5\}\Big\} \\ &= \{-1,-0.8,-0.6\}.\end{aligned}$$

Thus, $(\mu \tilde{\vee} \lambda)_{\vee}(b) = -0.5$, $(\mu \tilde{\wedge} \lambda)_{\vee}(b) = -0.6$ *and* $(\mu \tilde{\vee} \lambda)_{\wedge}(b) = -1 = (\mu \tilde{\wedge} \lambda)_{\wedge}(b)$.

Proposition 4. *Let X be an initial universe set with a binary operation $*$. If μ and λ are $\mathcal{N}$-hyper sets over X, then*

$$(\forall x \in X) \left((\mu \tilde{\vee} \lambda)_{\vee}(x) = \bigvee \{\mu_{\vee}(x), \lambda_{\vee}(x)\} \right) \tag{18}$$

and

$$(\forall x \in X) \left((\mu \tilde{\vee} \lambda)_{\wedge}(x) = \bigvee \{\mu_{\wedge}(x), \lambda_{\wedge}(x)\} \right). \tag{19}$$

Proof. For any $x \in X$, let $\alpha := \mu_{\vee}(x)$ and $\beta := \lambda_{\vee}(x)$. Then,

$$\begin{aligned}(\mu \tilde{\vee} \lambda)_{\vee}(x) &= \bigvee \{(\mu \tilde{\vee} \lambda)(x)\} \\ &= \bigvee \Big\{ \bigvee\{a, b\} \in [-1, 0] \mid a \in \mu(x), b \in \lambda(x) \Big\} \\ &= \bigvee \Big\{ \bigvee\{\alpha, b \mid b \in \lambda(x)\}, \bigvee\{a, b \mid a \in \mu(x), b \in \lambda(x)\}, \\ &\qquad \bigvee\{a, \beta \mid a \in \mu(x)\}, \bigvee\{\alpha, \beta\} \Big\} \\ &= \bigvee\{\alpha, \beta\} = \bigvee \{\mu_{\vee}(x), \lambda_{\vee}(x)\}.\end{aligned}$$

Thus, Equation (18) is valid. Similarly, we can prove Equation (19). □

Similarly, we have the following property.

Proposition 5. *Let X be an initial universe set with a binary operation $*$. If μ and λ are $\mathcal{N}$-hyper sets over X, then*

$$(\forall x \in X)\left((\mu \barwedge \lambda)_{\vee}(x) = \bigwedge\{\mu_{\vee}(x), \lambda_{\vee}(x)\}\right) \tag{20}$$

and

$$(\forall x \in X)\left((\mu \barwedge \lambda)_{\wedge}(x) = \bigwedge\{\mu_{\wedge}(x), \lambda_{\wedge}(x)\}\right). \tag{21}$$

Definition 5. *Let X be an initial universe set with a binary operation $*$. An $\mathcal{N}$-hyper set $\mu : X \to \mathcal{P}^*([-1,0])$ is called: an $\mathcal{N}$-hyper subset of $(X,*)$ with type (i,j) for $i,j \in \{1,2,3,4\}$ (briefly, $\mathcal{N}_{(i,j)}$-substructure of $(X,*)$) if $(X,\mu_{\vee})$ is an $\mathcal{N}_i$-substructure of $(X,*)$ and $(X,\mu_{\wedge})$ is an $\mathcal{N}_j$-substructure of $(X,*)$.*

Given an $\mathcal{N}$-hyper set $\mu : X \to \mathcal{P}^*([-1,0])$, we consider the set

$$\mathcal{N}_{(i,j)}(\mu) := \{(X,*) \in \mathcal{H}(X) \mid \mu \text{ is an } \mathcal{N}_{(i,j)}\text{-substructure of } (X,*)\}$$

for $i,j \in \{1,2,3,4\}$.

Theorem 3. *Let X be an initial universe set with a binary operation $*$. For any $\mathcal{N}$-hyper set $\mu : X \to \mathcal{P}^*([-1,0])$, we have*

$$(X,*) \in \mathcal{N}_{(3,4)}(\mu) \Rightarrow (X,*) \in \mathcal{N}_4(\mu_{\xi}). \tag{22}$$

Proof. Let $(X,*) \in \mathcal{N}_{(3,4)}(\mu)$. Then, $(X,\mu_{\vee})$ is an $\mathcal{N}_3$-substructure of $(X,*)$ and $(X,\mu_{\wedge})$ is an $\mathcal{N}_4$-substructure of $(X,*)$, that is,

$$\mu_{\vee}(x*y) \geq \bigvee\{\mu_{\vee}(x), \mu_{\vee}(y)\}$$

and

$$\mu_{\wedge}(x*y) \leq \bigwedge\{\mu_{\wedge}(x), \mu_{\wedge}(y)\}$$

for all $x,y \in X$. It follows that

$$\mu_{\xi}(x*y) = \mu_{\wedge}(x*y) - \mu_{\vee}(x*y) \leq \mu_{\wedge}(x) - \mu_{\vee}(x) = \mu_{\xi}(x).$$

Similarly, we get $\mu_{\xi}(x*y) \leq \mu_{\xi}(y)$. Hence, $\mu_{\xi}(x*y) \leq \bigwedge\{\mu_{\xi}(x), \mu_{\xi}(y)\}$, and so $(X,*) \in \mathcal{N}_4(\mu_{\xi})$. □

Corollary 1. *Let X be an initial universe set with a binary operation $*$. For any $\mathcal{N}$-hyper set $\mu : X \to \mathcal{P}^*([-1,0])$, we have*

$$(X,*) \in \mathcal{N}_{(3,4)}(\mu) \Rightarrow (X,*) \in \mathcal{N}_1(\mu_{\xi}).$$

Theorem 4. *Let X be an initial universe set with a binary operation $*$. For any $\mathcal{N}$-hyper set $\mu : X \to \mathcal{P}^*([-1,0])$, we have*

$$(X,*) \in \mathcal{N}_{(4,3)}(\mu) \Rightarrow (X,*) \in \mathcal{N}_3(\mu_{\xi}). \tag{23}$$

Proof. It is similar to the proof of Theorem 3. □

Corollary 2. *Let X be an initial universe set with a binary operation* $*$*. For any* $\mathcal{N}$*-hyper set* $\mu : X \to \mathcal{P}^*([-1,0])$*, we have*

$$(X,*) \in \mathcal{N}_{(4,3)}(\mu) \Rightarrow (X,*) \in \mathcal{N}_2(\mu_\xi).$$

Theorem 5. *Let X be an initial universe set with a binary operation* $*$*. For any* $\mathcal{N}$*-hyper set* $\mu : X \to \mathcal{P}^*([-1,0])$*, we have*

$$(X,*) \in \mathcal{N}_{(1,3)}(\mu) \Rightarrow (X,*) \in \mathcal{N}_3(\mu_\xi). \tag{24}$$

Proof. Let $(X,*) \in \mathcal{N}_{(1,3)}(\mu)$. Then, $(X,\mu_\vee)$ is an $\mathcal{N}_1$-substructure of $(X,*)$ and $(X,\mu_\wedge)$ is an $\mathcal{N}_3$-substructure of $(X,*)$, that is,

$$\mu_\vee(x*y) \leq \bigvee\{\mu_\vee(x), \mu_\vee(y)\} \tag{25}$$

and

$$\mu_\wedge(x*y) \geq \bigvee\{\mu_\wedge(x), \mu_\wedge(y)\}$$

for all $x,y \in X$. Equation (25) implies that

$$\mu_\vee(x*y) \leq \mu_\vee(x) \text{ or } \mu_\vee(x*y) \leq \mu_\vee(y).$$

If $\mu_\vee(x*y) \leq \mu_\vee(x)$, then

$$\mu_\xi(x*y) = \mu_\wedge(x*y) - \mu_\vee(x*y) \geq \mu_\wedge(x) - \mu_\vee(x) = \mu_\xi(x).$$

If $\mu_\vee(x*y) \leq \mu_\vee(y)$, then

$$\mu_\xi(x*y) = \mu_\wedge(x*y) - \mu_\vee(x*y) \geq \mu_\wedge(y) - \mu_\vee(y) = \mu_\xi(y).$$

It follows that $\mu_\xi(x*y) \geq \vee\{\mu_\xi(x), \mu_\xi(y)\}$, and so $(X,*) \in \mathcal{N}_3(\mu_\xi)$. □

Corollary 3. *Let X be an initial universe set with a binary operation* $*$*. For any* $\mathcal{N}$*-hyper set* $\mu : X \to \mathcal{P}^*([-1,0])$*, we have*

$$(X,*) \in \mathcal{N}_{(1,3)}(\mu) \Rightarrow (X,*) \in \mathcal{N}_2(\mu_\xi). \tag{26}$$

Theorem 6. *Let X be an initial universe set with a binary operation* $*$*. For any* $\mathcal{N}$*-hyper set* $\mu : X \to \mathcal{P}^*([-1,0])$*, we have*

$$(X,*) \in \mathcal{N}_{(3,1)}(\mu) \Rightarrow (X,*) \in \mathcal{N}_1(\mu_\xi). \tag{27}$$

Proof. It is similar to the proof of Theorem 5. □

Theorem 7. *Let X be an initial universe set with a binary operation* $*$*. For any* $\mathcal{N}$*-hyper set* $\mu : X \to \mathcal{P}^*([-1,0])$*, we have*

$$(X,*) \in \mathcal{N}_{(2,4)}(\mu) \Rightarrow (X,*) \in \mathcal{N}_1(\mu_\xi). \tag{28}$$

Proof. Let $(X,*) \in \mathcal{N}_{(2,4)}(\mu)$. Then, $(X,\mu_\vee)$ is an $\mathcal{N}_2$-substructure of $(X,*)$ and $(X,\mu_\wedge)$ is an $\mathcal{N}_4$-substructure of $(X,*)$, that is,

$$\mu_\vee(x*y) \geq \bigwedge\{\mu_\vee(x), \mu_\vee(y)\} \tag{29}$$

and

$$\mu_\wedge(x*y) \leq \bigwedge\{\mu_\wedge(x), \mu_\wedge(y)\}$$

for all $x,y \in X$. Then, $\mu_\vee(x*y) \geq \mu_\vee(x)$ or $\mu_\vee(x*y) \geq \mu_\vee(y)$ by Equation (29). If $\mu_\vee(x*y) \geq \mu_\vee(x)$, then

$$\mu_\xi(x*y) = \mu_\wedge(x*y) - \mu_\vee(x*y) \leq \mu_\wedge(x) - \mu_\vee(x) = \mu_\xi(x).$$

If $\mu_\vee(x*y) \geq \mu_\vee(y)$, then

$$\mu_\xi(x*y) = \mu_\wedge(x*y) - \mu_\vee(x*y) \leq \mu_\wedge(y) - \mu_\vee(y) = \mu_\xi(y).$$

It follows that $\mu_\xi(x*y) \leq \bigvee\{\mu_\xi(x), \mu_\xi(y)\}$, that is, $(X,*) \in \mathcal{N}_1(\mu_\xi)$. □

Theorem 8. *Let X be an initial universe set with a binary operation $*$. For any $\mathcal{N}$-hyper set $\mu : X \to \mathcal{P}^*([-1,0])$, if $(X,*) \in \mathcal{N}_{(4,2)}(\mu)$, then*

$$(\forall x,y \in X) \left(\mu_\ell(x*y) \leq \bigvee\{\mu_\ell(x), \mu_\ell(y)\}\right). \tag{30}$$

Proof. If $(X,*) \in \mathcal{N}_{(4,2)}(\mu)$, then $(X,\mu_\vee)$ is an $\mathcal{N}_4$-substructure of $(X,*)$ and $(X,\mu_\wedge)$ is an $\mathcal{N}_2$-substructure of $(X,*)$, that is,

$$\mu_\vee(x*y) \leq \bigwedge\{\mu_\vee(x), \mu_\vee(y)\}$$

and

$$\mu_\wedge(x*y) \geq \bigwedge\{\mu_\wedge(x), \mu_\wedge(y)\} \tag{31}$$

for all $x,y \in X$. Then, $\mu_\wedge(x*y) \geq \mu_\wedge(x)$ or $\mu_\wedge(x*y) \geq \mu_\wedge(y)$ by Equation (31). If $\mu_\wedge(x*y) \geq \mu_\wedge(x)$, then

$$\mu_\ell(x*y) = \mu_\vee(x*y) - \mu_\wedge(x*y) \leq \mu_\vee(x) - \mu_\wedge(x) = \mu_\ell(x).$$

If $\mu_\wedge(x*y) \geq \mu_\wedge(y)$, then

$$\mu_\ell(x*y) = \mu_\vee(x*y) - \mu_\wedge(x*y) \leq \mu_\vee(y) - \mu_\wedge(y) = \mu_\ell(y).$$

It follows that $\mu_\ell(x*y) \leq \bigvee\{\mu_\ell(x), \mu_\ell(y)\}$ for all $x,y \in X$. □

4. Conclusions

Fuzzy set theory has been considered by Zadeh as one of the mathematical tools to deal with the uncertainties. Because fuzzy set theory could not deal with negative information, Jun et al. have introduced a new function, which is called negative-valued function, and constructed $\mathcal{N}$-structures in 2009 as a mathematical tool to deal with negative information. Since then, $\mathcal{N}$-structures have been applied to algebraic structures and soft sets, etc. Using the $\mathcal{N}$-structures, in this article, we have studied the notions of (extended) $\mathcal{N}$-hyper sets, $\mathcal{N}$-substructures of type 1, 2, 3 and 4, and have been investigated several related properties.

Author Contributions: All authors contributed equally and significantly to the study and preparation of the article. They have read and approved the final manuscript.

Acknowledgments: The authors thank the anonymous reviewers for their valuable comments and suggestions.

Conflicts of Interest: The authors declare no conflict of interest.

References

1. Zadeh, L.A. Fuzzy sets. *Inf. Control* **1965**, *8*, 338–353. [CrossRef]
2. Jun, Y.B.; Lee, K.J.; Song, S.Z. $\mathcal{N}$-ideals of BCK/BCI-algebras. *J. Chungcheong Math. Soc.* **2009**, *22*, 417–437.
3. Ceven, Y. $\mathcal{N}$-ideals of rings. *Int. J. Algebra* **2012**, *6*, 1227–1232.
4. Jun, Y.B.; Öztürk, M.A.; Roh, E.H. $\mathcal{N}$-structures applied to closed ideals in BCH-algebras. *Int. J. Math. Math. Sci.* **2010**, 943565. [CrossRef]
5. Khan, A.; Jun, Y.B.; Shabir, M. $\mathcal{N}$-fuzzy quasi-ideals in ordered semigroups. *Quasigroups Relat. Syst.* **2009**, *17*, 237–252.
6. Khan, A.; Jun, Y.B.; Shabir, M. $\mathcal{N}$-fuzzy ideals in ordered semigroups. *Int. J. Math. Math. Sci.* **2009**, 814861. [CrossRef]
7. Khan, A.; Jun, Y.B.; Shabir, M. $\mathcal{N}$-fuzzy filters in ordered semigroups. *Fuzzy Syst. Math.* **2010**, *24*, 1–5.
8. Khan, A.; Jun, Y.B.; Shabir, M. $\mathcal{N}$-fuzzy bi-ideals in ordered semigroups. *J. Fuzzy Math.* **2011**, *19*, 747–762.
9. Jun, Y.B.; Alshehri, N.O.; Lee, K.J. Soft set theory and $\mathcal{N}$-structures applied to BCH-algebras. *J. Comput. Anal. Appl.* **2014**, *16*, 869–886.
10. Jun, Y.B.; Lee, K.J.; Kang, M.S. Ideal theory in BCK/BCI-algebras based on soft sets and $\mathcal{N}$-structures. *Discret. Dyn. Nat. Soc.* **2012**, 910450. [CrossRef]
11. Jun, Y.B.; Song, S.Z.; Lee, K.J. The combination of soft sets and $\mathcal{N}$-structures with applications. *J. Appl. Math.* **2013**, 420312. [CrossRef]
12. Bucolo, M.; Fortuna, L.; la Rosa, M. Complex dynamics through fuzzy chains. *IEEE Trans. Fuzzy Syst.* **2004**, *12*, 289–295. [CrossRef]

mathematics

MDPI

Article

Hypergraphs in m-Polar Fuzzy Environment

Muhammad Akram [1,*] **and Gulfam Shahzadi** [1]

[1] Department of Mathematics, University of the Punjab, New Campus, Lahore 54590, Pakistan; gulfamshahzadi22@gmail.com

* Correspondence: m.akram@pucit.edu.pk

Received: 7 February 2018; Accepted: 16 February 2018; Published: 20 February 2018

Abstract: Fuzzy graph theory is a conceptual framework to study and analyze the units that are intensely or frequently connected in a network. It is used to study the mathematical structures of pairwise relations among objects. An m-polar fuzzy (mF, for short) set is a useful notion in practice, which is used by researchers or modelings on real world problems that sometimes involve multi-agents, multi-attributes, multi-objects, multi-indexes and multi-polar information. In this paper, we apply the concept of mF sets to hypergraphs, and present the notions of regular mF hypergraphs and totally regular mF hypergraphs. We describe the certain properties of regular mF hypergraphs and totally regular mF hypergraphs. We discuss the novel applications of mF hypergraphs in decision-making problems. We also develop efficient algorithms to solve decision-making problems.

Keywords: regular mF hypergraph; totally regular mF hypergraph; decision-making; algorithm; time complexity

1. Introduction

Graph theory has interesting applications in different fields of real life problems to deal with the pairwise relations among the objects. However, this information fails when more than two objects satisfy a certain common property or not. In several real world applications, relationships are more problematic among the objects. Therefore, we take into account the use of hypergraphs to represent the complex relationships among the objects. In case of a set of multiarity relations, hypergraphs are the generalization of graphs, in which a hypergraph may have more than two vertices. Hypergraphs have many applications in different fields including biological science, computer science, declustering problems and discrete mathematics.

In 1994, Zhang [1] proposed the concept of bipolar fuzzy set as a generalization of fuzzy set [2]. In many problems, bipolar information are used, for instance, common efforts and competition, common characteristics and conflict characteristics are the two-sided knowledge. Chen et al. [3] introduced the concept of m-polar fuzzy (mF, for short) set as a generalization of a bipolar fuzzy set and it was shown that 2-polar and bipolar fuzzy set are cryptomorphic mathematical notions. The framework of this theory is that "multipolar information" (unlike the bipolar information which gives two-valued logic) arise because information for a natural world are frequently from n factors ($n \geq 2$). For example, 'Pakistan is a good country'. The truth value of this statement may not be a real number in $[0,1]$. Being a good country may have several properties: good in agriculture, good in political awareness, good in regaining macroeconomic stability etc. Each component may be a real number in $[0,1]$. If n is the number of such components under consideration, then the truth value of the fuzzy statement is a n-tuple of real numbers in $[0,1]$, that is, an element of $[0,1]^n$. The perception of fuzzy graphs based on Zadeh's fuzzy relations [4] was introduced by Kauffmann [5]. Rosenfeld [6] described the fuzzy graphs structure. Later, some remarks were given by Bhattacharya [7] on fuzzy graphs. Several concepts on fuzzy graphs were introduced by Mordeson and Nair [8]. In 2011, Akram introduced the notion of bipolar fuzzy graphs in [9]. Li et al. [10] considered different algebraic operations on mF graphs.

In 1977, Kauffmann [5] proposed the fuzzy hypergraphs. Chen [11] studied the interval-valued fuzzy hypergraph. Generalization and redefinition of the fuzzy hypergraph were explained by Lee-Kwang and Keon-Myung [12]. Parvathi et al. [13] introduced the concept of intuitionistic fuzzy hypergraphs. Samanta and Pal [14] dealt with bipolar fuzzy hypergraphs. Later on, Akram et al. [15] considered certain properties of the bipolar fuzzy hypergraph. Bipolar neutrosophic hypergraphs with applications were presented by Akram and Luqman [16]. Sometimes information is multipolar, that is, a communication channel may have various signal strengths from the others due to various reasons including atmosphere, device distribution, mutual interference of satellites etc. The accidental mixing of various chemical substances could cause toxic gases, fire or explosion of different degrees. All these are components of multipolar knowledge which are fuzzy in nature. This idea motivated researchers to study mF hypergraphs [17]. Akram and Sarwar [18] considered transversals of mF hypergraphs with applications. In this research paper, we introduce the idea of regular and totally regular mF hypergraphs and investigate some of their properties. We discuss the new applications of mF hypergraphs in decision-making problems. We develop efficient algorithms to solve decision-making problems and compute the time complexity of algorithms. For other notations, terminologies and applications not mentioned in the paper, the readers are referred to [19–31].

In this paper, we will use the notations defined in Table 1.

Table 1. Notations.

Symbol	Definition
$H^* = (A^*, B^*)$	Crisp hypergraph
$H = (A, B)$	mF hypergraph
$H^D = (A^*, B^*)$	Dual mF hypergraph
$N(x)$	Open neighourhood degree of a vertex in H
$N[x]$	Closed neighourhood degree of a vertex in H
$\gamma(x_1, x_2)$	Adjacent level of two vertices
$\sigma(T_1, T_2)$	Adjacent level of two hyperedges

2. Notions of mF Hypergraph

Definition 1. *An mF set on a non-empty crisp set X is a function $A : X \to [0,1]^m$. The degree of each element $x \in X$ is denoted by $A(x) = (P_1oA(x), P_2oA(x), ..., P_moA(x))$, where $P_ioA : [0,1]^m \to [0,1]$ is the i-th projection mapping [3].*

Note that $[0,1]^m$ (m-th-power of $[0,1]$) is considered as a poset with the point-wise order $\leq$, where m is an arbitrary ordinal number (we make an appointment that $m = \{n | n < m\}$ when $m > 0$), $\leq$ is defined by $x < y \Leftrightarrow p_i(x) \leq p_i(y)$ for each $i \in m$ ($x, y \in [0,1]^m$), and $P_i : [0,1]^m \to [0,1]$ is the i-th projection mapping ($i \in m$). $\mathbf{1} = (1, 1, ..., 1)$ is the greatest value and $\mathbf{0} = (0, 0, ..., 0)$ is the smallest value in $[0,1]^m$.

Definition 2. *Let A be an mF subset of a non-empty fuzzy subset of a non-empty set X. An mF relation on A is an mF subset B of $X \times X$ defined by the mapping $B : X \times X \to [0,1]^m$ such that for all $x, y \in X$*

$$P_ioB(xy) \leq \inf\{P_ioA(x), P_ioA(y)\}$$

$1 \leq i \leq m$, where $P_ioA(x)$ denotes the i-th degree of membership of a vertex x and $P_ioB(xy)$ denotes the i-th degree of membership of the edge xy.

Definition 3. *An mF graph is a pair $G = (A, B)$, where $A : X \to [0,1]^m$ is an mF set in X and $B : X \times X \to [0,1]^m$ is an mF relation on X such that*

$$P_ioB(xy) \leq \inf\{P_ioA(x), P_ioA(y)\}$$

$1 \leq i \leq m$, *for all* $x, y \in X$ *and* $P_i oB(xy) = 0$ *for all* $xy \in X \times X - E$ *for all* $i = 1, 2, ..., m$. *A is called the mF vertex set of G and B is called the mF edge set of G, respectively* [3].

Definition 4. *An mF hypergraph on a non-empty set X is a pair* $H = (A, B)$ [17], *where* $A = \{\zeta_1, \zeta_2, \zeta_3, ..., \zeta_r\}$ *is a family of mF subsets on X and B is an mF relation on the mF subsets* ζ_j *such that*

1. $B(E_j) = B(\{x_1, x_2, ..., x_r\}) \leq \inf\{\zeta_j(x_1), \zeta_j(x_2), ..., \zeta_j(x_s)\}$, *for all* $x_1, x_2, ..., x_s \in X$.
2. $\bigcup_k supp(\zeta_k) = X$, *for all* $\zeta_k \in A$.

Example 1. *Let* $A = \{\zeta_1, \zeta_2, \zeta_3, \zeta_4, \zeta_5\}$ *be a family of 4-polar fuzzy subsets on* $X = \{a, b, c, d, e, f, g\}$ *given in Table 2. Let B be a 4-polar fuzzy relation on* $\zeta_j^{'}s$, $1 \leq j \leq 5$, *given as,* $B(\{a, c, e\}) = (0.2, 0.4, 0.1, 0.3)$, $B(\{b, d, f\}) = (0.2, 0.1, 0.1, 0.1)$, $B(\{a, b\}) = (0.3, 0.1, 0.1, 0.6)$, $B(\{e, f\}) = (0.2, 0.4, 0.3, 0.2)$, $B(\{b, e, g\}) = (0.2, 0.1, 0.2, 0.4)$. *Thus, the 4-polar fuzzy hypergraph is shown in Figure 1.*

Table 2. 4-polar fuzzy subsets.

$x \in X$	ζ_1	ζ_2	ζ_3	ζ_4	ζ_5
a	(0.3,0.4,0.5,0.6)	(0,0,0,0)	(0.3,0.4,0.5,0.6)	(0,0,0,0)	(0,0,0,0)
b	(0,0,0,0)	(0.4,0.1,0.1,0.6)	(0.4,0.1,0.1,0.6)	(0,0,0,0)	(0.4,0.1,0.1,0.6)
c	(0.3,0.5,0.1,0.3)	(0,0,0,0)	(0,0,0,0)	(0,0,0,0)	(0,0,0,0)
d	(0,0,0,0)	(0.4,0.2,0.5,0.1)	(0,0,0,0)	(0,0,0,0)	(0,0,0,0)
e	(0.2,0.4,0.6,0.8)	(0,0,0,0)	(0,0,0,0)	(0.2,0.4,0.6,0.8)	(0.2,0.4,0.6,0.8)
f	(0,0,0,0)	(0.2,0.5,0.3,0.2)	(0,0,0,0)	(0.2,0.5,0.3,0.2)	(0,0,0,0)
g	(0,0,0,0)	(0,0,0,0)	(0,0,0,0)	(0,0,0,0)	(0.3,0.5,0.1,0.4)

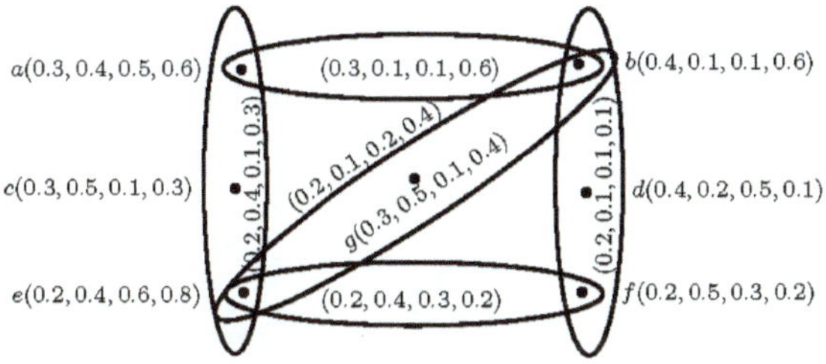

Figure 1. 4-Polar fuzzy hypergraph.

Example 2. *Consider a 5-polar fuzzy hypergraph with vertex set {a, b, c, d, e, f, g} whose degrees of membership are given in Table 3 and three hyperedges {a,b,c}, {b,d,e}, {b,f,g} such that* $B(\{a, b, c\}) = (0.2, 0.1, 0.3, 0.1, 0.2)$, $B(\{b, d, e\}) = (0.1, 0.2, 0.3, 0.4, 0.2)$, $B(\{b, f, g\}) = (0.2, 0.2, 0.3, 0.3, 0.2)$. *Hence, the 5-polar fuzzy hypergraph is shown in Figure 2.*

Table 3. 5-polar fuzzy subsets.

$x \in X$	ζ_1	ζ_2	ζ_3
a	(0.2,0.1,0.3,0.1,0.3)	(0,0,0,0,0)	(0,0,0,0,0)
b	(0.2,0.3,0.5,0.6,0.2)	(0.2,0.3,0.5,0.6,0.2)	(0.2,0.3,0.5,0.6,0.2)
c	(0.3,0.2,0.4,0.5,0.2)	(0,0,0,0,0)	(0,0,0,0,0)
d	(0,0,0,0,0)	(0.6,0.2,0.2,0.3,0.3)	(0,0,0,0,0)
e	(0,0,0,0,0)	(0.4,0.5,0.6,0.7,0.3)	(0,0,0,0,0)
f	(0,0,0,0,0)	(0,0,0,0,0)	(0.1,0.2,0.3,0.4,0.4)
g	(0,0,0,0,0)	(0,0,0,0,0)	(0.2,0.4,0.6,0.8,0.4)

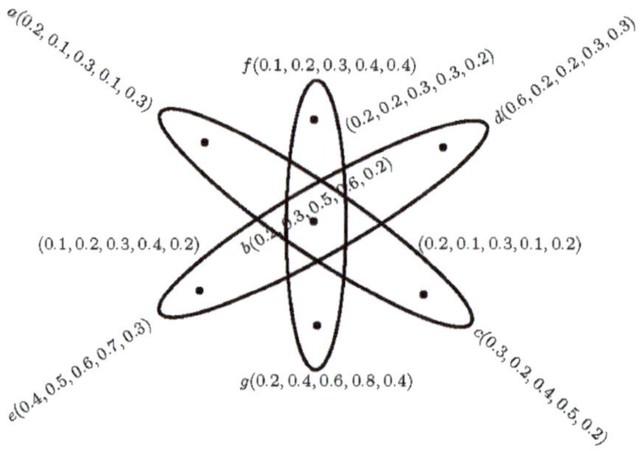

Figure 2. 5-Polar fuzzy hypergraph.

Definition 5. *Let $H = (A, B)$ be an mF hypergraph on a non-empty set X [17]. The dual mF hypergraph of H, denoted by $H^D = (A^*, B^*)$, is defined as*

1. *$A^* = B$ is the mF set of vertices of H^D.*
2. *If $|X| = n$ then, B^* is an mF set on the family of hyperedges $\{X_1, X_2, ..., X_n\}$ such that, X_i={$E_j \mid x_j \in E_j, E_j$ is a hyperedge of H}, i.e., X_i is the mF set of those hyperedges which share the common vertex x_i and $B^*(X_i) = \inf\{E_j \mid x_j \in E_j\}$.*

Example 3. *Consider the example of a 3-polar fuzzy hypergraph $H = (A, B)$ given in Figure 3, where $X = \{x_1, x_2, x_3, x_4, x_5, x_6\}$ and $E = \{E_1, E_2, E_3, E_4\}$. The dual 3-polar fuzzy hypergraph is shown in Figure 4 with dashed lines with vertex set $E = \{E_1, E_2, E_3, E_4\}$ and set of hyperedges $\{X_1, X_2, X_3, X_4, X_5, X_6\}$ such that $X_1 = X_3$.*

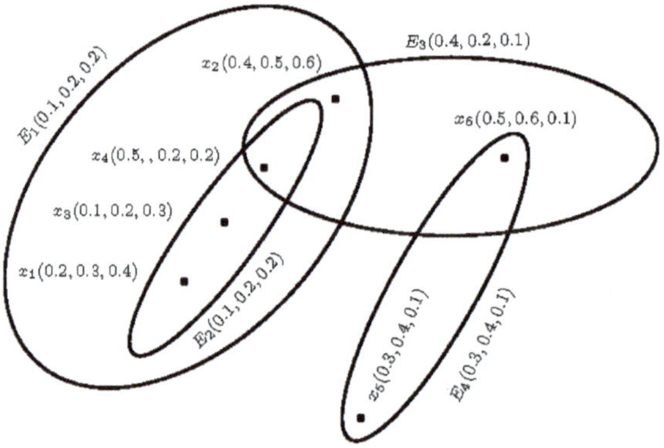

Figure 3. 3-Polar fuzzy hypergraph.

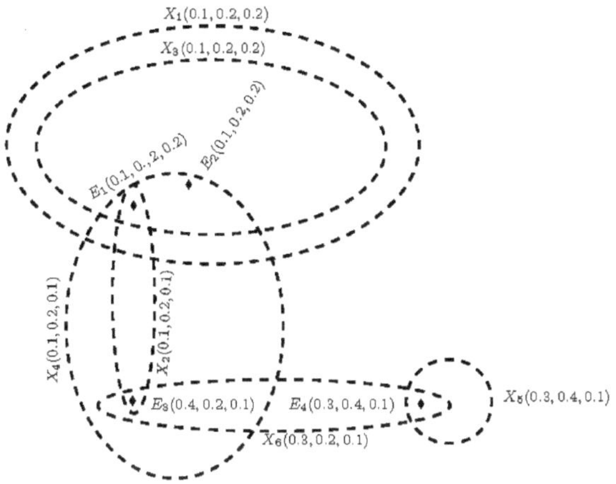

Figure 4. Dual 3-polar fuzzy hypergraph.

Definition 6. *The open neighourhood of a vertex x in the mF hypergraph is the set of adjacent vertices of x excluding that vertex and it is denoted by* $N(x)$.

Example 4. *Consider the 3-polar fuzzy hypergraph* $H = (A, B)$, *where* $A = \{\zeta_1, \zeta_2, \zeta_3, \zeta_4\}$ *is a family of 3-polar fuzzy subsets on* $X = \{a, b, c, d, e\}$ *and B is a 3-polar fuzzy relation on the 3-polar fuzzy subsets* ζ_i, *where* $\zeta_1 = \{(a, 0.3, 0.4, 0.5), (b, 0.2, 0.4, 0.6)\}$, $\zeta_2 = \{(c, 0.2, 0.1, 0.4), (d, 0.5, 0.1, 0.1), (e, 0.2, 0.3, 0.1)\}$, $\zeta_3 = \{(b, 0.1, 0.2, 0.4), (c, 0.4, 0.5, 0.6)\}$, $\zeta_4 = \{(a, 0.1, 0.3, 0.2), (d, 0.3, 0.4, 0.4)\}$. *In this example, open neighourhood of the vertex a is b and d, as shown in Figure 5.*

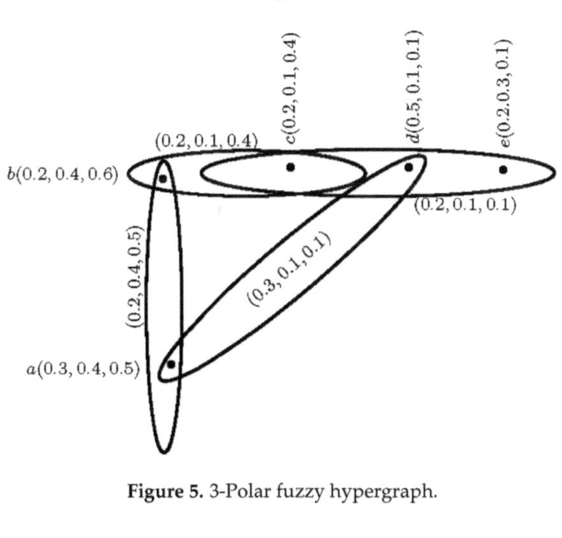

Figure 5. 3-Polar fuzzy hypergraph.

Definition 7. *The closed neighourhood of a vertex x in the mF hypergraph is the set of adjacent vertices of x including that vertex and it is denoted by* $N[x]$.

Example 5. *Consider the 3-polar fuzzy hypergraph* $H = (A, B)$, *where* $A = \{\zeta_1, \zeta_2, \zeta_3, \zeta_4\}$ *is a family of 3-polar fuzzy subsets on* $X = \{a, b, c, d, e\}$ *and B is a 3-polar fuzzy relation on the 3-polar fuzzy subsets* ζ_j, *where* $\zeta_1 = \{(a, 0.3, 0.4, 0.5), (b, 0.2, 0.4, 0.6)\}$, $\zeta_2 = \{(c, 0.2, 0.1, 0.4), (d, 0.5, 0.1, 0.1), (e, 0.2, 0.3, 0.1)\}$,

$\zeta_3 = \{(b, 0.1, 0.2, 0.4), (c, 0.4, 0.5, 0.6)\}$, $\zeta_4 = \{(a, 0.1, 0.3, 0.2), (d, 0.3, 0.4, 0.4)\}$. *In this example, closed neighourhood of the vertex a is a, b and d, as shown in Figure 5.*

Definition 8. *Let* $H = (A, B)$ *be an mF hypergraph on crisp hypergraph* $H^* = (A^*, B^*)$. *If all vertices in A have the same open neighbourhood degree n, then H is called n-regular mF hypergraph.*

Definition 9. *The open neighbourhood degree of a vertex x in H is denoted by* $deg(x)$ *and defined by* $deg(x) = (deg^{(1)}(x), deg^{(2)}(x), deg^{(3)}(x), \ldots, deg^{(m)}(x))$, *where*

$$deg^{(1)}(x) = \Sigma_{x \in N(x)} P_1 \circ \zeta_j(x),$$

$$deg^{(2)}(x) = \Sigma_{x \in N(x)} P_2 \circ \zeta_j(x),$$

$$deg^{(3)}(x) = \Sigma_{x \in N(x)} P_3 \circ \zeta_j(x),$$

$$\vdots$$

$$deg^{(m)}(x) = \Sigma_{x \in N(x)} P_m \circ \zeta_j(x).$$

Example 6. *Consider the 3-polar fuzzy hypergraph* $H = (A, B)$, *where* $A = \{\zeta_1, \zeta_2, \zeta_3, \zeta_4\}$ *is a family of 3-polar fuzzy subsets on* $X = \{a, b, c, d, e\}$ *and B is a 3-polar fuzzy relation on the 3-polar fuzzy subsets* ζ_j, *where* $\zeta_1 = \{(a, 0.3, 0.4, 0.5), (b, 0.2, 0.4, 0.6)\}$, $\zeta_2 = \{(c, 0.2, 0.1, 0.4), (d, 0.5, 0.1, 0.1), (e, 0.2, 0.3, 0.1)\}$, $\zeta_3 = \{(b, 0.1, 0.2, 0.4), (c, 0.4, 0.5, 0.6)\}$, $\zeta_4 = \{(a, 0.1, 0.3, 0.2), (d, 0.3, 0.4, 0.4)\}$. *The open neighbourhood degree of a vertex a is* $deg(a) = (0.5, 0.8, 1)$.

Definition 10. *Let* $H = (A, B)$ *be an mF hypergraph on crisp hypergraph* $H^* = (A^*, B^*)$. *If all vertices in A have the same closed neighbourhood degree m, then H is called m-totally regular mF hypergraph.*

Definition 11. *The closed neighbourhood degree of a vertex x in H is denoted by* $deg[x]$ *and defined by* $deg[x] = (deg^{(1)}[x], deg^{(2)}[x], deg^{(3)}[x], \ldots, deg^{(m)}[x])$, *where*

$$deg^{(1)}[x] = deg^{(1)}(x) + \wedge_j P_1 \circ \zeta_j(x),$$

$$deg^{(2)}[x] = deg^{(2)}(x) + \wedge_j P_2 \circ \zeta_j(x),$$

$$deg^{(3)}[x] = deg^{(3)}(x) + \wedge_j P_3 \circ \zeta_j(x),$$

$$\vdots$$

$$deg^{(m)}[x] = d_G^{(m)}(x) + \wedge_j P_m \circ \zeta_j(x).$$

Example 7. *Consider the 3-polar fuzzy hypergraph* $H = (A, B)$, *where* $A = \{\zeta_1, \zeta_2, \zeta_3, \zeta_4\}$ *is a family of 3-polar fuzzy subsets on* $X = \{a, b, c, d, e\}$ *and B is a 3-polar fuzzy relation on the 3-polar fuzzy subsets* ζ_j, *where* $\zeta_1 = \{(a, 0.3, 0.4, 0.5), (b, 0.2, 0.4, 0.6)\}$, $\zeta_2 = \{(c, 0.2, 0.1, 0.4), (d, 0.5, 0.1, 0.1), (e, 0.2, 0.3, 0.1)\}$, $\zeta_3 = \{(b, 0.1, 0.2, 0.4), (c, 0.4, 0.5, 0.6)\}$, $\zeta_4 = \{(a, 0.1, 0.3, 0.2), (d, 0.3, 0.4, 0.4)\}$. *The closed neighbourhood degree of a vertex a is* $deg[a] = (0.6, 1.1, 1.2)$.

Example 8. *Consider the 3-polar fuzzy hypergraph* $H = (A, B)$, *where* $A = \{\zeta_1, \zeta_2, \zeta_3\}$ *is a family of 3-polar fuzzy subsets on* $X = \{a, b, c, d, e\}$ *and B is a 3-polar fuzzy relation on the 3-polar fuzzy subsets* ζ_j, *where*

$$\zeta_1\{(a, 0.5, 0.4, 0.1), (b, 0.3, 0.4, 0.1), (c, 0.4, 0.4, 0.3)\},$$

$$\zeta_2 = \{(a, 0.3, 0.1, 0.1), (d, 0.2, 0.3, 0.2), (e, 0.4, 0.6, 0.1)\},$$

$$\zeta_3 = \{(b, 0.3, 0.4, 0.3), (d, 0.4, 0.3, 0.4), (e, 0.4, 0.3, 0.1)\}.$$

By routine calculations, we can show that the above 3-polar fuzzy hypergraph is neither regular nor totally regular.

Example 9. *Consider the 4-polar fuzzy hypergraph $H = (A, B)$; define $X = \{a, b, c, d, e, f, g, h, i\}$ and $A = \{\zeta_1, \zeta_2, \zeta_3, \zeta_4, \zeta_5, \zeta_6\}$, where*

$$\zeta_1 = \{(a, 0.4, 0.4, 0.4, 0.4), (b, 0.4, 0.4, 0.4, 0.4), (c, 0.4, 0.4, 0.4, 0.4)\},$$

$$\zeta_2 = \{(d, 0.4, 0.4, 0.4, 0.4), (e, 0.4, 0.4, 0.4, 0.4), (f, 0.4, 0.4, 0.4, 0.4)\},$$

$$\zeta_3 = \{(g, 0.4, 0.4, 0.4, 0.4), (h, 0.4, 0.4, 0.4, 0.4), (i, 0.4, 0.4, 0.4, 0.4)\},$$

$$\zeta_4 = \{(a, 0.4, 0.4, 0.4, 0.4), (d, 0.4, 0.4, 0.4, 0.4), (g, 0.4, 0.4, 0.4, 0.4)\},$$

$$\zeta_5 = \{(b, 0.4, 0.4, 0.4, 0.4), (e, 0.4, 0.4, 0.4, 0.4), (h, 0.4, 0.4, 0.4, 0.4)\},$$

$$\zeta_6 = \{(c, 0.4, 0.4, 0.4, 0.4), (f, 0.4, 0.4, 0.4, 0.4), (i, 0.4, 0.4, 0.4, 0.4)\}.$$

By routine calculations, we see that the 4-polar fuzzy hypergraph as shown in Figure 6 is both regular and totally regular.

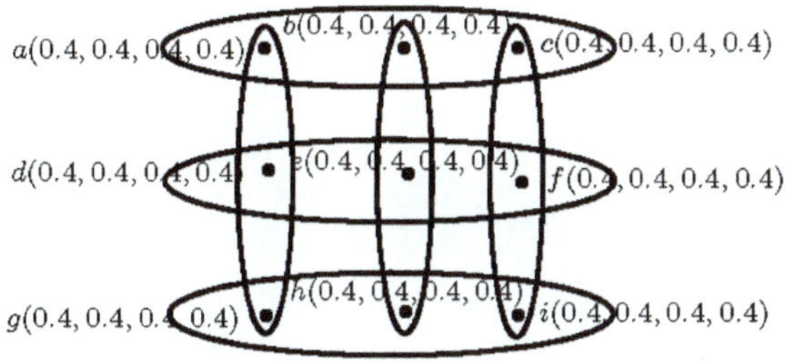

Figure 6. 4-Polar regular and totally regular fuzzy hypergraph.

Remark 1.
1. *For an mF hypergraph $H = (A, B)$ to be both regular and totally regular, the number of vertices in each hyperedge B_j must be the same. Suppose that $|B_j| = k$ for every j, then H is said to be k-uniform.*
2. *Each vertex lies in exactly the same number of hyperedges.*

Definition 12. *Let $H = (A, B)$ be a regular mF hypergraph. The order of a regular fuzzy hypergraph H is*

$$O(H) = (\Sigma_{x \in X} \wedge P_1 \circ \zeta_j(x), \Sigma_{x \in X} \wedge P_2 \circ \zeta_j(x), \cdots, \Sigma_{x \in X} \wedge P_m \circ \zeta_j(x)),$$

for every $x \in X$. The size of a regular mF hypergraph is $S(H) = \sum_j S(B_j)$, where

$$S(B_j) = (\Sigma_{x \in B_j} P_1 \circ \zeta_j(x), \Sigma_{x \in B_j} P_2 \circ \zeta_j(x), \cdots, \Sigma_{x \in B_j} P_m \circ \zeta_j(x)).$$

Example 10. *Consider the 4-polar fuzzy hypergraph $H = (A, B)$; define $X = \{a, b, c, d, e, f, g, h, i\}$ and $A = \{\zeta_1, \zeta_2, \zeta_3, \zeta_4, \zeta_5, \zeta_6\}$, where*

$$\zeta_1 = \{(a, 0.4, 0.4, 0.4, 0.4), (b, 0.4, 0.4, 0.4, 0.4), (c, 0.4, 0.4, 0.4, 0.4)\},$$

$$\zeta_2 = \{(d, 0.4, 0.4, 0.4, 0.4), (e, 0.4, 0.4, 0.4, 0.4), (f, 0.4, 0.4, 0.4, 0.4)\},$$

$$\zeta_3 = \{(g, 0.4, 0.4, 0.4, 0.4), (h, 0.4, 0.4, 0.4, 0.4), (i, 0.4, 0.4, 0.4, 0.4)\},$$

$$\zeta_4 = \{(a, 0.4, 0.4, 0.4, 0.4), (d, 0.4, 0.4, 0.4, 0.4), (g, 0.4, 0.4, 0.4, 0.4)\},$$

$$\zeta_5 = \{(b, 0.4, 0.4, 0.4, 0.4), (e, 0.4, 0.4, 0.4, 0.4), (h, 0.4, 0.4, 0.4, 0.4)\},$$

$$\zeta_6 = \{(c, 0.4, 0.4, 0.4, 0.4), (f, 0.4, 0.4, 0.4, 0.4), (i, 0.4, 0.4, 0.4, 0.4)\}.$$

The order of H is, $O(H) = (3.6, 3.6, 3.6, 3.6)$ *and* $S(H) = (7.2, 7.2, 7.2, 7.2)$.

We state the following propositions without proof.

Proposition 1. *The size of a n-regular mF hypergraph is* $\frac{nk}{2}$, $|X| = k$.

Proposition 2. *If H is both n-regular and m-totally regular mF hypergraph , then* $O(H) = k(m - n)$, *where* $|X| = K$.

Proposition 3. *If H is both m-totally regular mF hypergraph , then* $2S(H) + O(H) = mk, |X| = K$.

Theorem 1. *Let* $H = (A, B)$ *be an mF hypergraph of a crisp hypergraph* H^*. *Then* $A : X \longrightarrow [0, 1]^m$ *is a constant function if and only if the following are equivalent:*
(a) H is a regular mF hypergraph,
(b) H is a totally regular mF hypergraph.

Proof. Suppose that $A : X \longrightarrow [0, 1]^m$, where $A = \{\zeta_1, \zeta_2, ..., \zeta_r\}$ is a constant function. That is, $P_i o \zeta_j(x) = c_i$ for all $x \in \zeta_j$, $1 \leq i \leq m$, $1 \leq j \leq r$.

$(a) \Rightarrow (b)$: Suppose that H is n-regular mF hypergraph. Then $deg(x) = n_i$, for all $x \in \zeta_j$. By using definition 11, $deg[x] = n_i + k_i$ for all $x \in \zeta_j$. Hence, H is a totally regular mF hypergraph.

$(b) \Rightarrow (a)$: Suppose that H is a m-totally regular mF hypergraph. Then $deg[x] = k_i$, for all $x \in \zeta_j$, $1 \leq j \leq r$.

$$\Rightarrow deg(x) + \wedge_j P_i o \zeta_j(x) = k_i \text{ for all } x \in \zeta_j,$$

$$\Rightarrow deg(x) + c_i = k_i \text{ for all } x \in \zeta_j,$$

$$\Rightarrow deg(x) = k_i - c_i \text{ for all } x \in \zeta_j.$$

Thus, H is a regular mF hypergraph. Hence (1) and (2) are equivalent.

Conversely, suppose that (1) and (2) are equivalent, i.e. H is regular if and only if H is totally regular. On contrary, suppose that A is not constant, that is, $P_i o \zeta_j(x) \neq P_i o \zeta_j(y)$ for some x and y in A. Let $H = (A, B)$ be a n-regular mF hypergraph; then

$$deg(x) = n_i \text{ for all } x \in \zeta_j(x).$$

Consider,

$$\begin{aligned} deg[x] &= deg(x) + \wedge_j P_i o \zeta_j(x) = n_i + \wedge_j P_i o \zeta_j(x), \\ deg[y] &= deg(y) + \wedge_j P_i o \zeta_j(y) = n_i + \wedge_j P_i o \zeta_j(y). \end{aligned}$$

Since $P_i o \zeta_j(x)$ and $P_i o \zeta_j(y)$ are not equal for some x and y in X, hence $deg[x]$ and $deg[y]$ are not equal, thus H is not a totally regular m-poalr fuzzy hypergraph, which is again a contradiction to our assumption.

Next, let H be a totally regulr mF hypergraph, then $deg[x] = deg[y]$.

That is,

$$\begin{aligned} deg(x) + \wedge_j P_i o\zeta_j(x) &= deg(y) + \wedge_j P_i o\zeta_j(y), \\ deg(x) - deg(y) &= \wedge_j P_i o\zeta_j(y) - \wedge_j P_i o\zeta_j(x). \end{aligned}$$

Since the right hand side of the above equation is nonzero, the left hand side of the above equation is also nonzero. Thus $deg(x)$ and $deg(y)$ are not equal, so H is not a regular mF hypergraph, which is again contradiction to our assumption. Hence, A must be constant and this completes the proof. □

Theorem 2. *If an mF hypergraph is both regular and totally regular, then* $A : X \longrightarrow [0,1]^m$ *is constant function.*

Proof. Let H be a regular and totally regular mF hypergraph. Then

$$deg(x) = n_i \text{ for all } x \in \zeta_j(x),$$

and

$$\begin{aligned} deg[x] &= k_i \text{ for all } x \in \zeta_j(x), \\ \Leftrightarrow deg(x) + \wedge_j P_i o\zeta_j(x) &= k_i, \text{ for all } x \in \zeta_j(x), \\ \Leftrightarrow n_1 + \wedge_j P_i o\zeta_j(x) &= k_i, \text{ for all } x \in \zeta_j(x), \\ \Leftrightarrow \wedge_j P_i o\zeta_j(x) &= k_i - n_i, \text{ for all } x \in \zeta_j(x), \\ \Leftrightarrow P_i o\zeta_j(x) &= k_i - n_i, \text{ for all } x \in \zeta_j(x). \end{aligned}$$

Hence, $A : X \longrightarrow [0,1]^m$ is a constant function. □

Remark 2. *The converse of Theorem 1 may not be true, in general. Consider a 3-polar fuzzy hypergraph* $H = (A, B)$, *define* $X = \{a, b, c, d, e\}$,

$$\zeta_1 = \{(a, 0.2, 0, 2, 0.2), (b, 0.2, 0.2, 0.2), (c, 0.2, 0.2, 0.2)\},$$

$$\zeta_2 = \{(a, 0.2, 0, 2, 0.2), (d, 0.2, 0.2, 0.2)\},$$

$$\zeta_3 = \{(b, 0.2, 0.2, 0.2), (e, 0.2, 0.2, 0.2)\},$$

$$\zeta_4 = \{(c, 0.2, 0.2, 0.2), (e, 0.2, 0.2, 0.2)\}.$$

Then $A : X \longrightarrow [0,1]^m$, *where* $A = \{\zeta_1, \zeta_2, ..., \zeta_r\}$ *is a constant function. But* $deg(a) = (0.6, 0.6, 0.6) \neq (0.4, 0.4, 0.4) = deg(e)$. *Also* $(deg[a] = (0.8, 0.8, 0.8) \neq (0.6, 0.6, 0.6) = deg[e])$. *So H is neither regular nor totally regular mF hypergraph.*

Definition 13. *An mF hypergraph* $H = (A, B)$ *is called complete if for every* $x \in X$, $N(x) = \{x \mid x \in X - x\}$ *that is,* $N(x)$ *contains all the remaining vertices of X except x.*

Example 11. *Consider a 3-polar fuzzy hypergraph* $H = (A, B)$ *as shown in Figure 7, where* $X = \{a, b, c, d\}$ *and* $A = \{\zeta_1, \zeta_2, \zeta_3\}$, *where* $\zeta_1 = \{(a, 0.3, 0.4, 0.6), (c, 0.3, 0.4, 0.6)\}$, $\zeta_2 = \{(a, 0.3, 0.4, 0.6), (b, 0.3, 0.4, 0.6), (d, 0.3, 0.4, 0.6)\}$, $\zeta_3 = \{(b, 0.3, 0.4, 0.6), (c, 0.3, 0.4, 0.6), (d, 0.3, 0.4, 0.6)\}$. *Then* $N(a) = \{b, c, d\}$, $N(b) = \{a, c, d\}$, $N(c) = \{a, b, d\}$.

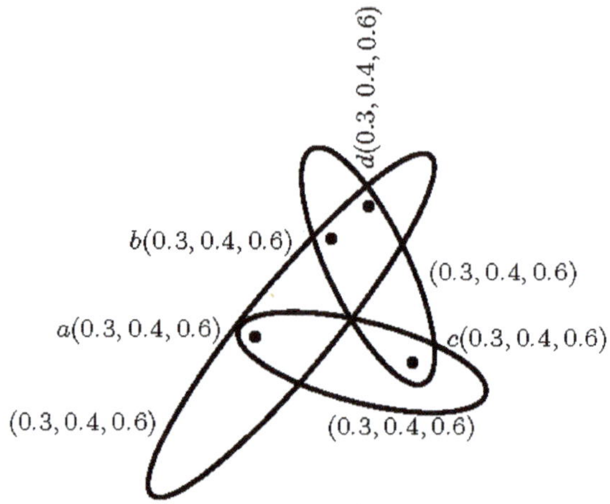

Figure 7. 3-Polar fuzzy hypergraph.

Remark 3. *For a complete mF hypergraph, the cardinality of* $N(x)$ *is the same for every vertex.*

Theorem 3. *Every complete mF hyprgraph is a totally regular mF hypergraph.*

Proof. Since given mF hypergraph H is complete, each vertex lies in exactly the same number of hyperedges and each vertex has the same closed neighborhood degree m. That is, $deg[x_1] = deg[x_2]$ for all $x_1, x_2 \in X$. Hence H is m-totally regular. □

3. Applications to Decision-Making Problems

Analysis of human nature and its culture has been entangled with the assessment of social networks for many years. Such networks are refined by designating one or more relations on the set of individuals and the relations can be taken from efficacious relationships, facets of some management and from a large range of others means. For super-dyadic relationships between the nodes, network models represented by simple graph are not sufficient. Natural presence of hyperedges can be found in co-citation, e-mail networks, co-authorship, web log networks and social networks etc. Representation of these models as hypergraphs maintain the dyadic relationships.

3.1. Super-Dyadic Managements in Marketing Channels

In marketing channels, dyadic correspondence organization has been a basic implementation. Marketing researchers and managers have realized that their common engagement in marketing channels is a central key for successful marketing and to yield benefits for the company. mF hypergraphs consist of marketing managers as vertices and hyperedges show their dyadic communication involving their parallel thoughts, objectives, plans, and proposals. The more powerful close relation in the research is more beneficial for the marketing strategies and the production of an organization. A 3-polar fuzzy network model showing the dyadic communications among the marketing managers of an organization is given in Figure 8.

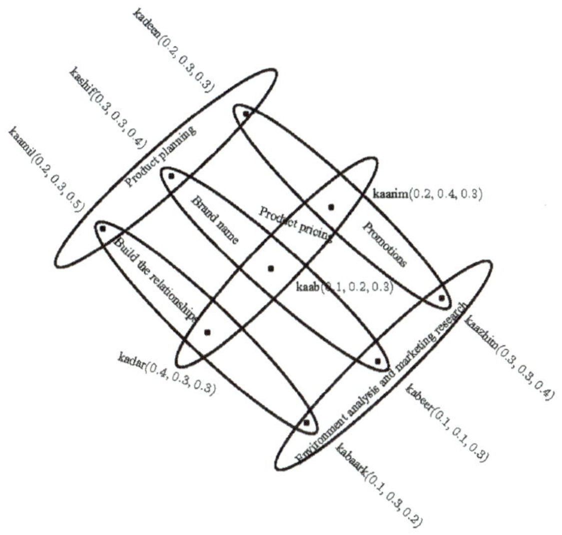

Figure 8. Super-dyadic managements in marketing channels.

The membership degrees of each person symbolize the percentage of its dyadic behaviour towards the other people of the same dyad group. The adjacent level between any pair of vertices illustrates how proficient their dyadic relationship is. The adjacent levels are given in Table 4.

Table 4. Adjacent levels.

Dyad pairs	Adjacent level	Dyad pairs	Adjacent level
γ(Kadeen, Kashif)	(0.2,0.3,0.3)	γ(Kaarim, Kaazhim)	(0.2,0.3,0.3)
γ(Kadeen, Kaamil)	(0.2,0.3,0.3)	γ(Kaarim, Kaab)	(0.1,0.2,0.3)
γ(Kadeen, Kaarim)	(0.2,0.3,0.3)	γ(Kaarim, Kadar)	(0.2,0.3,0.3)
γ(Kadeen, Kaazhim)	(0.2,0.3,0.3)	γ(Kaab, Kadar)	(0.1,0.2,0.3)
γ(Kashif, Kaamil)	(0.2,0.3,0.4)	γ(Kaab, Kabeer)	(0.1,0.1,0.3)
γ(Kashif, Kaab)	(0.1,0.2,0.3)	γ(Kadar, Kabaark)	(0.1,0.3,0.2)
γ(Kashif, Kabeer)	(0.1,0.1,0.3)	γ(Kaazhim, Kabeer)	(0.1,0.1,0.3)
γ(Kaamil, Kadar))	(0.2,0.2,0.3)	γ(Kaazhim, Kabaark)	(0.1,0.3,0.2)
γ(Kaamil, Kabaark)	(0.1,0.3,0.2)	γ(Kabeer, Kabaark)	(0.1,0.1,0.2)

It can be seen that the most capable dyadic pair is (Kashif, Kaamil). 3-polar fuzzy hyperedges are taken as different digital marketing strategies adopted by the different dyadic groups of the same organization. The vital goal of this model is to determine the most potent dyad of digital marketing techniques. The six different groups are made by the marketing managers and the digital marketing strategies adopted by these six groups are represented by hyperedges. i.e., the 3-polar fuzzy hyperedges $\{T_1, T_2, T_3, T_4, T_5, T_6\}$ show the following strategies {Product pricing, Product planning, Environment analysis and marketing research, Brand name, Build the relationships, Promotions}, respectively. The exclusive effects of membership degrees of each marketing strategy towards the achievements of an organization are given in Table 5.

Table 5. Effects of marketing strategies.

Marketing Strategy	Profitable growth	Instruction manual for company success	Create longevity of the business
Product pricing	0.1	0.2	0.3
Product planning	0.2	0.3	0.3
Environment analysis and marketing research	0.1	0.2	0.2
Brand name	0.1	0.3	0.3
Build the relationships	0.1	0.3	0.2
Promotions	0.2	0.3	0.3

Effective dyads of market strategies enhance the performance of an organization and discover the better techniques to be adopted. The adjacency of all dyadic communication managements is given in Table 6.

Table 6. Adjacency of all dyadic communication managements.

Dyadic strategies	Effects
σ(Product pricing, Product planning)	(0.1,0.2,0.3)
σ(Product pricing, Environment analysis and marketing research)	(0.1,0.2,0.2)
σ(Product pricing, Brand name)	(0.1,0.2,0.3)
σ(Product pricing, Build the relationships)	(0.1,0.2,0.2)
σ(Product pricing, Promotions)	(0.1,0.2,0.3)
σ(Product planning, Environment analysis and marketing research)	(0.1,0.2,0.2)
σ(Product planning, Brand name)	(0.1,0.3,0.3)
σ(Product planning, Build the relationships)	(0.1,0.3,0.2)
σ(Product planning, Promotions)	(0.2,0.3,0.3)
σ(Environment analysis and marketing research, Brand name)	(0.1,0.2,0.2)
σ(Environment analysis and marketing research, Build the relationships)	(0.1,0.2,0.2)
σ(Environment analysis and marketing research, Promotions)	(0.1,0.2,0.2)
σ(Brand name, Build the relationships)	(0.1,0.3,0.2)
σ(Brand name, Promotions)	(0.1,0.3,0.3)
σ(Build the relationships, Promotions)	(0.1,0.3,0.2)

The most dominant and capable marketing strategies adopted mutually are Product planning and Promotions. Thus, to increase the efficiency of an organization, dyadic managements should make powerful planning for products and use the promotions skill to attract customers to purchase their products. The membership degrees of this dyad is (0.2, 0.3, 0.3) which shows that the amalgamated effect of this dyad will increase the profitable growth of an organization up to 20%, instruction manual for company success up to 30%, create longevity of the business up to 30% . Thus, to promote the performance of an organization, super dyad marketing communications are more energetic. The method of determining the most effective dyads is explained in the following algorithm.

Algorithm 1

1. Input: The membership values $A(x_i)$ of all nodes (marketing managers) $x_1, x_2, ..., x_n$.
2. Input: The membership values $B(T_i)$ of all hyperedges $T_1, T_2, ..., T_r$.
3. Find the adjacent level between nodes x_i and x_j as,
4. **do** i from $1 \rightarrow n-1$
5.   **do** j from $i+1 \rightarrow n$
6.   **do** k from $1 \rightarrow r$
7.   **if** $x_i, x_j \in E_k$ **then**
8. $\gamma(x_i, x_j) = \max_k \inf\{A(x_i), A(x_j)\}$.
9.   **end if**
10.   **end do**
11.   **end do**
12. **end do**

13. Find the best capable dyadic pair as $\max_{i,j} \gamma(x_i, x_j)$.
14. **do** i from $1 \to r-1$
15. **do** j from $i+1 \to r$
16. **do** k from $1 \to r$
17. **if** $x_k \in T_i \cap T_j$ **then**
18. $\sigma(T_i, T_j) = \max_k \inf\{B(T_i), B(T_j)\}$.
19. **end if**
20. **end do**
21. **end do**
22. **end do**
23. Find the best effective super dyad management as $\max_{i,j} \sigma(T_i, T_j)$.

Description of Algorithm 1: Lines 1 and 2 pass the input of m-polar fuzzy set A on n vertices $x_1, x_2, \ldots, x_n$ and m-polar fuzzy relation B on r edges $T_1, T_2, ..., T_r$. Lines 3 to 12 calculate the adjacent level between each pair of nodes. Line 14 calculates the best capable dyadic pair. The loop initializes by taking the value $i = 1$ of do loop which is always true, i.e., the loop runs for the first iteration. For any *ith* iteration of do loop on line 3, the do loop on line 4 runs $n - i$ times and, the do loop on line 5 runs r times. If there exists a hyperedge E_k containing x_i and x_j then, line 7 is executed otherwise the if conditional terminates. For every *ith* iteration of the loop on line 3, this process continues n times and then increments i for the next iteration maintaining the loop throughout the algorithm. For $i = n - 1$, the loop calculates the adjacent level for every pair of distinct vertices and terminates successfully at line 12. Similarly, the loops on lines 13, 14 and 15 maintain and terminate successfully.

3.2. *m-Polar Fuzzy Hypergraphs in Work Allotment Problem*

In customer care centers, availability of employees plays a vital role in solving customer problems. Such a department should ensure that the system has been managed carefully to overcome practical difficulties. A lot of customers visit such centers to find a solution of their problems. In this part, focus is given to alteration of duties for the employees taking leave. The problem is that employees are taking leave without proper intimation and alteration. We now show the importance of m-polar fuzzy hypergraphs for the allocation of duties to avoid any difficulties.

Consider the example of a customer care center consisting of 30 employees. Assuming that six workers are necessary to be available at their duties. We present the employees as vertices and the degree of membership of each employee represents the work load, percentage of available time and number of workers who are also aware of the employee's work type. The range of values for present time and the workers, knowing the type of work is given in Tables 7 and 8.

Table 7. Range of membership values of table time.

Time	Membership value
5 h	0.40
6 h	0.50
8 h	0.70
10 h	0.90

Table 8. Workers knowing the work type.

Workers	Membership value
3	0.40
4	0.60
5	0.80
6	0.90

The degree of membership of each edge represents the common work load, percentage of available time and number of workers who are also aware of the employee's work type. This phenomenon can be represented by a 3-polar fuzzy graph as shown in Figure 9.

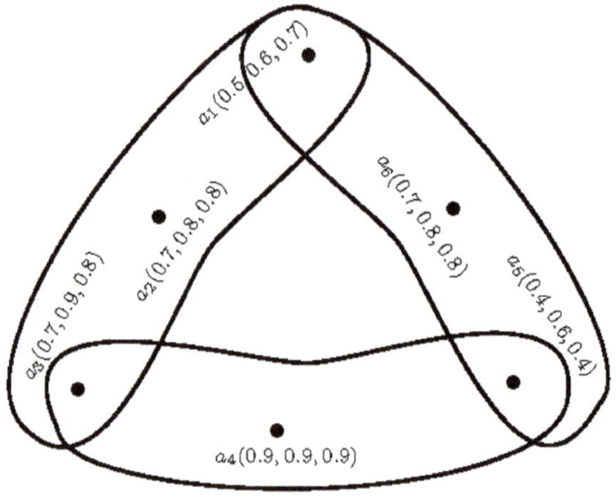

Figure 9. 3-Polar fuzzy graph.

Using Algorithm 2, the strength of allocation and alteration of duties among employees is given in Table 9.

Table 9. Alteration of duties.

Workers	$A(a_i, a_j)$	$S(a_i, a_j)$
a_1, a_2	(0.7,0.8,0.8)	0.77
a_1, a_3	(0.7,0.9,0.8)	0.80
a_2, a_3	(0.5,0.7,0.7)	0.63
a_3, a_4	(0.7,0.6,0.8)	0.70
a_3, a_5	(0.7,0.9,0.8)	0.80
a_4, a_5	(0.9,0.9,0.9)	0.90
a_5, a_6	(0.7,0.8,0.8)	0.77
a_5, a_1	(0.5,0.6,0.7)	0.60
a_1, a_6	(0.6,0.8,0.5)	0.63

Column 3 in Table 9 shows the percentage of alteration of duties. For example, in case of leave, duties of a_1 can be given to a_3 and similarly for other employees.

The method for the calculation of alteration of duties is given in Algorithm 2.

Algorithm 2

Input: The n number of employees $a_1, a_2, \ldots, a_n$.
Input: The number of edges $E_1, E_2, \ldots, E_r$.
Input: The incident matrix B_{ij} where, $1 \leq i \leq n, 1 \leq j \leq r$.
Input the membership values of edges $\xi_1, \xi_2, \ldots, \xi_r$
do i from $1 \to n$
  do j from $1 \to n$
    do k from $1 \to r$
      if $a_i, a_j \in E_k$ **then**
        do t from $1 \to m$
          $P_t \circ A(a_i, a_j) = |P_t \circ B_{ik} - P_t \circ B_{jk}| + P_t \circ \xi_k$
        end do

12. **end if**
13. **end do**
14. **end do**
15. **end do**
16. **do** i from $1 \rightarrow n$
17. **do** j from $1 \rightarrow n$
18. **if** $A(a_i, a_j) > 0$ **then**
19. $S(a_i, a_j) = \dfrac{P_1 \circ A(a_i, a_j) + P_2 \circ A(a_i, a_j) + \ldots + P_m \circ A(a_i, a_j)}{m}$
20. **end if**
21. **end do**
22. **end do**

Description of Algorithm 2: Lines 1, 2, 3 and 4 pass the input of membership values of vertices, hyperedges and an m-polar fuzzy adjacency matrix B_{ij}. The nested loops on lines 5 to 15 calculate the $rth, 1 \leq r \leq m$, strength of allocation and alteration of duties between each pair of employees. The nested loops on lines 16 to 22 calculate the strength of allocation and alteration of duties between each pair of employees. The net time complexity of the algorithm is $O(n^2 rm)$.

3.3. *Availability of Books in Library*

A library in a college is a collection of sources of information and similar resources, made accessible to the student community for reference and examination preparation. A student preparing for a given examination will use the knowledge sources such as

1. Prescribed textbooks (A)
2. Reference books in syllabus (B)
3. Other books from library (C)
4. Knowledgeable study materials (D)
5. E-gadgets and internet (E)

It is important to consider the maximum availability of the sources which students mostly use. This phenomenon can be discussed using m-polar fuzzy hypergraphs. We now calculate the importance of each source in the student community.

Consider the example of five library resources $\{A, B, C, D, E\}$ in a college. We represent these sources as vertices in a 3-polar fuzzy hypergraph. The degree of membership of each vertex represents the percentage of students using a particular source for exam preparation, percentage of faculty members using the sources and number of sources available. The degree of membership of each edge represents the common percentage. The 3-polar fuzzy hypergraph is shown in Figure 10.

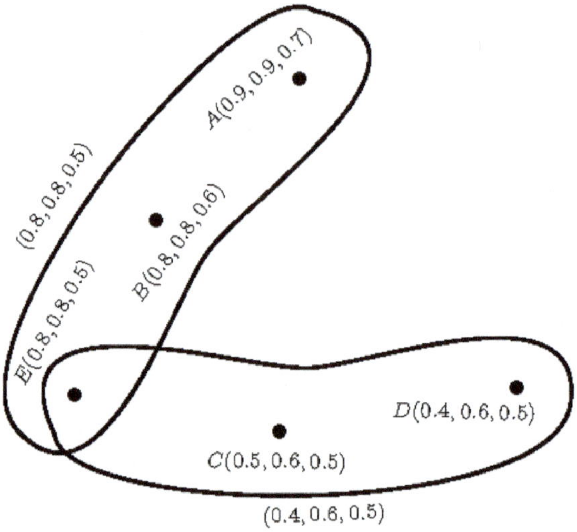

Figure 10. 3-Polar fuzzy hypergraph.

Using Algorithm 3, the strength of each library source in given in Table 10.

Table 10. Library sources.

Sources s_i	$T(s_i)$	$S(a_i, a_j)$
A	(1.7,1.7,1.4)	1.60
B	(1.6,1.6,1.1)	1.43
E	(1.6,1.6,1.0)	1.40
C	(0.9,1.2,1.0)	1.03
D	(0.8,1.2,1.0)	1.0

Column 3 in Table 10 shows that sources A and B are mostly used by students and faculty. Therefore, these should be available in maximum number. There is also a need to confirm the availability of source E to students and faculty.

The method for the calculation of percentage importance of the sources is given in Algorithm 3 whose net time complexity is $O(nrm)$.

Algorithm 3

Input: The n number of sources $s_1, s_2, \ldots, s_n$.
Input: The number of edges $E_1, E_2, \ldots, E_r$.
Input: The incident matrix B_{ij} where, $1 \leq i \leq n, 1 \leq j \leq r$.
Input: The membership values of edges $\xi_1, \xi_2, \ldots, \xi_r$
do i from $1 \to n$
    $A(s_i) = \mathbf{1}$
    $C(s_i) = \mathbf{1}$
    do k from $1 \to r$
        if $s_i \in E_k$ **then**
            $A(s_i) = \max\{A(s_i), \xi_k\}$
            $C(s_i) = \min\{C(s_i), B_{ik}\}$
        end if
    end do
    $T(s_i) = C(s_i) + A(s_i)$
end do

16. **do** i from $1 \to n$
17. **if** $T(s_i) > 0$ **then**
18. $S(s_i) = \dfrac{P_1 \circ T(s_i) + P_2 \circ T(s_i) + \ldots + P_m \circ T(s_i)}{m}$
19. **end if**
20. **end do**

Description of Algorithm 3: Lines 1, 2, 3 and 4 pass the input of membership values of vertices, hyperedges and an m-polar fuzzy adjacency matrix B_{ij}. The nested loops on lines 5 to 15 calculate the degree of usage and availability of library sources. The nested loops on lines 16 to 20 calculate the strength of each library source.

4. Conclusions

Hypergraphs are generalizations of graphs. Many problems which cannot be handled by graphs can be solved using hypergraphs. mF graph theory has numerous applications in various fields of science and technology including artificial intelligence, operations research and decision making. An mF hypergraph constitutes a generalization of the notion of an mF fuzzy graph. mF hypergraphs play an important role in discussing multipolar uncertainty among several individuals. In this research article, we have conferred certain concepts of regular mF hypergraphs and applications of mF hypergraphs in decision-making problems. We aim to generalize our notions to (1) mF soft hypergraphs, (2) soft rough mF hypergraphs, (3) soft rough hypergraphs, and (4) intuitionistic fuzzy rough hypergraphs.

Author Contributions: Muhammad Akram and Gulfam Shahzadi conceived and designed the experiments; Muhammad Akram performed the experiments; Muhammad Akram and Gulfam Shahzadi analyzed the data; Gulfam Shahzadi wrote the paper.

Conflicts of Interest: The authors declare no conflict of interest regarding the publication of this research article.

References

1. Zhang, W.R. Bipolar fuzzy sets and relations: a computational framework forcognitive modeling and multiagent decision analysis. In Proceedings of the Fuzzy Information Processing Society Biannual Conference, Industrial Fuzzy Control and Intelligent Systems Conference and the NASA Joint Technology Workshop on Neural Networks and Fuzzy Logic, San Antonio, TX, USA, 18–21 December 1994; pp. 305–309, doi: 10.1109/IJCF.1994.375115.
2. Zadeh, L.A. Fuzzy sets. *Inf. Control* **1965**, *8*, 338–353.
3. Chen, J.; Li, S.; Ma, S.; Wang, X. m-Polar fuzzy sets: An extension of bipolar fuzzy sets. *Sci. World J.* **2014**, 8.
4. Zadeh, L.A. Similarity relations and fuzzy orderings. *Inf. Sci.* **1971**, *3*, 177–200.
5. Kaufmann, A. *Introduction la Thorie des Sous-Ensembles Flous Lusage des Ingnieurs (Fuzzy Sets Theory)*; Masson: Paris, French, 1973.
6. Rosenfeld, A. *Fuzzy Graphs, Fuzzy Sets and Their Applications*; Academic Press: New York, NY, USA, 1975; pp. 77–95.
7. Bhattacharya, P. Remark on fuzzy graphs. *Pattern Recognit. Lett.* **1987**, *6*, 297–302.
8. Mordeson, J.N.; Nair, P.S. Fuzzy graphs and fuzzy hypergraphs. In *Studies in Fuzziness and Soft Computing*; Springer-Verlag: Berlin Heidelberg, Germany, 2000.
9. Akram, M. Bipolar fuzzy graphs. *Inf. Sci.* **2011**, *181*, 5548–5564.
10. Li, S.; Yang, X.; Li, H.; Ma, M. Operations and decompositions of m-polar fuzzy graphs. *Basic Sci. J. Text. Univ.* **2017**, *30*, 149–162.
11. Chen, S.M. Interval-valued fuzzy hypergraph and fuzzy partition. *IEEE Trans. Syst. Man Cybern. B Cybern.* **1997**, *27*, 725–733.
12. Lee-Kwang, H.; Lee, K.M. Fuzzy hypergraph and fuzzy partition. *IEEE Trans. Syst. Man Cybern.* **1995**, *25*, 196–201.
13. Parvathi, R.; Thilagavathi, S.; Karunambigai, M.G. Intuitionistic fuzzy hypergraphs. *Cybern. Inf. Technol.* **2009**, *9*, 46–48.

14. Samamta, S.; Pal, M. Bipolar fuzzy hypergraphs. *Int. J. Fuzzy Log. Intell. Syst.* **2012**, *2*, 17–28.
15. Akram, M.; Dudek, W.A.; Sarwa, S. Properties of bipolar fuzzy hypergraphs. *Ital. J. Pure Appl. Math.* **2013**, *31*, 141–160.
16. Akram, M.; Luqman, A. Bipolar neutrosophic hypergraphs with applications. *J. Intell. Fuzzy Syst.* **2017**, *33*, 1699–1713, doi:10.3233/JIFS-162207.
17. Akram, M.; Sarwar, M. Novel application of m-polar fuzzy hypergraphs. *J. Intell. Fuzzy Syst.* **2017**, *32*, 2747–2762.
18. Akram, M.; Sarwar, M. Transversals of m-polar fuzzy hypergraphs with applications. *J. Intell. Fuzzy Syst.* **2017**, *33*, 351–364.
19. Akram, M.; Adeel, M. mF labeling graphs with application. *Math. Comput. Sci.* **2016**, *10*, 387–402.
20. Akram, M.; Waseem, N. Certain metrics in m-polar fuzzy graphs. *N. Math. Nat. Comput.* **2016**, *12*, 135–155.
21. Akram, M.; Younas, H.R. Certain types of irregular m-polar fuzzy graphs. *J. Appl. Math. Comput.* **2017**, *53*, 365–382.
22. Akram, M.; Sarwar, M. Novel applications of m-polar fuzzy competition graphs in decision support system. *Neural Comput. Appl.* **2017**, 1–21, doi:10.1007/s00521-017-2894-y.
23. Chen, S.M. A fuzzy reasoning approach for rule-based systems based on fuzzy logics. *IEEE Trans. Syst. Man Cybern. B Cybern.* **1996**, *26*, 769–778.
24. Chen, S.M.; Lin, T.E.; Lee, L.W. Group decision making using incomplete fuzzy preference relations based on the additive consistency and the order consistency. *Inf. Sci.* **2014**, *259*, 1–15.
25. Chen, S.M.; Lee, S.H.; Lee, C.H. A new method for generating fuzzy rules from numerical data for handling classification problems. *Appl. Artif. Intell.* **2001**, *15*, 645–664.
26. Horng, Y.J.; Chen, S.M.; Chang, Y.C.; Lee, C.H. A new method for fuzzy information retrieval based on fuzzy hierarchical clustering and fuzzy inference techniques. *IEEE Trans. Fuzzy Syst.* **2005**, *13*, 216–228.
27. Narayanamoorthy, S.; Tamilselvi, A.; Karthick, P.; Kalyani, S.; Maheswari, S. Regular and totally regular fuzzy hypergraphs. *Appl. Math. Sci.* **2014**, *8*, 1933–1940.
28. Radhamani, C.; Radhika, C. Isomorphism on fuzzy hypergraphs. *IOSR J. Math.* **2012**, *2*, 24–31.
29. Sarwar, M.; Akram, M. Certain algorithms for computing strength of competition in bipolar fuzzy graphs. *Int. J. Uncertain. Fuzz.* **2017**, *25*, 877–896.
30. Sarwar, M.; Akram, M. Representation of graphs using m-polar fuzzy environment. *Ital. J. Pure Appl. Math.* **2017**, *38*, 291–312.
31. Sarwar, M.; Akram, M. Novel applications of m-polar fuzzy concept lattice. *N. Math. Nat. Comput.* **2017**, *13*, 261–287.

mathematics

MDPI

Article

On Generalized Roughness in LA-Semigroups

Noor Rehman [1], Choonkil Park [2,*], Syed Inayat Ali Shah [3] and Abbas Ali [1]

1 Department of Mathematics and Statistics, Riphah International University, Hajj Complex I-14, Islamabad 44000, Pakistan; noorrehman82@yahoo.com (N.R.); abbasali5068@gmail.com (A.A.)
2 Department of Mathematics, Research Institute of Natural Sciences, Hanyang University, Seoul 04763, Korea
3 Department of Mathematics, Islamia College University, Peshawar 25120, Pakistan; inayat64@gmail.com
* Correspondence: baak@hanyang.ac.kr; Tel.: +82-2-2220-0892; Fax: +82-2-2281-0019

Received: 30 May 2018; Accepted: 25 June 2018; Published: 27 June 2018

Abstract: The generalized roughness in LA-semigroups is introduced, and several properties of lower and upper approximations are discussed. We provide examples to show that the lower approximation of a subset of an LA-semigroup may not be an LA-subsemigroup/ideal of LA-semigroup under a set valued homomorphism.

Keywords: roughness; generalized roughness; LA-semigroup

MSC: 08A72; 54A40; 03B52; 20N25

1. Introduction

The algebraic structure of a left almost semigroup, abbreviated as an LA-semigroup, has been introduced by Naseerudin and Kazim in [1]. Later, Mushtaq and others investigated the structure of LA-semigroups and added some important results related to LA-semigroups (see [2–7]). LA-semigroups are also called AG-groupoids. Ideal theory, which was introduced in [8], plays a basic role in the study of LA-semigroups. Pawlak was the first to discuss rough sets with the help of equivalence relation among the elements of a set, which is a key point in discussing the uncertainty [9]. There are at least two methods for the development of rough set theory, the constructive and axiomatic approaches. In constructive methods, lower and upper approximations are constructed from the primitive notions, such as equivalence relations on a universe and neighborhood system. In rough sets, equivalence classes play an important role in the construction of both lower and upper approximations (see [10]). But sometimes in algebraic structures, as is the case in LA-semigroups, finding equivalence relations is too difficult. Many authors have worked on this to initiate rough sets without equivalence relations. Couso and Dubois in [11] initiated a generalized rough set or a T-rough set with the help of a set valued mapping. It is a more generalized rough set compared with the Pawlak rough set.

In this paper, we initiate the study of generalized roughness in LA-semigroups and of generalized rough sets applied in the crisp form of LA-semigroups. Approximations of LA-subsemigroups and approximations of ideals in LA-semigroups are given.

2. Preliminaries

A groupoid $(S, *)$ is called an LA-semigroup if it satisfies the left invertive law

$$(a * b) * c = (c * b) * a \text{ for all } a, b, c \in S.$$

Throughout the paper, S and R will denote LA-semigroups unless stated otherwise. Let S be an LA-semigroup and A be a subset of S. Then A is called an LA-subsemigroup of S if $A^2 \subseteq A$, that is, $ab \in A$ for all $a, b \in A$. A subset A of S is called left ideal (or right ideal) of S if $SA \subseteq A$ (or $AS \subseteq A$).

An LA-subsemigroup A of S is called bi-ideal of S if $(AS)A \subseteq A$. An LA-subsemigroup A of S is called an interior ideal of S if $(SA)S \subseteq A$. An element a of S is called idempotent, if $a^2 = a$ for all $a \in S$. If every element of S is an idempotent, then S is idempotent.

3. Rough Sets

In this section, we study Pawlak roughness and generalized roughness in LA-semigroups.

3.1. Pawlak Approximations in LA-Semigroups

The concept of a rough set was introduced by Pawlak in [9]. According to Pawlak, rough set theory is based on the approximations of a set by a pair of sets called lower approximation and upper approximation of that set. Let U be a nonempty finite set with an equivalence relation R. We say (U, R) is the approximation space. If $A \subseteq U$ can be written as the union of some classes obtained from R, then A is called definable; otherwise, it is not definable. Therefore, the approximations of A are as follows:

$$\underline{R}(A) = \{x \in U : [x]_R \subseteq A\}$$

$$\overline{R}(A) = \{x \in U : [x]_R \cap A \neq \emptyset\}.$$

The pair $(\underline{R}(A), \overline{R}(A))$ is a rough set, where $\underline{R}(A) \neq \overline{R}(A)$.

Definition 1. *[5] Let ρ be an equivalence relation on S. Then ρ is called a congruence relation on S if*

$$(a, b) \in \rho \text{ implies that } (ay, by) \in \rho \text{ and } (ya, yb) \in \rho \text{ for all } a, b, y \in S.$$

Definition 2. *[8] Let ρ be a congruence relation on S. Then the approximation of S is defined by $\rho(A) = \left(\underline{\rho}(A), \overline{\rho}(A)\right)$ for every $A \in \mathcal{P}(S)$, where $\mathcal{P}(S)$ is the power set of S, and*

$$\underline{\rho}(A) = \left\{x \in U : [x]_\rho \subseteq A\right\}$$

and

$$\overline{\rho}(A) = \left\{x \in U : [x]_\rho \cap A \neq \emptyset\right\}.$$

3.2. Generalized Roughness or T-Roughness in LA-Semigroups

A generalized rough set is the generalization of Pawlak's rough set. In this case, we use set valued mappings instead of congruence classes.

Definition 3. *[11] Let X and Y be two nonempty sets and $B \subseteq Y$. Let $T : X \to \mathcal{P}(Y)$ be a set valued (SV) mapping, where $\mathcal{P}(Y)$ denotes the set of all nonempty subsets of Y. The upper approximation and the lower approximation of B with respect to T are defined by*

$$\overline{T}(B) = \{x \in X \mid T(x) \cap B \neq \emptyset\}$$

and

$$\underline{T}(B) = \{x \in X \mid T(x) \subseteq B\}.$$

Definition 4. *[12] Let X and Y be two nonempty sets and $B \subseteq Y$. Let $T : X \to \mathcal{P}(Y)$ be an SV mapping, where $\mathcal{P}(Y)$ denotes the set of all nonempty subsets of Y. Then $(\underline{T}(B), \overline{T}(B))$ is called a T-rough set.*

Definition 5. *Let R and S be two LA-semigroups and $T : R \to \mathcal{P}(S)$ be an SV mapping. Then T is called an SV homomorphism if $T(a)T(b) \subseteq T(ab)$ for all $a, b \in R$.*

Example 1. *Let $R = \{a, b, c\}$ with the following multiplication table:*

$\cdot$	a	b	c
a	a	a	a
b	c	c	c
c	a	a	c

Then R is an LA-semigroup. Define an SV mapping $T : R \to \mathcal{P}(R)$ by $T(a) = T(c) = \{a, b, c\}$ and $T(b) = \{b, c\}$. Then clearly T is an SV homomorphism.

Example 2. *Let $S = \{a, b, c, d, e\}$ with the following multiplication table:*

$\cdot$	a	b	c	d	e
a	e	b	a	b	c
b	b	b	b	b	b
c	c	b	e	b	a
d	b	b	b	b	b
e	a	b	c	b	e

Then S is an LA-semigroup. Define an SV mapping $T : S \to \mathcal{P}(S)$ by $T(a) = T(b) = T(c) = T(e) = \{a, b, c, d, e\}$ and $T(d) = \{b, d\}$. Clearly T is an SV homomorphism.

Definition 6. *Let R and S be two LA-semigroups and $T : R \to \mathcal{P}(S)$ be an SV mapping. Then T is called a strong set valued (SSV) homomorphism if $T(a) T(b) = T(ab)$ for all $a, b \in R$.*

Example 3. *Let $R = \{a, b, c\}$ with the following multiplication table:*

$\cdot$	a	b	c
a	a	a	a
b	c	c	c
c	a	a	c

Then R is an LA-semigroup and $S = \{a, b, c\}$ with the following multiplication table:

$\cdot$	a	b	c
a	b	c	b
b	c	c	c
c	c	c	c

Then S is an LA-semigroup. Define an SV mapping $T : R \to \mathcal{P}(S)$ by $T(a) = T(c) = \{c\}$ and $T(b) = \{b, c\}$. Then T is an SSV homomorphism.

Proposition 1. *Let $T : R \to \mathcal{P}(S)$ be an SV homomorphism. If $\emptyset \neq A, B \subseteq S$, then $\overline{T}(A)\overline{T}(B) \subseteq \overline{T}(AB)$.*

Proof. Let $x \in \overline{T}(A)\overline{T}(B)$. Then $x = ab$, where $a \in \overline{T}(A)$ and $b \in \overline{T}(B)$. Then $T(a) \cap A \neq \emptyset$ and $T(b) \cap B \neq \emptyset$. Therefore, there exist $y, z \in S$ such that $y \in T(a) \cap A$ and $z \in T(b) \cap B$, which implies that $y \in T(a), y \in A, z \in T(b)$, and $z \in B$. It follows that $yz \in T(a) T(b) \subseteq T(ab)$ and $yz \in AB$. Thus, $yz \in T(ab) \cap AB$, so $T(ab) \cap AB \neq \emptyset$. It follows that $ab \in \overline{T}(AB)$. Hence, $x \in \overline{T}(AB)$; therefore, $\overline{T}(A)\overline{T}(B) \subseteq \overline{T}(AB)$. □

The following example shows that equality in Proposition 1 may not hold.

Example 4. *Consider the LA-semigroup R of Example 1.*

Define an SV mapping $T : R \to \mathcal{P}(R)$ *by* $T(a) = T(c) = \{a, b, c\}$ *and* $T(b) = \{b, c\}$. *Then* T *is an SV homomorphism. Let* $A = \{a, b\}$ *and* $B = \{b\}$. *Then* $\overline{T}(A) = \{a, b, c\}$ *and* $\overline{T}(B) = \{a, b, c\}$. *Therefore,* $\overline{T}(A)\overline{T}(B) = \{a, b, c\}\{a, b, c\} = \{a, c\}$, *and* $AB = \{a, b\}\{b\} = \{a, c\}$. *Thus,* $\overline{T}(AB) = \{a, b, c\}$. *Hence,* $\overline{T}(AB) \not\subseteq \overline{T}(A)\overline{T}(B)$.

Proposition 2. *Let* $T : R \to \mathcal{P}(S)$ *be an SSV homomorphism. If* $\emptyset \neq A, B \subseteq S$, *then* $\underline{T}(A)\underline{T}(B) \subseteq \underline{T}(AB)$.

Proof. Let $x \in \underline{T}(A)\underline{T}(B)$. Then $x = ab$, where $a \in \underline{T}(A)$ and $b \in \underline{T}(B)$. Therefore, $T(a) \subseteq A$ and $T(b) \subseteq B$. Thus, $T(a)T(b) \subseteq AB$. Therefore, $T(ab) \subseteq AB$, which implies $ab \in \underline{T}(AB)$. It follows that $x \in \underline{T}(AB)$. Hence $\underline{T}(A)\underline{T}(B) \subseteq \underline{T}(AB)$. □

The following example shows that equality in Proposition 2 may not hold.

Example 5. *Consider the LA-semigroups* R *and* S *of Example 3. Define an SV mapping* $T : R \to \mathcal{P}(S)$ *by* $T(a) = T(c) = \{c\}$ *and* $T(b) = \{b, c\}$. *Then,* T *is an SSV homomorphism. Let* $A = \{a, c\}$ *and* $B = \{b, c\} \subseteq S$. *Then* $\underline{T}(A) = \{a, c\}$ *and* $\underline{T}(B) = \{a, b, c\}$. *Thus,* $\underline{T}(A)\underline{T}(B) = \{a, c\}\{a, b, c\} = \{a, c\}$, *and* $AB = \{a, c\}\{b, c\} = \{b, c\}$. *Thus,* $\underline{T}(AB) = \{a, b, c\}$. *Hence,* $\underline{T}(AB) \not\subseteq \underline{T}(A)\underline{T}(B)$.

The fact that considered groupoids are LA-semigroups is important in Propositions 3 and 4 and examples.

Proposition 3. *Let* $T : R \to \mathcal{P}(S)$ *be an SV homomorphism. If* H *is an LA-subsemigroup of* S, *then* $\overline{T}(H)$ *is an LA-subsemigroup of* R.

Proof. Let $x, y \in \overline{T}(H)$. Then $\overline{T}(x) \cap H \neq \emptyset$ and $\overline{T}(y) \cap H \neq \emptyset$. Thus, there exist $a, b \in S$ such that $a \in T(x) \cap H$ and $b \in T(y) \cap H$. Thus, $a \in T(x), a \in H$ and $b \in T(y), b \in H$. Therefore, $ab \in T(x)T(y) \subseteq T(xy)$ and $ab \in H$. Hence, $ab \in T(xy) \cap H$, and $T(xy) \cap H \neq \emptyset$. Therefore, $xy \in \overline{T}(H)$. Hence, $\overline{T}(H)$ is an LA-subsemigroup of R. □

Proposition 4. *Let* $T : R \to \mathcal{P}(S)$ *be an SSV homomorphism. If* H *is an LA-subsemigroup of* S, *then* $\underline{T}(H)$ *is an LA-subsemigroup of* R.

Proof. Let $x, y \in \underline{T}(H)$. Then $T(x) \subseteq H$ and $T(y) \subseteq H$. Therefore, $T(x)T(y) \subseteq HH = H^2$. Thus, $T(xy) \subseteq H^2$, so $T(xy) \subseteq H$, which implies $xy \in \underline{T}(H)$. Hence, $\underline{T}(H)$ is an LA-subsemigroup of R. □

The following example shows that, in the case of an SV homomorphism, $\underline{T}(A)$ may not be an LA-subsemigroup.

Example 6. *Consider the LA-semigroup* S *of Example 3.*

Define an SV mapping $T : S \to \mathcal{P}(S)$ *by* $T(b) = T(c) = \{a, b, c\}$ *and* $T(a) = \{b, c\}$. *Then* T *is an SV homomorphism. Let* $A = \{b, c\} \subseteq S$. *Then* A *is an LA-subsemigroup of* S, *and* $\underline{T}(A) = \{a\}$. *It follows that* $\underline{T}(A)\underline{T}(A) = \{a\}\{a\} = \{b\} \not\subseteq \underline{T}(A)$. *Hence,* $\underline{T}(A)$ *is not an LA-subsemigroup of* S.

Proposition 5. *Let* $T : R \to \mathcal{P}(S)$ *be an SV homomorphism. If* A *is a left ideal of* S, *then* $\overline{T}(A)$ *is a left ideal of* R.

Proof. Let x and r be elements of $\overline{T}(A)$ and R, respectively. Then $T(x) \cap A \neq \emptyset$, so there exists $a \in S$ such that $a \in T(x) \cap A$. Thus, $a \in T(x)$ and $a \in A$. Since $r \in R$, there exists a $y \in S$ such that $y \in T(r)$. Hence, $ya \in T(r)a \subseteq SA \subseteq A$. Thus, $ya \in A$ and $ya \in T(r)T(x) \subseteq T(rx)$. Hence, $ya \in T(rx) \cap A$. It follows that $T(rx) \cap A \neq \emptyset$. Therefore, $rx \in \overline{T}(A)$. Therefore, $\overline{T}(A)$ is a left ideal of R. □

Corollary 1. *Let $T : R \to \mathcal{P}(S)$ be an SV homomorphism. If A is a right ideal of S, then $\overline{T}(A)$ is a right ideal of R.*

Corollary 2. *Let $T : R \to \mathcal{P}(S)$ be an SV homomorphism. If A is an ideal of S, then $\overline{T}(A)$ is an ideal of R.*

Proposition 6. *Let $T : R \to \mathcal{P}(S)$ be an SSV homomorphism. If A is a left ideal of S, then $\underline{T}(A)$ is a left ideal of R.*

Proof. Let $x \in \underline{T}(A)$ and $r \in R$. Then $T(x) \subseteq A$. Since $r \in R$, $T(r) \subseteq S$. Thus, $T(r)T(x) \subseteq SA \subseteq A$. Thus, $T(r)T(x) \subseteq A$, and $T(rx) \subseteq A$. It follows that $rx \in \underline{T}(A)$. Hence, $\underline{T}(A)$ is a left ideal of R. □

The following example shows that, in the case of an SV homomorphism, $\underline{T}(A)$ may not be a left ideal.

Example 7. *Consider the LA-semigroup S of Example 2.*

Define an SV mapping $T : S \to \mathcal{P}(S)$ by $T(a) = T(b) = T(c) = T(e) = \{a,b,c,d,e\}$ and $T(d) = \{b,d\}$. Clearly T is an SV homomorphism. Let $A = \{b,d\}$ be a subset of S. Then A is a left ideal of S, and $\underline{T}(A) = \{d\}$. Hence, $S\underline{T}(A) = \{b\} \nsubseteq \underline{T}(A)$. Therefore, $\underline{T}(A)$ is not a left ideal of S.

Corollary 3. *Let $T : R \to \mathcal{P}(S)$ be an SSV homomorphism. If A is a right ideal of S, then $\underline{T}(A)$ is a right ideal of R.*

Corollary 4. *Let $T : R \to \mathcal{P}(S)$ be an SSV homomorphism. If A is an ideal of S, then $\underline{T}(A)$ is an ideal of R.*

Proposition 7. *Let R and S be two idempotent LA-semigroups and $T : R \to \mathcal{P}(S)$ be an SV homomorphism. If A, B are ideals of S, then*

$$\overline{T}(A) \cap \overline{T}(B) = \overline{T}(AB).$$

Proof. Since $AB \subseteq AS \subseteq A$, $AB \subseteq A$. Thus, $\overline{T}(AB) \subseteq \overline{T}(A)$, and $AB \subseteq SB \subseteq B$. It follows that $AB \subseteq B$. Thus, $\overline{T}(AB) \subseteq \overline{T}(B)$. Hence, $\overline{T}(AB) \subseteq \overline{T}(A) \cap \overline{T}(B)$.

Let $c \in \overline{T}(A) \cap \overline{T}(B)$. Then $c \in \overline{T}(A)$ and $c \in \overline{T}(B)$. Thus, $T(c) \cap A \neq \emptyset$, and $T(c) \cap B \neq \emptyset$, so there exist $x, y \in S$ such that $x \in T(c) \cap A$ and $y \in T(c) \cap B$. It follows that $x \in T(c), x \in A$, and $y \in T(c), y \in B$. Thus, $xy \in T(c)T(c) \subseteq T(cc) = T(c)$, and $x \in A$ and $y \in B$. Hence, $xy \in AB$, so $xy \in T(c) \cap AB$. Thus, $T(c) \cap AB \neq \emptyset$. Hence, $c \in \overline{T}(AB)$. Thus, $\overline{T}(A)\overline{T}(B) \subseteq \overline{T}(AB)$. Therefore,

$$\overline{T}(A) \cap \overline{T}(B) = \overline{T}(AB),$$

as desired. □

Proposition 8. *Let R and S be two idempotent LA-semigroups and $T : R \to \mathcal{P}(S)$ be an SSV homomorphism. If A and B are ideals of S, then*

$$\underline{T}(A) \cap \underline{T}(B) = \underline{T}(AB).$$

Proof. Let $AB \subseteq AS \subseteq A$. Then $AB \subseteq A$. Therefore, $\underline{T}(AB) \subseteq \underline{T}(A)$, and $AB \subseteq SB \subseteq B$. Hence, $\underline{T}(AB) \subseteq \underline{T}(B)$. Therefore,

$$\underline{T}(AB) \subseteq \underline{T}(A) \cap \underline{T}(B).$$

Let $c \in \underline{T}(A) \cap \underline{T}(B)$. Then $c \in \underline{T}(A)$ and $c \in \underline{T}(B)$. Hence, $T(c) \subseteq A$ and $T(c) \subseteq B$, so $T(c)T(c) \subseteq AB$. Thus, $T(cc) \subseteq AB$. Thus, $T(c) \subseteq AB$. Hence, $c \in \underline{T}(AB)$. This implies that $\underline{T}(A) \cap \underline{T}(B) \subseteq \underline{T}(AB)$. Therefore,

$$\underline{T}(A) \cap \underline{T}(B) = \underline{T}(AB),$$

as desired. □

Proposition 9. *Let $T : R \to \mathcal{P}(S)$ be an SV homomorphism. If A is a bi-ideal of S, then $\overline{T}(A)$ is a bi-ideal of R.*

Proof. Let $x, y \in \overline{T}(A)$ and $r \in R$. Then $T(x) \cap A \neq \emptyset$ and $T(y) \cap A \neq \emptyset$. Hence, there exist $a, b \in S$ such that $a \in T(x) \cap A$ and $b \in T(y) \cap B$, so $a \in T(x), a \in A$, and $b \in T(y), b \in A$. Since $r \in R$, there is a $c \in S$ such that $c \in T(r)$. Now, $(ac)b \in (T(x)T(r))T(y) \subseteq T(xr)T(y) \subseteq T((xr)y)$. Thus, $(ac)b \in T((xr)y)$ and $(ac)b \in A$, so $(ac)b \in T((xr)y) \cap A$. Hence, $T((xr)y) \cap A \neq \emptyset$. Thus, $(xr)y \in \overline{T}(A)$. Therefore, $\overline{T}(A)$ is a bi-ideal of R. □

Proposition 10. *Let $T : R \to \mathcal{P}(S)$ be an SSV homomorphism. If A is a bi-ideal of S, then $\underline{T}(A)$ is a bi-ideal of R.*

Proof. Let $x, y \in \underline{T}(A)$ and $r \in R$. Then $T(x) \subseteq A$ and $T(y) \subseteq A$. Since $r \in R$, $T(r) \subseteq S$. Now, $T((xr)y) = T(xr)T(y) = (T(x)T(r))T(y) \subseteq (AS)A \subseteq A$. Therefore, $T((xr)y) \subseteq A$. Thus, $(xr)y \subseteq \underline{T}(A)$. Hence, $\underline{T}(A)$ is a bi-ideal of R. □

The following example shows that, in the case of an SV homomorphism, $\underline{T}(A)$ may not be a bi-ideal.

Example 8. *Consider the LA-semigroup S of Example 2.*

Define an SV mapping $T : S \to \mathcal{P}(S)$ by $T(a) = T(b) = T(c) = T(e) = \{a, b, c, d, e\}$ and $T(d) = \{b\}$. Then T is an SV homomorphism. Let $A = \{b, d\}$. Then A is a bi-ideal of S, and $\underline{T}(A) = \{d\}$. Now, $(\underline{T}(A)S)\underline{T}(A) = \{b\} \nsubseteq \underline{T}(A)$. Hence, $\underline{T}(A)$ is not a bi-ideal of S.

Proposition 11. *Let $T : R \to \mathcal{P}(S)$ be an SV homomorphism. If A is an interior ideal of S, then $\overline{T}(A)$ is an interior ideal of R.*

Proof. Let $r \in \overline{T}(A)$, and $a, b \in R$. Then $T(r) \cap A \neq \emptyset$. Thus, there exists a $c \in S$ such that $c \in T(r) \cap A$. This implies that $c \in T(r)$ and $c \in A$. Since $a, b \in R$, there exist $x, y \in S$ such that $x \in T(a)$ and $y \in T(b)$. It follows that $(xc)y \in (T(a)T(r))T(b) \subseteq T((ar)b)$, and $(xc)y \in A$. Therefore, $(xc)y \in T((ar)b) \cap A$. Thus, $T((ar)b) \cap A \neq \emptyset$, so $(ar)b \in \overline{T}(A)$. Hence, $\overline{T}(A)$ is an interior of R. □

Proposition 12. *Let $T : R \to \mathcal{P}(S)$ be an SSV homomorphism. If A is an interior ideal of S, then $\underline{T}(A)$ is an interior ideal of R.*

Proof. Let $r \in \underline{T}(A)$ and $a, b \in R$. Then $T(r) \subseteq A$. Since $a, b \in R$, $T(a) \subseteq S, T(b) \subseteq S$. It follows that $T((ar)b) = T(ar)T(b) = (T(a)T(r))T(b) \subseteq (SA)S \subseteq A$. Therefore, $T((ar)b) \subseteq A$. Thus, $(ar)b \in \underline{T}(A)$. Hence, $\underline{T}(A)$ is an interior ideal of R. □

Definition 7. *A subset A of an LA-semigroup S is called a quasi-ideal of S if $SA \cap AS \subseteq A$.*

Proposition 13. *Let $T : R \to \mathcal{P}(S)$ be an SSV homomorphism. If A is a quasi-ideal of S, then $\underline{T}(A)$ is a quasi-ideal of R.*

Proof. Let A be a quasi-ideal of S. We prove $\underline{T}(AS \cap SA) \subseteq \underline{T}(A)$. Let $x \in \underline{T}(AS \cap SA)$. Then $T(x) \subseteq AS \cap SA \subseteq A$. Therefore, $T(x) \subseteq A$. Therefore, $x \in \underline{T}(A)$. Thus, $\underline{T}(AS \cap SA) \subseteq \underline{T}(A)$. Hence, $\underline{T}(A)$ is a quasi-ideal of R. □

Proposition 14. *Let $T : R \to \mathcal{P}(S)$ be an SV homomorphism. If A is a quasi-ideal of S, then $\overline{T}(A)$ is a quasi-ideal of R.*

Proof. Let A be a quasi-ideal of S. Then we have to show that $\overline{T}(AS \cap SA) \subseteq \overline{T}(A)$. Let $x \in \overline{T}(AS \cap SA)$. Then $T(x) \cap (AS \cap SA) \neq \emptyset$. Thus, there exists a $y \in S$ such that $y \in T(x) \cap (AS \cap SA)$. This implies that $y \in T(x)$ and $y \in (AS \cap SA) \subseteq A$, so $y \in T(x)$ and $y \in A$. Thus, $y \in T(x) \cap A$. Therefore, $x \in \overline{T}(A)$. Hence, $\overline{T}(AS \cap SA) \subseteq \overline{T}(A)$. Therefore, $\overline{T}(A)$ is a quasi-ideal of R. □

Definition 8. *An ideal P of an LA-semigroup S with left identity e is said to be prime if $AB \subseteq P$ implies either $A \subseteq P$ or $B \subseteq P$ for all ideals A, B of S.*

Proposition 15. *Let $T : R \to \mathcal{P}(S)$ be an SSV homomorphism. If A is a prime ideal of S, then $\overline{T}(A)$ is a prime ideal of R.*

Proof. Since A is an ideal of S, by Corollary 2, $\overline{T}(A)$ is an ideal of R. Let $xy \in \overline{T}(A)$. Then $T(xy) \cap A \neq \emptyset$. Thus, there exists a $z \in S$ such that $z \in T(xy) \cap A$, so $z \in T(xy) = T(x)T(y)$, and $z \in A$. Since $z = ab \in T(x)T(y)$, $ab \in A$, and A is a prime ideal of S, $a \in A$ or $b \in A$, which implies that $a \in T(x)$ and $a \in A$ or that $b \in T(y)$ and $b \in A$. Therefore, $a \in T(x) \cap A$ or $b \in T(y) \cap A$. Thus, $T(x) \cap A \neq \emptyset$ or $T(y) \cap A \neq \emptyset$. It follows that $x \in \overline{T}(A)$ or $y \in \overline{T}(A)$. Hence, $\overline{T}(A)$ is a prime ideal of R. □

Proposition 16. *Let $T : R \to \mathcal{P}(S)$ be an SSV homomorphism. If A is a prime ideal of S, then $\underline{T}(A)$ is a prime ideal of R.*

Proof. Since A is an ideal of S, by Corollary 4, $\underline{T}(A)$ is an ideal of R. Let $xy \in \underline{T}(A)$. Then $T(xy) \subseteq A$. Let $z \in T(xy) = T(x)T(y)$, where $z = ab \in T(x)T(y)$. Then $a \in T(x)$, $b \in T(y)$, and $ab \in A$. Since A is a prime ideal of S, $a \in A$ or $b \in A$. Thus, $a \in T(x) \subseteq A$ or $b \in T(y) \subseteq A$. Thus, $x \in \underline{T}(A)$ or $y \in \underline{T}(A)$. Hence, $\underline{T}(A)$ is a prime ideal of R. □

Remark 1. *The algebraic approach—in particular, the semigroup theory—can be introduced in the area of genetic algorithms and to the evolutionary based procedure for optimization and clustering* (see [13]).

4. Conclusions

In this paper, we discussed the generalized roughness in (crisp) LA-subsemigroups or ideals of LA-semigroups with the help of set valued/strong set valued homomorphisms. We have provided examples showing that the lower approximations of a subset of an LA-semigroup may not be an LA-subsemigroup or ideal of LA-semigroup, under a set valued homomorphism.

Author Contributions: Conceptualization: N.R. and C.P.; Methodology: A.A.; Software: N.R.; Validation: A.A. and C.P.; Formal Analysis: S.I.A.S.; Investigation: C.P.; Resources: N.R.; Data Curation: C.P.; Writing—Original Draft Preparation: A.A.; Writing—Review & Editing: N.R.; Visualization: S.I.A.S.; Supervision: N.R.; Project Administration: A.A.; Funding Acquisition: C.P.

Funding: This work was supported by Basic Science Research Program through the National Research Foundation of Korea funded by the Ministry of Education, Science and Technology (NRF-2017R1D1A1B04032937).

Conflicts of Interest: The authors declare that they have no competing interests.

References

1. Kazim, M.A.; Naseerudin, M. On almost-semigroup. *Alig. Bull. Math.* **1972**, 2, 1–7.
2. Dudek, W.A.; Jun, Y.B. Rough subalgebras of some binary algebras connected with logics. *Int. J. Math. Math. Sci.* **2005**, *2005*, 437–443. [CrossRef]

3. Dudek, W.A.; Jun, Y.B.; Kim, H.S. Rough set theory applied to BCI-algebras. *Quasigroups Relat. Syst.* **2002**, *9*, 45–54.
4. Mushtaq, Q. Abelian groups defined by LA-semigroups. *Stud. Sci. Math. Hungar.* **1983**, *18*, 427–428.
5. Mushtaq, Q.; Iqbal, M. Partial ordering and congruences on LA-semigroups. *Ind. J. Pure Appl. Math.* **1991**, *22*, 331–336.
6. Mushtaq, Q.; Khan, M. M-Systems in LA-semigroups. *SEA Bull. Math.* **2009**, *33*, 321–327.
7. Mushtaq, Q.; Yusuf, S.M. On LA-semigroups. *Alig. Bull. Math.* **1978**, *8*, 65–70.
8. Aslam, M.; Shabir, M.; Yaqoob, N. Roughness in left almost semigroups. *J. Adv. Pure Math.* **2011**, *3*, 70–88. [CrossRef]
9. Pawlak, Z. Rough set. *Int. J. Comput. Sci.* **1982**, *11*, 341–356.
10. Wang, J.; Zhang, Q.; Abdel-Rahman, H.; Abdel-Monem, M.I. A rough set approach to feature selection based on scatter search metaheuristic. *J. Syst. Sci. Complex.* **2014**, *27*, 157–168. [CrossRef]
11. Couois, I.; Dubois, D. Rough set, coverings and incomplete information. *Fund. Inf.* **2001**, *XXI*, 1001–1025.
12. Davvaz, B. A short note on algebraic T-rough sets. *Inf. Sci.* **2008**, *178*, 3247–3252. [CrossRef]
13. Fortuna, L.; Graziani, S.; Xibilia, M.G. Genetic algorithms and applications in system engeering: A survey. *Trans. Inst. Meas. Control* **1993**, *15*, 143–156.

mathematics

MDPI

Article

Fuzzy Semi-Metric Spaces

Hsien-Chung Wu

Department of Mathematics, National Kaohsiung Normal University, Kaohsiung 802, Taiwan; hcwu@nknucc.nknu.edu.tw

Received: 27 May 2018; Accepted: 19 June 2018; Published: 22 June 2018

Abstract: The T_1-spaces induced by the fuzzy semi-metric spaces endowed with the special kind of triangle inequality are investigated in this paper. The limits in fuzzy semi-metric spaces are also studied to demonstrate the consistency of limit concepts in the induced topologies.

Keywords: fuzzy semi-metric space; T_1-space; triangle inequality; triangular norm

1. Introduction

Given a universal set X, for any $x, y \in X$, let $\tilde{d}(x,y)$ be a fuzzy subset of $\mathbb{R}_+$ with membership function $\xi_{\tilde{d}(x,y)} : \mathbb{R}_+ \to [0,1]$, where the value $\xi_{\tilde{d}(x,y)}(t)$ means that the membership degree of the distance between x and y is equal to t. Kaleva and Seikkala [1] proposed the fuzzy metric space by defining a function $M^* : X \times X \times [0,\infty) \to [0,1]$ as follows:

$$M^*(x,y,t) = \xi_{\tilde{d}(x,y)}(t). \tag{1}$$

On the other hand, inspired by the Menger space that is a special kind of probabilistic metric space (by referring to Schweizer and Sklar [2–4], Hadžić and Pap [5] and Chang et al. [6]), Kramosil and Michalek [7] proposed another concept of fuzzy metric space.

Let X be a nonempty universal set, let $*$ be a t-norm, and let M be a mapping defined on $X \times X \times [0,\infty)$ into $[0,1]$. The 3-tuple $(X, M, *)$ is called a fuzzy metric space if and only if the following conditions are satisfied:

- for any $x, y \in X$, $M(x,y,t) = 1$ for all $t > 0$ if and only if $x = y$;
- $M(x,y,0) = 0$ for all $x, y \in X$;
- $M(x,y,t) = M(y,x,t)$ for all $x, y \in X$ and $t \geq 0$;
- $M(x,y,t) * M(y,z,s) \leq M(x,z,t+s)$ for all $x, y, z \in X$ and $s, t \geq 0$ (the so-called triangle inequality).

The mapping M in fuzzy metric space $(X, M, *)$ can be regarded as a membership function of a fuzzy subset of $X \times X \times [0,\infty)$. Sometimes, M is called a fuzzy metric of the space $(X, M, *)$. According to the first and second conditions of fuzzy metric space, the mapping $M(x,y,t)$ can be interpreted as the membership degree of the distance that is less than t between x and y. Therefore, the meanings of M and M^* defined in Equation (1) are different.

George and Veeramani [8,9] studied some properties of fuzzy metric spaces. Gregori and Romaguera [10–12] also extended their research to study the properties of fuzzy metric spaces and fuzzy quasi-metric spaces. In particular, Gregori and Romaguera [11] proposed the fuzzy quasi-metric spaces in which the symmetric condition was not assumed. In this paper, we study the so-called fuzzy semi-metric space without assuming the symmetric condition. The main difference is that four forms of triangle inequalities that were not addressed in Gregori and Romaguera [11] are considered in this paper. Another difference is that the t-norm in Gregori and Romaguera [11] was assumed to be continuous. However, the assumption of continuity for t-norm is relaxed in this paper.

The Hausdorff topology induced by the fuzzy metric space was studied in Wu [13], and the concept of fuzzy semi-metric space was considered in Wu [14]. In this paper, we shall extend to study the T_1-spaces induced by the fuzzy semi-metric spaces that is endowed with special kind of triangle inequality. Roughly speaking, the fuzzy semi-metric space does not assume the symmetric condition $M(x,y,t) = M(y,x,t)$. In this case, there are four kinds of triangle inequalities that can be considered, which will be presented in Definition 2. We shall induce the T_1-spaces from the fuzzy semi-metric space based on a special kind of triangle inequality, which will generalize the results obtained in Wu [13]. On the other hand, since the symmetric condition is not satisfied in the fuzzy semi-metric space, three kinds of limit concepts will also be considered in this paper. Furthermore, we shall prove the consistency of limit concepts in the induced topologies.

This paper is organized as follows. In Section 2, the basic properties of t-norm are presented that will be used for the further discussion. In Section 3, we propose the fuzzy semi-metric space that is endowed with four kinds of triangle inequalities. In Section 4, we induce the T_1-space from a given fuzzy semi-metric space endowed with a special kind of triangle inequality. In Section 5, three kinds of limits in fuzzy semi-metric space will be considered. We also present the consistency of limit concepts in the induced topologies.

2. Properties of t-Norm

We first recall the concept of triangular norm (i.e., t-norm). We consider the function $* : [0,1] \times [0,1] \to [0,1]$ from the product space $[0,1] \times [0,1]$ of unit intervals into the unit interval $[0,1]$. The function $*$ is called a *t-norm* if and only if the following conditions are satisfied:

- (boundary condition) $a * 1 = a$;
- (commutativity) $a * b = b * a$;
- (increasing property) if $b \leq c$, then $a * b \leq a * c$;
- (associativity) $(a * b) * c = a * (b * c)$.

From the third condition, it follows that, for any $a \in [0,1]$, we have $0 * a \leq 0 * 1$. From the first condition, we also have $0 * 1 = 0$, which implies $0 * a = 0$. The following proposition from Wu [13] will be useful for further study

Proposition 1. *By the commutativity of t-norm, if the t-norm is continuous with respect to the first component (resp. second component), then it is also continuous with respect to the second component (resp. first component). In other words, for any fixed $a \in [0,1]$, if the function $f(x) = a * x$ (resp. $f(x) = x * a$) is continuous, then the function $g(x) = x * a$ (resp. $g(x) = a * x$) is continuous. Similarly, if the t-norm is left-continuous (resp. right-continuous) with respect to the first or second component, then it is also left-continuous (resp. right-continuous) with respect to each component.*

We first provide some properties that will be used in the subsequent discussion.

Proposition 2. *We have the following properties:*

(i) *Given any fixed $a, b \in [0,1]$, suppose that the t-norm $*$ is continuous at a and b with respect to the first or second component. If $\{a_n\}_{n=1}^{\infty}$ and $\{b_n\}_{n=1}^{\infty}$ are two sequences in $[0,1]$ such that $a_n \to a$ and $b_n \to b$ as $n \to \infty$, then $a_n * b_n \to a * b$ as $n \to \infty$.*

(ii) *Given any fixed $a, b \in (0,1]$, suppose that the t-norm $*$ is left-continuous at a and b with respect to the first or second component. If $\{a_n\}_{n=1}^{\infty}$ and $\{b_n\}_{n=1}^{\infty}$ are two sequences in $[0,1]$ such that $a_n \to a-$ and $b_n \to b-$ as $n \to \infty$, then $a_n * b_n \to a * b$ as $n \to \infty$.*

(iii) *Given any fixed $a, b \in [0,1)$, suppose that the t-norm $*$ is right-continuous at a and b with respect to the first or second component. If $\{a_n\}_{n=1}^{\infty}$ and $\{b_n\}_{n=1}^{\infty}$ are two sequences in $[0,1]$ such that $a_n \to a+$ and $b_n \to b+$ as $n \to \infty$, then $a_n * b_n \to a * b$ as $n \to \infty$.*

Proof. To prove part (i), since $a_n \to a$ as $n \to \infty$, there exist an increasing sequence $\{p_n\}_{n=1}^{\infty}$ and a decreasing sequence $\{q_n\}_{n=1}^{\infty}$ such that $p_n \uparrow a$ and $q_n \downarrow a$ satisfying $p_n \leq a_n \leq q_n$. In addition, there exists an increasing sequence $\{r_n\}_{n=1}^{\infty}$ and a decreasing sequence $\{s_n\}_{n=1}^{\infty}$ such that $r_n \uparrow b$ and $s_n \downarrow b$ satisfying $r_n \leq b_n \leq s_n$. By Remark 1, we see that the t-norm is continuous with respect to each component. Given any $\epsilon > 0$, using the continuity of t-norm at b with respect to the second component, there exists $n_0 \in \mathbb{N}$ such that

$$a * b - \frac{\epsilon}{2} < a * r_{n_0} \text{ and } a * s_{n_0} < a * b + \frac{\epsilon}{2}. \tag{2}$$

In addition, using the continuity of t-norm at a with respect to the first component, there exists $n_1 \in \mathbb{N}$ such that

$$a * r_{n_0} - \frac{\epsilon}{2} < p_{n_1} * r_{n_0} \text{ and } q_{n_1} * s_{n_0} < a * s_{n_0} + \frac{\epsilon}{2}. \tag{3}$$

According to Equation (2) and using the increasing property of t-norm, for $n \geq n_0$, we have

$$a * b - \frac{\epsilon}{2} < a * r_{n_0} \tag{4}$$

$$\leq a * r_n \leq a * b_n \leq a * s_n \leq a * s_{n_0} < a * b + \frac{\epsilon}{2}. \tag{5}$$

In addition, according to Equation (3), for $m \geq n_1$ and $n \geq n_0$, we have

$$a * r_{n_0} - \frac{\epsilon}{2} < p_{n_1} * r_{n_0} \leq p_m * r_n \leq a_m * b_n \leq q_m * s_n \leq q_{n_1} * s_{n_0} < a * s_{n_0} + \frac{\epsilon}{2}. \tag{6}$$

By taking $n_2 = \max\{n_0, n_1\}$, from Equations (4) and (6), we obtain that $n \geq n_2$ implies

$$\begin{aligned} a * b - \epsilon &< a * r_{n_0} - \frac{\epsilon}{2} \text{ (by Equation (4))} \\ &< a_n * b_n < a * s_{n_0} + \frac{\epsilon}{2} \text{ (from Equation (6) by taking } m = n) \\ &< a * b + \epsilon, \text{ (by Equation (5))}, \end{aligned}$$

which says that $|a * b - a_n * b_n| < \epsilon$. This shows the desired convergence.

To prove part (ii), we note that there exist two increasing sequences $\{p_n\}_{n=1}^{\infty}$ and $\{r_n\}_{n=1}^{\infty}$ such that $p_n \uparrow a$ and $r_n \uparrow b$ satisfying $p_n \leq a_n$ and $r_n \leq b_n$. By Remark 1, we see that the t-norm is left continuous with respect to each component. Given any $\epsilon > 0$, using the left-continuity of t-norm at b with respect to the second component, there exists $n_0 \in \mathbb{N}$ such that

$$a * b - \frac{\epsilon}{2} < a * r_{n_0}.$$

In addition, using the left-continuity of t-norm at a with respect to the first component, there exists $n_1 \in \mathbb{N}$ such that

$$a * r_{n_0} - \frac{\epsilon}{2} < p_{n_1} * r_{n_0}.$$

Using the increasing property of t-norm, for $m \geq n_1$ and $n \geq n_0$, we have

$$a * r_{n_0} - \frac{\epsilon}{2} < p_{n_1} * r_{n_0} \leq p_m * r_n \leq a_m * b_n.$$

Since $a_n \to a-$ and $b_n \to b-$, we see that $a_n \leq a$ and $b_n \leq b$ for all n. By taking $n_2 = \max\{n_0, n_1\}$, for $n \geq n_2$, we obtain

$$a * b - \epsilon < a * r_{n_0} - \frac{\epsilon}{2} < a_n * b_n \leq a * b < a * b + \epsilon,$$

which says that $|a * b - a_n * b_n| < \epsilon$. This shows the desired convergence. Part (iii) can be similarly proved, and the proof is complete. □

The associativity of t-norm says that the operation $a_1 * a_2 * \cdots * a_p$ is well-defined for $p \geq 2$. The following proposition from Wu [13] will be useful for further study.

Proposition 3. *Suppose that the t-norm $*$ is left-continuous at 1 with respect to the first or second component. We have the following properties:*

(i) *For any $a, b \in (0,1)$ with $a > b$, there exists $r \in (0,1)$ such that $a * r \geq b$.*
(ii) *For any $a \in (0,1)$ and any $n \in \mathbb{N}$ with $n \geq 2$, there exists $r \in (0,1)$ such that $r * r * \cdots * r > a$ for n-times.*

3. Fuzzy Semi-Metric Space

In the sequel, we shall define the concept of fuzzy semi-metric space without considering the symmetric condition. Because of lacking symmetry, the concept of triangle inequality should be carefully interpreted. Therefore, we propose four kinds of triangle inequalities.

Definition 1. *Let X be a nonempty universal set, and let M be a mapping defined on $X \times X \times [0,\infty)$ into $[0,1]$. Then, (X, M) is called a fuzzy semi-metric space if and only if the following conditions are satisfied:*

- *for any $x, y \in X$, $M(x,y,t) = 1$ for all $t > 0$ if and only if $x = y$;*
- *$M(x,y,0) = 0$ for all $x, y \in X$ with $x \neq y$.*

We say that M satisfies the symmetric condition if and only if $M(x,y,t) = M(y,x,t)$ for all $x, y \in X$ and $t > 0$. We say that M satisfies the strongly symmetric condition if and only if $M(x,y,t) = M(y,x,t)$ for all $x, y \in X$ and $t \geq 0$.

We remark that the first condition says that $M(x,x,t) = 1$ for all $t > 0$. However, the value of $M(x,x,0)$ is free. Recall that the mapping $M(x,y,t)$ is interpreted as the membership degree of the distance that is less than t between x and y. Therefore, $M(x,x,t) = 1$ for all $t > 0$ means that the distance that is less than $t > 0$ between x and x is always true. The second condition says that $M(x,y,0) = 0$ for $x \neq y$, which can be similarly realized that the distance that is less than 0 between two distinct elements x and y is impossible.

Definition 2. *Let X be a nonempty universal set, let $*$ be a t-norm, and let M be a mapping defined on $X \times X \times [0,\infty)$ into $[0,1]$.*

- *We say that M satisfies the $\bowtie$-triangle inequality if and only if the following inequality is satisfied:*

$$M(x,y,t) * M(y,z,s) \leq M(x,z,t+s) \text{ for all } x,y,z \in X \text{ and } s,t > 0.$$

- *We say that M satisfies the $\triangleright$-triangle inequality if and only if the following inequality is satisfied:*

$$M(x,y,t) * M(z,y,s) \leq M(x,z,t+s) \text{ for all } x,y,z \in X \text{ and } s,t > 0.$$

- *We say that M satisfies the $\triangleleft$-triangle inequality if and only if the following inequality is satisfied:*

$$M(y,x,t) * M(y,z,s) \leq M(x,z,t+s) \text{ for all } x,y,z \in X \text{ and } s,t > 0.$$

- *We say that M satisfies the $\diamond$-triangle inequality if and only if the following inequality is satisfied:*

$$M(y,x,t) * M(z,y,s) \leq M(x,z,t+s) \text{ for all } x,y,z \in X \text{ and } s,t > 0.$$

We say that M satisfies the strong $\circ$*-triangle inequality for* $\circ \in \{\bowtie, \triangleright, \triangleleft, \diamond\}$ *when* $s,t > 0$ *is replaced by* $s,t \geq 0$.

Remark 1. *It is obvious that if the mapping M satisfies the symmetric condition, then the concepts of* $\bowtie$*-triangle inequality,* $\triangleright$*-triangle inequality,* $\triangleleft$*-triangle inequality and* $\diamond$*-triangle inequality are all equivalent.*

Example 1. *Let X be a universal set, and let* $d : X \times X \to \mathbb{R}_+$ *satisfy the following conditions:*

- $d(x,y) \geq 0$ *for any* $x,y \in X$;
- $d(x,y) = 0$ *if and only if* $x = y$ *for any* $x,y \in X$;
- $d(x,y) + d(y,z) \geq d(x,z)$ *for any* $x,y,z \in X$.

Note that we do not assume $d(x,y) = d(y,x)$. *For example, let* $X = [0,1]$. *We define*

$$d(x,y) = \begin{cases} y - x, & \text{if } y \geq x, \\ 1, & \text{otherwise.} \end{cases}$$

Then, $d(x,y) \neq d(y,x)$ *and the above three conditions are satisfied. Now, we take t-norm* $*$ *as* $a * b = ab$ *and define*

$$M(x,y,t) = \begin{cases} \dfrac{t}{t+d(x,y)}, & \text{if } t > 0, \\ 1, & \text{if } t = 0 \text{ and } d(x,y) = 0, \\ 0, & \text{if } t = 0 \text{ and } d(x,y) > 0, \end{cases} = \begin{cases} \dfrac{t}{t+d(x,y)} & \text{if } t > 0, \\ 1, & \text{if } t = 0 \text{ and } x = y, \\ 0, & \text{if } t = 0 \text{ and } x \neq y. \end{cases}$$

It is clear to see that $M(x,y,t) \neq M(y,x,t)$ *for* $t > 0$, *since* $d(x,y) \neq d(y,x)$. *We are going to claim that* $(X, M, *)$ *is a fuzzy semi-metric space satisfying the* $\bowtie$*-triangle inequality. For* $t > 0$ *and* $M(x,y,t) = 1$, *we have* $t = t + d(x,y)$, *which says that* $d(x,y) = 0$, *i.e.,* $x = y$. *Next, we are going to check the* $\bowtie$*-triangle inequality. For* $s > 0$ *and* $t > 0$, *we first have*

$$\frac{1}{t}d(x,y) + \frac{1}{s}d(y,z) \geq \frac{1}{s+t}\left[d(x,y) + d(y,z)\right] \geq \frac{1}{t+s}d(x,z).$$

Then, we obtain

$$\begin{aligned} M(x,y,t) * M(y,z,s) &= \frac{t}{t+d(x,y)} \cdot \frac{s}{s+d(y,z)} = \frac{ts}{ts + td(y,z) + sd(x,y) + d(x,y)d(y,z)} \\ &\leq \frac{ts}{ts + td(y,z) + sd(x,y)} = \frac{1}{1 + \frac{1}{s}d(y,z) + \frac{1}{t}d(x,y)} \\ &\leq \frac{1}{1 + \frac{1}{t+s}d(x,z)} = \frac{t+s}{t+s+d(x,z)} = M(x,z,t+s). \end{aligned}$$

This shows that $(X, M, *)$ *defined above is indeed a fuzzy semi-metric space satisfying the* $\bowtie$*-triangle inequality.*

Given a fuzzy semi-metric space (X, M), when we say that the mapping M satisfies some kinds of (strong) triangle inequalities, it implicitly means that the t-norm is considered in (X, M).

- Suppose that M satisfies the (strong) $\triangleright$-triangle inequality. Then,

$$M(x,y,t) * M(z,y,s) \leq M(x,z,t+s) \text{ and } M(z,y,t) * M(x,y,s) \leq M(z,x,t+s).$$

Since the t-norm is commutative, it follows that

$$M(x,y,t) * M(z,y,s) = M(z,y,t) * M(x,y,s) \leq \min\{M(x,z,t+s), M(z,x,t+s)\}.$$

- Suppose that M satisfies the (strong) $\triangleleft$-triangle inequality. Then, we similarly have
$$M(y,x,t) * M(y,z,s) = M(y,z,t) * M(y,x,s) \leq \min\{M(x,z,t+s), M(z,x,t+s)\}.$$

Definition 3. *Let (X, M) be a fuzzy semi-metric space.*

- *We say that M is nondecreasing if and only if, given any fixed $x, y \in X$, $M(x,y,t_1) \geq M(x,y,t_2)$ for $t_1 > t_2 > 0$. We say that M is strongly nondecreasing if and only if, given any fixed $x, y \in X$, $M(x,y,t_1) \geq M(x,y,t_2)$ for $t_1 > t_2 \geq 0$.*
- *We say that M is symmetrically nondecreasing if and only if, given any fixed $x, y \in X$, $M(x,y,t_1) \geq M(y,x,t_2)$ for $t_1 > t_2 > 0$. We say that M is symmetrically strongly nondecreasing if and only if, given any fixed $x, y \in X$, $M(x,y,t_1) \geq M(y,x,t_2)$ for $t_1 > t_2 \geq 0$.*

Proposition 4. *Let (X, M) be a fuzzy semi-metric space. Then, we have the following properties:*

(i) *If M satisfies the $\bowtie$-triangle inequality, then M is nondecreasing.*
(ii) *If M satisfies the $\triangleright$-triangle inequality or the $\triangleleft$-triangle inequality, then M is both nondecreasing and symmetrically nondecreasing.*
(iii) *If M satisfies the $\diamond$-triangle inequality, then M is symmetrically nondecreasing.*

Proof. Given any fixed $x, y \in X$, for $t_1 > t_2 > 0$, we have the following inequalities.

- Suppose that M satisfies the $\bowtie$-triangle inequality. Then,
$$M(x,y,t_1) \geq M(x,y,t_2) * M(y,y,t_1 - t_2) = M(x,y,t_2) * 1 = M(x,y,t_2).$$
- Suppose that M satisfies the $\triangleright$-triangle inequality. Then,
$$M(x,y,t_1) \geq M(x,y,t_2) * M(y,y,t_1 - t_2) = M(x,y,t_2) * 1 = M(x,y,t_2)$$
and
$$M(x,y,t_1) \geq M(x,x,t_1 - t_2) * M(y,x,t_2) = 1 * M(y,x,t_2) = M(y,x,t_2).$$
- Suppose that M satisfies the $\triangleleft$-triangle inequality. Then,
$$M(x,y,t_1) \geq M(x,x,t_1 - t_2) * M(x,y,t_2) = 1 * M(x,y,t_2) = M(x,y,t_2)$$
and
$$M(x,y,t_1) \geq M(y,x,t_2) * M(y,y,t_1 - t_2) = M(y,x,t_2) * 1 = M(y,x,t_2).$$
- Suppose that M satisfies the $\diamond$-triangle inequality. Then,
$$M(x,y,t_1) \geq M(x,x,t_1 - t_2) * M(y,x,t_2) = 1 * M(y,x,t_2) = M(y,x,t_2).$$

This completes the proof. □

Definition 4. *Let (X, M) be a fuzzy semi-metric space.*

- *We say that M is left-continuous with respect to the distance at $t_0 > 0$ if and only if, for any fixed $x, y \in X$, given any $\epsilon > 0$, there exists $\delta > 0$ such that $0 < t_0 - t < \delta$ implies $|M(x,y,t) - M(x,y,t_0)| < \epsilon$; that is, the mapping $M(x,y,\cdot) : (0,\infty) \to [0,1]$ is left-continuous at t_0. We say that M is left-continuous with respect to the distance on $(0,\infty)$ if and only if the mapping $M(x,y,\cdot)$ is left-continuous on $(0,\infty)$ for any fixed $x, y \in X$.*

- *We say that M is right-continuous with respect to the distance at $t_0 \geq 0$ if and only if, for any fixed $x, y \in X$, given any $\epsilon > 0$, there exists $\delta > 0$ such that $0 < t - t_0 < \delta$ implies $|M(x,y,t) - M(x,y,t_0)| < \epsilon$; that is, the mapping $M(x,y,\cdot) : (0,\infty) \to [0,1]$ is right-continuous at t_0. We say that M is right-continuous with respect to the distance on $[0,\infty)$ if and only if the mapping $M(x,y,\cdot)$ is left-continuous on $[0,\infty)$ for any fixed $x, y \in X$.*
- *We say that M is continuous with respect to the distance at $t_0 \geq 0$ if and only if, for any fixed $x, y \in X$, given any $\epsilon > 0$, there exists $\delta > 0$ such that $|t - t_0| < \delta$ implies $|M(x,y,t) - M(x,y,t_0)| < \epsilon$; that is, the mapping $M(x,y,\cdot) : (0,\infty) \to [0,1]$ is continuous at t_0. We say that M is continuous with respect to the distance on $[0,\infty)$ if and only if the mapping $M(x,y,\cdot)$ is continuous on $[0,\infty)$ for any fixed $x, y \in X$.*
- *We say that M is symmetrically left-continuous with respect to the distance at $t_0 > 0$ if and only if, for any fixed $x, y \in X$, given any $\epsilon > 0$, there exists $\delta > 0$ such that $0 < t_0 - t < \delta$ implies $|M(x,y,t) - M(y,x,t_0)| < \epsilon$. We say that M is symmetrically left-continuous with respect to the distance on $(0,\infty)$ if and only if it is symmetrically left-continuous with respect to the distance at each $t > 0$.*
- *We say that M is symmetrically right-continuous with respect to the distance at $t_0 \geq 0$ if and only if, for any fixed $x, y \in X$, given any $\epsilon > 0$, there exists $\delta > 0$ such that $0 < t - t_0 < \delta$ implies $|M(x,y,t) - M(y,x,t_0)| < \epsilon$. We say that M is symmetrically right-continuous with respect to the distance on $[0,\infty)$ if and only if it is symmetrically right-continuous with respect to the distance at each $t \geq 0$.*
- *We say that M is symmetrically continuous with respect to the distance at $t_0 \geq 0$ if and only if, for any fixed $x, y \in X$, given any $\epsilon > 0$, there exists $\delta > 0$ such that $|t - t_0| < \delta$ implies $|M(x,y,t) - M(y,x,t_0)| < \epsilon$. We say that M is symmetrically continuous with respect to the distance on $[0,\infty)$ if and only if it is symmetrically continuous with respect to the distance at each $t \geq 0$.*

Proposition 5. *Let (X, M) be a fuzzy semi-metric space such that the $\circ$-triangle inequality is satisfied for $\circ \in \{\triangleright, \triangleleft, \diamond\}$. Then, we have the following properties:*

(i) *Suppose that M is left-continuous or symmetrically left-continuous with respect to the distance at $t > 0$. Then $M(x,y,t) = M(y,x,t)$. In other words, if M is left-continuous or symmetrically left-continuous with respect to the distance on $(0,\infty)$. Then M satisfies the symmetric condition.*

(ii) *Suppose that M is right-continuous or symmetrically right-continuous with respect to the distance at $t \geq 0$. Then $M(x,y,t) = M(y,x,t)$. In other words, if M is right-continuous or symmetrically right-continuous with respect to the distance on $[0,\infty)$. Then M satisfies the strongly symmetric condition.*

Proof. To prove part (i), given any $t > 0$, there exists $n_t \in \mathbb{N}$ satisfying $t - \frac{1}{n_t} > 0$. We consider the following cases:

- Suppose that the $\triangleright$-triangle inequality is satisfied. Then,

$$M\left(y,x,t-\frac{1}{n_t}\right) = 1 * M\left(y,x,t-\frac{1}{n_t}\right) = M\left(x,x,\frac{1}{n_t}\right) * M\left(y,x,t-\frac{1}{n_t}\right) \leq M(x,y,t)$$

and

$$M\left(x,y,t-\frac{1}{n_t}\right) = 1 * M\left(x,y,t-\frac{1}{n_t}\right) = M\left(y,y,\frac{1}{n_t}\right) * M\left(x,y,t-\frac{1}{n_t}\right) \leq M(y,x,t).$$

Using the left-continuity of M, it follows that $M(y,x,t) \leq M(x,y,t)$ and $M(x,y,t) \leq M(y,x,t)$ by taking $n_t \to \infty$. This shows that $M(x,y,t) = M(y,x,t)$ for all $t > 0$. On the other hand, we also have

$$M\left(x,y,t-\frac{1}{n_t}\right) = M\left(x,y,t-\frac{1}{n_t}\right) * 1 = M\left(x,y,t-\frac{1}{n_t}\right) * M\left(y,y,\frac{1}{n_t}\right) \leq M(x,y,t)$$

and

$$M\left(y,x,t-\frac{1}{n_t}\right)=M\left(y,x,t-\frac{1}{n_t}\right)*1=M\left(y,x,t-\frac{1}{n_t}\right)*M\left(x,x,\frac{1}{n_t}\right)\leq M(y,x,t).$$

Using the symmetric left-continuity of M, it follows that $M(y,x,t)\leq M(x,y,t)$ and $M(x,y,t)\leq M(y,x,t)$ by taking $n_t\to\infty$. This shows that $M(x,y,t)=M(y,x,t)$ for all $t>0$.

- Suppose that the $\triangleleft$-triangle inequality is satisfied. Then,

$$M\left(y,x,t-\frac{1}{n_t}\right)=M\left(y,x,t-\frac{1}{n_t}\right)*1=M\left(y,x,t-\frac{1}{n_t}\right)*M\left(y,y,\frac{1}{n_t}\right)\leq M(x,y,t)$$

and

$$M\left(x,y,t-\frac{1}{n_t}\right)=M\left(x,y,t-\frac{1}{n_t}\right)*1=M\left(x,y,t-\frac{1}{n_t}\right)*M\left(x,x,\frac{1}{n_t}\right)\leq M(y,x,t).$$

The left-continuity of M shows that $M(x,y,t)=M(y,x,t)$ for all $t>0$. We can similarly obtain the desired result using the symmetric left-continuity of M.

- Suppose that the $\diamond$-triangle inequality is satisfied. Then, this is the same situation as the $\triangleright$-triangle inequality.

To prove part (ii), given any $t\geq 0$ and $n\in\mathbb{N}$, we consider the following cases.

- Suppose that the $\triangleright$-triangle inequality is satisfied. Then,

$$M\left(y,x,t+\frac{1}{n}\right)=1*M\left(y,x,t+\frac{1}{n}\right)=M\left(x,x,\frac{1}{n}\right)*M\left(y,x,t+\frac{1}{n}\right)\leq M\left(x,y,t+\frac{2}{n}\right)$$

and

$$M\left(x,y,t+\frac{1}{n}\right)=1*M\left(x,y,t+\frac{1}{n}\right)=M\left(y,y,\frac{1}{n}\right)*M\left(x,y,t+\frac{1}{n}\right)\leq M\left(y,x,t+\frac{2}{n}\right).$$

The right-continuity of M shows that $M(x,y,t)=M(y,x,t)$ for all $t\geq 0$. We can similarly obtain the desired result using the symmetric right-continuity of M.

- Suppose that the $\triangleleft$-triangle inequality is satisfied. Then,

$$M\left(y,x,t+\frac{1}{n}\right)=M\left(y,x,t+\frac{1}{n}\right)*1=M\left(y,x,t+\frac{1}{n}\right)*M\left(y,y,\frac{1}{n}\right)\leq M\left(x,y,t+\frac{2}{n}\right)$$

and

$$M\left(x,y,t+\frac{1}{n}\right)=M\left(x,y,t+\frac{1}{n}\right)*1=M\left(x,y,t+\frac{1}{n}\right)*M\left(x,x,\frac{1}{n}\right)\leq M\left(y,x,t+\frac{2}{n}\right).$$

The right-continuity of M shows that $M(x,y,t)=M(y,x,t)$ for all $t\geq 0$. We can similarly obtain the desired result using the symmetric right-continuity of M.

- Suppose that the $\diamond$-triangle inequality is satisfied. Then, this is the same situation as the $\triangleright$-triangle inequality.

This completes the proof. □

From Proposition 5, if M is left-continuous or symmetrically left-continuous with respect to the distance on $(0,\infty)$, or right-continuous and or symmetrically right-continuous with respect to the distance on on $(0,\infty]$, then we can just consider the $\bowtie$-triangle inequality.

Proposition 6. *Let (X, M) be a fuzzy semi-metric space such that M is left-continuous or symmetrically left-continuous with respect to the distance on $(0, \infty)$, or right-continuous and or symmetrically right-continuous with respect to the distance on $(0, \infty]$. Suppose that $M(x, x, 0) = 1$ for any $x \in X$. Then, M satisfies the $\circ$-triangle inequality if and only if M satisfies the strong $\circ$-triangle inequality for $\circ \in \{\triangleright, \triangleleft, \diamond\}$.*

Proof. We first note that the converse is obvious. Now, we assume that M satisfies the $\triangleleft$-triangle inequality.

- Suppose that $s = t = 0$. If $x \neq y$ or $y \neq z$, then $M(y, x, 0) = 0$ or $M(y, z, 0) = 0$, which implies
$$M(y, x, 0) * M(y, z, 0) = 0 \leq M(x, z, 0).$$
If $x = y = z$, then $M(y, x, 0) = 1 = M(y, z, 0) = M(x, z, 0)$, which implies
$$M(y, x, 0) * M(y, z, 0) = 1 * 1 = 1 = M(x, z, 0).$$
- Suppose that $s > 0$ and $t = 0$. If $x \neq y$, then $M(y, x, 0) = 0$, which implies
$$M(y, x, t) * M(y, z, s) = M(y, x, 0) * M(y, z, s) = 0 \leq M(x, z, t + s).$$
If $x = y$, then $M(y, x, t) = M(x, x, 0) = 1$, which implies
$$M(y, x, t) * M(y, z, s) = 1 * M(y, z, s) = M(y, z, s) = M(x, z, t + s).$$
- Suppose that $s = 0$ and $t > 0$. If $y \neq z$, then $M(y, z, 0) = 0$, which implies
$$M(y, x, t) * M(y, z, s) = M(y, x, t) * M(y, z, 0) = 0 \leq M(x, z, t + s).$$
If $y = z$, then $M(y, z, s) = M(y, y, 0) = 1$. Using Proposition 5, we have
$$M(y, x, t) * M(y, z, s) = M(y, x, t) * 1 = M(y, x, t) = M(x, y, t) = M(x, z, t + s).$$

We can similarly obtain the desired results for $\circ \in \{\triangleleft, \diamond\}$. This completes the proof. □

Proposition 7. *Let (X, M) be a fuzzy semi-metric space. Suppose that M satisfies the $\bowtie$-triangle inequality, and that M is left-continuous with respect to the distance at $t > 0$. Given any fixed $x, y \in X$, if $M(x, y, t) > 1 - r$, then there exists t_0 with $0 < t_0 < t$ such that $M(x, y, t_0) > 1 - r$.*

Proof. Let $\epsilon = M(x, y, t) - (1 - r) > 0$. Using the left-continuity of M, there exists t_0 with $0 < t_0 < t$ such that $|M(x, y, t) - M(x, y, t_0)| < \epsilon$. From part (i) of Proposition 4, we also have $0 \leq M(x, y, t) - M(x, y, t_0) < \epsilon$, which implies $M(x, y, t_0) > 1 - r$. This completes the proof. □

4. T_1-Spaces

Let $(X, M, *)$ be a fuzzy metric space, i.e., the symmetric condition is satisfied. Given $t > 0$ and $0 < r < 1$, the (r, t)-ball of x is defined by
$$B(x, r, t) = \{y \in X : M(x, y, t) = M(y, x, t) > 1 - r\}$$
by referring to Wu [13]. In this paper, since the symmetric condition is not satisfied, two different concepts of open ball will be proposed below. Therefore, the T_1-spaces generated from these two different open balls will generalize the results obtained in Wu [13].

Definition 5. *Let (X, M) be a fuzzy semi-metric space. Given $t > 0$ and $0 < r < 1$, the (r,t)-*balls *centered at x are denoted and defined by*

$$B^{\triangleleft}(x,r,t) = \{y \in X : M(x,y,t) > 1 - r\}$$

and

$$B^{\triangleright}(x,r,t) = \{y \in X : M(y,x,t) > 1 - r\}.$$

Let $\mathcal{B}^{\triangleleft}$ denote the family of all (r,t)-balls $B^{\triangleleft}(x,r,t)$, and let $\mathcal{B}^{\triangleright}$ denote the family of all (r,t)-balls $B^{\triangleright}(x,r,t)$.

It is clearly that if the symmetric condition for M is satisfied, then

$$B^{\triangleleft}(x,r,t) = B^{\triangleright}(x,r,t).$$

In this case, we simply write $B(x,r,t)$ to denote the (r,t)-balls centered at x, and write $\mathcal{B}$ to denote the family of all (r,t)-balls $B(x,r,t)$.

We also see that $B^{\triangleleft}(x,r,t) \neq \emptyset$ and $B^{\triangleright}(x,r,t) \neq \emptyset$, since $x \in B^{\triangleleft}(x,r,t)$ and $x \in B^{\triangleright}(x,r,t)$ by the fact of $M(x,x,t) = 1$ for all $t > 0$. Since $0 < r < 1$, it is obvious that if $M(x,y,t) = 0$, then $y \notin B^{\triangleleft}(x,r,t)$. In other words, if $y \in B^{\triangleleft}(x,r,t)$, then $M(x,y,t) > 0$. Similarly, if $y \in B^{\triangleright}(x,r,t)$, then $M(y,x,t) > 0$.

Proposition 8. *Let (X, M) be a fuzzy semi-metric space.*

(i) *For each $x \in X$, we have $x \in B^{\triangleleft}(x,r,t) \in \mathcal{B}^{\triangleleft}$ and $x \in B^{\triangleright}(x,r,t) \in \mathcal{B}^{\triangleright}$.*
(ii) *If $x \neq y$, then there exist $B^{\triangleleft}(x,r,t)$ and $B^{\triangleright}(x,r,t)$ such that $y \notin B^{\triangleleft}(x,r,t)$ and $y \notin B^{\triangleright}(x,r,t)$.*

Proof. Part (i) is obvious. To prove part (ii), since $x \neq y$, there exists $t_0 > 0$ such that $M(x,y,t_0) < 1$. There also exists r_0 such that $M(x,y,t_0) < r_0 < 1$. Suppose that $y \in B^{\triangleleft}(x, 1 - r_0, t_0)$. Then, we have

$$M(x,y,t_0) > r_0 > M(x,y,t_0).$$

This contradiction says that $y \notin B^{\triangleleft}(x, 1 - r_0, t_0)$, and the proof is complete. □

Proposition 9. *Let (X, M) be a fuzzy semi-metric space.*

(i) *Suppose that M satisfies the $\circ$-triangle for $\circ \in \{\bowtie, \triangleright, \triangleleft\}$. Then, the following statements hold true:*
   - *Given any $B^{\triangleleft}(x,r,t) \in \mathcal{B}^{\triangleleft}$, there exists $n \in \mathbb{N}$ such that $B^{\triangleleft}(x, 1/n, 1/n) \subseteq B^{\triangleleft}(x,r,t)$.*
   - *Given any $B^{\triangleright}(x,r,t) \in \mathcal{B}^{\triangleright}$, there exists $n \in \mathbb{N}$ such that $B^{\triangleright}(x, 1/n, 1/n) \subseteq B^{\triangleright}(x,r,t)$.*

(ii) *Suppose that M satisfies the $\circ$-triangle for $\circ \in \{\triangleright, \triangleleft, \diamond\}$. Then, the following statements hold true:*
   - *Given any $B^{\triangleleft}(x,r,t) \in \mathcal{B}^{\triangleleft}$, there exists $n \in \mathbb{N}$ such that $B^{\triangleright}(x, 1/n, 1/n) \subseteq B^{\triangleleft}(x,r,t)$.*
   - *Given any $B^{\triangleright}(x,r,t) \in \mathcal{B}^{\triangleright}$, there exists $n \in \mathbb{N}$ such that $B^{\triangleleft}(x, 1/n, 1/n) \subseteq B^{\triangleright}(x,r,t)$.*

Proof. To prove part (i), it suffices to prove the first case. We take $n \in \mathbb{N}$ such that $1/n \leq \min\{r, t\}$. Then, for $y \in B^{\triangleleft}(x, 1/n, 1/n)$, using parts (i) and (ii) of Proposition 4, we have

$$M(x,y,t) \geq M\left(x,y,\frac{1}{n}\right) > 1 - \frac{1}{n} \geq 1 - r,$$

which says that $y \in B^{\triangleleft}(x,r,t)$. Part (ii) can be similarly obtained by using parts (ii) and (iii) of Proposition 4, and the following inequalities:

$$M(x,y,t) \geq M\left(y,x,\frac{1}{n}\right) > 1 - \frac{1}{n} \geq 1 - r.$$

This completes the proof. □

Proposition 10. *(Left-Continuity for M) Let (X, M) be a fuzzy semi-metric space along with a t-norm $*$ such that the following conditions are satisfied:*

- *M is left-continuous with respect to the distance on $(0, \infty)$;*
- *the t-norm $*$ is left-continuous at 1 with respect to the first or second component.*

Suppose that M satisfies the $\bowtie$-triangle inequality. Then, we have the following inclusions:

(i) *Given any $y \in B^{\triangleleft}(x,r,t)$, there exists $B^{\triangleleft}(y,\bar{r},\bar{t})$ such that $B^{\triangleleft}(y,\bar{r},\bar{t}) \subseteq B^{\triangleleft}(x,r,t)$.*
(ii) *Given any $y \in B^{\triangleright}(x,r,t)$, there exists $B^{\triangleright}(y,\bar{r},\bar{t})$ such that $B^{\triangleright}(y,\bar{r},\bar{t}) \subseteq B^{\triangleright}(x,r,t)$.*

Proof. For $y \in B^{\triangleleft}(x,r,t)$, we have $M(x,y,t) > 1-r$. By part (i) of Proposition 7, there exists t_0 with $0 < t_0 < t$ such that $M(x,y,t_0) > 1-r$. Let $r_0 = M(x,y,t_0)$. Then, we have $r_0 > 1-r$. There exists s with $0 < s < 1$ such that $r_0 > 1-s > 1-r$. By part (i) of Proposition 3, there exists r_1 with $0 < r_1 < 1$ such that $r_0 * r_1 \geq 1-s$. Let $\bar{r} = 1-r_1$ and $\bar{t} = t-t_0$. Similarly, for $y \in B^{\triangleright}(x,r,t)$, we have $M(y,x,t) > 1-r$. In this case, let $r_0 = M(y,x,t_0)$.

To prove part (i), for $y \in B^{\triangleleft}(x,r,t)$ and $z \in B^{\triangleleft}(y,\bar{r},\bar{t})$, we have

$$M(y,z,t-t_0) = M(y,z,\bar{t}) > 1-\bar{r} = r_1.$$

By the $\bowtie$-triangle inequality, we also have

$$M(x,z,t) \geq M(x,y,t_0) * M(y,z,t-t_0) = r_0 * M(y,z,t-t_0) \geq r_0 * r_1 \geq 1-s > 1-r.$$

This shows that $z \in B^{\triangleleft}(x,r,t)$. Therefore, we obtain the inclusion $B^{\triangleleft}(y,\bar{r},\bar{t}) \subseteq B^{\triangleleft}(x,r,t)$.

To prove part (ii), for $y \in B^{\triangleright}(x,r,t)$ and $z \in B^{\triangleright}(y,\bar{r},\bar{t})$, we have

$$M(z,y,t-t_0) = M(z,y,\bar{t}) > 1-\bar{r} = r_1.$$

By the $\bowtie$-triangle inequality, we also have

$$M(z,x,t) \geq M(z,y,t-t_0) * M(y,x,t_0) = M(z,y,t-t_0) * r_0 \geq r_0 * r_1 \geq 1-s > 1-r.$$

This shows that $z \in B^{\triangleright}(x,r,t)$. Therefore, we obtain the inclusion $B^{\triangleright}(y,\bar{r},\bar{t}) \subseteq B^{\triangleright}(x,r,t)$. This completes the proof. □

According to Proposition 5, since M is assumed to be left-continuous with respect to the distance on $(0, \infty)$, it is not necessarily to consider the $\circ$-triangle inequality for $\circ \in \{\triangleright, \triangleleft, \diamond\}$ in Proposition 10.

Proposition 11. *(Symmetric Left-Continuity for M) Let (X, M) be a fuzzy semi-metric space along with a t-norm $*$ such that the following conditions are satisfied:*

- *M is symmetrically left-continuous with respect to the distance on $(0, \infty)$;*
- *the t-norm $*$ is left-continuous at 1 with respect to the first or second component.*

Suppose that M satisfies the $\bowtie$-triangle inequality. Then, we have the following inclusions:

(i) *Given any $y \in B^{\triangleleft}(x,r,t)$, there exists $B^{\triangleright}(y,\bar{r},\bar{t})$ such that $B^{\triangleright}(y,\bar{r},\bar{t}) \subseteq B^{\triangleright}(x,r,t)$.*
(ii) *Given any $y \in B^{\triangleright}(x,r,t)$, there exists $B^{\triangleleft}(y,\bar{r},\bar{t})$ such that $B^{\triangleleft}(y,\bar{r},\bar{t}) \subseteq B^{\triangleleft}(x,r,t)$.*

Proof. For $y \in B^{\triangleleft}(x,r,t)$, we have $M(x,y,t) > 1-r$. By part (ii) of Proposition 7, there exists t_0 with $0 < t_0 < t$ such that $M(y,x,t_0) > 1-r$. Let $r_0 = M(y,x,t_0)$. Then, we have $r_0 > 1-r$. There exists s with $0 < s < 1$ such that $r_0 > 1-s > 1-r$. By part (i) of Proposition 3, there exists r_1 with

$0 < r_1 < 1$ such that $r_0 * r_1 \geq 1 - s$. Let $\bar{r} = 1 - r_1$ and $\bar{t} = t - t_0$. Similarly, for $y \in B^{\triangleright}(x, r, t)$, we have $M(y, x, t) > 1 - r$. In this case, let $r_0 = M(x, y, t_0)$.

To prove part (i), for $y \in B^{\triangleleft}(x, r, t)$ and $z \in B^{\triangleright}(y, \bar{r}, \bar{t})$, we have

$$M(z, y, t - t_0) = M(z, y, \bar{t}) > 1 - \bar{r} = r_1.$$

By the $\bowtie$-triangle inequality, we have

$$M(z, x, t) \geq M(z, y, t - t_0) * M(y, x, t_0) = M(z, y, t - t_0) * r_0 \geq r_1 * r_0 \geq 1 - s > 1 - r.$$

This shows that $z \in B^{\triangleright}(x, r, t)$. Therefore, we obtain the inclusion $B^{\triangleright}(y, \bar{r}, \bar{t}) \subseteq B^{\triangleright}(x, r, t)$.
To prove part (ii), for $y \in B^{\triangleright}(x, r, t)$ and $z \in B^{\triangleleft}(y, \bar{r}, \bar{t})$, we have

$$M(y, z, t - t_0) = M(y, z, \bar{t}) > 1 - \bar{r} = r_1.$$

By the $\bowtie$-triangle inequality, we have

$$M(x, z, t) \geq M(x, y, t_0) * M(y, z, t - t_0) = r_0 * M(y, z, t - t_0) \geq r_1 * r_0 \geq 1 - s > 1 - r.$$

This shows that $z \in B^{\triangleleft}(x, r, t)$. Therefore, we obtain the inclusion $B^{\triangleleft}(y, \bar{r}, \bar{t}) \subseteq B^{\triangleleft}(x, r, t)$.
This completes the proof. □

According to Proposition 5, since M is assumed to be symmetrically left-continuous with respect to the distance on $(0, \infty)$, it is not necessarily to consider the $\circ$-triangle inequality for $\circ \in \{\triangleright, \triangleleft, \diamond\}$ in Proposition 11.

Proposition 12. *(Left-Continuity for M) Let (X, M) be a fuzzy semi-metric space along with a t-norm $*$ such that the following conditions are satisfied:*

- *M is left-continuous with respect to the distance on $(0, \infty)$;*
- *the t-norm $*$ is left-continuous at 1 with respect to the first or second component.*

Suppose that M satisfies the $\bowtie$-triangle inequality. We have the following inclusions:

(i) *If $x \in B^{\triangleleft}(x_1, r_1, t_1) \cap B^{\triangleleft}(x_2, r_2, t_2)$, then there exists $B^{\triangleleft}(x, r_3, t_3)$ such that*

$$B^{\triangleleft}(x, r_3, t_3) \subseteq B^{\triangleleft}(x_1, r_1, t_1) \cap B^{\triangleleft}(x_2, r_2, t_2). \tag{7}$$

(ii) *If $x \in B^{\triangleright}(x_1, r_1, t_1) \cap B^{\triangleright}(x_2, r_2, t_2)$, then there exists $B^{\triangleright}(x, r_3, t_3)$ such that*

$$B^{\triangleright}(x, r_3, t_3) \subseteq B^{\triangleright}(x_1, r_1, t_1) \cap B^{\triangleright}(x_2, r_2, t_2). \tag{8}$$

Proof. Using part (i) of Proposition 10, there exist $\bar{t}_1, \bar{t}_2, \bar{r}_1, \bar{r}_2$ such that

$$B^{\triangleleft}(x, \bar{r}_1, \bar{t}_1) \subseteq B^{\triangleleft}(x_1, r_1, t_1) \text{ and } B^{\triangleleft}(x, \bar{r}_2, \bar{t}_2) \subseteq B^{\triangleleft}(x_2, r_2, t_2).$$

We take $t_3 = \min\{\bar{t}_1, \bar{t}_2\}$ and $r_3 = \min\{\bar{r}_1, \bar{r}_2\}$. Then, for $y \in B^{\triangleleft}(x, r_3, t_3)$, using part (i) of Proposition 4, we have

$$M(x, y, \bar{t}_1) \geq M(x, y, t_3) > 1 - r_3 \geq 1 - \bar{r}_1$$

and

$$M(x, y, \bar{t}_2) \geq M(x, y, t_3) > 1 - r_3 \geq 1 - \bar{r}_2,$$

which say that

$$y \in B^{\triangleleft}(x, \bar{r}_1, \bar{t}_1) \cap B^{\triangleleft}(x, \bar{r}_2, \bar{t}_2) \subseteq B^{\triangleleft}(x_1, r_1, t_1) \cap B^{\triangleleft}(x_2, r_2, t_2).$$

Therefore, we obtain the inclusion of Equation (7). The second inclusion of Equation (8) can be similarly obtained. This completes the proof. □

Proposition 13. *(Symmetric Left-Continuity for M) Let (X, M) be a fuzzy semi-metric space along with a t-norm $*$ such that the following conditions are satisfied:*

- *M is symmetrically left-continuous with respect to the distance on $(0, \infty)$;*
- *the t-norm $*$ is left-continuous at 1 with respect to the first or second component.*

Suppose that M satisfies the $\bowtie$-triangle inequality. Then, we have the following inclusions:

(i) *If $x \in B^{\triangleleft}(x_1, r_1, t_1) \cap B^{\triangleleft}(x_2, r_2, t_2)$, then there exists $B^{\triangleleft}(x, r_3, t_3)$ such that*

$$B^{\triangleleft}(x, r_3, t_3) \subseteq B^{\triangleleft}(x_1, r_1, t_1) \cap B^{\triangleleft}(x_2, r_2, t_2). \tag{9}$$

(ii) *If $x \in B^{\triangleright}(x_1, r_1, t_1) \cap B^{\triangleright}(x_2, r_2, t_2)$, then there exists $B^{\triangleleft}(x, r_3, t_3)$ such that*

$$B^{\triangleright}(x, r_3, t_3) \subseteq B^{\triangleright}(x_1, r_1, t_1) \cap B^{\triangleright}(x_2, r_2, t_2). \tag{10}$$

Proof. Using part (iv) of Proposition 11, there exist $\bar{t}_1, \bar{t}_2, \bar{r}_1, \bar{r}_2$ such that

$$B^{\triangleright}(x, \bar{r}_1, \bar{t}_1) \subseteq B^{\triangleright}(x_1, r_1, t_1) \text{ and } B^{\triangleright}(x, \bar{r}_2, \bar{t}_2) \subseteq B^{\triangleright}(x_2, r_2, t_2).$$

We take $t_3 = \min\{\bar{t}_1, \bar{t}_2\}$ and $r_3 = \min\{\bar{r}_1, \bar{r}_2\}$. Then, for $y \in B^{\triangleleft}(x, r_3, t_3)$, using part (i) of Proposition 4, we have

$$M(y, x, \bar{t}_1) \geq M(y, x, t_3) > 1 - r_3 \geq 1 - \bar{r}_1$$

and

$$M(y, x, \bar{t}_2) \geq M(y, x, t_3) > 1 - r_3 \geq 1 - \bar{r}_2,$$

which say that

$$y \in B^{\triangleright}(x, \bar{r}_1, \bar{t}_1) \cap B^{\triangleright}(x, \bar{r}_2, \bar{t}_2) \subseteq B^{\triangleright}(x_1, r_1, t_1) \cap B^{\triangleright}(x_2, r_2, t_2).$$

Therefore, we obtain the inclusion of Equation (9). The second inclusion of Equation (10) can be similarly obtained. This completes the proof. □

The following proposition does not assume the left-continuity or symmetric left-continuity for M. Therefore, we can consider the different $\circ$-triangle inequality for $\circ \in \{\bowtie, \triangleright, \triangleleft, \diamond\}$.

Proposition 14. *Let (X, M) be a fuzzy semi-metric space along with a t-norm $*$ that is left-continuous at 1 with respect to the first or second component. Suppose that $x \neq y$. We have the following properties.*

(i) *Suppose that M satisfies the $\bowtie$-triangle inequality or the $\diamond$-triangle inequality. Then,*

$$B^{\triangleleft}(x, r, t) \cap B^{\triangleright}(y, r, t) = \emptyset \text{ and } B^{\triangleright}(x, r, t) \cap B^{\triangleleft}(y, r, t) = \emptyset$$

for some $r \in (0, 1)$ and $t > 0$.

(ii) *Suppose that M satisfies the $\triangleright$-triangle inequality. Then,*

$$B^{\triangleleft}(x, r, t) \cap B^{\triangleleft}(y, r, t) = \emptyset$$

for some $r \in (0, 1)$ and $t > 0$.

(iii) *Suppose that M satisfies the $\triangleleft$-triangle inequality. Then,*

$$B^{\triangleright}(x, r, t) \cap B^{\triangleright}(y, r, t) = \emptyset$$

for some $r \in (0,1)$ *and* $t > 0$.

Proof. Since $x \neq y$, there exists $t_0 > 0$ such that $M(x,y,t_0) < 1$. There also exists r_0 such that $M(x,y,t_0) < r_0 < 1$. By part (ii) of Proposition 3, there exists $\hat{r}$ with $0 < \hat{r} < 1$ such that $\hat{r} * \hat{r} > r_0$.

- Suppose that M satisfies the $\triangleright$-triangle inequality. We are going to prove that

$$B^{\triangleleft}\left(x,1-\hat{r},\frac{t_0}{2}\right) \cap B^{\triangleleft}\left(y,1-\hat{r},\frac{t_0}{2}\right) = \emptyset$$

by contradiction. Suppose that

$$z \in B^{\triangleleft}\left(x,1-\hat{r},\frac{t_0}{2}\right) \cap B^{\triangleleft}\left(y,1-\hat{r},\frac{t_0}{2}\right).$$

Since M satisfies the $\triangleright$-triangle inequality, it follows that

$$M(x,y,t_0) \geq M\left(x,z,\frac{t_0}{2}\right) * M\left(y,z,\frac{t_0}{2}\right) \geq \hat{r} * \hat{r} > r_0 > M(x,y,t_0),$$

which is a contradiction.

- Suppose that M satisfies the $\triangleleft$-triangle inequality for

$$z \in B^{\triangleright}\left(x,1-\hat{r},\frac{t_0}{2}\right) \cap B^{\triangleright}\left(y,1-\hat{r},\frac{t_0}{2}\right).$$

Since M satisfies the $\triangleleft$-triangle inequality, it follows that

$$M(x,y,t_0) \geq M\left(z,x,\frac{t_0}{2}\right) * M\left(z,y,\frac{t_0}{2}\right) \geq \hat{r} * \hat{r} > r_0 > M(x,y,t_0),$$

which is a contradiction.

- Suppose that M satisfies the $\bowtie$-triangle inequality for

$$z \in B^{\triangleleft}\left(x,1-\hat{r},\frac{t_0}{2}\right) \cap B^{\triangleright}\left(y,1-\hat{r},\frac{t_0}{2}\right).$$

Since M satisfies the $\bowtie$-triangle inequality, it follows that

$$M(x,y,t_0) \geq M\left(x,z,\frac{t_0}{2}\right) * M\left(z,y,\frac{t_0}{2}\right) \geq \hat{r} * \hat{r} > r_0 > M(x,y,t_0),$$

which is a contradiction. On the other hand, for

$$z \in B^{\triangleright}\left(x,1-\hat{r},\frac{t_0}{2}\right) \cap B^{\triangleleft}\left(y,1-\hat{r},\frac{t_0}{2}\right).$$

Since M satisfies the $\bowtie$-triangle inequality, it follows that

$$M(y,x,t_0) \geq M\left(y,z,\frac{t_0}{2}\right) * M\left(z,x,\frac{t_0}{2}\right) \geq \hat{r} * \hat{r} > r_0 > M(x,y,t_0),$$

which is a contradiction.

- Suppose that M satisfies the $\diamond$-triangle inequality. For

$$z \in B^{\triangleright}\left(x,1-\hat{r},\frac{t_0}{2}\right) \cap B^{\triangleleft}\left(y,1-\hat{r},\frac{t_0}{2}\right).$$

Since M satisfies the $\diamond$-triangle inequality, it follows that

$$M(x,y,t_0) \geq M\left(z,x,\frac{t_0}{2}\right) * M\left(y,z,\frac{t_0}{2}\right) \geq \hat{r} * \hat{r} > r_0 > M(x,y,t_0),$$

which is a contradiction. On the other hand, for

$$z \in B^{\triangleleft}\left(x,1-\hat{r},\frac{t_0}{2}\right) \cap B^{\triangleright}\left(y,1-\hat{r},\frac{t_0}{2}\right).$$

Since M satisfies the $\diamond$-triangle inequality, it follows that

$$M(y,x,t_0) \geq M\left(z,y,\frac{t_0}{2}\right) * M\left(x,z,\frac{t_0}{2}\right) \geq \hat{r} * \hat{r} > r_0 > M(x,y,t_0),$$

which is a contradiction.

This completes the proof. □

Theorem 1. *Let (X, M) be a fuzzy semi-metric space along with a t-norm $*$ that is left-continuous at 1 with respect to the first or second component. Suppose that M is left-continuous or symmetrically left-continuous with respect to the distance on $(0, \infty)$, and that M satisfies the $\bowtie$-triangle inequality.*

(i) *We define*

$$\tau^{\triangleleft} = \{O^{\triangleleft} \subseteq X : x \in O^{\triangleleft} \text{ if and only if there exist } t > 0 \text{ and } r \in (0,1) \text{ such that } B^{\triangleleft}(x,r,t) \subseteq O^{\triangleleft}\}.$$

Then, the family $\mathcal{B}^{\triangleleft}$ induces a T_1-space $(X, \tau^{\triangleleft})$ such that $\mathcal{B}^{\triangleleft}$ is a base for the topology $\tau^{\triangleleft}$, in which $O^{\triangleleft} \in \tau^{\triangleleft}$ if and only if, for each $x \in O^{\triangleleft}$, there exist $t > 0$ and $r \in (0,1)$ such that $B^{\triangleleft}(x,r,t) \subseteq O^{\triangleleft}$.

(ii) *We define*

$$\tau^{\triangleright} = \{O^{\triangleright} \subseteq X : x \in O^{\triangleright} \text{ if and only if there exist } t > 0 \text{ and } r \in (0,1) \text{ such that } B^{\triangleright}(x,r,t) \subseteq O^{\triangleright}\}.$$

Then, the family $\mathcal{B}^{\triangleright}$ induces a T_1-space $(X, \tau^{\triangleright})$ such that $\mathcal{B}^{\triangleright}$ is a base for the topology $\tau^{\triangleright}$, in which $O^{\triangleright} \in \tau^{\triangleright}$ if and only if, for each $x \in O^{\triangleright}$, there exist $t > 0$ and $r \in (0,1)$ such that $B^{\triangleright}(x,r,t) \subseteq O^{\triangleright}$.

Moreover, the T_1-spaces $(X, \tau^{\triangleleft})$ and $(X, \tau^{\triangleright})$ satisfy the first axiom of countability.

Proof. Using part (i) of Proposition 8, part (i) of Proposition 12 and part (i) of Proposition 13, we see that $\tau^{\triangleleft}$ is a topology such that $\mathcal{B}^{\triangleleft}$ is a base for $\tau^{\triangleleft}$. Part (ii) of Proposition 8 says that $(X, \tau^{\triangleleft})$ is a T_1-space. Part (i) of Proposition 9 says that there exist countable local bases at each $x \in X$ for $\tau^{\triangleright}$ and $\tau^{\triangleleft}$, respectively, which also says that $\tau^{\triangleright}$ and $\tau^{\triangleleft}$ satisfy the first axiom of countability. We can similarly obtain the desired results regarding the topology $\tau^{\triangleright}$. This completes the proof. □

According to Proposition 5, since M is assumed to be left-continuous or symmetrically left-continuous with respect to the distance on $(0, \infty)$, it follows that the topologies obtained in Wu [13] are still valid when we consider the $\circ$-triangle inequality for $\circ \in \{\triangleright, \triangleleft, \diamond\}$.

Proposition 15. *Let (X, M) be a fuzzy semi-metric space along with a t-norm $*$ that is left-continuous at 1 with respect to the first or second component. Suppose that M is left-continuous with respect to the distance on $(0, \infty)$, and that M satisfies the $\bowtie$-triangle inequality. Then, regarding the T_1-spaces $(X, \tau^{\triangleleft})$ and $(X, \tau^{\triangleright})$, $B^{\triangleleft}(x,r,t)$ is a $\tau^{\triangleleft}$-open set and $B^{\triangleright}(x,r,t)$ is a $\tau^{\triangleright}$-open set.*

Proof. Using part (i) of Proposition 10, we see that $B^{\triangleleft}(x,r,t)$ is a $\tau^{\triangleleft}$-open set and $B^{\triangleright}(x,r,t)$ is a $\tau^{\triangleright}$-open set. This completes the proof. □

5. Limits in Fuzzy Semi-Metric Space

Since the symmetric condition is not satisfied in the fuzzy semi-metric space, three kinds of limit concepts will also be considered in this paper by referring to Wu [14]. In this section, we shall study the consistency of limit concepts in the induced topologies, which was not addressed in Wu [13].

Let (X,d) be a metric space. If the sequence $\{x_n\}_{n=1}^{\infty}$ in (X,d) converges to x, i.e., $d(x_n,x) \to 0$ as $n \to \infty$, then it is denoted by $x_n \stackrel{d}{\longrightarrow} x$ as $n \to \infty$. In this case, we also say that x is a *d-limit* of the sequence $\{x_n\}_{n=1}^{\infty}$.

Definition 6. *Let (X,M) be a fuzzy semi-metric space, and let $\{x_n\}_{n=1}^{\infty}$ be a sequence in X.*

- *We write $x_n \stackrel{M^{\triangleright}}{\longrightarrow} x$ as $n \to \infty$ if and only if*

$$\lim_{n\to\infty} M(x_n,x,t) = 1 \text{ for all } t > 0.$$

In this case, we call x a $M^{\triangleright}$-limit of the sequence $\{x_n\}_{n=1}^{\infty}$.

- *We write $x_n \stackrel{M^{\triangleleft}}{\longrightarrow} x$ as $n \to \infty$ if and only if*

$$\lim_{n\to\infty} M(x,x_n,t) = 1 \text{ for all } t > 0.$$

In this case, we call x a $M^{\triangleleft}$-limit of the sequence $\{x_n\}_{n=1}^{\infty}$.

- *We write $x_n \stackrel{M}{\longrightarrow} x$ as $n \to \infty$ if and only if*

$$\lim_{n\to\infty} M(x_n,x,t) = \lim_{n\to\infty} M(x,x_n,t) = 1 \text{ for all } t > 0.$$

In this case, we call x a M-limit of the sequence $\{x_n\}_{n=1}^{\infty}$.

Proposition 16. *Let (X,M) be a fuzzy semi-metric space along with a t-norm $*$ that is left-continuous at 1 with respect to the first or second component, and let $\{x_n\}_{n=1}^{\infty}$ be a sequence in X.*

(i) *Suppose that M satisfies the $\bowtie$-triangle inequality or the $\circ$-triangle inequality. Then, we have the following properties:*

- *If $x_n \stackrel{M^{\triangleleft}}{\longrightarrow} x$ and $x_n \stackrel{M^{\triangleright}}{\longrightarrow} y$, then $x = y$.*
- *If $x_n \stackrel{M^{\triangleright}}{\longrightarrow} x$ and $x_n \stackrel{M^{\triangleleft}}{\longrightarrow} y$, then $x = y$.*

(ii) *Suppose that M satisfies the $\triangleleft$-triangle inequality. If $x_n \stackrel{M^{\triangleright}}{\longrightarrow} x$ and $x_n \stackrel{M^{\triangleright}}{\longrightarrow} y$, then $x = y$. In other words, the $M^{\triangleright}$-limit is unique.*

(iii) *Suppose that M satisfies the $\triangleright$-triangle inequality. If $x_n \stackrel{M^{\triangleleft}}{\longrightarrow} x$ and $x_n \stackrel{M^{\triangleleft}}{\longrightarrow} y$, then $x = y$. In other words, the $M^{\triangleleft}$-limit is unique.*

Proof. To prove the first case of part (i), we first assume that M satisfies the $\bowtie$-triangle inequality. For any $t > 0$, using the left-continuity of t-norm at 1, we have

$$M(x,y,t) \geq M\left(x,x_n,\frac{t}{2}\right) * M\left(x_n,y,\frac{t}{2}\right) \to 1 * 1 = 1,$$

which says that $x = y$. To prove the second case of part (i), we have

$$M(y,x,t) \geq M\left(y, x_n, \frac{t}{2}\right) * M\left(x_n, x, \frac{t}{2}\right) \to 1 * 1 = 1,$$

which says that $x = y$. Now suppose that M satisfies the $\diamond$-triangle inequality. Then, we have

$$M(y,x,t) \geq M\left(x_n, y, \frac{t}{2}\right) * M\left(x, x_n, \frac{t}{2}\right) \to 1 * 1 = 1,$$

and

$$M(x,y,t) \geq M\left(x_n, x, \frac{t}{2}\right) * M\left(y, x_n, \frac{t}{2}\right) \to 1 * 1 = 1.$$

Therefore, we can similarly obtain the desired result.

To prove part (ii), we have

$$M(x,y,t) \geq M\left(x_n, x, \frac{t}{2}\right) * M\left(x_n, y, \frac{t}{2}\right) \to 1 * 1 = 1,$$

which says that $x = y$. To prove part (iii), we have

$$M(x,y,t) \geq M\left(x, x_n, \frac{t}{2}\right) * M\left(y, x_n, \frac{t}{2}\right) \to 1 * 1 = 1,$$

which says that $x = y$. This completes the proof. □

Let (X, τ) be a topological space. The sequence $\{x_n\}_{n=1}^{\infty}$ in X converges to $x \in X$ with respect to the topology τ is denoted by $x_n \xrightarrow{\tau} x$ as $n \to \infty$, where the limit is unique when τ is a Hausdorff topology.

Remark 2. *Let (X, M) be a fuzzy semi-metric space along with a t-norm $*$ and be endowed with a topology $\tau^{\triangleright}$ given in Theorem 1. Let $\{x_n\}_{n=1}^{\infty}$ be a sequence in X. Since $\mathcal{B}^{\triangleright}$ is a base for $\tau^{\triangleright}$, it follows that $x_n \xrightarrow{\tau^{\triangleright}} x$ as $n \to \infty$, if and only if, given any $t > 0$ and $0 < r < 1$, there exists $n_{r,t}$ such that $x_n \in B^{\triangleright}(x,r,t)$ for all $n \geq n_{r,t}$. Since $x_n \in B^{\triangleright}(x,r,t)$ means $M(x_n, x, t) > 1 - r$, it says that $x_n \xrightarrow{\tau^{\triangleright}} x$ as $n \to \infty$, if and only if, given any $t > 0$ and $0 < r < 1$, there exists $n_{r,t}$ such that $M(x_n, x, t) > 1 - r$ for all $n \geq n_{r,t}$.*

Proposition 17. *Let (X, M) be a fuzzy semi-metric space along with a t-norm $*$. Suppose that M is left-continuous or symmetrically left-continuous with respect to the distance on $(0, \infty)$, and that M satisfies the strong $\bowtie$-triangle inequality. Then, the following statements hold true:*

(i) *Let $\tau^{\triangleright}$ be the topology induced by $(X, M, *)$, and let $\{x_n\}_{n=1}^{\infty}$ be a sequence in X. Then, $x_n \xrightarrow{\tau^{\triangleright}} x$ as $n \to \infty$ if and only if $x_n \xrightarrow{M^{\triangleright}} x$ as $n \to \infty$.*
(ii) *Let $\tau^{\triangleleft}$ be the topology induced by $(X, M, *)$, and let $\{x_n\}_{n=1}^{\infty}$ be a sequence in X. Then, $x_n \xrightarrow{\tau^{\triangleleft}} x$ as $n \to \infty$ if and only if $x_n \xrightarrow{M^{\triangleleft}} x$ as $n \to \infty$.*

Proof. Under the assumptions, Theorem 1 says that we can induce two topologies $\tau^{\triangleright}$ and $\tau^{\triangleleft}$. It suffices to prove part (i). Suppose that $x_n \xrightarrow{\tau^{\triangleright}} x$ as $n \to \infty$. Fixed $t > 0$, given any $\epsilon \in (0,1)$, there exists $n_{\epsilon,t} \in \mathbb{N}$ such that $x_n \in B^{\triangleright}(x, \epsilon, t)$ for all $n \geq n_{\epsilon,t}$, which says that $M(x_n, x, t) > 1 - \epsilon$, i.e., $0 \leq 1 - M(x_n, x, t) < \epsilon$ for all $n \geq n_{\epsilon,t}$. Therefore, we obtain $M(x_n, x, t) \to 1$ as $n \to \infty$. Conversely, given any $t > 0$, if $M(x_n, x, t) \to 1$ as $n \to \infty$, then, given any $\epsilon \in (0,1)$, there exists $n_{\epsilon,t} \in \mathbb{N}$ such that $1 - M(x_n, x, t) < \epsilon$, i.e., $M(x_n, x, t) > 1 - \epsilon$ for all $n \geq n_{\epsilon,t}$, which says that $x_n \in B^{\triangleright}(x, \epsilon, t)$ for all $n \geq n_{\epsilon,t}$. This shows that $x_n \xrightarrow{\tau^{\triangleright}} x$ as $n \to \infty$, and the proof is complete. □

Let (X, M) be a fuzzy semi-metric space. We consider the following sets

$$\bar{B}^{\triangleleft}(x,r,t) = \{y \in X : M(x,y,t) \geq 1 - r\}$$

and

$$\bar{B}^{\triangleright}(x,r,t) = \{y \in X : M(y,x,t) \geq 1 - r\}.$$

If the symmetric condition is satisfied, then we simply write $\bar{B}(x,r,t)$. We are going to consider the closeness of $\bar{B}^{\triangleleft}(x,r,t)$ and $\bar{B}^{\triangleright}(x,r,t)$.

Proposition 18. *Let (X, M) be a fuzzy semi-metric space along with a t-norm $*$ that is left-continuous at 1 with respect to the first or second component. Suppose that M is continuous or symmetrically continuous with respect to the distance on $(0, \infty)$. If M satisfies the $\bowtie$-triangle inequality, then $\bar{B}^{\triangleleft}(x,r,t)$ and $\bar{B}^{\triangleright}(x,r,t)$ are $\tau^{\triangleright}$-closed and $\tau^{\triangleleft}$-closed, respectively. In other words, we have*

$$\tau^{\triangleright}\text{-cl}(\bar{B}^{\triangleleft}(x,r,t)) = \bar{B}^{\triangleleft}(x,r,t) \text{ and } \tau^{\triangleleft}\text{-cl}(\bar{B}^{\triangleright}(x,r,t)) = \bar{B}^{\triangleright}(x,r,t).$$

Proof. Under the assumptions, Theorem 1 says that we can induce two topologies $\tau^{\triangleright}$ and $\tau^{\triangleleft}$ satisfying the first axiom of countability. To prove the first case, for $y \in \tau^{\triangleright}\text{-cl}(\bar{B}^{\triangleleft}(x,r,t))$, since $(X, \tau^{\triangleright})$ satisfies the first axiom of countability, there exists a sequence $\{y_n\}_{n=1}^{\infty}$ in $\bar{B}^{\triangleleft}(x,r,t)$ such that $y_n \xrightarrow{\tau^{\triangleright}} y$ as $n \to \infty$. We also have $M(x, y_n, t) \geq 1 - r$ for all n. By Proposition 17, we have $M(y_n, y, t) \to 1$ as $n \to \infty$ for all $t > 0$. Given any $\epsilon > 0$, the $\bowtie$-triangle inequality says that

$$M(x,y,t+\epsilon) \geq M(x,y_n,t) * M(y_n,y,\epsilon) \geq (1-r) * M(y_n,y,\epsilon).$$

Since the t-norm $*$ is left-continuous at 1 with respect to each component by Remark 1, we obtain

$$M(x,y,t+\epsilon) \geq (1-r) * \left(\lim_{n\to\infty} M(y_n,y,\epsilon)\right) = (1-r) * 1 = 1 - r.$$

By the right-continuity of M, we also have

$$M(x,y,t) = \lim_{\epsilon \to 0+} M(x,y,t+\epsilon) \geq 1 - r,$$

which says that $y \in \bar{B}^{\triangleleft}(x,r,t)$.

To prove the second case, for $y \in \tau^{\triangleleft}\text{-cl}(\bar{B}^{\triangleright}(x,r,t))$, since $(X, \tau^{\triangleleft})$ satisfies the first axiom of countability, there exists a sequence $\{y_n\}_{n=1}^{\infty}$ in $\bar{B}^{\triangleright}(x,r,t)$ such that $y_n \xrightarrow{\tau^{\triangleleft}} y$ as $n \to \infty$. We also have $M(y_n, x, t) \geq 1 - r$ for all n. By Proposition 17, we have $M(y, y_n, t) \to 1$ as $n \to \infty$ for all $t > 0$. Given any $\epsilon > 0$, the $\bowtie$-triangle inequality says that

$$M(y,x,t+\epsilon) \geq M(y,y_n,\epsilon) * M(y_n,x,t) \geq M(y,y_n,\epsilon) * (1-r).$$

Since the t-norm $*$ is left-continuous at 1 with respect to each component by Remark 1, we obtain

$$M(y,x,t+\epsilon) \geq \left(\lim_{n\to\infty} M(y,y_n,\epsilon)\right) * (1-r) = 1 * (1-r) = 1 - r.$$

By the right-continuity of M, we also have

$$M(y,x,t) = \lim_{\epsilon \to 0+} M(y,x,t+\epsilon) \geq 1 - r,$$

which says that $y \in \bar{B}^{\triangleright}(x,r,t)$. This completes the proof. □

6. Conclusions

In fuzzy metric space, the triangle inequality plays an important role. In general, since the symmetric condition is not necessarily to be satisfied, the so-called fuzzy semi-metric space is proposed in this paper. In this situation, four different types of triangle inequalities are proposed and studied. The main purpose of this paper is to establish the T_1-spaces that are induced by the fuzzy semi-metric spaces along with the special kind of triangle inequality.

On the other hand, the limit concepts in fuzzy semi-metric space are also proposed and studied in this paper. Since the symmetric condition is not satisfied, three kinds of limits in fuzzy semi-metric space are considered. The concepts of uniqueness for the limits are also studied. Finally, we present the consistency of limit concepts in the induced T_1-spaces.

Funding: This research received no external funding.

Acknowledgments: The author would like to thank the reviewers for providing the useful suggestions that improve the presentation of this paper.

Conflicts of Interest: The author declares no conflict of interest.

References

1. Kaleva, O.; Seikkala, S. On Fuzzy Metric Spaces. *Fuzzy Sets Syst.* **1984**, *12*, 215–229. [CrossRef]
2. Schweizer, B.; Sklar, A. Statistical Metric Spaces. *Pac. J. Math.* **1960**, *10*, 313–334. [CrossRef]
3. Schweizer, B.; Sklar, A.; Thorp, E. The Metrization of Statistical Metric Spaces. *Pac. J. Math.* **1960**, *10*, 673–675. [CrossRef]
4. Schweizer, B.; Sklar, A. Triangle Inequalities in a Class of Statistical Metric Spaces. *J. Lond. Math. Soc.* **1963**, *38*, 401–406. [CrossRef]
5. Hadžić, O.; Pap, E. *Fixed Point Theory in Probabilistic Metric Spaces*; Klumer Academic Publishers: Norwell, MA, USA, 2001.
6. Chang, S.S.; Cho, Y.J.; Kang, S.M. *Nonlinear Operator Theory in Probabilistic Metric Space*; Nova Science Publishers: New York, NY, USA, 2001.
7. Kramosil, I.; Michalek, J. Fuzzy Metric and Statistical Metric Spaces. *Kybernetika* **1975**, *11*, 336–344.
8. George, A.; Veeramani, P. On Some Results in Fuzzy Metric Spaces. *Fuzzy Sets Syst.* **1994**, *64*, 395–399. [CrossRef]
9. George, A.; Veeramani, P. On Some Results of Analysis for Fuzzy Metric Spaces. *Fuzzy Sets Syst.* **1997**, *90*, 365–368. [CrossRef]
10. Gregori, V.; Romaguera, S. Some Properties of Fuzzy Metric Spaces. *Fuzzy Sets Syst.* **2002**, *115*, 399–404. [CrossRef]
11. Gregori, V.; Romaguera, S. Fuzzy Quasi-Metric Spaces. *Appl. Gen. Topol.* **2004**, *5*, 129–136. [CrossRef]
12. Gregori, V.; Romaguera, S.; Sapena, A. A Note on Intuitionistic Fuzzy Metric Spaces. *Chaos Solitons Fractals* **2006**, *28*, 902–905. [CrossRef]
13. Wu, H.-C. Hausdorff Topology Induced by the Fuzzy Metric and the Fixed Point Theorems in Fuzzy Metric Spaces. *J. Korean Math. Soc.* **2015**, *52*, 1287–1303. [CrossRef]
14. Wu, H.-C. Common Coincidence Points and Common Fixed Points in Fuzzy Semi-Metric Spaces. *Mathematics* **2018**, *6*, 29. [CrossRef]

mathematics

MDPI

Article

Nilpotent Fuzzy Subgroups

Elaheh Mohammadzadeh [1], Rajab Ali Borzooei [2,*]

[1] Department of Mathematics, Payame Noor University, Tehran 19395-3697, Iran; mohamadzadeh36@gmail.com
[2] Department of Mathematics, Shahid Beheshti University, G. C., Tehran 19395-3697, Iran
* Correspondence: borzooei@sbu.ac.ir

Received: 18 December 2017; Accepted: 13 February 2018; Published: 19 February 2018

Abstract: In this paper, we introduce a new definition for nilpotent fuzzy subgroups, which is called the good nilpotent fuzzy subgroup or briefly g-nilpotent fuzzy subgroup. In fact, we prove that this definition is a good generalization of abstract nilpotent groups. For this, we show that a group G is nilpotent if and only if any fuzzy subgroup of G is a g-nilpotent fuzzy subgroup of G. In particular, we construct a nilpotent group via a g-nilpotent fuzzy subgroup. Finally, we characterize the elements of any maximal normal abelian subgroup by using a g-nilpotent fuzzy subgroup.

Keywords: nilpotent group; nilpotent fuzzy subgroup; generalized nilpotent fuzzy subgroup

1. Introduction

Applying the concept of fuzzy sets of Zadeh [1] to group theory, Rosenfeld [2] introduced the notion of a fuzzy subgroup as early as 1971. Within a few years, it caught the imagination of algebraists like wildfire and there seems to be no end to its ramifications. With appropriate definitions in the fuzzy setting, most of the elementary results of group theory have been superseded with a startling generalized effect (see [3–5]). In [6] Dudek extended the concept of fuzzy sets to the set with one n-ary operation i.e., to the set G with one operation on $f : G \longrightarrow G$, where $n \geq 2$. Such defined groupoid will be denoted by (G, f). Moreover, he introduced the notion of a fuzzy subgroupoid of an n-ary groupoid. Specially, he proved that if every fuzzy subgroupoid μ defined on (G, f) has the finite image, then every descending chain of subgroupoids of (G, f) terminates at finite step. One of the important concept in the study of groups is the notion of nilpotency. In [7] Kim proposed the notion of a nilpotent fuzzy subgroup. There, he attached to a fuzzy subgroup an ascending series of subgroups of the underlying group to define nilpotency of the fuzzy subgroup. With this definition, the nilpotence of a group can be completely characterized by the nilpotence of its fuzzy subgroups. Then, in [8] Guptaa and Sarmahas, defined the commutator of a pair of fuzzy subsets of a group to generate the descending central chain of fuzzy subgroups of a given fuzzy subgroup and they proposed a new definition of a nilpotent fuzzy subgroup through its descending central chain. Specially, They proved that every Abelian (see [9]) fuzzy subgroup is nilpotent. There are many natural generalizations of the notion of a normal subgroup. One of them is subnormal subgroup. The new methods are important to guarantee some properties of the fuzzy sets; for example, see [10]. In [3] Kurdachenko and et all formulated this concept for fuzzy subgroups to prove that if every fuzzy subgroup of γ is subnormal in γ with defect at most d, then γ is nilpotent ([3] Corollary 4.6). Finally in [11,12] Borzooei et. al. defind the notions of Engel fuzzy subgroups (subpolygroups) and investigated some related results. Now, in this paper we define the ascending series differently with Kim's definition. We then propose a definition of a nilpotent fuzzy subgroup through its ascending central series and call it g-nilpotent fuzzy subgroups. Also, we show that each g-nilpotent fuzzy subgroup is nilpotent. Moreover, we get the main results of nilpotent fuzzy subgroups with our definition. Basically this definition help us with the fuzzification of much more properties of nilpotent groups. Furthermore, we prove that for a fuzzy subgroup μ of G, $\{x \in G \mid \mu([x, y_1, ..., y_n]) = \mu(e) \ for \ any \ y_1, ..., y_n \in G\}$ is equal to the $n-$th term of ascending series where $[x, y_1] = x^{-1}y_1^{-1}xy_1$ and $[x, y_1, ..., y_n] = [[x, y_1, ..., y_{n-1}], y_n]$. Therefore, we have a complete analogy concept of nilpotent groups of an abstract group. Specially, we prove that a finite maximal normal subgroup can control the g-nilpotent fuzzy subgroup and makes it finite.

2. Preliminary

Let G be any group and $x, y \in G$. Define the n-commutator $[x,_n y]$, for any $n \in \mathbb{N}$ and $x, y \in G$, by $[x,_0 y] = x$, $[x,_1 y] = x^{-1}y^{-1}xy$ and $[x,_n y] = [[x,_{n-1} y], y]$ also, for any $y_1, ..., y_n \in G$, $[x, y_1, ..., y_n] = [[x, y_1, ..., y_{n-1}], y_n]$. For any $x, g \in G$, we consider $x^g = g^{-1}xg$ and $[x, y] = [x,_1 y]$.

Theorem 1. [13] *Let G be a group and $x, y, z \in G$. Then*
(1) $[x, y] = [y, x]^{-1}$,
(2) $[x.y, z] = [x, z]^y.[y, z]$ and $[x, y.z] = [x, z].[x, y]^z$,
(3) $[x, y^{-1}] = ([x, y]^{y^{-1}})^{-1}$ and $[x^{-1}, y] = ([x, y]^{x^{-1}})^{-1}$.
Note that $x^g = x.[x, g]$.

Definition 1. [13] *Let $X_1, X_2, ...$ be nonempty subsets of a group G. Define the commutator subgroup of X_1 and X_2 by*

$$[X_1, X_2] = \langle [x_1, x_2] \mid x_1 \in X_1, x_2 \in X_2 \rangle.$$

More generally, define

$$[X_1, ..., X_n] = [[X_1, ..., X_{n-1}], X_n]$$

where $n \geq 2$ and $[X_1] = \langle X_1 \rangle$. Also recall that $X_1^{X_2} = \langle x_1^{x_2} \mid x_1 \in X_1, x_2 \in X_2 \rangle$

Definition 2. [1] *A fuzzy subset μ of X is a function $\mu : X \to [0, 1]$.*

Also, for fuzzy subsets μ_1 and μ_2 of X, then μ_1 is *smaller* than μ_2 and write $\mu_1 \leq \mu_2$ iff for all $x \in X$, we have $\mu_1(x) \leq \mu_2(x)$. Also, $\mu_1 \vee \mu_2$ and $\mu_1 \wedge \mu_2$, for any μ_1, μ_2 are defined as follows:
$(\mu_1 \vee \mu_2)(x) = max\{\mu_1(x), \mu_2(x)\}$, $(\mu_1 \wedge \mu_2)(x) = min\{\mu_1(x), \mu_2(x)\}$, for any $x \in X$.

Definition 3. [14] *Let f be a function from X into Y, and μ be a fuzzy subset of X. Define the fuzzy subset $f(\mu)$ of Y, for any $y \in Y$, by*

$$(f(\mu))(y) = \begin{cases} \bigvee_{x \in f^{-1}(y)} \mu(x), & f^{-1}(y) \neq \phi \\ 0, & otherwise \end{cases}$$

Definition 4. [2] *Let μ be a fuzzy subset of a group G. Then μ is called a fuzzy subgroup of G if for any $x, y \in G$; $\mu(xy) \geq \mu(x) \wedge \mu(y)$, and $\mu(x^{-1}) \geq \mu(x)$. A fuzzy subgroup μ of G is called normal if $\mu(xy) = \mu(yx)$, for any x, y in G. It is easy to prove that a fuzzy subgroup μ is normal if and only if $\mu(x) = \mu(y^{-1}xy)$, for any $x, y \in G$ (See [14]).*

Theorem 2. [14] *Let μ be a fuzzy subgroup of G. Then for any $x, y \in G$, $\mu(x) \neq \mu(y)$, implies $\mu(xy) = \mu(x) \wedge \mu(y)$. Moreover, for a normal subgroup N of G, fuzzy subset ξ of $\frac{G}{N}$ as the following definition:*

$$\xi(xN) = \bigvee_{z \in xN} \mu(z), \ for \ any \ x \in G$$

is a fuzzy subgroup of $\frac{G}{N}$.

Definition 5. [14] *Let* μ *be a fuzzy subset of a semigroup* G. *Then* $Z(\mu)$ *is define as follows:*

$$Z(\mu) = \{x \in G \mid \mu(xy) = \mu(yx) \text{ and } \mu(xyz) = \mu(yxz), \text{ for any } y, z \in G\}$$

If $Z(\mu) = G$, *then* μ *is called a commutative fuzzy subset of* G.

Note that since $\mu(xy) = \mu(yx)$ then we have $\mu(xyz) = \mu(x(yz)) = \mu((yz)x) = \mu(yzx)$.

Theorem 3. [14] *Let* μ *be a fuzzy subset of a semigroup* G. *If* $Z(\mu)$ *is nonempty, then* $Z(\mu)$ *is a subsemigroup of* G. *Moreover, if* G *is a group, then* $Z(\mu)$ *is a normal subgroup of* G.

We recall the notion of the ascending central series of a fuzzy subgroup and a nilpotent fuzzy subgroup of a group [14]. Let μ be a fuzzy subgroup of a group G and $Z^0(\mu) = \{e\}$. Clearly $\{e\}$ is a normal subgroup of G. Let π_0 be the natural homomorphism of G onto $\frac{G}{Z^0(\mu)}$. It is clear that $\pi_0 = I$. Suppose that $Z^1(\mu) = \pi_0^{-1}(Z(\pi_0(\mu)))$. Since $Z(\pi_0(\mu))$ is a normal subgroup of $\frac{G}{Z^0(\mu)}$, then it is clear that $Z^1(\mu)$ is a normal subgroup of G. Also we see that $Z^1(\mu) = Z(\mu)$. Now let π_1 be the natural homomorphism of G onto $\frac{G}{Z^1(\mu)}$ and $Z^2(\mu) = \pi_1^{-1}(Z(\pi_1(\mu)))$. Since $\pi_1(\mu)$ is a fuzzy subgroup of $\frac{G}{Z^1(\mu)}$, then $Z(\pi_1(\mu))$ is a normal subgroup of $\frac{G}{Z^1(\mu)}$, which implies that $Z^2(\mu)$ is a normal subgroup of G. Similarly suppose that $Z^i(\mu)$ has been defined and so $Z^i(\mu)$ is a normal subgroup of G, for $i \in \mathbb{N} \cup \{0\}$. Let π_i be the natural homomorphism of G onto $\frac{G}{Z^i(\mu)}$ and $Z^{i+1}(\mu) = \pi_i^{-1}(Z(\pi_i(\mu)))$. Then $Z^{i+1}(\mu)$ is a normal subgroup of G. Since $1_{\frac{G}{Z^i(\mu)}} \subseteq Z(\pi_i(\mu)$, then $\pi_i^{-1}(1_{\frac{G}{Z^i(\mu)}}) \subseteq \pi_i^{-1}(Z(\pi_i(\mu)))$. Therefore, $Ker(\pi_i) = Z^i(\mu) \subseteq Z^{i+1}(\mu)$, for $i = 0, 1, \ldots$.

Definition 6. [14] *Let* μ *be a fuzzy subgroup of a group* G. *The ascending central series of* μ *is defined to be the ascending chain of normal subgroups of* G *as follows:*

$$Z^0(\mu) \subseteq Z^1(\mu) \subseteq Z^2(\mu) \ldots$$

Now the fuzzy subgroup μ *of* G *is called nilpotent if there exists a nonnegative integer* m, *such that* $Z^m(\mu) = G$. *The smallest such integer is called the class of* μ.

Theorem 4. [14] *Let* μ *be a fuzzy subgroup of a group* G, $i \in \mathbb{N}$ *and* $x \in G$. *If* $xyx^{-1}y^{-1} \in Z^{i-1}(\mu)$, *for any* $y \in G$, *then* $x \in Z^i(\mu)$. *Moreover, if* $T = \{x \in G \mid \mu(xyx^{-1}y^{-1}) = \mu(e), \text{for any } y \in G\}$, *then* $T = Z(\mu)$.

Let G be a group. We know that $Z(G)$ is a normal subgroup of G. Let $Z_2(G)$ be the inverse image of $Z(\frac{G}{Z(G)})$, under the canonical projection $G \longrightarrow \frac{G}{Z(G)}$. Then $Z_2(G)$ is normal in G and contains $Z(G)$. Continue this process by defining inductively, $Z_1(G) = Z(G)$ and $Z_i(G)$ is the inverse image of $Z(\frac{G}{Z_{i-1}(G)})$ under the canonical projection $G \longrightarrow \frac{G}{Z_{i-1}(G)}$ for any $i \in \mathbb{N}$. Thus we obtain a sequence of normal subgroups of G, called the ascending central series of G that is, $\{e\} = Z_0(G) \subseteq Z_1(G) \subseteq Z_2(G) \subseteq \ldots$. The other definition is as follows [13]: Let G be a group and $Z_0(G) = \{e\}$. It is clear that $\{e\}$ is a normal subgroup of G. Put $\frac{Z_1(G)}{\{e\}} = Z(\frac{G}{\{e\}})$. Then $Z_1(G) = Z(G)$ is a normal subgroup of G. Similarly for any integer $n > 1$, put $\frac{Z_n(G)}{Z_{n-1}(G)} = Z(\frac{G}{Z_{n-1}(G)})$. Then $Z_i(\mu)$ is called the *i-th center* of group G. Moreover, $\{e\} = Z_0(G) \subseteq Z_1(G) \subseteq Z_2(G) \subseteq \ldots$ is called upper central series of G. These two definitions are equivalent since, $\pi(Z_2(G)) = \pi(\pi^{-1}(Z(\frac{G}{Z(G)}))) = Z(\frac{G}{Z(G)})$. Thus $\frac{Z_2(G)}{Z(G)} = Z(\frac{G}{Z(G)})$. Similarly we get the result for any $n \in \mathbb{N}$.

Theorem 5. [13] *Let G be a group and $n \in \mathbb{N}$. Then*
(i) $x \in Z_n(G)$ if and only if for any $y_i \in G$ where $1 \le i \le n$, $[x, y_1, ..., y_n] = e$,
(ii) $[Z_n(G), G] \subseteq Z_{n-1}(G)$.
(iii) Class of nilpotent groups is closed with respect to subgroups and homomorphic images.

Notation. From now on, in this paper we let G be a group.

3. Good Nilpotent Fuzzy Subgroups

One of the important concept in the study of groups is the notion of nilpotency. It was introduced for fuzzy subgroups, too (See [14]). Now, in this section we give a new definition of nilpotent fuzzy subgroups which is similar to one in the abstract group theory. It is a good generation of the last one. With this nilpotency we get some new main results.

Let μ be a fuzzy subgroup of G. Put $Z_0(\mu) = \{e\}$. Clearly $Z_0(\mu) \trianglelefteq G$. Let $Z_1(\mu) = \{x \in G \mid \mu([x,y]) = \mu(e), \text{for any } y \in G\}$. Now using Theorems 4, we have $Z_1(\mu) = Z(\mu)$ is a normal subgroup of G. We define a subgroup $Z_2(\mu)$ of G such that $\frac{Z_2(\mu)}{Z_1(\mu)} = Z(\frac{G}{Z_1(\mu)})$; Since $Z_1(\mu) \trianglelefteq G$ then $Z_1(\mu) \trianglelefteq Z_2(\mu)$. We show that $[Z_2(\mu), G] \subseteq Z_1(\mu)$. For this let $x \in Z_2(\mu)$ and $g \in G$. Thus $xZ_1(\mu) \in \frac{Z_2(\mu)}{Z_1(\mu)} = Z(\frac{G}{Z_1(\mu)})$, which implies that $[xZ_1(\mu), gZ_1(\mu)] = Z_1(\mu)$ for any $g \in G$. Therefore $[x, g] \in Z_1(\mu)$. Hence $[Z_2(\mu), G] \subseteq Z_1(\mu)$. Therefore $x^g = x[x, g] \in Z_2(\mu)$. Thus $Z_2(\mu) \trianglelefteq G$. Similarly for $k \ge 2$ we define a normal subgroup $Z_k(\mu)$ such that $\frac{Z_k(\mu)}{Z_{k-1}(\mu)} = Z(\frac{G}{Z_{k-1}(\mu)})$. It is clear that $Z_0(\mu) \subseteq Z_1(\mu) \subseteq Z_2(\mu) \subseteq ....$

Definition 7. *A fuzzy subgroup μ of G is called a good nilpotent fuzzy subgroup of G or briefly g-nilpotent fuzzy subgroup of G if there exists a none negative integer n, such that $Z_n(\mu) = G$. The smallest such integer is called the class of μ.*

Example 1. *Let $D_3 = \langle a, b; a^3 = b^2 = e, ba = a^2b\rangle$ be the dihedral group with six element and $t_0, t_1 \in [0, 1]$ such that $t_0 > t_1$. Define a fuzzy subgroup μ of D_3 as follows:*

$$\mu(x) = \begin{cases} t_0 & if \quad x \in < a > \\ t_1 & if \quad x \notin < a > \end{cases}$$

Then $(D_3 \backslash \langle a\rangle)(D_3 \backslash \langle a\rangle) = \langle a\rangle$, $(\langle a\rangle)(D_3 \backslash \langle a\rangle) = (D_3 \backslash \langle a\rangle)$, $(D_3 \backslash \langle a\rangle)(\langle a\rangle) = (D_3 \backslash \langle a\rangle)$ and $(\langle a\rangle)(\langle a\rangle) = (\langle a\rangle)$. Now, we show that $Z_1(\mu) = D_3$. If $x \in \langle a\rangle$ and $y \notin \langle a\rangle$, then $xy \notin \langle a\rangle$. Thus by the above relations, we have $[x, y] = x^{-1}y^{-1}xy = (yx)^{-1}(xy) \in \langle a\rangle$, which implies that $\mu[x, y] = t_0 = \mu(e)$. Similarly, for the cases $x \notin \langle a\rangle$ and $y \in \langle a\rangle$ or $x, y \in \langle a\rangle$ or $x, y \notin \langle a\rangle$, we have $\mu[x, y] = \mu(e)$. Hence for any $x, y \in D_3$, $\mu[x, y] = \mu(e)$ and so by Theorem 4, $Z(\mu) = D_3$. Now, since $Z_1(\mu) = Z(\mu)$, we get μ is g-nilpotent fuzzy subgroup.

In the following we see that for $n \in \mathbb{N}$, each normal subgroup $Z_n(\mu)$, in which is defined by $\frac{Z_{n+1}(\mu)}{Z_n(\mu)} = Z(\frac{G}{Z_n(\mu)})$ is equal to $\{x \in G \mid \mu([x, y_1, ..., y_n]) = \mu(e), \ for \ any \ y_1, y_2, ..., y_n \in G\}$.

Lemma 1. *Let μ be a fuzzy subgroup of G. Then for $k \in \mathbb{N}$*

$$Z_k(\mu) = \{x \in G \mid \mu([x, y_1, ..., y_k]) = \mu(e), \ for \ any \ y_1, y_2, ..., y_k \in G\}.$$

Proof. We prove it by induction on k. If $k = 1$, then by definition of $Z_1(\mu)$ we have $Z_1(\mu) = \{x \in G \mid \mu([x,y]) = \mu(e)$ for any $y \in G\}$. Now let $k = n+1$, and the result is true for $k \le n$. Then

$$\begin{aligned} x \in Z_{n+1}(\mu) &\iff xZ_n(\mu) \in \frac{Z_{n+1}(\mu)}{Z_n(\mu)} = Z(\frac{G}{Z_n(\mu)}) \\ &\iff [xZ_n(\mu), y_1 Z_n(\mu)] = Z_n(\mu),\ for\ any\ y_1 \in G \\ &\iff [x,y_1]Z_n(\mu) = Z_n(\mu),\ for\ any\ y_1 \in G \\ &\iff [x,y_1] \in Z_n(\mu),\ for\ any\ y_1 \in G \\ &\iff \mu([[x,y_1],y_2,...,y_{n+1}]) = \mu(e),\ for\ any\ y_1,...,y_{n+1} \in G. \end{aligned}$$

This complete the proof. □

Theorem 6. *Any g-nilpotent fuzzy subgroup of G is a nilpotent fuzzy subgroup.*

Proof. Let fuzzy subgroup μ of G be g-nilpotent. Since $Z_1(\mu) = Z(\mu) = Z^1(\mu)$, for $n = 1$, the proof is true. Now let $Z_{n+1}(\mu) = G$. Then by Lemma 1, $\{x \mid \mu([x,y_1,...,y_{n+1}]) = \mu(e)\ for\ any\ y_1,y_2,...,y_{n+1} \in G\} = G$. We should prove that $Z^{n+1}(\mu) = G$. Let $x \in G$. Then $\mu([x,y_1,...,y_{n+1}]) = \mu(e)$, for any $y_1,...,y_{n+1} \in G$. Therefore by Theorem 4, $[x,y_1,...,y_n] \in Z(\mu)$. Consequently, by Theorem 4, $[x,y_1,...,y_{n-1}] \in Z^2(\mu)$. Similarly, by using k-times Theorem 4, we have $x \in Z^{n+1}(\mu)$ and so $Z^{n+1}(\mu) = G$. Therefore μ is a nilpotent fuzzy subgroup of G. □

Theorem 7. *Let μ be a fuzzy subgroup of G. Then μ is commutative if and only if μ is g-nilpotent fuzzy subgroup of class 1.*

Proof. ($\Rightarrow$) Let μ be commutative. Then $Z(\mu) = G$. Since $Z_1(\mu) = Z(\mu)$, then $Z_1(\mu) = G$ which implies that μ is g-nilpotent of class 1.
($\Leftarrow$) If μ is g-nilpotent of class 1, then $Z_1(\mu) = G$. Hence $Z_1(\mu) = Z(\mu) = G$. Therefore, μ is commutative. □

Notation. If μ is a fuzzy subgroup of G, then $Z_{k-1}(\frac{G}{Z(\mu)})$ means the $(k-1)$-th center of $\frac{G}{Z(\mu)}$ ([15]).

Next we see that a g-nilpotent fuzzy subgroup of G makes the g-nilpotent fuzzy subgroup of $\frac{G}{Z(\mu)}$. For this, we need the following two Lemmas.

Lemma 2. *Let μ be a fuzzy subgroup of G. Then for any $k \in \mathbb{N}$, $\frac{Z_k(\mu)}{Z(\mu)} = Z_{k-1}(\frac{G}{Z(\mu)})$.*

Proof. First we recall that for $i \in \mathbb{N}$, $x \in Z_i(G)$ if and only if $[x,y_1,...,y_i] = e$, for any $y_1,y_2,...,y_i \in G$ (See [13]). Hence

$$\begin{aligned} xZ(\mu) \in Z_{k-1}(\frac{G}{Z(\mu)}) &\iff [xZ(\mu), y_1Z(\mu),...,y_{k-1}Z(\mu)] = Z(\mu),\ for\ any\ y_1,...,y_{k-1} \in G \\ &\iff [x,y_1,...,y_{k-1}]Z(\mu) = Z(\mu),\ for\ any\ y_1,...,y_{k-1} \in G \\ &\iff [x,y_1,...,y_{k-1}] \in Z(\mu),\ for\ any\ y_1,...,y_{k-1} \in G \\ &\iff \mu[x,y_1,...,y_{k-1},y_k] = \mu(e),\ for\ any\ y_k \in G\ (by\ Theorem\ 4) \\ &\iff x \in Z_k(\mu)\ (by\ Lemma\ 1) \\ &\iff xZ(\mu) \in \frac{Z_k(\mu)}{Z(\mu)}. \end{aligned}$$

Therefore $\frac{Z_k(\mu)}{Z(\mu)} = Z_{k-1}(\frac{G}{Z(\mu)})$. □

Lemma 3. *Let μ be a fuzzy subgroup of G, $H = \frac{G}{Z(\mu)}$, $\overline{\mu}$ be a fuzzy subgroup of H and $N = Z(\overline{\mu})$. If H is nilpotent, then $\frac{H}{N}$ is nilpotent, too.*

Proof. Let H be nilpotent of class n, that is $Z_n(H) = H$. We will prove that there exist $m \leq n$ such that $Z_m(\frac{H}{N}) = \frac{H}{N}$. For this by Theorem 5, since $\frac{H}{N}$ is a homomorphic image of H, we get $\frac{H}{N}$ is nilpotent of class at most m. □

Theorem 8. *Let μ be a fuzzy subgroup of G and $\overline{\mu}$ be a fuzzy subgroup of $\frac{G}{Z(\mu)}$. If μ is a g-nilpotent fuzzy subgroup of class n, then $\overline{\mu}$ is a g-nilpotent fuzzy subgroup of class m, where $m \leq n$.*

Proof. Let μ be a g-nilpotent fuzzy subgroup of class n. Then $Z_n(\mu) = G$. Now we show that there exists $m \leq n$, such that $Z_m(\overline{\mu}) = \frac{G}{Z(\mu)}$. By Lemma 2, $Z_n(\mu) = G \Longleftrightarrow \frac{G}{Z(\mu)} = \frac{Z_n(\mu)}{Z(\mu)} = Z_{n-1}(\frac{G}{Z(\mu)})$, and similarly (put m instead of n and $\overline{\mu}$ instead of μ),

$$Z_m(\overline{\mu}) = \frac{G}{Z(\mu)} \Longleftrightarrow Z_{m-1}(\frac{\frac{G}{Z(\mu)}}{Z(\overline{\mu})}) = \frac{\frac{G}{Z(\mu)}}{Z(\overline{\mu})}$$

Consequently, it is enough to show that if $Z_{n-1}(\frac{G}{Z(\mu)}) = \frac{G}{Z(\mu)}$, then

$$Z_{m-1}(\frac{\frac{G}{Z(\mu)}}{Z(\overline{\mu})}) = \frac{\frac{G}{Z(\mu)}}{Z(\overline{\mu})}$$

It follows by Lemma 3 (put $H = \frac{G}{Z(\mu)}$ in Lemma 3).

□

We now consider homomorphic images and the homomorphic pre-image of g-nilpotent fuzzy subgroups.

Theorem 9. *Let H be a group, $f : G \longrightarrow H$ be an epimorphism and μ be a fuzzy subgroup of G. If μ is a g-nilpotent fuzzy subgroup, then $f(\mu)$ is a g-nilpotent fuzzy subgroup.*

Proof. First, we show that $f(Z_i(\mu)) \subseteq Z_i(f(\mu))$, for any $i \in \mathbb{N}$. Let $i \in \mathbb{N}$. Then $x \in f(Z_i(\mu))$ implies that $x = f(u)$, for some $u \in Z_i(\mu)$. Since f is epimorphism, hence for any $y_1, ..., y_n \in H$ we get $y_i = f(v_i)$ for some $v_i \in G$ where $1 \leq i \leq n$. Therefore $[x, y_1, ..., y_n] = [f(u), f(v_1), ..., f(v_n)]$ which implies that

$$(f(\mu))([x, y_1, ..., y_n]) = \bigvee_{f(z)=[x,y_1,...,y_n]} \mu(z) = \bigvee_{f(z)=f([u,v_1,...,v_n])} \mu(z)$$

Now, since $u \in Z_i(\mu)$, by Lemma 1, we get $\mu([u, v_1, ..., v_n]) = \mu(e_G)$. Therefore,

$$(f(\mu))([x, y_1, ..., y_n]) = \mu(e_G) = (f(\mu))(e_H)$$

Hence by Lemma 1, $x \in Z_i(f(\mu))$. Consequently, $f(Z_i(\mu)) \subseteq Z_i(f(\mu))$. Hence if μ is g-nilpotent, then there exists nonnegative integer n such that $Z_n(\mu) = G$ which implies that $f(Z_n(\mu)) = f(G)$. Therefore $Z_n(f(\mu)) = H$ which implies that $f(\mu)$ is g-nilpotent. □

Theorem 10. *Let H be a group, $f : G \longrightarrow H$ be an epimorphism and v be a fuzzy subgroup of H. Then v is a g-nilpotent fuzzy subgroup if and only if $f^{-1}(v)$ is a g-nilpotent fuzzy subgroup.*

Proof. First, we show that $Z_i(f^{-1}(\nu)) = f^{-1}(Z_i(\nu))$, for any $i \in \mathbb{N}$. Now, let $i \in \mathbb{N}$. Then by Lemma 1,

$$\begin{aligned} x \in Z_i(f^{-1}(\nu)) &\Longleftrightarrow (f^{-1}(\nu))([x, x_1, ..., x_i]) = (f^{-1}(\nu))(e), \ \ for \ any \ x_1, x_2, ..., x_i \in G \\ &\Longleftrightarrow \nu([f(x), ..., f(x_i)]) = \nu(e), \ \ for \ any \ x_1, x_2, ..., x_i \in G \\ &\Longleftrightarrow f(x) \in Z_i(\nu), \\ &\Longleftrightarrow x \in f^{-1}(Z_i(\nu)). \end{aligned}$$

Hence ν is g-nilpotent if and only if there exists nonnegative integer n such that $Z_n(\nu) = H$ if and only if $f^{-1}(Z_n(\nu)) = f^{-1}(H)$ if and only if $Z_n(f^{-1}(\nu)) = G$ if and only if, $f^{-1}(\nu)$ is g-nilpotent. □

Proposition 1. *Let μ and ν be two fuzzy subgroups of G such that $\mu \subseteq \nu$ and $\mu(e) = \nu(e)$. Then $Z(\mu) \subseteq Z(\nu)$.*

Proof. Let $x \in Z(\mu)$. Then $\mu([x, y]) = \mu(e)$, for any $y \in G$. Since

$$\nu(e) = \mu(e) = \mu([x, y]) \leq \nu([x, y]) \leq \nu(e).$$

hence $\nu(e) = \nu([x, y])$ and so $x \in Z(\nu)$. Therefore $Z(\mu) \subseteq Z(\nu)$. □

Lemma 4. *Let μ be a fuzzy subgroup of G and $i > 1$. Then for any $y \in G$, $[x, y] \in Z_{i-1}(\mu)$ if and only if $x \in Z_i(\mu)$.*

Proof. ($\Longrightarrow$) Let $[x, y] \in Z_{i-1}(\mu)$. Then by Lemma 1, $\mu([[x, y], y_1, ..., y_{i-1}]) = \mu(e)$ for any $y, y_1, ..., y_{i-1} \in G$. Hence $x \in Z_i(\mu)$.
($\Longleftarrow$) The proof is similar. □

In the following we see a relation between nilpotency of a group and its fuzzy subgroups.

Theorem 11. *G is nilpotent if and only if any fuzzy subgroup μ of G is a g-nilpotent fuzzy subgroup.*

Proof. ($\Longrightarrow$) Let G be nilpotent of class n and μ be a fuzzy subgroup of G. Since $Z_n(G) = G$, it is enough to prove that for any nonnegative integer i, $Z_i(G) \subseteq Z_i(\mu)$. For i=0 or 1, the proof is clear. Let for $i > 1$, $Z_i(G) \subseteq Z_i(\mu)$ and $x \in Z_{i+1}(G)$. Then for any $y \in G$, $[x, y] \in Z_i(G) \subseteq Z_i(\mu)$ and so by Lemma 4, $x \in Z_{i+1}(\mu)$. Hence $Z_{i+1}(G) \subseteq Z_{i+1}(\mu)$, for any $i \geq 0$, and this implies that $Z_n(\mu) = G$. Therefore, μ is g-nilpotent.
($\Longleftarrow$) Let any fuzzy subgroups of G be g-nilpotent. Suppose that fuzzy set μ on G is defined as follows:

$$\mu(x) = \begin{cases} 1 & if \ \ x \in Z_0(G) \\ \frac{1}{i+1} & if \ \ x \in Z_i(G) - Z_{i-1}(G) \\ 0 & otherwise \end{cases}$$

We show that $Z_i(\mu) \subseteq Z_i(G)$, for any nonnegative integer i. For $i = 0$, the result is immediate. If $i = 1$ and $x \in Z_1(\mu)$, then $\mu([x, y]) = \mu(e) = 1$ for any $y \in G$. By definition of μ, $[x, y] \in Z_0(G) = \{e\}$ and so $x \in Z_1(G)$. Now let $Z_{i-1}(\mu) \subseteq Z_{i-1}(G)$, for $i \geq 2$. Then by Lemma 4, $x \in Z_i(\mu)$ implies that for any $y \in G$; $[x, y] \in Z_{i-1}(\mu) \subseteq Z_{i-1}(G)$. Hence, for any $y, y_1, ..., y_{i-1} \in G$, $[x, y, y_1, ..., y_{i-1}] = e$ which implies that $x \in Z_i(G)$. Thus by induction on i, $Z_i(\mu) \subseteq Z_i(G)$, for any nonnegative integer i. Now since $Z_i(G) \subseteq Z_i(\mu)$ for any nonnegative integer i, then $Z_i(\mu) = Z_i(G)$. Now by the hypotheses there exist $n \in \mathbb{N}$ such that $G = Z_n(\mu) = Z_n(G)$. Hence, G is nilpotent. □

Theorem 12. *Let fuzzy subgroups μ_1 and μ_2 of G be g-nilpotent fuzzy subgroups. Then the fuzzy set $\mu_1 \times \mu_2$ of $G \times G$ is a g-nilpotent fuzzy subgroup, too.*

Proof. Let $\mu = \mu_1 \times \mu_2$. It is clear that μ is fuzzy subgroup of G. So we show that μ is g-nilpotent. It is enough to show that $Z_n(\mu_1 \times \mu_2) = G \times G$, for $n \in \mathbb{N}$. Suppose that $(x, y) \in G \times G$. Then there exist $n_1, n_2 \in \mathbb{N}$ such that $Z_{n_1}(\mu_1) = G$ and $Z_{n_2}(\mu_2) = G$. Hence for any $x_1..., x_n, y_1..., y_n \in G$, $\mu_1([x, x_1..., x_n]) = \mu(e)$ and $\mu_2([y, y_1..., y_n]) = \mu(e)$ for $n = max\{n_2, n_1\}$. Then

$$(\mu_1 \times \mu_2)([(x,y), ..., (x_n, y_n)]) = min\{\mu_1[x, x_1..., x_n], \mu_2[y, y_1..., y_n]\} = (\mu_1 \times \mu_2)(e,e).$$

Therefore, $Z_n(\mu_1 \times \mu_2) = G \times G$. □

Definition 8. *Let μ be a normal fuzzy subgroup of G. For any $x, y \in G$, define a binary relation on G as follows*

$$x \sim y \Longleftrightarrow \mu(xy^{-1}) = \mu(e)$$

Lemma 5. *Binary relation $\sim$ in Definition 8, is a congruence relation.*

Proof. The proof of reflexivity and symmetrically is clear. Hence, we prove the transitivity. Let $x \sim y$ and $y \sim z$, for $x, y, z \in G$. Then $\mu(xy^{-1}) = \mu(yz^{-1}) = \mu(e)$. Since μ is a fuzzy subgroup of G, then $\mu(xz^{-1}) \geq min\{\mu(xy^{-1}), \mu(yz^{-1})\} = \mu(e)$. Hence $\mu(xz^{-1}) = \mu(e)$ and so $x \sim z$. Therefore $\sim$ is an equivalence relation. Now let $x \sim y$ and $z \in G$. Then $\mu((xz)(yz)^{-1}) = \mu(xy^{-1}) = \mu(e)$ and so $xz \sim yz$. Since μ is normal, we get $\mu((zx)(zy)^{-1}) = \mu((zy)^{-1}(zx)) = \mu(y^{-1}x) = \mu(xy^{-1}) = \mu(e)$ and so $zx \sim zy$. Therefore, $\sim$ is a congruence relation on G. □

Notation. For the congruence relation in Definition 8, for any $x \in G$, the equivalence class containing x is denoted by $x\mu$, and $\frac{G}{\mu} = \{x\mu \mid x \in G\}$. It is easy to prove that $\frac{G}{\mu}$ by the operation $(x\mu).(y\mu) = xy\mu$ for any $x\mu, y\mu \in \frac{G}{\mu}$ is a group, where $e\mu$ is unit of $\frac{G}{\mu}$ and $(x\mu)^{-1} = x^{-1}\mu$, for any $x\mu \in \frac{G}{\mu}$.

Theorem 13. *Let μ be a normal fuzzy subgroup of G. Then μ is a g-nilpotent fuzzy subgroup if and only if $\frac{G}{\mu}$ is a nilpotent group.*

Proof. ($\Longrightarrow$) Let μ be a g-nilpotent fuzzy subgroup of G. First we show that for any $n \in \mathbb{N}$ and $x_1, ..., x_n \in G$, $[x, x_1, ..., x_n]\mu = [x\mu, x_1\mu, ..., x_n\mu]$. For $n = 1$, we have

$$[x, x_1]\mu = (x^{-1}\mu).((x_1)^{-1}\mu).(x\mu).(x_1\mu) = [x\mu, x_1\mu]$$

Now assume that it is true for $n - 1$. By hypotheses of induction, we have

$$\begin{aligned}[x, x_1, ..., x_n]\mu &= ([x, x_1, ..., x_{n-1}]^{-1}\mu).(x_n^{-1}\mu).([x, x_1, ..., x_{n-1}]\mu).(x_n\mu) \\ &= ([x\mu, x_1\mu..., x_{n-1}\mu]^{-1}).(x_n^{-1}\mu).([x\mu, x_1\mu..., x_{n-1}\mu]).(x_n\mu) \\ &= [x\mu, x_1\mu, ..., x_n\mu].\end{aligned}$$

Therefore, if μ is a g-nilpotent fuzzy subgroup then there exist $n \in \mathbb{N}$; $Z_n(\mu) = G$, which implies by Lemma 1, that

$$\{x \in G \mid \mu[x, x_1, ..., x_n] = \mu(e) \ \ for \ \ any \ \ x_1, x_2, x_3, ..., x_n \in G\} = G\,, (I)$$

Also $\mu(x) = \mu(e)$ if and only if $x \sim e$ if and only if $x\mu = e\mu$, (II). Thus, by (I) and (II) we have

$$\begin{aligned} \frac{G}{\mu} &= \{x\mu \mid x \in G\} = \{x\mu \mid \mu[x, x_1, ..., x_n] = \mu(e), \forall\, x_1, x_2, x_3, ..., x_n \in G\} \\ &= \{x\mu \mid [x\mu, x_1\mu, ..., x_n\mu] = e\mu, \forall x_1, x_2, x_3, ..., x_n \in G\} = Z_n(\frac{G}{\mu}) \end{aligned}$$

Consequently $\frac{G}{\mu}$ is a nilpotent group of class n.

($\Longleftarrow$) If $\frac{G}{\mu}$ is a nilpotent group of class n, then

$$\frac{G}{\mu} = Z_n(\frac{G}{\mu}) = \{x\mu \mid [x\mu, x_1\mu, ..., x_n\mu] = e\mu, \forall\, x_1, x_2, x_3, ..., x_n \in G\}\}$$

Thus for $x \in G$ we have $x\mu \in \frac{G}{\mu}$. Therefore $[x\mu, x_1\mu, ..., x_n\mu] = e\mu$ for any $x_1, x_2, x_3, ..., x_n \in G$ which implies by (II) that $\mu[x, x_1, ..., x_n] = \mu(e)$. Thus, by Lemma 1, $x \in Z_n(\mu)$. Thus $G = Z_n(\mu)$ and so μ is g-nilpotent. □

Theorem 14. *Let μ be a fuzzy subgroup of G and $\mu_* = \{x \mid \mu(x) = \mu(e)\}$ be a normal subgroup of G. If $\frac{G}{\mu_*}$ is a nilpotent group, then μ is a g-nilpotent fuzzy subgroup.*

Proof. Let $\frac{G}{\mu_*}$ be a nilpotent group and $\pi : G \longrightarrow \frac{G}{\mu_*}$ be the natural epimomorphism. Since

$$\begin{aligned} z \in \pi^{-1}(\pi(x)) &\Longleftrightarrow \pi(z) = \pi(x) \Longleftrightarrow \pi(z^{-1}x) = e \Longleftrightarrow z^{-1}x \in ker\pi = \mu_* \\ &\Longleftrightarrow \mu(z^{-1}x) = \mu(e) \Longleftrightarrow \mu(z) = \mu(x). \end{aligned}$$

hence for any $x \in G$,

$$\pi^{-1}(\pi(\mu))(x) = \pi(\mu)(\pi(x)) = \bigvee_{z \in \pi^{-1}(\pi(x))} \mu(z) = \bigvee_{\mu(z)=\mu(x)} \mu(z) = \mu(x),$$

and so $\pi^{-1}(\pi(\mu)) = \mu$. Now since $\frac{G}{\mu_*}$ is a nilpotent group and $\pi(\mu)$ is a fuzzy subgroup of $\frac{G}{\mu_*}$, then by Theorem 11, $\pi(\mu)$ is g-nilpotent and by Theorem 10, $\pi^{-1}(\pi(\mu)) = \mu$ is g-nilpotent. □

Example 2. *In Example 1, $\mu(e) = t_0$ and so $\mu_* = \{x \mid \mu(x) = \mu(e)\} = \langle a \rangle$. Thus μ_* is a normal subgroup of D_3. Also $\frac{D_3}{\mu_*} \approx \mathbb{Z}_2$. Since $\mathbb{Z}_2$ is Abelian hence it is nilpotent and so by Theorem 14, μ is a g-nilpotent fuzzy subgroup.*

Theorem 15. *Let μ and ν be two fuzzy subgroups of G such that $\mu \subseteq \nu$ and $\mu(e) = \nu(e)$. If μ is a g-nilpotent fuzzy subgroup of class m , then ν is a g-nilpotent fuzzy subgroup of class n, where $n \leq m$.*

Proof. Let μ and ν be two fuzzy subgroups of G where $\mu \subseteq \nu$ and $\mu(e) = \nu(e)$. First, we show that for any $i \in \mathbb{N}$, $Z_i(\mu) \subseteq Z_i(\nu)$. By Theorem 1, for $i = 1$ the proof is clear. Let for $i \geq 2$, $Z_i(\mu) \subseteq Z_i(\nu)$ and $x \in Z_{i+1}(\mu)$. Then by Lemma 4, for any $y \in G$, $[x, y] \in Z_i(\mu) \subseteq Z_i(\nu)$. Thus, by Lemma 4, $x \in Z_{i+1}(\nu)$. Hence $Z_{i+1}(\mu) \subseteq Z_{i+1}(\nu)$. Now let μ be g-nilpotent of class m. Then $G = Z_m(\mu) \subseteq Z_m(\nu) \subseteq G$. Thus $G = Z_m(\nu)$, which implies that ν is g-nilpotent of class at most m. □

Definition 9. *[4] Let μ be a fuzzy set of a set S. Then the lower level subset is*

$$\overline{\mu}_t = \{x \in S; \mu(x) \leq t\}, \text{ where } t \in [0, 1].$$

Now fuzzification of $\overline{\mu}_t$ is the fuzzy set $A_{\overline{\mu}_t}$ defined by

$$A_{\overline{\mu}_t}(x) = \begin{cases} \mu(x) & if \quad x \in \overline{\mu}_t \\ 0 & otherwise \end{cases}$$

Clearly, $A_{\overline{\mu}_t} \subseteq \mu$ and $\overline{(A_{\overline{\mu}_t})_t} = \overline{\mu}_t$.

Corollary 1. *Let μ be a nilpotent fuzzy subgroup of G. Then $A_{\overline{\mu}_t}$ is nilpotent too.*

Proof. Let μ be a nilpotent fuzzy subgroup of G, since $A_{\overline{\mu}_t} \subseteq \mu$ then by Theorem 15, $A_{\overline{\mu}_t}$ is nilpotent. □

In the following we see that our definition for terms of $Z_k(\mu)$, is equivalent to an important relation, which will be used in the main Lemma 7.

Lemma 6. *Let μ be a fuzzy subgroup of G. For $k \geq 2$, $\frac{Z_k(\mu)}{Z_{k-1}(\mu)} = Z(\frac{G}{Z_{k-1}(\mu)})$ if and only if $[Z_k(\mu), G] \subseteq Z_{k-1}(\mu)$.*

Proof. ($\Longrightarrow$) Let for $k \geq 2$, $\frac{Z_k(\mu)}{Z_{k-1}(\mu)} = Z(\frac{G}{Z_{k-1}(\mu)})$ and $w \in [Z_k(\mu), G]$. Then there exist $x \in Z_k(\mu)$ and $g \in G$ such that $w = [x, g]$. Since

$$\begin{aligned} x \in Z_k(\mu) &\Longrightarrow xZ_{k-1}(\mu) \in \frac{Z_k(\mu)}{Z_{k-1}(\mu)} = Z(\frac{G}{Z_{k-1}(\mu)}) \\ &\Longrightarrow [xZ_{k-1}(\mu), gZ_{k-1}(\mu)] = Z_{k-1}(\mu), \ for \ any \ g \in G \\ &\Longrightarrow [x,g]Z_{k-1}(\mu) = Z_{k-1}(\mu), \ for \ any \ g \in G \\ &\Longrightarrow [x,g] \in Z_{k-1}(\mu). \end{aligned}$$

hence $w \in Z_{k-1}(\mu)$.
($\Longleftarrow$) Let for $k \geq 2$, $[Z_k(\mu), G] \subseteq Z_{k-1}(\mu)$ and $xZ_{k-1}(\mu) \in \frac{Z_k(\mu)}{Z_{k-1}(\mu)}$. Hence $x \in Z_k(\mu)$. Since $[Z_k(\mu), G] \subseteq Z_{k-1}(\mu)$, for any $g \in G$, we have $[x, g] \in Z_{k-1}(\mu)$ which implies that $[xZ_{k-1}(\mu), gZ_{k-1}(\mu)] = Z_{k-1}(\mu)$ and so $xZ_{k-1}(\mu) \in Z(\frac{G}{Z_{k-1}(\mu)})$. Hence $\frac{Z_k(\mu)}{Z_{k-1}(\mu)} \subseteq Z(\frac{G}{Z_{k-1}(\mu)})$. Now, let $xZ_{k-1}(\mu) \in Z(\frac{G}{Z_{k-1}(\mu)})$. Then for any $g \in G$ we have, $[xZ_{k-1}(\mu), gZ_{k-1}(\mu)] = Z_{k-1}(\mu)$ which implies that $[x,g]Z_{k-1}(\mu) = Z_{k-1}(\mu)$ and so $[x, g] \in Z_{k-1}(\mu)$. Now by Lemma 1, $\mu([x, g, y_1, y_2 ..., y_{k-1}]) = \mu(e)$, for any $g, y_1, y_2 ..., y_{k-1} \in G$. Hence $x \in Z_k(\mu)$ and this implies that $xZ_{k-1}(\mu) \in \frac{Z_k(\mu)}{Z_{k-1}(\mu)}$. So $\frac{Z_k(\mu)}{Z_{k-1}(\mu)} \supseteq Z(\frac{G}{Z_{k-1}(\mu)})$. Therefore, $\frac{Z_k(\mu)}{Z_{k-1}(\mu)} = Z(\frac{G}{Z_{k-1}(\mu)})$. □

Lemma 7. *Let μ be a g-nilpotent fuzzy subgroup of G of class $n \geq 2$ and N, be a nontrivial normal subgroup of G (i.e $1 \neq N \trianglelefteq G$). Then $N \cap Z(\mu) \neq 1$.*

Proof. Since μ is g-nilpotent, so there exist $n \geq 2$ such that $Z_n(\mu) = G$. Thus

$$1 = Z_0(\mu) \subseteq Z_1(\mu) \subseteq ... \subseteq Z_n(\mu) = G$$

Since $N \cap Z_n(\mu) = N \cap G = N \neq 1$, then there is $j \in \mathbb{N}$ such that $N \cap Z_j(\mu) \neq 1$. Let i be the smallest index such that $N \cap Z_i(\mu) \neq 1$ (so $N \cap Z_{i-1}(\mu) = 1$). Then we claim that $[N \cap Z_i(\mu), G] \subseteq N$. For this let $w \in [N \cap Z_i(\mu), G]$. Then there exists $x \in N \cap Z_i(\mu)$ and $g \in G$ such that $w = [x, g] = x^{-1}g^{-1}xg$. Since $N \trianglelefteq G$, then $w = x^{-1}x^g \in N$. Thus $[N \cap Z_i(\mu), G] \subseteq N$. Also since $x \in N \cap Z_i(\mu)$, by Lemma 6, $[x, g] \in [Z_i(\mu), G] \subseteq Z_{i-1}(\mu)$. Thus $[N \cap Z_i(\mu), G] \subseteq Z_{i-1}(\mu)$. Hence $[N \cap Z_i(\mu), G] \subseteq N \cap Z_{i-1}(\mu) = 1$. Therefore $N \cap Z_i(\mu) \leq Z(G) \leq Z(\mu)$ and so $N \cap Z_i(\mu) \leq N \cap Z(\mu) = 1$. Hence $N \cap Z_i(\mu) = 1$ which is a contradiction. Consequently $N \cap Z(\mu) \neq 1$. □

The following theorem shows that for a g-nilpotent fuzzy subgroup μ each minimal normal subgroup of G is contained in $Z(\mu)$.

Theorem 16. *Let μ be a g-nilpotent fuzzy subgroup of G of class $n \geq 2$. If N is a minimal normal subgroup of G, then $N \leq Z(\mu)$.*

Proof. Since N and $Z(\mu)$ are normal subgroups of G, we get $N \cap Z(\mu) \trianglelefteq G$. Now since N is a minimal normal subgroup of G, $N \cap Z(\mu) \leq N$ and by Lemma 7, $1 \neq N \cap Z(\mu)$. we get $N \cap Z(\mu) = N$. Therefore $N \leq Z(\mu)$. □

Theorem 17. *Let μ be a g-nilpotent fuzzy subgroup of G and A be a maximal normal Abelian subgroup of G. If $\mu(x) = \mu(e)$ for any $x \in A$, and $\mu(x) \neq \mu(e)$ for any $x \in G - A$, then*

$$A = C_G(A) = \{x \in G \mid [x, a] = e, \; for \;\; any \;\; a \in A\}.$$

Proof. First, we prove that $C_G(A) \trianglelefteq G$. For this, let $x \in C_G(A)$ and $g \in G$. Then for all $a \in A$ we have $[x^g, a] = [x, a^{g^{-1}}]^g$. Since A is Abelian, then $a^{g^{-1}} = a$. Hence $x \in C_G(A)$ implies that $[x^g, a] = [x, a^{g^{-1}}]^g = [x, a]^g = e$ and so $x^g \in C_G(A)$. Thus $C_G(A) \trianglelefteq G$. Suppose $A \subsetneq C_G(A)$. Then $1 \neq \frac{C_G(A)}{A} \trianglelefteq \frac{G}{A}$. Let $\overline{\mu}$ be the fuzzy subgroup of $\frac{G}{A}$. Then by Lemma 7, $\frac{C_G(A)}{A} \cap Z(\overline{\mu}) \neq 1$. So there exists $A \neq gA \in \frac{C_G(A)}{A} \cap Z(\overline{\mu})$. Hence $g \in C_G(A)$ and $\overline{\mu}[gA, xA] = \overline{\mu}(eA)$ for any $x \in G$. Thus by Theorem 2, $\bigvee_{a \in A} \mu([g, x]a) = \mu(e)$. Now if for some $a \in A$, $\mu([g, x]) = \mu(a)$, then by definition of μ, $[g, x] \in A$ and if for any $a \in A$, $\mu([g, x]) \neq \mu(a)$, then by Theorem 2, $\bigvee_{a \in A} \mu([g, x]a) = \bigvee_{a \in A} (\mu([g, x]) \wedge \mu(a)) = \mu([g, x])$. Thus $[g, x] \in A$. Now let $B = \langle A, g \rangle$. Then $A \subsetneq B \trianglelefteq G (B \trianglelefteq G$ since $A \trianglelefteq G$, and for $x \in G$ we have $g^x = g[g, x] \in B$). Moreover, since $g \in C_G(A)$, then B is Abelian. Therefore, B is a normal Abelian subgroup of G, which is a contradiction. Thus $A = C_G(A)$.

□

Now we show that with some conditions every g-nilpotent fuzzy subgroup is finite.

Corollary 2. *Let μ be a g-nilpotent fuzzy subgroup of G and A be a finite maximal normal Abelian subgroup of G. If $\mu(x) = \mu(e)$ for any $x \in A$, and $\mu(x) \neq \mu(e)$ for any $x \in G - A$, then μ is finite, too.*

Proof. Since $A \trianglelefteq G$, for $g \in G$ and $x \in A$ we have $x^g \in A$. Now let

$$\begin{aligned} \theta \;\; : \;\; & G \longrightarrow Aut(A) \\ & g \longrightarrow \theta_g : A \longrightarrow A \\ & \qquad\qquad x \longrightarrow x^g. \end{aligned}$$

We prove that θ is a homomorphism. Let $g_1, g_2 \in G$. Then $\theta(g_1 g_2) = \theta_{g_1 g_2}$. Thus for $x \in A$,

$$(\theta(g_1 g_2))(x) = (\theta_{g_1 g_2})(x) = x^{g_1 g_2} = x^{g_1}.x^{g_2} = (\theta(g_1))(x).(\theta(g_2))(x).$$

But $Ker(\theta) = \{g \in G \mid \theta(g) = I\}$, in which I is the identity homomorphism. Thus for any $x \in A$, $(\theta(g))(x) = I(x)$ which implies that $x^g = x$. Hence $g \in C_G(A)$. Therefore, $Ker(\theta) = C_G(A)$. By Theorem 17, $A = C_G(A)$. Thus $\frac{G}{Ker(\theta)} = \frac{G}{A}$ is embeded in $Aut(A)$. Now since A and so $Aut(A)$ are finite we get G is finite which implies that μ is finite. □

4. Conclusions

By the notion of a g-nilpotent fuzzy subgroup we can investigate on fuzzification of nilpotent groups. Moreover, since this is similar to group theory' definition, it is much easier than before to study the properties of nilpotent fuzzy groups. Moreover, if we accept the definition of a g-nilpotent fuzzy subgroup, then one can verify, as we have done in Theorem 16, that for a g-nilpotent fuzzy subgroup μ each minimal normal subgroup of G is contained in the center of μ. We hope that these results inspire other papers on nilpotent fuzzy subgroups.

Author Contributions: Elaheh Mohammadzadeh and Rajab Ali Borzooei conceived and designed the paper structure. Then Elaheh Mohammadzadeh performed the first reaserch and Rajab Ali Borzooei completed the research.

Conflicts of Interest: The authors declare no conflicts of interest.

References

1. Zadeh, L.A. Fuzzy sets. *Inf. Control* **1965**, *8*, 338–353, doi:10.1016/S0019-9958(65)90241-X.
2. Rosenfeld, A. Fuzzy groups. *J. Math. Anal. Appl.* **1971**, *35*, 512–517, doi:10.1016/0022-247X(71)90199-5.
3. Kurdachenko, L.A.; Subbotin, I.Y.; Grin, K.O. On some properties of normal and subnormal fuzzy subgros. *Southeast Asian Bull. Math.* **2014**, *38*, 401–421.
4. Biswas, R. Fuzzy subgroups and anti fuzzy subgroups. *Fuzzy Sets Syst.* **1990**, *35*, 121–124, doi:10.1016/0165-0114(90)90025-2.
5. Bhattacharya, P.; Mukherjee, N.P. Fuzzy relations and fuzzy groups. *Inf. Sci.* **1985**, *36*, 267–282, doi:10.1016/0020-0255(85)90057-X.
6. Dudek, W.A. Fuzzification of n-ary groupoids. *Quasigroups Related Syst.* **2000**, *7*, 45–66.
7. Kim, J.G. Commutative fuzzy sets and nilpotent fuzzy groups. *Inf. Sci.* **1995**, *83*, 161–174, doi:10.1016/0020-0255(94)00082-M.
8. Guptaa, K.C.; Sarma, B.K. Nilpotent fuzzy groups. *Fuzzy Sets Syst.* **1999**, *101*, 167–176, doi:10.1016/S0165-0114(97)00067-5.
9. Ray, S. Generated and cyclic fuzzy groups. *Inf. Sci.* **1993**, *69*, 185–200, doi:10.1016/0020-0255(93)90119-7.
10. Bucolo, M.; Fazzino, S.; la Rosa, M.; Fortuna, L. Small-world networks of fuzzy chaotic oscillators. *Chaos Solitons Fractals* **2003**, *17*, 557–565, doi:10.1016/S0960-0779(02)00398-3.
11. Ameri, R.; Borzooei, R.A.; Mohammadzadeh, E. Engel fuzzy subgroups. *Ital. J. Pure Appl. Math.* **2015**, *34* , 251–262.
12. Borzooei, R.A.; Mohammadzadeh, E.; Fotea, V. On Engel Fuzzy Subpolygroups. *New Math. Nat. Comput.* **2017**, *13*, 195–206, doi:10.1142/S1793005717500089.
13. Robinson, D.J.S. *A Course in the Theory of Groups*; Springer-Verlag: New York, NY, USA, 1980; pp. 121–158; ISBN 978-1-4612-643-9.
14. Mordeson, J.N.; Bhutani, K.R.; Rosenfeld, A. *Fuzzy Group Theory*; Studies in Fuzziness and Soft Computing; Springer-Verlag: Berlin/Heidelberg, Germany, 2005; pp. 61–89; ISBN 978-3-540-25072-2.
15. Hungerford, T.W. *Algebra*; Springer-Verlag: New York, NY, USA, 1974; pp. 23–69; ISBN 978-1-4612-6103-2.

mathematics

MDPI

Article

Neutrosophic Triplet G-Module

Florentin Smarandache [1], Mehmet Şahin [2] and Abdullah Kargın [2,*]

[1] Department of Mathematics, University of Mexico, 705-Ave., Gallup, NM 87301, USA; smarand@unm.edu
[2] Department of Mathematics, Gaziantep University, Gaziantep 27310, Turkey; mesahin@gantep.edu.tr
* Correspondence: abdullahkargin27@gmail.com

Received: 19 March 2018; Accepted: 2 April 2018; Published: 5 April 2018

Abstract: In this study, the neutrosophic triplet G-module is introduced and the properties of neutrosophic triplet G-module are studied. Furthermore, reducible, irreducible, and completely reducible neutrosophic triplet G-modules are defined, and relationships of these structures with each other are examined. Also, it is shown that the neutrosophic triplet G-module is different from the G-module.

Keywords: neutrosophic triplet G-module; neutrosophic triplet group; neutrosophic triplet vector space

1. Introduction

Neutrosophy is a branch of philosophy, firstly introduced by Smarandache in 1980. Neutrosophy [1] is based on neutrosophic logic, probability, and set. Neutrosophic logic is a generalized form of many logics such as fuzzy logic, which was introduced by Zadeh [2], and intuitionistic fuzzy logic, which was introduced by Atanassov [3]. Furthermore, Bucolo et al. [4] studied complex dynamics through fuzzy chains; Chen [5] introduced MAGDM based on intuitionistic 2–Tuple linguistic information, and Chen [6] obtain some q–Rung Ortopair fuzzy aggregation operators and their MAGDM. Fuzzy set has function of membership; intuitionistic fuzzy set has function of membership and function of non-membership. Thus; they do not explain the indeterminancy states. However, the neutrosophic set has a function of membership, a function of indeterminacy, and a function of non-membership. Also, many researchers have studied the concept of neutrosophic theory in [7–12]. Recently, Olgun et al. [13] studied the neutrosophic module; Şahin et al. [14] introduced Neutrosophic soft lattices; Şahin et al. [15] studied the soft normed ring; Şahin et al. [16] introduced the centroid single-valued neutrosophic triangular number and its applications; Şahin et al. [17] introduced the centroid points of transformed single-valued neutrosophic number and its applications; Ji et al. [18] studied multi-valued neutrosophic environments and their applications. Also, Smarandache et al. [19] studied neutrosophic triplet (NT) theory and [20,21] neutrosophic triplet groups. A NT has a form <x, neut(x), anti(x)>, in which neut(x) is neutral of "x" and anti(x) is opposite of "x". Furthermore, neut(x) is different from the classical unitary element. Also, the neutrosophic triplet group is different from the classical group. Recently, Smarandache et al. [22] studied the NT field and [23] the NT ring; Şahin et al. [24] introduced the NT metric space, the NT vector space, and the NT normed space; Şahin et al. [25] introduced the NT inner product.

The concept of G-module [26] was introduced by Curties. G-modules are algebraic structures constructed on groups and vector spaces. The concept of group representation was introduced by Frobenious in the last two decades of the 19th century. The representation theory is an important algebraic structure that makes the elements, which are abstract concepts, more evident. Many important results could be proved only for representations over algebraically closed fields. The module theoretic approach is better suited to deal with deeper results in representation theory. Moreover, the module theoretic approach adds more elegance to the theory. In particular, the G-module structure has been extensively used for the study of representations of finite groups. Also, the representation theory of groups describes all the ways in which group G may be embedded in any linear group GL (V). The G-module also holds an important place in the representation theory of groups. Recently some researchers have been dealing with the G-module. For example, Fernandez [27] studied fuzzy G-modules. Sinho and Dewangan [28] studied isomorphism theory for fuzzy submodules of G-modules. Şahin et al. [29] studied soft G-modules. Sharma and Chopra [30] studied the injectivity of intuitionistic fuzzy G-modules.

In this paper, we study neutrosophic triplet G-Modules in order to obtain a new algebraic constructed on neutrosophic triplet groups and neutrosophic triplet vector spaces. Also we define the reducible neutrosophic triplet G-module, the irreducible neutrosophic triplet G-module, and the completely reducible neutrosophic triplet G-module. In this study, in Section 2, we give some preliminary results for neutrosophic triplet sets, neutrosophic triplet groups, the neutrosophic triplet field, the neutrosophic triplet vector space, and G-modules. In Section 3, we define the neutrosophic triplet G-module, and we introduce some properties of a neutrosophic triplet G-module. We show that the neutrosophic triplet G-module is different from the G-module, and we show that if certain conditions are met, every neutrosophic triplet vector space or neutrosophic triplet group can be a neutrosophic triplet G-module at the same time. Also, we introduce the neutrosophic triplet G-module homomorphism and the direct sum of neutrosophic triplet vector space. In Section 4, we define the reducible neutrosophic triplet G-module, the irreducible neutrosophic triplet G-module, and the completely reducible neutrosophic triplet G-module, and we give some properties and theorems for them. Furthermore, we examine the relationships of these structures with each other, and we give some properties and theorems. In Section 5, we give some conclusions.

2. Preliminaries

Definition 1. *Let N be a set together with a binary operation *. Then, N is called a neutrosophic triplet set if for any a ∈ N there exists a neutral of "a" called neut(a) that is different from the classical algebraic unitary element and an opposite of "a" called anti(a) with neut(a) and anti(a) belonging to N, such that [21]:*

*a*neut(a) = neut(a)* a = a,*
and
*a*anti(a) = anti(a)* a = neut(a).*

Definition 2. *Let (N,*) be a neutrosophic triplet set. Then, N is called a neutrosophic triplet group if the following conditions are satisfied [21].*

(1) *If (N,*) is well-defined, i.e., for any a, b ∈ N, one has a*b ∈ N.*
(2) *If (N,*) is associative, i.e., (a*b)*c = a*(b*c) for all a, b, c ∈ N.*

Theorem 1. *Let (N,*) be a commutative neutrosophic triplet group with respect to * and a, b ∈ N, in which a and b are both cancellable [21],*

(i) *neut(a)*neut(b) = neut(a*b).*
(ii) *anti(a)*anti(b) = anti(a*b).*

Definition 3. *Let (NTF,*, #) be a neutrosophic triplet set together with two binary operations * and #. Then (NTF,*, #) is called neutrosophic triplet field if the following conditions hold [22].*

1. *(NTF,*) is a commutative neutrosophic triplet group with respect to *.*
2. *(NTF, #) is a neutrosophic triplet group with respect to #.*
3. *a#(b*c) = (a#b)*(a#c) and (b*c)#a = (b#a)*(c#a) for all a, b, c ∈ NTF.*

Theorem 2. *Let (N,*) be a neutrosophic triplet group with respect to *. For (left or right) cancellable a ∈ N, one has the following [24]:*

(i) $neut(neut(a)) = neut(a)$;
(ii) $anti(neut(a)) = neut(a)$;
(iii) $anti(anti(a)) = a$;
(iv) $neut(anti(a)) = neut(a)$.

Definition 4. *Let $(NTF, *_1, \#_1)$ be a neutrosophic triplet field, and let $(NTV, *_2, \#_2)$ be a neutrosophic triplet set together with binary operations "$*_2$" and "$\#_2$". Then $(NTV, *_2, \#_2)$ is called a neutrosophic triplet vector space if the following conditions hold. For all $u, v \in NTV$, and for all $k \in NTF$, such that $u *_2 v \in NTV$ and $u \#_2 k \in NTV$ [24];*

(1) $(u *_2 v) *_2 t = u *_2 (v *_2 t)$; $u, v, t \in NTV$;
(2) $u *_2 v = v *_2 u$; $u, v \in NTV$;
(3) $(v *_2 u) \#_2 k = (v \#_2 k) *_2 (u \#_2 k)$; $k \in NTF$ *and* $u, v \in NTV$;
(4) $(k *_1 t) \#_2 u = (k \#_2 v) *_1 (u \#_2 v)$; $k, t \in NTF$ *and* $u \in NTV$;
(5) $(k \#_1 t) \#_2 u = k \#_1 (t \#_2 u)$; $k, t \in NTF$ *and* $u \in NTV$;
(6) *There exists any* $k \in NTF$ *such that* $u \#_2 neut(k) = neut(k) \#_2 u = u$; $u \in NTV$.

Definition 5. *Let G be a finite group. A vector space M over a field K is called a G-module if for every $g \in G$ and $m \in M$ there exists a product (called the action of G on M) $m.g \in M$ satisfying the following axioms [26]:*

(i) $m.1_G = m, \forall\, m \in M$ *(1_G being the identity element in G);*
(ii) $m.(g.h) = (m.g).h, \forall\, m \in M; g, h \in G$;
(iii) $(k_1 m_1 + k_2 m_2).g = k_1(m_1.g) + k_2(m_2.g)$; $k_1, k_2 \in K$; $m_1, m_2 \in M$, *and* $g \in G$.

Definition 6. *Let M be a G-module. A vector subspace N of M is a G-submodule if N is also a G-module under the same action of G [26].*

Definition 7. *Let M and M^* be G-modules. A mapping ϕ [26]: $M \to M^*$ is a G-module homomorphism if*

(i) $\phi(k_1.m_1 + k_2.m_2) = k_1.\phi(m_1) + k_2.\phi(m_2)$;
(ii) $\phi(m.g) = \phi(m).g$; $k_1, k_2 \in K$; $m, m_1, m_2 \in M$; $g \in G$.

Further, if ϕ is 1-1, then ϕ is an isomorphism. The G-modules M and M^* are said to be isomorphic if there exists an isomorphism ϕ of M onto M^*. Then we write $M \cong M^*$.

Definition 8. *Let M be a nonzero G-module. Then, M is irreducible if the only G-submodules of M are M and {0}. Otherwise, M is reducible [26].*

Definition 9. *Let $M_1, M_2, M_3, \ldots, M_n$ be vector spaces over a field K [31]. Then, the set $\{m_1 + m_2 + \ldots + m_n;\ m_i \in M_i\}$ becomes a vector space over K under the operations*

$$(m_1 + m_2 + \ldots + m_n) + (m_1{}' + m_2{}' + \ldots + m_n{}') = (m_1 + m_1{}') + (m_2 + m_2{}') + \ldots + (m_n + m_n{}') \text{ and}$$

$$\alpha(m_1 + m_2 + \ldots + m_n) = \alpha m_1 + \alpha m_2 + \ldots + \alpha m_n;\ \alpha \in K, m_n{}' \in M_i$$

It is the called direct sum of the vector spaces $M_1, M_2, M_3, \ldots, M_n$ and is denoted by $\oplus_{i=1}^{n} M_i$.

Remark 1. *The direct sum $M = \oplus_{i=1}^{n} M_i$ of vector spaces M_i has the following properties [31]:*

(i) *Each element $m \in M$ has a unique expression as the sum of elements of M_i.*
(ii) *The vector subspaces $M_1, M_2, M_3, \ldots, M_n$ of M are independent.*
(iii) *For each $1 \le i \le n$, $M_j \cap (M_1 + M_2 + \ldots + M_{j-1} + M_{j+1} + \ldots + M_n) = \{0\}$.*

Definition 10. *A nonzero G-module M is completely reducible if for every G-submodule N of M there exists a G-submodule N^* of M such that $M = N \oplus N^*$ [26].*

Proposition 1. *A G-submodule of a completely reducible G-module is completely reducible [26].*

3. Neutrosophic Triplet G-Module

Definition 11. *Let (G, *) be a neutrosophic triplet group, (NTV,$*_1$, $\#_1$) be a neutrosophic triplet vector space on a neutrosophic triplet field (NTF,$*_2$, $\#_2$), and g*m ϵ NTV for g ϵ G, m ϵ NTV. If the following conditions are satisfied, then (NTV,$*_1$, $\#_1$) is called neutrosophic triplet G-module.*

(a) *There exists g ∈ G such that m*neut(g) = neut(g)*m = m for every m ∈ NTV;*
(b) $m *_1 (g *_1 h) = (m *_1 g) *_1 h$, $\forall\, m \in NTV; g, h \in G$;
(c) $(k_1\#_1 m_1 *_1 k_2\, \#_1 m_2)^* g = k_1\#_1\, (m_1{}^* g) *_1 k_2 \#_1\, (m_2{}^* g)$, $\forall k_1, k_2 \epsilon\, NTF; m_1, m_2\, \epsilon\, NTV; g\, \epsilon\, G$.

Corollary 1. *Neutrosophic G-modules are generally different from the classical G-modules, since there is a single unit element in classical G-module. However, the neutral element neut(g) in neutrosophic triplet G-module is different from the classical one. Also, neutrosophic triplet G-modules are different from fuzzy G-modules, intuitionistic fuzzy G-modules, and soft G-modules, since neutrosophic triplet set is a generalized form of fuzzy set, intuitionistic fuzzy set, and soft set.*

Example 1. *Let X be a nonempty set and let P(X) be set of all subset of X. From Definition 4, (P(X), ∪, ∩) is a neutrosophic triplet vector space on the (P(X), ∪, ∩) neutrosophic triplet field, in which the neutrosophic triplets with respect to ∪; neut(A) = A and anti(A) = B, such that A, B ∈ P(X); A ⊆ B; and the neutrosophic triplets with respect to ∩; neut(A) = A and anti(A) = B, such that A, B ∈ P(X); B ⊇ A. Furthermore, (P(X), ∪) is a neutrosophic triplet group with respect to ∪, in which neut(A) = A and anti(A) = B such that A, B ⊂ P(X); A ⊆ B. We show that (P(X), ∪, ∩) satisfies condition of neutrosophic triplet G-module. From Definition 11:*

(a) *It is clear that if A = B, there exists any A ∈ P(X) for every B ∈ P(X), such that B ∪ neut(A) = neut(A) ∪ B = A.*
(b) *It is clear that A ∪ (B ∪ C) = (A ∪ B) ∪ C, ∀ A ∈ P(X); B, C ∈ P(X).*
(c) *It is clear that*

$((A_1 \cap B_1) \cup (A_2 \cap B_2)) \cup C = (A_1 \cap B_1) \cup C)) \cup (A_2 \cap B_2) \cup C)), \forall A_1, A_2\, \epsilon\, P(X); B_1, B_2\, \epsilon\, P(X); C\, \epsilon\, P(X)$.
Thus, (P(X), ∪, ∩) is a neutrosophic triplet G-module.

Corollary 2. *If G = NTV, * = $*_1$, then each (NTV,$*_1$, $\#_1$) neutrosophic triplet vector space is a neutrosophic triplet G-module at the same time. Thus, if G = NTV and * = $*_1$, then every neutrosophic triplet vector space or neutrosophic triplet group can be a neutrosophic triplet G-module at the same time. It is not provided by classical G-module.*

Proof of Corollary 1. *If G = NTV, * = $*_1$;*

(a) *There exists a g ϵ NTV such that m*neut(g) = neut(g)*m = m, ∀m ϵ NTV;*
(b) *It is clear that m*(g*h) = (m*g)*h, as (NTV,*) is a neutrosophic triplet group; ∀ m, g, h ∈ NTV;*
(c) *It is clear that* $(k_1\#_1 m_1 *_1 k_2\, \#_1 m_2)^* g = k_1\#_1 (m_1{}^* g) *_1 k_2 \#_1\, (m_2{}^* g)$, *since (NTV,$*_1$, $\#_1$) is a neutrosophic triplet vector space;* $\forall\, g, k_1, k_2\, \epsilon\, NTF; m_1, m_2\, \epsilon\, NTV$.

Definition 12. *Let (NTV,$*_1$, $\#_1$) be a neutrosophic triplet G-module. A neutrosophic triplet subvector space (N, $*_1$, $\#_1$) of (NTV,$*_1$, $\#_1$) is a neutrosophic triplet G-submodule if (N,$*_1$, $\#_1$) is also a neutrosophic triplet G-module.*

Example 2. *From Example 1; for N ⊆ X, (P(N), ∪, ∩) is a neutrosophic triplet subvector space of (P(X), ∪, ∩). Also, (P(N), ∪, ∩) is a neutrosophic triplet G-module. Thus, (P(N), ∪, ∩) is a neutrosophic triplet G-submodule of (P(X), ∪, ∩).*

Example 3. *Let $(NTV,*_1, \#_1)$ be a neutrosophic triplet G-module. N = {neut(x)} ∈ NTV is a neutrosophic triplet subvector space of $(NTV,*_1, \#_1)$. Also, N = {neut(x) = x} ∈ NTV is a neutrosophic triplet G-submodule of $(NTV,*_1, \#_1)$.*

Definition 13. *Let $(NTV,*_1, \#_1)$ and $(NTV^*,*_3, \#_3)$ be neutrosophic triplet G-modules on neutrosophic triplet field $(NTF,*_2, \#_2)$ and (G, *) be a neutrosophic triplet group. A mapping ϕ: $NTV \rightarrow NTV^*$ is a neutrosophic triplet G-module homomorphism if*

(i) $\phi(neut(m)) = neut(\phi(m))$
(ii) $\phi(anti(m)) = anti(\phi(m))$
(iii) $\phi((k_1\#_1 m_1) *_1 (k_2\#_1 m_2)) = (k_1\#_3\phi(m_1)) *_3 (k_2\#_3\phi(m_2))$
(iv) $\phi(m^*g) = \phi(m)^*g$; $\forall\, k_1, k_2 \in NTF$; $m, m_1, m_2 \in M$; $g \in G$.

Further, if ϕ is 1-1, then ϕ is an isomorphism. The neutrosophic triplet G-modules $(NTV,*_1, \#_1)$ and $(NTV^*,*_3, \#_3)$ are said to be isomorphic if there exists an isomorphism ϕ: $NTV \rightarrow NTV^*$. Then, we write $NTV \cong NTV^*$.

Example 4. *From Example 1, (P(X), ∪, ∩) is neutrosophic triplet vector space on neutrosophic triplet field (P(X), ∪, ∩). Furthermore, (P(X), ∪, ∩) is a neutrosophic triplet G-module. We give a mapping ϕ: P(X) → P(X), such that ϕ(A) = neut(A). Now, we show that ϕ is a neutrosophic triplet G-module homomorphism.*

(i) $\phi(neut(A)) = neut(neut(A)) = neut(\phi(A))$
(ii) $\phi(anti(A)) = neut(anti(A))$; *from Theorem 2*, $neut(anti(A)) = neut(A)$.

$anti(\phi(A)) = anti(neut(A))$; *from Theorem 2*, $anti(neut(A)) = neut(A)$. *Then* $\phi(anti(A)) = anti(\phi(A))$.

(iii) $\phi((A_1 \cap B_1) \cup (A_2 \cap B_2)) = neut(A_1 \cap B_1) \cup (A_2 \cap B_2))$; *from Theorem 1, as* $neut(a)^*neut(b) = neut(a^*b)$;

$neut(A_1 \cap B_1) \cup (A_2 \cap B_2)) = neut(A_1 \cap B_1) \cup neut(A_2 \cap B_2) =$

$((neut(A_1) \cap neut(B_1)) \cup ((neut(A_2) \cap neut(B_2))$. *From Example 1, as* $neut(A) = A$,

$((neut(A_1) \cap neut(B_1)) \cup ((neut(A_2) \cap neut(B_2)) = (A_1 \cap neut(B_1)) \cup (A_2 \cap neut(B_2)) =$

$(A_1 \cap neut(B_1)) \cup (A_2 \cap neut(B_2)) = (A_1 \cap \phi(B_1)) \cup (A_2 \cap \phi(B_2))$.

(iv) $\phi(A^*B) = neut(A^*B)$; *from Theorem 1, as* $neut(a)^*neut(b) = neut(a^*b)$, $neut(A^*B) = neut(A)^* neut(B)$. *From Example 1, as* $neut(A) = A$, $neut(A)^* neut(B) = A^* neut(B) = A^* \phi(B)$.

4. Reducible, Irreducible, and Completely Reducible Neutrosophic Triplet G-Modules

Definition 14. *Let $(NTV,*_1, \#_1)$ be neutrosophic triplet G-modules on neutrosophic triplet field $(NTF,*_2, \#_2)$. Then, $(NTV,*_1, \#_1)$ is irreducible neutrosophic triplet G-modules if the only neutrosophic triplet G-submodules of $(NTV,*_1, \#_1)$ are $(NTV,*_1, \#_1)$ and {neut(x) = x}, x ∈ NTV. Otherwise, $(NTV,*_1, \#_1)$ is reducible neutrosophic triplet G-module.*

Example 5. *From Example 2, for N = {1,2} ⊆ {1,2,3} = X, (P(N), ∪, ∩) is a neutrosophic triplet subvector space of (P(X), ∪, ∩). Also, (P(N), ∪, ∩) is a neutrosophic triplet G-module. Thus, (P(N), ∪, ∩) is a neutrosophic triplet G-submodule of (P(X), ∪, ∩). Also, from Definition 14, (P(X), ∪, ∩) is a reducible neutrosophic triplet G-module.*

Example 6. *Let X = G = {1, 2} and P(X) be power set of X. Then, (P(X), *, ∩) is a neutrosophic triplet vector space on the (P(X), *, ∩) neutrosophic triplet field and (G, *) is a neutrosophic triplet group, in which*

$$A^*B = \begin{cases} B\backslash A, s(A) < s(B) \wedge B \supset A \wedge A' = B \\ A\backslash B, s(B) < s(A) \wedge A \supset B \wedge B' = A \\ (A\backslash B)', s(A) > s(B) \wedge A \supset B \wedge B\prime \neq A \\ (B\backslash A)', s(B) > s(A) \wedge B \supset A \wedge A\prime \neq B \\ X, s(A) = s(B) \wedge A \neq B \\ \varnothing, A = B \end{cases}$$

Here, s(A) means the cardinal of A, and A′ means the complement of A.
*The neutrosophic triplets with respect to *;*
neut(∅) = ∅, anti(∅) = ∅; neut({1}) = {1, 2}, anti({1}) = {2}; neut({2}) = {1, 2}, anti({2}) = {1}; neut({1, 2}) = ∅, anti({1,2}) = {1, 2};
The neutrosophic triplets with respect to ∩;
neut(A) = A and anti(A) = B, in which B ⊃ ⋂ A.
*Also, (P(X), *, ∩) is a neutrosophic triplet G-module. Here, only neutrosophic triplet G-submodules of (P(X), *, ∩) are (P(X), *, ∩) and {neut(∅) = ∅}. Thus, (P(X),*, ∩) is a irreducible neutrosophic triplet G-module.*

Definition 15. *Let* $(NTV_1, *_1, \#_1)$, $(NTV_2, *_1, \#_1)$, …, $(NTV_n, *_1, \#_1)$ *be neutrosophic triplet vector spaces on* $(NTF, *_2, \#_2)$. *Then, the set* $\{m_1 + m_2 + \ldots + m_n;\ m_i \in NTV_i\}$ *becomes a neutrosophic triplet vector space on* $(NTF, *_2, \#_2)$, *such that*

$$(m_1 *_1 m_2 *_1 \ldots *_1 m_n) *_1 (m_1{}' *_1 m_2{}' *_1 \ldots *_1 m_n{}') = (m_1 *_1 m_1{}') *_1 (m_2 *_1 m_2{}') *_1 \ldots *_1 (m_n *_1 m_n{}') \text{ and}$$

$$\alpha\#_1(m_1 *_1 m_2 *_1 \ldots *_1 m_n) = (\alpha\#_1 m_1) *_1 \alpha\#_1 m_2) *_1 \ldots *_1 (\alpha\#_1 m_n);\ \alpha \in NTf,\ m_n{}' \in NTV_i.$$

It is called the direct sum of the neutrosophic triplet vector spaces $NTV_1, NTV_2, NTV_3, \ldots, NTV_n$ and is denoted by ${}_{i=1}^{n}\oplus NTV_i$.

Remark 2. *The direct sum* $NTV = {}_{i=1}^{n}\oplus NTV_i$ *of neutrosophic triplet vector spaces* NTV_i *has the following properties.*

(i) *Each element* $m \in NTV$ *has a unique expression as the sum of elements of* NTV_i.
(ii) *For each* $1 \leq i \leq n$, $NTV_j \cap (NTV_1 + NTV_2 + \ldots + NTV_{j-1} + NTV_{j+1} + \ldots + NTV_n) = \{x: neut(x) = x\}$.

Definition 16. *Let* $(NTV, *_1, \#_1)$ *be neutrosophic triplet G-modules on neutrosophic triplet field* $(NTF, *_2, \#_2)$, *such that* $NTV \neq \{neut(x) = x\}$. *Then,* $(NTV, *_1, \#_1)$ *is a completely reducible neutrosophic triplet G-module if for every neutrosophic triplet G-submodule* $(N_1, *_1, \#_1)$ *of* $(NTV, *_1, \#_1)$ *there exists a neutrosophic triplet G-submodule* $(N_2, *_1, \#_1)$ *of* $(NTV, *_1, \#_1)$, *such that* $NTV = N_1 \oplus N_2$.

Example 7. *From Example 5, for N = {1, 2}, (P(N), ∪, ∩) is a neutrosophic triplet vector space on (P(N), ∪, ∩) and a neutrosophic triplet G-module. Also, the neutrosophic triplet G-submodules of (P(N), ∪, ∩) are (P(N), ∪, ∩), (P(M), ∪, ∩), (P(K), ∪, ∩), and (P(L), ∪, ∩). Here, M = {1}, K = {2}, and T = {∅}, in which P(M)⊕P(K) = P(N), P(K)⊕P(M) = P(N), P(N)⊕P(T) = P(N), and P(T)⊕P(N) = P(N). Thus, (P(N), ∪, ∩) is a completely reducible neutrosophic triplet G-module.*

Theorem 3. *A neutrosophic triplet G-submodule of a completely reducible neutrosophic triplet G-module is completely neutrosophic triplet G-module.*

Proof of Theorem 1. *Let* $(NTV, *_1, \#_1)$ *is a completely reducible neutrosophic triplet G-module on neutrosophic triplet field* $(NTF, *_2, \#_2)$. *Assume that* $(N, *_1, \#_1)$ *is a neutrosophic triplet G-submodule of* $(NTV, *_1, \#_1)$ *and* $(M, *_1, \#_1)$ *is a neutrosophic triplet G-submodule of* $(N, *_1, \#_1)$. *Then,* $(M, *_1, \#_1)$ *is a neutrosophic triplet*

G-submodule of (NTV, $*_1$, $\#_1$). There exists a neutrosophic triplet G-submodule (T, $*_1$, $\#_1$), such that NTV = M⊕T, since (NTV, $*_1$, $\#_1$) is a completely reducible neutrosophic triplet G-module. Then, we take N′ = T ∩ N. From Remark 2,

$$N' \cap M \subset M \cap T = \{x : \text{neut}(x) = x\} \tag{1}$$

Then, we take y ∈ N. If y ∈ N, y ∈ NTV and y = m $*_1$ t, in which m ∈ M; t ∈ T. Therefore, we obtain t ∈ N. Thus,

$$tN' = T \cap N \text{ and } y = m*_1\, tN' \oplus M \tag{2}$$

From (i) and (ii), we obtain N = N′⊕M. Thus, (N, $*_1$, $\#_1$) is completely reducible neutrosophic triplet G-module.

Theorem 4. *Let (NTV, $*_1$, $\#_1$) be a completely reducible neutrosophic triplet G-module on neutrosophic triplet field (NTF, $*_2$, $\#_2$). Then, there exists a irreducible neutrosophic triplet G-submodule of (NTV, $*_1$, $\#_1$).*

Proof of Theorem 2. *Let (NTV, $*_1$, $\#_1$) be a completely reducible neutrosophic triplet G-module and (N, $*_1$, $\#_1$) be a neutrosophic triplet G-submodule of (NTV, $*_1$, $\#_1$). We take y ≠ neut(y) ∈ N, and we take collection sets of neutrosophic triplet G-submodules of (N, $*_1$, $\#_1$) such that do not contain element y. This set is not empty, because there is {x: x = neut(x)} neutrosophic triplet G-submodule of (N, $*_1$, $\#_1$). From Zorn's Lemma, the collection has maximal element (M, $*_1$, $\#_1$). From Theorem 3, (N, $*_1$, $\#_1$) is a completely reducible neutrosophic triplet G-module, and there exists a (N_1, $*_1$, $\#_1$) neutrosophic triplet G-submodule, such that N = M⊕N_1. We show that ($N_1 *_1$, $\#_1$) is a irreducible neutrosophic triplet G–submodule. Assume that (N_1, $*_1$, $\#_1$) is a reducible neutrosophic triplet G–submodule. Then, there exists (K_1, $*_1$, $\#_1$) and (K_2, $*_1$, $\#_1$) neutrosophic triplet G-submodules of (N_1, $*_1$, $\#_1$), such that y ∈ N_1, N_2, and from Theorem 3, N_1 = $K_1 \oplus K_2$, in which, as N = M⊕N_1, N = M⊕$K_1 \oplus K_2$. From Remark 2, ($M*_1K_1$) ∩ K_2 = {neut(x) = x} or ($M*_1K_2$) ∩ K_1 = neut(x) = x}. Then, y ∉ ($M*_1K_1$) ∩ K_2 or y ∉ ($M*_1K_2$) ∩ K_1. Hence, y ∉ ($M*_1K_1$) or y ∉ ($M*_1K_2$). This is a contraction. Thus, (N_1*_1, $\#_1$) is an irreducible neutrosophic triplet G-submodule.*

Theorem 5. *Let (NTV, $*_1$, $\#_1$) be a completely reducible neutrosophic triplet G-module. Then, (NTV, $*_1$, $\#_1$) is a direct sum of irreducible neutrosophic triplet G-modules of (NTV, $*_1$, $\#_1$).*

Proof of Theorem 3. *From Theorem 3, (N_i,$*_1$, $\#_1$) (i = 1, 2, ... , n), neutrosophic triplet G-submodules of (NTV, $*_1$, $\#_1$) are completely reducible neutrosophic triplet G-modules, such that NTV = $N_{i-k} \oplus N_k$ (k = 1, 2, ... , $i-1$). From Theorem 4, there exists (M_i, $*_1$, $\#_1$) irreducible neutrosophic triplet G-submodules of (N_i, $*_1$, $\#_1$). Also, from Theorem 3, (M_i, $*_1$, $\#_1$) are completely reducible neutrosophic triplet G-modules, such that N_i = $N_{i-k} \oplus N_k$ (k = 1, 2, ... , $i-1$). If these steps are followed, we obtained (NTV, $*_1$, $\#_1$), which is a direct sum of irreducible neutrosophic triplet G-modules of (NTV, $*_1$, $\#_1$).*

5. Conclusions

In this paper; we studied the neutrosophic triplet G-module. Furthermore, we showed that neutrosophic triplet G-module is different from the classical G-module. Also, we introduced the reducible neutrosophic triplet G-module, the irreducible neutrosophic triplet G-module, and the completely reducible neutrosophic triplet G-module. The neutrosophic triplet G-module has new properties compared to the classical G-module. By using the neutrosophic triplet G-module, a theory of representation of neutrosophic triplet groups can be defined. Thus, the usage areas of the neutrosophic triplet structures will be expanded.

Author Contributions: Florentin Smarandache defined and studied neutrosophic triplet G-module, Abdullah Kargın defined and studied reducible neutrosophic triplet G-module, irreducible neutrosophic triplet G-module, and the completely reducible neutrosophic triplet G-module. Mehmet Şahin provided the examples and organized the paper.

Conflicts of Interest: The authors declare no conflict of interest.

References

1. Smarandache, F. *Neutrosophy: Neutrosophic Probability, Set and Logic;* Research Press: Rehoboth, MA, USA, 1998.
2. Zadeh, A.L. Fuzzy sets. *Inf. Control* **1965**, *8*, 338–353. [CrossRef]
3. Atanassov, T.K. Intuitionistic fuzzy sets. *Fuzzy Sets Syst.* **1986**, *20*, 87–96. [CrossRef]
4. Bucolo, M.; Fortuna, L.; Rosa, M.L. Complex dynamics through fuzzy chains. *IEEE Trans. Fuzzy Syst.* **2004**, *12*, 289–295. [CrossRef]
5. Chen, S.M. Multi-attribute group decision making based on intuitionistic 2-Tuple linguistic information. *Inf. Sci.* **2018**, *430–431*, 599–619.
6. Chen, S.M. Some q–Rung Orthopair fuzzy aggregation operators and their applications to multiple-Attribute decision making. *Int. J. Intell. Syst.* **2018**, *33*, 259–280.
7. Sahin, M.; Deli, I.; Ulucay, V. Similarity measure of bipolar neutrosophic sets and their application to multiple criteria decision making. *Neural Comput. Appl.* **2016**. [CrossRef]
8. Liu, C.; Luo, Y. Power aggregation operators of simplifield neutrosophic sets and their use in multi-attribute group decision making. *İEE/CAA J. Autom. Sin.* **2017**. [CrossRef]
9. Sahin, R.; Liu, P. Some approaches to multi criteria decision making based on exponential operations of simplied neutrosophic numbers. *J. Intell. Fuzzy Syst.* **2017**, *32*, 2083–2099. [CrossRef]
10. Liu, P.; Li, H. Multi attribute decision-making method based on some normal neutrosophic bonferroni mean operators. *Neural Comput. Appl.* **2017**, *28*, 179–194. [CrossRef]
11. Broumi, S.; Bakali, A.; Talea, M.; Smarandache, F. Decision-Making Method Based on the Interval Valued Neutrosophic Graph. In Proceedings of the IEEE Future Technologie Conference, San Francisco, CA, USA, 6–7 December 2016; pp. 44–50.
12. Liu, P. The aggregation operators based on Archimedean t-conorm and t-norm for the single valued neutrosophic numbers and their application to Decision Making. *Int. J. Fuzzy Syst.* **2016**, *18*, 849–863. [CrossRef]
13. Olgun, N.; Bal, M. Neutrosophic modules. *Neutrosophic Oper. Res.* **2017**, *2*, 181–192.
14. Şahin, M.; Uluçay, V.; Olgun, N.; Kilicman, A. On neutrosophic soft lattices. *Afr. Mat.* **2017**, *28*, 379–388.
15. Şahin, M.; Uluçay, V.; Olgun, N. Soft normed rings. *Springerplus* **2016**, *5*, 1–6.
16. Şahin, M.; Ecemiş, O.; Uluçay, V.; Kargın, A. Some new generalized aggregation operator based on centroid single valued triangular neutrosophic numbers and their applications in multi- attribute decision making. *Assian J. Mat. Comput. Res.* **2017**, *16*, 63–84.
17. Şahin, M.; Olgun, N.; Uluçay, V.; Kargın, A.; Smarandache, F. A new similarity measure based on falsity value between single valued neutrosophic sets based on the centroid points of transformed single valued neutrosophic numbers with applications to pattern recognition. *Neutrosophic Sets Syst.* **2017**, *15*, 31–48. [CrossRef]
18. Ji, P.; Zang, H.; Wang, J. A projection–based TODIM method under multi-valued neutrosophic enviroments and its application in personnel selection. *Neutral Comput. Appl.* **2018**, *29*, 221–234. [CrossRef]
19. Smarandache, F.; Ali, M. Neutrosophic triplet as extension of matter plasma, unmatter plasma and antimatter plasma. In Proceedings of the APS Gaseous Electronics Conference, Bochum, Germany, 10–14 October 2016. [CrossRef]
20. Smarandache, F.; Ali, M. *The Neutrosophic Triplet Group and its Application to Physics;* Universidad National de Quilmes, Department of Science and Technology, Bernal: Buenos Aires, Argentina, 2014.
21. Smarandache, F.; Ali, M. Neutrosophic triplet group. *Neural Comput. Appl.* **2016**, *29*, 595–601. [CrossRef]
22. Smarandache, F.; Ali, M. Neutrosophic Triplet Field Used in Physical Applications, (Log Number: NWS17-2017-000061). In Proceedings of the 18th Annual Meeting of the APS Northwest Section, Pacific University, Forest Grove, OR, USA, 1–3 June 2017.
23. Smarandache, F.; Ali, M. Neutrosophic Triplet Ring and its Applications, (Log Number: NWS17-2017-000062). In Proceedings of the 18th Annual Meeting of the APS Northwest Section, Pacific University, Forest Grove, OR, USA, 1–3 June 2017.
24. Şahin, M.; Kargın, A. Neutrosophic triplet normed space. *Open Phys.* **2017**, *15*, 697–704. [CrossRef]
25. Şahin, M.; Kargın, A. Neutrosophic triplet inner product space. *Neutrosophic Oper. Res.* **2017**, *2*, 193–215.

26. Curties, C.W. *Representation Theory of Finite Group and Associative Algebra*; American Mathematical Society: Providence, RI, USA, 1962.
27. Fernandez, S. A Study of Fuzzy G-Modules. Ph.D. Thesis, Mahatma Gandhi University, Kerala, India, April 2004.
28. Sinho, A.K.; Dewangan, K. Isomorphism Theory for Fuzzy Submodules of G–modules. *Int. J. Eng.* **2013**, *3*, 852–854.
29. Şahin, M.; Olgun, N.; Kargın, A.; Uluçay, V. Soft G-Module. In Proceedings of the Eighth International Conference on Soft Computing, Computing with Words and Perceptions in System Analysis, Decision and Control (ICSCCW-2015), Antalya, Turkey, 3–4 September 2015.
30. Sharma, P.K.; Chopra, S. Injectivity of intuitionistic fuzzy G-modules. *Ann. Fuzzy Math. Inform.* **2016**, *12*, 805–823.
31. Keneth, H.; Ray, K. *Linear Algebra, Eastern Economy*, 2nd ed.; Pearson: New York, NY, USA, 1990.

mathematics

MDPI

Article

Some Types of Subsemigroups Characterized in Terms of Inequalities of Generalized Bipolar Fuzzy Subsemigroups

Pannawit Khamrot [1] and Manoj Siripitukdet [1,2,*]

[1] Departament of Mathematics, Faculty of Science, Naresuan University, Phitsanulok 65000, Thailand; pannawit.k@gmail.com

[2] Research Center for Academic Excellence in Mathematics, Faculty of Science, Naresuan University, Phitsanulok 65000, Thailand

* Correspondence: manojs@nu.ac.th

Received: 14 November2017; Accepted: 23 November 2017; Published: 27 November 2017

Abstract: In this paper, we introduce a generalization of a bipolar fuzzy (BF) subsemigroup, namely, a $(\alpha_1, \alpha_2; \beta_1, \beta_2)$-BF subsemigroup. The notions of $(\alpha_1, \alpha_2; \beta_1, \beta_2)$-BF quasi(generalized bi-, bi-) ideals are discussed. Some inequalities of $(\alpha_1, \alpha_2; \beta_1, \beta_2)$-BF quasi(generalized bi-, bi-) ideals are obtained. Furthermore, any regular semigroup is characterized in terms of generalized BF semigroups.

Keywords: generalized bipolar fuzzy (BF) semigroup; $(\alpha_1, \alpha_2; \beta_1, \beta_2)$-bipolar fuzzy subsemigroup; fuzzy quasi(generalized bi-, bi-) ideal; regular semigroup

1. Introduction

Most of the bipolarities separate positive and negative information response; positive information representations are compiled to be possible, while negative information representations are impossible [1]. The bipolar information of evaluation can help to evaluate decisions. Sometimes, decisions are not only influenced by the positive decision criterion, but also with the negative decision criterion, for example, environmental and social impact assessment. Evaluated alternative consideration should weigh the negative effects to select the optimal choice. Therefore bipolar information affects the effectiveness and efficiency of decision making. It is used in decision-making problems, organization problems, economic problems, evaluation, risk management, environmental and social impact assessment, and so forth. Thus, the concept of bipolar fuzzy (BF) sets are more relevant in mathematics.

In 1965, Zadeh [2] introduced the fuzzy set theory, which can be applied to many areas, such as mathematics, statistics, computers, electrical instruments, the industrial industry, business, engineering, social applications, and so forth. In 2003, Bucolo et al. [3] proposed small-world networks of fuzzy chaotic oscillators. The fuzzy set was used to establish the mathematical method for dealing with imprecise and uncertain environments. In 1971, Rosenfeld [4] applied fuzzy sets to group structures. Then, the fuzzy set was used in the theory of semigroups in 1979. Kuroki [5] initiated fuzzy semigroups based on the notion of fuzzy ideals in semigroups and introduced some properties of fuzzy ideals and fuzzy bi-ideals of semigroups. The fundamental concepts of BF sets were initiated by Zhang [6] in 1994. He innovated the BF set as BF logic, which has been widely applied to solve many real-world problems. In 2000, Lee [7] studied the notion of bipolar-valued fuzzy sets. Kim et al. [8] studied the notions of BF subsemigroups, BF left (right, bi-) ideals. He provided some necessary and sufficient conditions for a BF subsemigroup and BF left (right, bi-) ideals of semigroups.

In this paper, generalizations of BF semigroups are introduced. Definitions and properties of $(\alpha_1, \alpha_2; \beta_1, \beta_2)$-BF quasi (generalized bi-, bi-) ideals are obtained. Some inequalities of $(\alpha_1, \alpha_2; \beta_1, \beta_2)$-BF quasi (generalized bi-, bi-) ideals are obtained. Finally, we characterize a regular semigroup in terms of generalized BF semigroups.

2. Preliminaries

In this section, we give definitions and examples that are used in this paper. By a subsemigroup of a semigroup S we mean a non-empty subset A of S such that $A^2 \subseteq A$, and by a left (right) ideal of S we mean a non-empty subset A of S such that $SA \subseteq A (AS \subseteq A)$. By a two-sided ideal or simply an ideal, we mean a non-empty subset of a semigroup S that is both a left and a right ideal of S. A non-empty subset A of S is called an **interior ideal** of S if $SAS \subseteq A$, and a **quasi-ideal** of S if $AS \cap SA \subseteq A$. A subsemigroup A of S is called a **bi-ideal** of S if $ASA \subseteq A$. A non-empty subset A is called a **generalized bi-ideal** of S if $ASA \subseteq A$ [9].

By the definition of a left (right) ideal of a semigroup S, it is easy to see that every left (right) ideal of S is a quasi-ideal of S.

Definition 1. *A semigroup S is called **regular** if for all $a \in S$ there exists $x \in S$ such that $a = axa$.*

Theorem 1. *For a semigroup S, the following conditions are equivalent.*

(1) *S is regular.*
(2) *$R \cap L = RL$ for every right ideal R and every left ideal L of S.*
(3) *$ASA = A$ for every quasi-ideal A of S.*

Definition 2. *Let X be a set; a fuzzy set (or fuzzy subset) f on X is a mapping $f : X \to [0,1]$, where $[0,1]$ is the usual interval of real numbers.*

The symbols $f \wedge g$ and $f \vee g$ will denote the following fuzzy sets on S:

$$(f \wedge g)(x) = f(x) \wedge g(x)$$

$$(f \vee g)(x) = f(x) \vee g(x)$$

for all $x \in S$.

A product of two fuzzy sets f and g is denoted by $f \circ g$ and is defined as

$$(f \circ g)(x) = \begin{cases} \bigvee_{x=yz}\{f(y) \wedge g(z)\}, & \text{if } x = yz \text{ for some } y,z \in S \\ 0, & \text{otherwise} \end{cases}$$

Definition 3. *Let S be a non-empty set. A BF set f on S is an object having the following form:*

$$f := \{(x, f_p(x), f_n(x)) : x \in S\}$$

where $f_p : S \to [0,1]$ and $f_n : S \to [-1,0]$.

Remark 1. *For the sake of simplicity we use the symbol $f = (S; f_p, f_n)$ for the BF set $f = \{(x, f_p(x), f_n(x)) : x \in S\}$.*

Definition 4. *Given a BF set $f = (S; f_p, f_n)$, $\alpha \in [0,1]$ and $\beta \in [-1,0]$, the sets*

$$P(f;\alpha) := \{x \in S | f_p(x) \geq \alpha\}$$

and

$$N(f;\beta) := \{x \in S | f_n(x) \leq \beta\}$$

*are called the **positive** α**-cut** and **negative** β**-cut** of f, respectively. The set $C(f;(\alpha,\beta)) := P(f;\alpha) \cap N(f;\beta)$ is called the **bipolar** (α,β)**-cut** of f.*

We give the generalization of a BF subsemigroup, which is defined by Kim et al. (2011).

Definition 5. *A BF set $f = (S; f_p, f_n)$ on S is called a $(\alpha_1, \alpha_2; \beta_1, \beta_2)$**-BF subsemigroup** on S, where $\alpha_1, \alpha_2 \in [0,1], \beta_1, \beta_2 \in [-1,0]$ if it satisfies the following conditions:*

(1) $f_p(xy) \vee \alpha_1 \geq f_p(x) \wedge f_p(y) \wedge \alpha_2$
(2) $f_n(xy) \wedge \beta_2 \leq f_n(x) \vee f_n(y) \vee \beta_1$

for all $x, y \in S$.

We note that every BF subsemigroup is a $(0,1,-1,0)$-BF subsemigroup.

The following examples show that $f = (S; f_p, f_n)$ is a $(\alpha_1, \alpha_2; \beta_1, \beta_2)$-BF subsemigroup on S but $f = (S; f_p, f_n)$ is not a BF subsemigroup on S.

Example 1. *The set $S = \{2,3,4,...\}$ is a semigroup under the usual multiplication. Let $f = (S; f_p, f_n)$ be a BF set on S defined as follows:*

$$f_p(x) := \frac{1}{x+2} \text{ and } f_n(x) := -\frac{1}{x+2}$$

for all $x \in S$.

Let $x, y \in S$. Then

$$f_p(xy) = \frac{1}{xy+2} < \frac{1}{x+2} = f_p(x)$$

and

$$f_p(xy) = \frac{1}{xy+2} < \frac{1}{y+2} = f_p(y)$$

Thus, $f_p(xy) < f_p(x) \wedge f_p(y)$. Therefore $f = (S; f_p, f_n)$ is not a BF subsemigroup on S.

Let $\alpha_2 \in [0,1]$, $\beta_1 \in [-1,0]$, $\alpha_1 = \frac{1}{4}$ and $\beta_2 = -\frac{1}{4}$. Thus for all $x, y \in S$,

$$f_p(xy) \vee \frac{1}{4} \geq \frac{1}{x+2} \wedge \frac{1}{y+2} \geq f_p(x) \wedge f_p(y) \wedge \alpha_2$$

and

$$f_n(xy) \wedge -\frac{1}{4} \leq -\frac{1}{x+2} \vee -\frac{1}{y+2} \leq f_n(x) \vee f_n(y) \vee \beta_1$$

Hence $f = (S; f_p, f_n)$ is a $(\frac{1}{4}, \alpha_2; \beta_1, -\frac{1}{4})$-BF subsemigroup on S.

We note that $f = (S; f_p, f_n)$ is a $(\alpha_1, \alpha_2; \beta_1, \beta_2)$-BF subsemigroup on S for all $\alpha_1 \geq \dfrac{1}{4}$ and $\beta_2 \leq -\dfrac{1}{4}$.

Definition 6. *A BF set $f = (S; f_p, f_n)$ on S is called a $(\alpha_1, \alpha_2; \beta_1, \beta_2)$**-BF left (right) ideal** on S, where $\alpha_1, \alpha_2 \in [0,1]$, and $\beta_1, \beta_2 \in [-1,0]$ if it satisfies the following conditions:*

(1) $f_p(xy) \vee \alpha_1 \geq f_p(y) \wedge \alpha_2$ $(f_p(xy) \vee \alpha_1 \geq f_p(x) \wedge \alpha_2)$
(2) $f_n(xy) \wedge \beta_2 \leq f_n(y) \vee \beta_1$ $(f_n(xy) \wedge \beta_2 \leq f_n(x) \vee \beta_1)$

for all $x, y \in S$.

A BF set $f = (S; f_p, f_n)$ on S is called a $(\alpha_1, \alpha_2; \beta_1, \beta_2)$-**BF ideal** on S $(\alpha_1, \alpha_2 \in [0,1], \beta_1, \beta_2 \in [-1,0])$ if it is both a $(\alpha_1, \alpha_2; \beta_1, \beta_2)$-BF left ideal and a $(\alpha_1, \alpha_2; \beta_1, \beta_2)$-BF right ideal on S.

By Definition 6, every $(\alpha_1, \alpha_2; \beta_1, \beta_2)$-BF ideal on a semigroup S is a $(\alpha_1, \alpha_2; \beta_1, \beta_2)$-BF subsemigroup on S.

We note that a $(0,1,-1,0)$-BF left (right) ideal is a BF left (right) ideal.

Definition 7. *A* $(\alpha_1, \alpha_2; \beta_1, \beta_2)$*-BF subsemigroup* $f = (S; f_p, f_n)$ *on a subsemigroup* S *is called a* $(\alpha_1, \alpha_2; \beta_1, \beta_2)$***-BF bi-ideal*** *on* S*, where* $\alpha_1, \alpha_2 \in [0, 1]$*, and* $\beta_1, \beta_2 \in [-1, 0]$ *if it satisfies the following conditions:*

(1) $f_p(xay) \vee \alpha_1 \geq f_p(x) \wedge f_p(y) \wedge \alpha_2$

(2) $f_n(xay) \wedge \beta_2 \leq f_n(x) \vee f_n(y) \vee \beta_1$

for all $x, y \in S$.

We note that every $(\alpha_1, \alpha_2; \beta_1, \beta_2)$-BF bi-ideal on a semigroup is a $(\alpha_1, \alpha_2; \beta_1, \beta_2)$-BF subsemigroup on the semigroup.

3. Generalized Bi-Ideal and Quasi-Ideal

In this section, we introduce a product of BF sets and characterize a regular semigroup by generalized BF subsemigroups.

We let $f = (S : f_p, f_n)$ and $g = (S : g_p, g_n)$ be two BF sets on a semigroup S and let $\alpha_1, \alpha_2 \in [0, 1]$, and $\beta_1, \beta_2 \in [-1, 0]$. We define two fuzzy sets $f_p^{(\alpha_1,\alpha_2)}$and $f_n^{(\beta_1,\beta_2)}$ on S as follows:

$$f_p^{(\alpha_1,\alpha_2)}(x) = (f_p(x) \wedge \alpha_1) \vee \alpha_2$$

$$f_n^{(\beta_1,\beta_2)}(x) = (f_n(x) \vee \beta_2) \wedge \beta_1$$

for all $x \in S$.

We define two operations $\overset{(\alpha_1,\alpha_2)}{\wedge}$ and $\underset{(\beta_1,\beta_2)}{\vee}$ on S as follows:

$$(f_p \overset{(\alpha_1,\alpha_2)}{\wedge} g_p)(x) = ((f_p \wedge g_p)(x) \wedge \alpha_1) \vee \alpha_2$$

$$(f_n \underset{(\beta_1,\beta_2)}{\wedge} g_n)(x) = ((f_n \vee g_n)(x) \vee \alpha_2) \wedge \alpha_1$$

for all $x \in S$, and we define products $f_p \overset{(\alpha_1,\alpha_2)}{\circ} g_p$ and $f_n \underset{(\beta_1,\beta_2)}{\circ} g_n$ as follows:

For all $x \in S$,

$$(f_p \overset{(\alpha_1,\alpha_2)}{\circ} g_p)(x) = ((f_p \bar{\circ} g_p)(x) \wedge \alpha_1) \vee \alpha_2$$

$$(f_n \underset{(\beta_1,\beta_2)}{\circ} g_n)(x) = ((f_n \underline{\circ} g_n)(x) \vee \alpha_2) \wedge \alpha_1$$

where

$$(f_p \bar{\circ} g_p)(x) = \begin{cases} \bigvee_{x=yz}\{f_p(y) \wedge g_p(z)\} & \text{if } x = yz \text{ for some } y, z \in S \\ 0 & \text{otherwise} \end{cases}$$

$$(f_n \underline{\circ} g_n)(x) = \begin{cases} \bigwedge_{y=yz}\{f_n(y) \vee g_n(z)\} & \text{if } x = yz \text{ for some } y, z \in S \\ 0 & \text{otherwise} \end{cases}$$

We set

$$f \overset{(\alpha_1,\alpha_2)}{\underset{(\beta_1,\beta_2)}{\circ}} g := (S; f_p \overset{(\alpha_1,\alpha_2)}{\circ} g_p, f_n \underset{(\beta_1,\beta_2)}{\circ} g_n)$$

Then it is a BF set.

We note that

(1) $f_p^{(1,0)}(x) = f_p(x)$,

(2) $f_n^{(0,-1)}(x) = f_n(x)$,

(3) $f = (S; f_p, f_n) = (S; f_p^{(1,0)}, f_n^{(0,-1)})$,

(4) $(f \overset{(\alpha_1,\alpha_2)}{\underset{(\beta_1,\beta_2)}{\circ}} g)_p = f_p \overset{(\alpha_1,\alpha_2)}{\circ} g_p$ and $(f \overset{(\alpha_1,\alpha_2)}{\underset{(\beta_1,\beta_2)}{\circ}} g)_n = f_n \underset{(\beta_1,\beta_2)}{\circ} g_n$.

Definition 8. *A BF set $f = (S; f_p, f_n)$ on S is called a $(\alpha_1, \alpha_2; \beta_1, \beta_2)$-**BF generalized bi-ideal** on S, where $\alpha_1, \alpha_2 \in [0,1]$, and $\beta_1, \beta_2 \in [-1,0]$ if it satisfies the following conditions:*

(1) $f_p(xay) \vee \alpha_1 \geq f_p(x) \wedge f_p(y) \wedge \alpha_2$

(2) $f_n(xay) \wedge \beta_2 \leq f_n(x) \vee f_n(y) \vee \beta_1$

for all $x, y, a \in S$.

Definition 9. *A BF set $f = (S; f_p, f_n)$ on S is called a $(\alpha_1, \alpha_2; \beta_1, \beta_2)$-**BF quasi-ideal** on S, where $\alpha_1, \alpha_2 \in [0,1]$, and $\beta_1, \beta_2 \in [-1,0]$ if it satisfies the following conditions:*

(1) $f_p(x) \vee \alpha_1 \geq (f_p \bar{\circ} \mathcal{S}_p)(x) \wedge (\mathcal{S}_p \bar{\circ} f_p)(x) \wedge \alpha_2$

(2) $f_n(x) \wedge \beta_2 \leq (f_n \underline{\circ} \mathcal{S}_n)(x) \vee (\mathcal{S}_n \underline{\circ} f_n)(x) \vee \beta_1$

for all $x, y \in S$.

In the following theorem, we give a relation between a bipolar (α, β)-cut of f and a $(\alpha_1, \alpha_2; \beta_1, \beta_2)$-BF generalized bi-ideal on S.

Theorem 2. *Let $f = (S; f_p, f_n)$ be a BF set on a semigroup S with $Im(f_p) \subseteq \Delta^+ \subseteq [0,1]$ and $Im(f_n) \subseteq \Delta^- \subseteq [-1,0]$. Then $C(f; (\alpha, \beta))(\neq \emptyset)$ is a generalized bi-ideal of S for all $\alpha \in \Delta^+$ and $\beta \in \Delta^-$ if and only if f is a $(\alpha_1, \alpha_2; \beta_1, \beta_2)$-BF generalized bi-ideal on S for all $\alpha_1, \alpha_2 \in [0,1]$ and $\beta_1, \beta_2 \in [-1,0]$.*

Proof. Let $\alpha_1, \alpha_2 \in [0,1], \beta_1, \beta_2 \in [-1,0]$. Suppose on the contrary that f is not a $(\alpha_1, \alpha_2; \beta_1, \beta_2)$-BF generalized bi-ideal on S. Then there exists $x, y, a \in S$ such that

$$f_p(xay) \vee \alpha_1 < f_p(x) \wedge f_p(y) \wedge \alpha_2 \text{ or } f_n(xay) \wedge \beta_2 > f_n(x) \vee f_n(y) \vee \beta_1 \quad (1)$$

Let $\alpha' = f_p(x) \wedge f_p(y)$ and $\beta' = f_n(x) \vee f_n(y)$. Then $x, y \in C(f; (\alpha', \beta'))$. By assumption, we have $xay \in C(f; (\alpha', \beta'))$. By Equation (1), $f_p(xay) \leq f_p(xay) \wedge \alpha_1 < f_p(x) \wedge f_p(y) \wedge \alpha_2 \leq f_p(x) \wedge f_p(y) = \alpha'$ or $f_n(xay) \geq f_n(xay) \wedge \beta_2 > f_n(x) \vee f_n(y) \vee \beta_1 \geq f_n(x) \vee f_n(y) = \beta'$. Thus, $xay \notin C(f; (\alpha', \beta'))$. This is a contradiction. Therefore f is a $(\alpha_1, \alpha_2; \beta_1, \beta_2)$-BF generalized bi-ideal on S.

Conversely, let $\alpha \in \Delta^+$, and $\beta \in \Delta^-$, and suppose that $C(f; (\alpha, \beta)) \neq \emptyset$. Let $a \in S$ and $x, y \in C(f; (\alpha, \beta))$. Then $f_p(x) \geq \alpha, f_p(y) \geq \alpha, f_n(x) \leq \beta$ and $f_n(y) \leq \beta$. By assumption, f is a $(\alpha, f_p(xay); f_n(xay), \beta)$-BF generalized bi-ideal on S, and thus $f_p(xay) \vee f_p(xay) \geq f_p(x) \wedge f_p(y) \wedge \alpha$ and $f_n(xay) \wedge f_n(xay) \leq f_n(x) \vee f_n(y) \vee \beta$. Then $f_p(xay) \geq f_p(x) \wedge f_p(y) \wedge \alpha \geq \alpha \wedge \alpha = \alpha$ and $f_n(xay) \leq f_n(x) \vee f_n(x) \vee \beta \leq \beta \vee \beta = \beta$. Hence, $xay \in C(f; (\alpha, \beta))$. Therefore $C(f; (\alpha, \beta))$ is a generalized bi-ideal of S. □

Corollary 1. *Let $f = (S; f_n, f_p)$ be a BF set on a semigroup. Then the following statements hold:*

(1) *f is a $(\alpha_1, \alpha_2; \beta_1, \beta_2)$-BF generalized bi-ideal on S for all $\alpha_1, \alpha_2 \in [0,1]$ and $\beta_1, \beta_2 \in [-1,0]$ if and only if $C(f; (\alpha, \beta))(\neq \emptyset)$ is a generalized bi-ideal of S for all $\alpha \in Im(f_p)$, and $\beta \in Im(f_n)$;*

(2) *f is a $(\alpha_1, \alpha_2; \beta_1, \beta_2)$-BF generalized bi-ideal on S for all $\alpha_1, \alpha_2 \in [0,1]$, and $\beta_1, \beta_2 \in [-1,0]$ if and only if $C(f; (\alpha, \beta))(\neq \emptyset)$ is a generalized bi-ideal of S for all $\alpha \in [0,1]$, and $\beta \in [-1,0]$.*

Proof. (1) Set $\Delta^+ = [0,1]$ and $\Delta^- = [-1,0]$, and apply Theorem 2.

(2) Set $\Delta^+ = Im(f_p)$ and $\Delta^- = Im(f_n)$, and apply Theorem 2. □

Lemma 1. *Every $(\alpha_1, \alpha_2; \beta_1, \beta_2)$-BF generalized bi-ideal on regular semigroup S is a $(\alpha_1, \alpha_2; \beta_1, \beta_2)$-BF bi-ideal on S.*

Proof. Let S be a regular semigroup and $f = (S; f_p, f_n)$ be a $(\alpha_1, \alpha_2; \beta_1, \beta_2)$-BF generalized bi-ideal on S. Let $a, b \in S$; then there exists $x \in S$ such that $b = bxb$. Thus we have $f_p(ab) \vee \alpha_1 = f_p(a(bxb)) \vee \alpha_1 = f_p(a(bx)b) \vee \alpha_1 \geq f_p(a) \wedge f_p(b) \wedge \alpha_2$ and $f_n(ab) \wedge \beta_2 = f_n(a(bxb)) \wedge \beta_2 = f_n(a(bx)b) \wedge \beta_2 \leq f_n(a) \vee f_n(b) \vee \beta_1$. This shows that f is a $(\alpha_1, \alpha_2; \beta_1, \beta_2)$-BF subsemigroup on S, and thus f is a $(\alpha_1, \alpha_2; \beta_1, \beta_2)$-BF bi-ideal on S. □

Let S be a semigroup and $\emptyset \neq I \subseteq S$. A positive characteristic function and a negative characteristic function are respectively defined by

$$C_I^p : S \to [0,1], x \mapsto C_I^p(x) := \begin{cases} 1, & x \in I \\ 0, & x \notin I \end{cases}$$

and

$$C_I^n : S \to [-1,0], x \mapsto C_I^n(x) := \begin{cases} -1, & x \in I \\ 0, & x \notin I \end{cases}$$

Remark 2.

(1) *For the sake of simplicity, we use the symbol* $C_I = (S; C_I^p, C_I^n)$ *for the BF set. That is,* $C_I = (S; C_I^p, C_I^n) = (S; (C_I)_p, (C_I)_n)$. *We call this a bipolar characteristic function.*

(2) *If* $I = S$, *then* $C_S = (S; C_S^p, C_S^n)$. *In this case, we denote* $\mathcal{S} = (S, \mathcal{S}_p, \mathcal{S}_n)$.

In the following theorem, some necessary and sufficient conditions of $(\alpha_1, \alpha_2; \beta_1, \beta_2)$-BF generalized bi-ideals are obtained.

Theorem 3. *Let* $f = (S; f_p, f_n)$ *be a BF set on a semigroup* S. *Then the following statements are equivalent:*

(1) f *is a* $(\alpha_1, \alpha_2; \beta_1, \beta_2)$*-BF generalized bi-ideal on* S.

(2) $f_p \overset{(\alpha_2,\alpha_1)}{\circ} \mathcal{S}_p \overset{(\alpha_2,\alpha_1)}{\circ} f_p \leq f_p^{(\alpha_2,\alpha_1)}$ *and* $f_n \underset{(\beta_2,\beta_1)}{\circ} \mathcal{S}_n \underset{(\beta_2,\beta_1)}{\circ} f_n \geq f_n^{(\beta_2,\beta_1)}$.

Proof. ($\Rightarrow$) Let a be any element of S. In the case for which $(f_p \overset{(\alpha_2,\alpha_1)}{\circ} \mathcal{S}_p \overset{(\alpha_2,\alpha_1)}{\circ} f_p)(a) = 0$, it is clear that $f_p \overset{(\alpha_2,\alpha_1)}{\circ} \mathcal{S}_p \overset{(\alpha_2,\alpha_1)}{\circ} f_p \leq f_p^{(\alpha_2,\alpha_1)}$. Otherwise, there exist $x, y, r, s \in S$ such that $a = xy$ and $x = rs$. Because f is a $(\alpha_1, \alpha_2; \beta_1, \beta_2)$-BF generalized bi-ideal on S, we have $f_p(rsy) \vee \alpha_1 \geq f_p(r) \wedge f_p(s) \wedge \alpha_2$ and $f_n(rsy) \wedge \beta_2 \leq f_n(r) \vee f_n(s) \vee \beta_1$. Consider

$$\begin{aligned}
(f_p \overset{(\alpha_2,\alpha_1)}{\circ} \mathcal{S}_p \overset{(\alpha_2,\alpha_1)}{\circ} f_p)(a) &= ((f_p \bar{\circ} \mathcal{S}_p \bar{\circ} f_p)(a) \wedge \alpha_2) \vee \alpha_1 \\
&= (\bigvee_{a=xy} \{(f_p \bar{\circ} \mathcal{S}_p)(x) \wedge f_p(y)\} \wedge \alpha_2) \vee \alpha_1 \\
&= (\bigvee_{a=xy} \{\bigvee_{x=rs} \{f_p(r) \wedge \mathcal{S}_p(s)\} \wedge f_p(y)\} \wedge \alpha_2) \vee \alpha_1 \\
&= (\bigvee_{a=xy} \{\bigvee_{x=rs} \{f_p(r) \wedge 1\} \wedge f_p(y)\} \wedge \alpha_2) \vee \alpha_1 \\
&= (\bigvee_{a=rsy} \{f_p(r) \wedge f_p(y) \wedge \alpha_2\} \wedge \alpha_2) \vee \alpha_1 \\
&\leq (\bigvee_{a=rsy} \{f_p(rsy) \vee \alpha_1\} \wedge \alpha_2) \vee \alpha_1 \\
&\leq (\bigvee_{a=rsy} \{f_p(a) \vee \alpha_1\} \wedge \alpha_2) \vee \alpha_1 \\
&\leq ((f_p(a) \vee \alpha_1) \wedge \alpha_2) \vee \alpha_1 \\
&= (f_p(a) \wedge \alpha_2) \vee \alpha_1 \\
&= f_p^{(\alpha_2,\alpha_1)}(a)
\end{aligned}$$

Hence $f_p \overset{(\alpha_2,\alpha_1)}{\circ} \mathcal{S}_p \overset{(\alpha_2,\alpha_1)}{\circ} f_p \leq f_p^{(\alpha_2,\alpha_1)}$.

Similarly, we can show that $f_n \underset{(\beta_2,\beta_1)}{\circ} \mathcal{S}_n \underset{(\beta_2,\beta_1)}{\circ} f_n \geq f_n^{(\beta_2,\beta_1)}$.

($\Leftarrow$) Conversely, let $a, x, y, z \in S$ such that $a = xyz$. Then we have

$$\begin{aligned}
f_p(xyz) \vee \alpha_1 &\geq (f_p(a) \wedge \alpha_2) \vee \alpha_1 \\
&= f_p^{(\alpha_2,\alpha_1)}(a) \\
&\geq (f_p \overset{(\alpha_2,\alpha_1)}{\circ} \mathcal{S}_p \overset{(\alpha_2,\alpha_1)}{\circ} f_p)(a) \\
&= ((f_p \bar{\circ} \mathcal{S}_p \bar{\circ} f_p)(a) \wedge \alpha_2) \vee \alpha_1 \\
&= (\bigvee_{a=bc} \{(f_p \bar{\circ} \mathcal{S}_p)(b) \wedge f_p(c)\} \wedge \alpha_2) \vee \alpha_1 \\
&\geq ((f_p \bar{\circ} \mathcal{S}_p)(xy) \wedge f_p(z) \wedge \alpha_2) \vee \alpha_1 \\
&= (\bigvee_{xy=rs} \{f_p(r) \wedge \mathcal{S}_p(s)\} \wedge f_p(z) \wedge \alpha_2) \vee \alpha_1 \\
&\geq (f_p(x) \wedge \mathcal{S}_p(y) \wedge f_p(z) \wedge \alpha_2) \vee \alpha_1 \\
&\geq (f_p(x) \wedge f_p(z) \wedge \alpha_2) \vee \alpha_1 \\
&\geq f_p(x) \wedge f_p(z) \wedge \alpha_2
\end{aligned}$$

Similarly, we can show that $f_n(xyz) \wedge \beta_2 \leq f_n(x) \vee f_n(z) \vee \beta_1$ for all $x, y, z \in S$. Therefore f is a $(\alpha_1, \alpha_2; \beta_1, \beta_2)$-BF generalized bi-ideal on S for all $\alpha_1, \alpha_2 \in [0,1]$ and $\beta_1, \beta_2 \in [-1,0]$. □

Theorem 4. *Let $f = (S; f_p, f_n)$ be a BF set on a semigroup S. Then the following statements are equivalent:*

(1) *f is a $(\alpha_1, \alpha_2; \beta_1, \beta_2)$-BF bi-ideal on S .*

(2) $f_p \overset{(\alpha_2,\alpha_1)}{\circ} \mathcal{S}_p \overset{(\alpha_2,\alpha_1)}{\circ} f_p \leq f_p^{(\alpha_2,\alpha_1)}$ *and* $f_n \underset{(\beta_2,\beta_1)}{\circ} \mathcal{S}_n \underset{(\beta_2,\beta_1)}{\circ} f_n \geq f_n^{(\beta_2,\beta_1)}$.

Proof. The proof is similar to the proof of Theorem 3. □

In the following theorem, we give a relation between a bipolar (α, β)-cut of f and a $(\alpha_1, \alpha_2; \beta_1, \beta_2)$-BF quasi-ideal on S.

Theorem 5. *Let $f = (S; f_p, f_n)$ be a BF set on a semigroup S with $Im(f_p) \subseteq \Delta^+ \subseteq [0,1]$ and $Im(f_n) \subseteq \Delta^- \subseteq [-1,0]$. Then $C(f; (\alpha, \beta))(\neq\emptyset)$ is a quasi-ideal of S for all $\alpha \in \Delta^+$ and $\beta \in \Delta^-$ if and only if f is a $(\alpha_1, \alpha_2; \beta_1, \beta_2)$-BF quasi-ideal on S for all $\alpha_1, \alpha_2 \in [0,1]$ and $\beta_1, \beta_2 \in [-1,0]$.*

Proof. ($\Rightarrow$) Let $\alpha_1, \alpha_2 \in [0,1]$ and $\beta_1, \beta_2 \in [-1,0]$. Suppose on the contrary that f is not a $(\alpha_1, \alpha_2; \beta_1, \beta_2)$-BF quasi-ideal on S. Then there exists $x \in S$ such that

$$f_p(x) \vee \alpha_1 < (f_p \bar{\circ} \mathcal{S}_p)(x) \wedge (\mathcal{S}_p \bar{\circ} f_p)(x) \wedge \alpha_2$$

or

$$f_n(x) \wedge \beta_2 > (f_n \underline{\circ} \mathcal{S}_n)(x) \vee (\mathcal{S}_n \underline{\circ} f_n)(x) \vee \beta_1 \tag{2}$$

Case 1: $f_p(x) \vee \alpha_1 < (f_p \bar{\circ} \mathcal{S}_p)(x) \wedge (\mathcal{S}_p \bar{\circ} f_p)(x) \wedge \alpha_2$. Let $\alpha' = (f_p \bar{\circ} \mathcal{S}_p)(x) \wedge (\mathcal{S}_p \bar{\circ} f_p)(x)$. Then $\alpha' \leq (f_p \bar{\circ} \mathcal{S}_p)(x)$, $\alpha' \leq (\mathcal{S}_p \bar{\circ} f_p)(x)$. This implies that there exist $a, b, c, d \in S$ such that $x = ab = cd$. Then

$$\alpha' \leq (f_p \bar{\circ} \mathcal{S}_p)(x) = \bigvee_{x=yz} \{f_p(y) \wedge \mathcal{S}_p(z)\} \leq f_p(a) \wedge \mathcal{S}_p(b) = f_p(a)$$

$$\alpha' \leq (\mathcal{S}_p\bar{\circ}f_p)(x) = \bigvee_{x=yz}\{\mathcal{S}_p(z)\wedge f_p(y)\} \leq \mathcal{S}_p(c)\wedge f_p(d) = f_p(d)$$

Let $\beta' = f_n(a) \vee f_n(d)$. Then $f_n(a) \leq \beta'$ and $f_n(d) \leq \beta'$.

Thus $a,d \in C(f;(\alpha',\beta'))$, and so $ad \in C(f;(\alpha',\beta'))S$ and $ad \in SC(f;(\alpha',\beta'))$. Hence $x \in C(f;(\alpha',\beta'))S$ and $x \in SC(f;(\alpha',\beta'))$, and it follows that $x \in C(f;(\alpha',\beta'))S \cap SC(f;(\alpha',\beta'))$. By hypothesis, $x \in C(f;(\alpha',\beta'))$.

Case 2: $f_n(x) \wedge \beta_2 > (f_n\underline{\circ}\mathcal{S}_n)(x) \vee (\mathcal{S}_n\underline{\circ}f_n)(x) \vee \beta_1$. Let $\beta' = (f_n\underline{\circ}\mathcal{S}_n)(x) \vee (\mathcal{S}_n\underline{\circ}f_n)(x)$. Then $\beta' \geq (f_n\underline{\circ}\mathcal{S}_n)(x)$ and $\beta' \geq (\mathcal{S}_n\underline{\circ}f_n)(x)$. This implies that there exist $a',b',c',d' \in S$ such that $x = a'b' = c'd'$. Then

$$\beta' \geq (f_n\underline{\circ}\mathcal{S}_n)(x) = \bigwedge_{x=yz}\{f_n(y)\vee\mathcal{S}_n(z)\} \geq f_n(a')\vee\mathcal{S}_n(b') \geq f_n(a')$$

$$\beta' \geq (\mathcal{S}_n\underline{\circ}f_n)(x) = \bigwedge_{x=yz}\{\mathcal{S}_n(z)\vee f_n(y)\} \geq \mathcal{S}_n(c')\vee f_n(d') \geq f_n(d')$$

Let $\alpha' = f_p(a') \wedge f_p(d')$. Then $f_p(a') \geq \alpha'$ and $f_p(d') \geq \alpha'$. Thus $a',d' \in C(f;(\alpha',\beta'))$, and so $a'd' \in C(f;(\alpha',\beta'))S$ and $a'd' \in SC(f;(\alpha',\beta'))$. Hence $x \in C(f;(\alpha',\beta'))S$ and $x \in SC(f;(\alpha',\beta'))$, and it follows that $x \in C(f;(\alpha',\beta'))S \cap SC(f;(\alpha',\beta'))$. By hypothesis, $x \in C(f;(\alpha',\beta'))$. Therefore $x \in C(f;(\alpha',\beta'))$. By Equation (2),

$$f_p(x) \leq f_p(x)\vee\alpha_1 < (f_p\bar{\circ}\mathcal{S}_p)(x)\wedge(\mathcal{S}_p\bar{\circ}f_p)(x)\wedge\alpha_2 \leq (f_p\bar{\circ}\mathcal{S}_p)(x)\wedge(\mathcal{S}_p\bar{\circ}f_p)(x) = \alpha'$$

or

$$f_n(x) \geq f_n(x)\wedge\beta_2 > (f_n\underline{\circ}\mathcal{S}_n)(x)\vee(\mathcal{S}_n\underline{\circ}f_n)(x)\vee\beta_1 \geq (f_n\underline{\circ}\mathcal{S}_n)(x)\vee(\mathcal{S}_n\underline{\circ}f_n)(x) = \beta'$$

and it follows that $x \notin C(f;(\alpha',\beta'))$. This is a contradiction. Therefore f is a $(\alpha_1,\alpha_2;\beta_1,\beta_2)$-BF quasi-ideal on S.

($\Leftarrow$) Conversely, let $\alpha \in \Delta^+$ and $\beta \in \Delta^-$, and suppose that $C(f;(\alpha,\beta)) \neq \emptyset$. Let $x \in S$ be such that $x \in C(f;(\alpha,\beta))S \cap SC(f;(\alpha,\beta))$. Then $x \in C(f;(\alpha,\beta))S$ and $x \in SC(f;(\alpha,\beta))$. Thus there exist $y,z' \in C(f;(\alpha,\beta))$ and $z,y' \in S$ such that $x = yz$ and $x = y'z'$.

By assumption, f is a $(f_p(x),\alpha;\beta,f_n(x))$-BF quasi-ideal on S, and thus

$$\begin{aligned}
f_p(x) &= f_p(x)\vee f_p(x)\\
&\geq (f_p\bar{\circ}\mathcal{S}_p)(x)\wedge(\mathcal{S}_p\bar{\circ}f_p)(x)\wedge\alpha\\
&= \bigvee_{x=yz}\{f_p(y)\wedge\mathcal{S}_p(z)\}\wedge\bigvee_{x=y'z'}\{\mathcal{S}_p(y')\wedge f_p(z')\}\wedge\alpha\\
&= \bigvee_{x=yz}\{f_p(y)\wedge 1\}\wedge\bigvee_{x=y'z'}\{1\wedge f_p(z')\}\wedge\alpha\\
&= \bigvee_{x=yz}\{f_p(y)\}\wedge\bigvee_{x=y'z'}\{f_p(z')\}\wedge\alpha\\
&\geq f_p(y)\wedge f_p(z')\wedge\alpha
\end{aligned}$$

Because $y,z' \in C(f;(\alpha,\beta))$, we have $f_p(y) \geq \alpha$ and $f_p(z') \geq \alpha$. Then $f_p(x) \geq \alpha$. Similarly, we can show that $f_n(x) \leq \beta$. Hence, $x \in C(f;(\alpha,\beta))$. Therefore $C(f;(\alpha,\beta))$ is a quasi-ideal of S. □

Corollary 2. *Let $f = (S;f_n,f_p)$ be a BF set on a semigroup S. Then*

(1) *f is a $(\alpha_1,\alpha_2;\beta_1,\beta_2)$-BF quasi-ideal on S for all $\alpha_1,\alpha_2 \in [0,1]$ and $\beta_1,\beta_2 \in [-1,0]$ if and only if $C(f;(\alpha,\beta))(\neq\emptyset)$ is a quasi-ideal of S for all $\alpha \in Im(f_p)$ and $\beta \in Im(f_n)$;*

(2) *f is a $(\alpha_1,\alpha_2;\beta_1,\beta_2)$-BF quasi-ideal on S for all $\alpha_1,\alpha_2 \in [0,1]$ and $\beta_1,\beta_2 \in [-1,0]$ if and only if $C(f;(\alpha,\beta))(\neq\emptyset)$ is a quasi-ideal of S for all $\alpha \in [0,1]$ and $\beta \in [-1,0]$.*

Proof. (1) Set $\Delta^+ = [0,1]$ and $\Delta^- = [-1,0]$, and apply Theorem 5.
(2) Set $\Delta^+ = Im(f_p)$ and $\Delta^- = Im(f_n)$, and apply Theorem 5. □

In the following theorem, we discuss a quasi-ideal of a semigroup S in terms of the bipolar characteristic function being a $(\alpha_1, \alpha_2; \beta_1, \beta_2)$-BF quasi-ideal on S.

Theorem 6. *Let S be a semigroup. Then a non-empty subset I is a quasi-ideal of S if and only if the bipolar characteristic function $C_I = (S; C_I^p, C_I^n)$ is a $(\alpha_1, \alpha_2; \beta_1, \beta_2)$-BF quasi-ideal on S for all $\alpha_1, \alpha_2 \in [0,1]$ and $\beta_1, \beta_2 \in [-1,0]$.*

Proof. ($\Rightarrow$) Let I be a quasi-ideal of S and $x \in S$. Let $\alpha_1, \alpha_2 \in [0,1]$ and $\beta_1, \beta_2 \in [-1,0]$.
Case 1: $x, y \in I$. Then

$$C_I^p(x) \vee \alpha_1 = 1 \geq (C_I^p \bar{\circ} \mathcal{S}_p)(x) \wedge (\mathcal{S}_p \bar{\circ} C_I^p)(x) \wedge \alpha_2$$

$$C_I^n(x) \wedge \beta_2 = -1 \leq (C_I^n \underline{\circ} \mathcal{S}_n)(x) \vee (\mathcal{S}_n \underline{\circ} C_I^n)(x) \vee \beta_1$$

Case 2: $x \notin I$. Then $x \notin SI$ or $x \notin IS$. If $x \notin SI$, then $(C_I^p \bar{\circ} \mathcal{S}_p)(x) = 0$ and $(C_I^n \underline{\circ} \mathcal{S}_n)(x) = 0$. Thus

$$C_I^p(x) \vee \alpha_1 \geq 0 = (C_I^p \bar{\circ} \mathcal{S}_p)(x) \wedge (\mathcal{S}_p \bar{\circ} C_I^p)(x) \wedge \alpha_2$$

$$C_I^n(x) \wedge \beta_2 \leq 0 = (C_I^n \underline{\circ} \mathcal{S}_n)(x) \vee (\mathcal{S}_n \underline{\circ} C_I^n)(x) \vee \beta_1$$

Therefore $C_I = (S; C_I^p, C_I^n)$ is a $(\alpha_1, \alpha_2; \beta_1, \beta_2)$-BF quasi-ideal on S.
($\Leftarrow$) Conversely, let C_I be a $(\alpha_1, \alpha_2; \beta_1, \beta_2)$-BF quasi-ideal on S for all $\alpha_1, \alpha_2 \in [0,1]$ and $\beta_1, \beta_2 \in [-1,0]$. Let $a \in IS \cap SI$. Then there exist $a, b \in S$ and $x, y \in I$ such that $a = xb = cy$. Then $(C_I)_p(x) = 1 = (C_I)_p(y)$ and $(C_I)_n(x) = -1 = (C_I)_n(y)$. Hence $x, y \in C(C_I; (1,-1))$, and so $xb \in C(C_I; (1,-1))S$ and $cy \in SC(C_I; (1,-1))$. Hence $a \in C(C_I; (1,-1))S$ and $a \in SC(C_I; (1,-1))$, and it follows that $a \in C(C_I; (1,-1))S \cap SC(C_I; (1,-1))$. By Corollary 2, $C(C_I; (1,-1))$ is a quasi-ideal. Thus $a \in C(C_I; (1,-1))$, and so $C_I^p(a) \geq 1$. This implies that $a \in I$. Therefore I is a quasi-ideal on S. □

Theorem 7. *Let S be a semigroup. Then I is a generalized bi-ideal of S if and only if the bipolar characteristic function $C_I = (S; C_I^p, C_I^n)$ is a $(\alpha_1, \alpha_2; \beta_1, \beta_2)$-BF generalized bi-ideal on S for all $\alpha_1, \alpha_2 \in [0,1]$ and $\beta_1, \beta_2 \in [-1,0]$.*

Proof. ($\Rightarrow$) Let I be a generalized bi-ideal of S and $x, y, a \in S$. Let $\alpha_1, \alpha_2 \in [0,1]$ and $\beta_1, \beta_2 \in [-1,0]$.
Case 1: $x, y \in I$. Then $xay \in I$; thus

$$C_I^p(xay) \vee \alpha_1 = 1 \geq C_I^p(x) \wedge C_I^p(y) \wedge \alpha_2$$

and

$$C_I^n(xay) \wedge \beta_2 = -1 \leq C_I^n(x) \vee C_I^n(y) \vee \beta_1$$

Case 2: $x \notin I$ or $y \notin I$. Then

$$C_I^p(xay) \vee \alpha_1 \geq 0 = C_I^p(x) \wedge C_I^p(y) \wedge \alpha_2$$

and

$$C_I^n(xay) \wedge \beta_2 \leq 0 = C_I^n(x) \vee C_I^n(y) \vee \beta_1$$

Therefore $C_I = (S; C_I^p, C_I^n)$ is a $(\alpha_1, \alpha_2; \beta_1, \beta_2)$-BF generalized bi-ideal on S.
($\Leftarrow$) Conversely, let C_I be a $(\alpha_1, \alpha_2; \beta_1, \beta_2)$-BF generalized bi-ideal on S for all $\alpha_1, \alpha_2 \in [0,1]$ and $\beta_1, \beta_2 \in [-1,0]$. Let $a \in S$ and $x, y \in I$. Then $(C_I)_p(x) = 1 = (C_I)_p(y)$ and $(C_I)_n(x) = -1 = (C_I)_n(y)$. Hence, $x, y \in C(C_I; (1,-1))$. By Corollary 1, $C(C_I; (1,-1))$ is a generalized bi-ideal. Thus $xay \in C(C_I; (1,-1))$, and so $C_I^p(xay) \geq 1$. This implies that $xay \in I$. Therefore I is a generalized bi-ideal on S. □

Theorem 8. *Every* $(\alpha_1, \alpha_2; \beta_1, \beta_2)$*-BF left (right) ideal on a semigroup* S *is a* $(\alpha_1, \alpha_2; \beta_1, \beta_2)$*-BF quasi-ideal on* S.

Proof. Let $f = (S; f_p, f_n)$ be a $(\alpha_1, \alpha_2; \beta_1, \beta_2)$-BF left ideal on S and $x, y \in S$. Then

$$\begin{aligned}
(\mathcal{S}\bar{\circ} f)_p(xy) \wedge \alpha_2 &= (\mathcal{S}_p \bar{\circ} f_p)(xy) \wedge \alpha_2 \\
&= \Big(\bigvee_{x=yz} \{\mathcal{S}_p(y) \wedge f_p(z)\}\Big) \wedge \alpha_2 \\
&= \bigvee_{x=yz} \{f_p(z)\} \wedge \alpha_2 \\
&\leq \bigvee_{x=yz} \{f_p(yz)\} \vee \alpha_1 \\
&\leq f_p(x) \vee \alpha_1
\end{aligned}$$

Thus $f_p(x) \vee \alpha_1 \geq (\mathcal{S}_p \bar{\circ} f_p)(xy) \wedge \alpha_2$. Hence $f_p(x) \vee \alpha_1 \geq (\mathcal{S}_p \bar{\circ} f_p)(xy) \wedge \alpha_2 \geq (\mathcal{S}_p \bar{\circ} f_p)(xy) \wedge (f_p \bar{\circ} \mathcal{S}_p)(xy) \wedge \alpha_2$. Similarly, we can show that $f_n(x) \wedge \beta_2 \leq (f_n \underline{\circ} \mathcal{S}_n)(x) \vee (\mathcal{S}_n \underline{\circ} f_n)(x) \vee \beta_1$. Therefore f is a $(\alpha_1, \alpha_2; \beta_1, \beta_2)$-BF quasi-ideal on S. □

Lemma 2. *Every* $(\alpha_1, \alpha_2; \beta_1, \beta_2)$*-BF quasi-ideal on a semigroup* S *is a* $(\alpha_1, \alpha_2; \beta_1, \beta_2)$*-BF bi-ideal on* S.

Proof. Let $f = (S; f_p, f_n)$ be a $(\alpha_1, \alpha_2; \beta_1, \beta_2)$-BF quasi-ideal on S and $x, y, z \in S$. Then

$$\begin{aligned}
f_p(xy) \vee \alpha_1 &\geq (f_p \bar{\circ} \mathcal{S}_p)(xy) \wedge (\mathcal{S}_p \bar{\circ} f_p)(xy) \wedge \alpha_2 \\
&= \bigvee_{xy=ab} \{f_p(a) \wedge \mathcal{S}_p(b)\} \wedge \bigvee_{xy=rs} \{\mathcal{S}_p(r) \wedge f_p(s)\} \wedge \alpha_2 \\
&\geq f_p(x) \wedge \mathcal{S}_p(y) \wedge \mathcal{S}_p(x) \wedge f_p(y) \wedge \alpha_2 \\
&\geq f_p(x) \wedge 1 \wedge 1 \wedge f_p(y) \wedge \alpha_2 \\
&= f_p(x) \wedge f_p(y) \wedge \alpha_2
\end{aligned}$$

Hence, $f_p(xy) \vee \alpha_1 \geq f_p(x) \wedge f_p(y) \wedge \alpha_2$. Additionally,

$$\begin{aligned}
f_p(xyz) \vee \alpha_1 &\geq (f_p \bar{\circ} \mathcal{S}_p)(xyz) \wedge (\mathcal{S}_p \bar{\circ} f_p)(xyz) \wedge \alpha_2 \\
&= \bigvee_{xyz=ab} \{f_p(a) \wedge \mathcal{S}_p(b)\} \wedge \bigvee_{xyz=rs} \{\mathcal{S}_p(r) \wedge f_p(s)\} \wedge \alpha_2 \\
&\geq f_p(x) \wedge \mathcal{S}_p(yz) \wedge \mathcal{S}_p(xy) \wedge f_p(z) \wedge \alpha_2 \\
&\geq f_p(x) \wedge 1 \wedge 1 \wedge f_p(z) \wedge \alpha_2 \\
&= f_p(x) \wedge f_p(z) \wedge \alpha_2
\end{aligned}$$

Hence, $f_p(xyz) \vee \alpha_1 \geq f_p(x) \wedge f_p(z) \wedge \alpha_2$. Similarly, we can show that $f_n(xy) \wedge \beta_2 \leq f_n(x) \vee f_n(y) \vee \beta_1$ and $f_n(xyz) \wedge \beta_2 \leq f_n(x) \vee f_n(z) \vee \beta_1$. Therefore f is a $(\alpha_1, \alpha_2; \beta_1, \beta_2)$-BF bi-ideal on S. □

Lemma 3. *Let* A *and* B *be non-empty subsets of a semigroup* S*. Then the following conditions hold:*

(1) $(C_A)_p \overset{(\alpha_2,\alpha_1)}{\wedge} (C_B)_p = (C_{A\cap B})_p^{(\alpha_2,\alpha_1)}$.

(2) $(C_A)_n \overset{(\beta_2,\beta_1)}{\vee} (C_B)_n = (C_{A\cup B})_n^{(\beta_2,\beta_1)}$.

(3) $(C_A)_p \overset{(\alpha_2,\alpha_1)}{\circ} (C_B)_p = (C_{AB})_p^{(\alpha_2,\alpha_1)}$.

(4) $(C_A)_n \overset{(\beta_2,\beta_1)}{\circ} (C_B)_n = (C_{AB})_n^{(\beta_2,\beta_1)}$.

Lemma 4. *If $f = (S; f_p, f_n)$ is a $(\alpha_1, \alpha_2; \beta_1, \beta_2)$-BF left ideal and $g = (S; g_p, g_n)$ is a $(\alpha_1, \alpha_2; \beta_1, \beta_2)$-BF right ideal on a semigroup S, then $f_p \overset{(\alpha_2,\alpha_1)}{\circ} g_p \leq f_p \overset{(\alpha_2,\alpha_1)}{\wedge} g_p$ and $f_n \underset{(\beta_2,\beta_1)}{\circ} g_n \geq f_n \underset{(\beta_2,\beta_1)}{\vee} g_n$.*

Theorem 9. *For a semigroup S, the following are equivalent.*

(1) S is regular.

(2) $f_p \overset{(\alpha_2,\alpha_1)}{\wedge} g_p = f_p \overset{(\alpha_2,\alpha_1)}{\circ} g_p$ and $f_n \underset{(\beta_2,\beta_1)}{\vee} g_n = f_n \underset{(\beta_2,\beta_1)}{\circ} g_n$ for every $(\alpha_1, \alpha_2; \beta_1, \beta_2)$-BF right ideal $f = (S; f_p, f_n)$ and every $(\alpha_1, \alpha_2; \beta_1, \beta_2)$-BF left ideal $g = (S; g_p, g_n)$ on S.

Next, we characterize a regular semigroup by generalizations of BF subsemigroups.

Theorem 10. *For a semigroup S, the following are equivalent.*

(1) S is regular.

(2) $f_p \overset{(\alpha_2,\alpha_1)}{\wedge} h_p \overset{(\alpha_2,\alpha_1)}{\wedge} g_p \leq f_p \overset{(\alpha_2,\alpha_1)}{\circ} h_p \overset{(\alpha_2,\alpha_1)}{\circ} g_p$ and $f_n \underset{(\beta_2,\beta_1)}{\vee} h_n \underset{(\beta_2,\beta_1)}{\vee} g_n \geq f_n \underset{(\beta_2,\beta_1)}{\circ} h_n \underset{(\beta_2,\beta_1)}{\circ} g_n$ for every $(\alpha_1, \alpha_2; \beta_1, \beta_2)$-BF right ideal $f = (S; f_p, f_n)$, every $(\alpha_1, \alpha_2; \beta_1, \beta_2)$-BF generalized bi-ideal $h = (S; h_p, h_n)$ and every $(\alpha_1, \alpha_2; \beta_1, \beta_2)$-BF left ideal $g = (S; g_p, g_n)$ on S.

(3) $f_p \overset{(\alpha_2,\alpha_1)}{\wedge} h_p \overset{(\alpha_2,\alpha_1)}{\wedge} g_p \leq f_p \overset{(\alpha_2,\alpha_1)}{\circ} h_p \overset{(\alpha_2,\alpha_1)}{\circ} g_p$ and $f_n \underset{(\beta_2,\beta_1)}{\vee} h_n \underset{(\beta_2,\beta_1)}{\vee} g_n \geq f_n \underset{(\beta_2,\beta_1)}{\circ} h_n \underset{(\beta_2,\beta_1)}{\circ} g_n$ for every $(\alpha_1, \alpha_2; \beta_1, \beta_2)$-BF right ideal $f = (S; f_p, f_n)$, every $(\alpha_1, \alpha_2; \beta_1, \beta_2)$-BF bi-ideal $h = (S; h_p, h_n)$ and every $(\alpha_1, \alpha_2; \beta_1, \beta_2)$-BF left ideal $g = (S; g_p, g_n)$ on S.

(4) $f_p \overset{(\alpha_2,\alpha_1)}{\wedge} h_p \overset{(\alpha_2,\alpha_1)}{\wedge} g_p \leq f_p \overset{(\alpha_2,\alpha_1)}{\circ} h_p \overset{(\alpha_2,\alpha_1)}{\circ} g_p$ and $f_n \underset{(\beta_2,\beta_1)}{\vee} h_n \underset{(\beta_2,\beta_1)}{\vee} g_n \geq f_n \underset{(\beta_2,\beta_1)}{\circ} h_n \underset{(\beta_2,\beta_1)}{\circ} g_n$ for every $(\alpha_1, \alpha_2; \beta_1, \beta_2)$-BF right ideal $f = (S; f_p, f_n)$, every $(\alpha_1, \alpha_2; \beta_1, \beta_2)$-BF quasi-ideal $h = (S; h_p, h_n)$ and every $(\alpha_1, \alpha_2; \beta_1, \beta_2)$-BF left ideal $g = (S; g_p, g_n)$ on S.

Proof. (1 ⇒ 2). Let f, h and g be a $(\alpha_1, \alpha_2; \beta_1, \beta_2)$-BF right ideal, a $(\alpha_1, \alpha_2; \beta_1, \beta_2)$-BF generalized bi-ideal and a $(\alpha_1, \alpha_2; \beta_1, \beta_2)$-BF left ideal on S, respectively. Let $a \in S$. Because S is regular, there exists $x \in S$ such that $a = axa$. Thus

$$
\begin{aligned}
(f_p \overset{(\alpha_2,\alpha_1)}{\circ} h_p \overset{(\alpha_2,\alpha_1)}{\circ} g_p)(a) &= ((f_p \bar{\circ} h_p \bar{\circ} g_p)(a) \wedge \alpha_2) \vee \alpha_1 \\
&= (\bigvee_{a=yz} \{f_p(y) \wedge (h_p \bar{\circ} g_p)(z) \wedge \alpha_2\}) \vee \alpha_1 \\
&\geq (f_p(ax) \wedge (h_p \bar{\circ} g_p)(a) \wedge \alpha_2) \vee \alpha_1 \\
&\geq (f_p(ax) \vee \alpha_1) \wedge ((h_p \bar{\circ} g_p)(a) \vee \alpha_1) \wedge (\alpha_2 \vee \alpha_1) \\
&\geq (f_p(a) \wedge \alpha_2) \wedge (\bigvee_{a=rs} \{(h_p(r) \wedge g_p(s)) \vee \alpha_1\}) \wedge (\alpha_2 \vee \alpha_1) \\
&\geq (f_p(a) \wedge \alpha_2) \wedge ((h_p(a) \vee \alpha_1) \wedge (g_p(xa) \vee \alpha_1) \wedge (\alpha_2 \vee \alpha_1) \\
&\geq (f_p(a) \wedge \alpha_2) \wedge ((h_p(a) \vee \alpha_1) \wedge ((g_p(a) \wedge \alpha_2) \vee \alpha_1) \wedge (\alpha_2 \vee \alpha_1) \\
&\geq (f_p(a) \wedge h_p(a) \wedge g_p(a) \wedge \alpha_2) \wedge \alpha_1 \\
&= ((f_p \wedge h_p \wedge g_p)(a) \wedge \alpha_2) \wedge \alpha_1 \\
&= (f_p \overset{(\alpha_2,\alpha_1)}{\wedge} h_p \overset{(\alpha_2,\alpha_1)}{\wedge} g_p)(a)
\end{aligned}
$$

Similarly, we can show that $f_n \underset{(\beta_2,\beta_1)}{\vee} h_n \underset{(\beta_2,\beta_1)}{\vee} g_n \geq f_n \underset{(\beta_2,\beta_1)}{\circ} h_n \underset{(\beta_2,\beta_1)}{\circ} g_n$.

(2 ⇒ 3 ⇒ 4). This is straightforward, because every $(\alpha_1, \alpha_2; \beta_1, \beta_2)$-BF bi-ideal is a $(\alpha_1, \alpha_2; \beta_1, \beta_2)$-BF generalized bi-ideal and every $(\alpha_1, \alpha_2; \beta_1, \beta_2)$-BF quasi-ideal is a $(\alpha_1, \alpha_2; \beta_1, \beta_2)$-BF bi-ideal on S.

$(4 \Rightarrow 1)$. Let f and g be any $(\alpha_1, \alpha_2; \beta_1, \beta_2)$-BF right ideal and $(\alpha_1, \alpha_2; \beta_1, \beta_2)$-BF left ideal on S, respectively. Let $a \in S$. By Theorem 8, $\mathcal{S} = (S, \mathcal{S}_p, \mathcal{S}_n)$ is a $(\alpha_1, \alpha_2; \beta_1, \beta_2)$-BF quasi ideal, and we have

$$\begin{aligned}(f_p \overset{(\alpha_2,\alpha_1)}{\wedge} g_p)(a) &= ((f_p \wedge g_p)(a) \wedge \alpha_2) \vee \alpha_1 \\ &= ((f_p \wedge \mathcal{S}_p \wedge g_p)(a) \wedge \alpha_2) \vee \alpha_1 \\ &= (f_p \overset{(\alpha_2,\alpha_1)}{\wedge} \mathcal{S}_p \overset{(\alpha_2,\alpha_1)}{\wedge} g_p)(a) \\ &\leq (f_p \overset{(\alpha_2,\alpha_1)}{\circ} \mathcal{S}_p \overset{(\alpha_2,\alpha_1)}{\circ} g_p)(a) \\ &\leq (f_p \overset{(\alpha_2,\alpha_1)}{\circ} g_p)(a)\end{aligned}$$

Thus $f_p \overset{(\alpha_2,\alpha_1)}{\wedge} g_p \leq f_p \overset{(\alpha_2,\alpha_1)}{\circ} g_p$ for every $(\alpha_1, \alpha_2; \beta_1, \beta_2)$-BF right ideal f and every $(\alpha_1, \alpha_2; \beta_1, \beta_2)$-BF left ideal g on S. Similarly, we can show that $f_n \underset{(\beta_2,\beta_1)}{\circ} g_n \leq f_n \underset{(\beta_2,\beta_1)}{\vee} g_n$. By Lemma 4, $f_p \overset{(\alpha_2,\alpha_1)}{\circ} g_p \leq f_p \overset{(\alpha_2,\alpha_1)}{\wedge} g_p$ and $f_n \underset{(\beta_2,\beta_1)}{\circ} g_n \geq f_n \underset{(\beta_2,\beta_1)}{\vee} g_n$. Thus $f_p \overset{(\alpha_2,\alpha_1)}{\circ} g_p = f_p \overset{(\alpha_2,\alpha_1)}{\wedge} g_p$ and $f_n \underset{(\beta_2,\beta_1)}{\circ} g_n = f_n \underset{(\beta_2,\beta_1)}{\vee} g_n$. Therefore by Theorem 9, S is regular. □

Theorem 11. *For a semigroup S, the following are equivalent.*

(1) *S is regular.*

(2) $f_p^{(\alpha_2,\alpha_1)} = f_p \overset{(\alpha_2,\alpha_1)}{\circ} \mathcal{S}_p \overset{(\alpha_2,\alpha_1)}{\circ} f_p$ *and* $f_n^{(\beta_2,\beta_1)} = f_n \underset{(\beta_2,\beta_1)}{\circ} \mathcal{S}_n \underset{(\beta_2,\beta_1)}{\circ} f_n$ *for every $(\alpha_1, \alpha_2; \beta_1, \beta_2)$-BF generalized bi-ideal $f = (S; f_p, f_n)$ on S.*

(3) $f_p^{(\alpha_2,\alpha_1)} = f_p \overset{(\alpha_2,\alpha_1)}{\circ} \mathcal{S}_p \overset{(\alpha_2,\alpha_1)}{\circ} f_p$ *and* $f_n^{(\beta_2,\beta_1)} = f_n \underset{(\beta_2,\beta_1)}{\circ} \mathcal{S}_n \underset{(\beta_2,\beta_1)}{\circ} f_n$ *for every $(\alpha_1, \alpha_2; \beta_1, \beta_2)$-BF bi-ideal $f = (S; f_p, f_n)$ on S.*

(4) $f_p^{(\alpha_2,\alpha_1)} = f_p \overset{(\alpha_2,\alpha_1)}{\circ} \mathcal{S}_p \overset{(\alpha_2,\alpha_1)}{\circ} f_p$ *and* $f_n^{(\beta_2,\beta_1)} = f_n \underset{(\beta_2,\beta_1)}{\circ} \mathcal{S}_n \underset{(\beta_2,\beta_1)}{\circ} f_n$ *for every $(\alpha_1, \alpha_2; \beta_1, \beta_2)$-BF quasi-ideal $f = (S; f_p, f_n)$ on S.*

Proof. $(1 \Rightarrow 2)$. Let f be a $(\alpha_1, \alpha_2; \beta_1, \beta_2)$-BF generalized bi-ideal on S and $a \in S$. Because S is regular, there exists $x \in S$ such that $a = axa$. Hence we have

$$\begin{aligned}(f_p \overset{(\alpha_2,\alpha_1)}{\circ} \mathcal{S}_p \overset{(\alpha_2,\alpha_1)}{\circ} f_p)(a) &= ((f_p \bar{\circ} \mathcal{S}_p \bar{\circ} f_p)(a) \wedge \alpha_2) \vee \alpha_1 \\ &= (\bigvee_{a=yz} \{(f_p \bar{\circ} \mathcal{S}_p)(y) \wedge f_p(z)\} \wedge \alpha_2) \vee \alpha_1 \\ &\geq ((f_p \bar{\circ} \mathcal{S}_p)(ax) \wedge f_p(a)) \wedge \alpha_2) \vee \alpha_1 \\ &= ((\bigvee_{ax=rs} \{f_p(r) \wedge \mathcal{S}_p(s)\}) \wedge f_p(a)) \wedge \alpha_2) \vee \alpha_1 \\ &\geq (((f_p(a) \wedge \mathcal{S}_p(x)) \wedge f_p(a)) \wedge \alpha_2) \vee \alpha_1 \\ &= ((f_p(a) \wedge 1)) \wedge f_p(a)) \wedge \alpha_2) \vee \alpha_1 \\ &= (f_p(a) \wedge \alpha_2) \vee \alpha_1 \\ &= f_p^{(\alpha_2,\alpha_1)}(a)\end{aligned}$$

Thus $f_p \overset{(\alpha_2,\alpha_1)}{\circ} \mathcal{S}_p \overset{(\alpha_2,\alpha_1)}{\circ} f_p \geq f_p^{(\alpha_2,\alpha_1)}$. Similarly, we can show that $f_n \underset{(\beta_2,\beta_1)}{\circ} \mathcal{S}_n \underset{(\beta_2,\beta_1)}{\circ} f_n \leq f_n^{(\beta_2,\beta_1)}$. By Theorem 3, $f_p \overset{(\alpha_2,\alpha_1)}{\circ} \mathcal{S}_p \overset{(\alpha_2,\alpha_1)}{\circ} f_p \leq f_p^{(\alpha_2,\alpha_1)}$ and $f_n \underset{(\beta_2,\beta_1)}{\circ} \mathcal{S}_n \underset{(\beta_2,\beta_1)}{\circ} f_n \geq f_n^{(\beta_2,\beta_1)}$. Therefore, $f_p \overset{(\alpha_2,\alpha_1)}{\circ} \mathcal{S}_p \overset{(\alpha_2,\alpha_1)}{\circ} f_p = f_p^{(\alpha_2,\alpha_1)}$ and $f_n \underset{(\beta_2,\beta_1)}{\circ} \mathcal{S}_n \underset{(\beta_2,\beta_1)}{\circ} f_n = f_n^{(\beta_2,\beta_1)}$.

$(2 \Rightarrow 3 \Rightarrow 4)$. Obvious.

$(4 \Rightarrow 1)$. Let Q be any quasi-ideal of S. By Theorem 6 and Lemma 3, we have

$$\begin{aligned}(C_Q)_p^{(\alpha_2,\alpha_1)} &= (C_Q)_p \overset{(\alpha_2,\alpha_1)}{\circ} (\mathcal{S})_p \overset{(\alpha_2,\alpha_1)}{\circ} (C_Q)_p \\ &= (C_{QSQ})_p^{(\alpha_2,\alpha_1)}\end{aligned}$$

Thus, $Q = QSQ$. Therefore it follows from Theorem 1 that S is regular. □

Theorem 12. *For a semigroup S, the following are equivalent.*

(1) *S is regular.*

(2) $f_p \overset{(\alpha_2,\alpha_1)}{\wedge} g_p \leq f_p \overset{(\alpha_2,\alpha_1)}{\circ} g_p$ *and* $f_n \underset{(\beta_2,\beta_1)}{\vee} g_n \geq f_n \underset{(\beta_2,\beta_1)}{\circ} g_n$ *for every $(\alpha_1,\alpha_2;\beta_1,\beta_2)$-BF generalized bi-ideal $f = (S; f_p, f_n)$ and every $(\alpha_1,\alpha_2;\beta_1,\beta_2)$-BF left ideal $g = (S; g_p, g_n)$ on S.*

(3) $f_p \overset{(\alpha_2,\alpha_1)}{\wedge} g_p \leq f_p \overset{(\alpha_2,\alpha_1)}{\circ} g_p$ *and* $f_n \underset{(\beta_2,\beta_1)}{\vee} g_n \geq f_n \underset{(\beta_2,\beta_1)}{\circ} g_n$ *for every $(\alpha_1,\alpha_2;\beta_1,\beta_2)$-BF bi-ideal $f = (S; f_p, f_n)$ and every $(\alpha_1,\alpha_2;\beta_1,\beta_2)$-BF left ideal $g = (S; g_p, g_n)$ on S.*

(4) $f_p \overset{(\alpha_2,\alpha_1)}{\wedge} g_p \leq f_p \overset{(\alpha_2,\alpha_1)}{\circ} g_p$ *and* $f_n \underset{(\beta_2,\beta_1)}{\vee} g_n \geq f_n \underset{(\beta_2,\beta_1)}{\circ} g_n$ *for every $(\alpha_1,\alpha_2;\beta_1,\beta_2)$-BF quasi-ideal $f = (S; f_p, f_n)$ and every $(\alpha_1,\alpha_2;\beta_1,\beta_2)$-BF left ideal $g = (S; g_p, g_n)$ on S.*

Proof. $(1 \Rightarrow 2)$. Let f and g be any $(\alpha_1,\alpha_2;\beta_1,\beta_2)$-BF generalized bi-ideal and any $(\alpha_1,\alpha_2;\beta_1,\beta_2)$-BF left ideal on S, respectively. Let $a \in S$. Because S is regular, there exists $x \in S$ such that $a = axa$. Thus we have

$$\begin{aligned}(f_p \overset{(\alpha_2,\alpha_1)}{\circ} g_p)(a) &= ((f_p \bar{\circ} g_p)(a) \wedge \alpha_2) \vee \alpha_1 \\ &= (\bigvee_{a=yz} \{f_p(y) \wedge g_p(z)\} \wedge \alpha_2) \vee \alpha_1 \\ &\geq (f_p(a) \wedge g_p(xa)) \wedge \alpha_2) \vee \alpha_1 \\ &\geq (f_p(a) \vee \alpha_1) \wedge (g_p(xa) \vee \alpha_1) \wedge (\alpha_2 \vee \alpha_1) \\ &\geq (f_p(a) \vee \alpha_1) \wedge ((g_p(a) \wedge \alpha_2) \vee \alpha_1) \wedge (\alpha_2 \vee \alpha_1) \\ &= (f_p(a) \wedge g_p(a) \wedge \alpha_2) \vee \alpha_1 \\ &= ((f_p \wedge g_p)(a) \wedge \alpha_2) \vee \alpha_1 \\ &= (f_p \overset{(\alpha_2,\alpha_1)}{\wedge} g_p)(a)\end{aligned}$$

Hence $f_p \overset{(\alpha_2,\alpha_1)}{\circ} g_p \geq f_p \overset{(\alpha_2,\alpha_1)}{\wedge} g_p$. Similarly, we can show that $f_n \underset{(\beta_2,\beta_1)}{\circ} g_n \leq f_n \underset{(\beta_2,\beta_1)}{\vee} g_n$.

$(2 \Rightarrow 3 \Rightarrow 4)$. Obvious.

$(4 \Rightarrow 1)$. Let f and g be any $(\alpha_1,\alpha_2;\beta_1,\beta_2)$-BF right ideal and $(\alpha_1,\alpha_2;\beta_1,\beta_2)$-BF left ideal on S, respectively. By Theorem 8, f is a $(\alpha_1,\alpha_2;\beta_1,\beta_2)$-BF quasi ideal. Thus $f_p \overset{(\alpha_2,\alpha_1)}{\circ} g_p \geq f_p \overset{(\alpha_2,\alpha_1)}{\wedge} g_p$

and $f_n \underset{(\beta_2,\beta_1)}{\circ} g_n \leq f_n \underset{(\beta_2,\beta_1)}{\vee} g_n$. By Lemma 4, $f_p \overset{(\alpha_2,\alpha_1)}{\circ} g_p \leq f_p \overset{(\alpha_2,\alpha_1)}{\wedge} g_p$ and $f_n \underset{(\beta_2,\beta_1)}{\circ} g_n \geq f_n \underset{(\beta_2,\beta_1)}{\vee} g_n$. Thus $f_p \overset{(\alpha_2,\alpha_1)}{\circ} g_p = f_p \overset{(\alpha_2,\alpha_1)}{\wedge} g_p$ and $f_n \underset{(\beta_2,\beta_1)}{\circ} g_n = f_n \underset{(\beta_2,\beta_1)}{\vee} g_n$. Therefore by Theorem 9, S is regular. □

4. Conclusions

In this paper, we propose the generalizations of BF sets. In particular, we introduce several concepts of generalized BF sets and study the relationship between such sets and semigroups. In other words, we propose generalized BF subsemigroups. This under consideration, the results obtained in this paper are some inequalities of $(\alpha_1, \alpha_2; \beta_1, \beta_2)$-BF quasi(generalized bi-, bi-) ideals and characterize a regular semigroup in terms of generalized BF semigroups. The importance of BF sets has positive and negative components frequently found in daily life, for example, in organizations, economics, performance, development, evaluation, risk management or decisions, and so forth. Therefore we establish generalized BF sets on semigroups, which enhances the structure of the algebra. We hope that the study of some types of subsemigroups characterized in terms of inequalities of generalized BF subsemigroups is a useful mathematical tool.

Author Contributions: Both authors contributed equally to this manuscript.

Conflicts of Interest: The authors declare no conflict of interest.

References

1. Bloch, I. Bipolar Fuzzy Mathematical Morphology for Spatial Reasoning. *Int. J. Approx. Reason.* **2012**, *53*, 1031–1070.
2. Zadeh, L.A. Fuzzy sets. *Inf. Control* **1965**, *8*, 338–353.
3. Bucolo, M.; Fazzino, S.; La Rosa, M.; Fortuna, L. Small-world networks of fuzzy chaotic oscillators. *Chaos Solitons Fractals* **2003**, *17*, 557–565.
4. Rosenfeld, A. Fuzzy group. *J. Math. Anal. Appl.* **1971**, *35*, 338–353.
5. Kuroki, N. On fuzzy ideals and fuzzy bi-ideals in semigroups. *Fuzzy Sets Syst.* **1981**, *5*, 203–215.
6. Zhang, W.R. Bipolar Fuzzy Sets and Relations: A Computational Framework Forcognitive Modeling and Multiagent Decision Analysis. In Proceedings of the 1994 Industrial Fuzzy Control and Intelligent Systems Conference and the NASA Joint Technology Workshop on Neural Networks and Fuzzy Logic Fuzzy Information Processing Society Biannual Conference, San Antonio, TX, USA, 18–21 December 1994; pp. 305–309.
7. Lee, K.M. Bipolar-Valued Fuzzy Sets and Their Operations. In Proceedings of the International Conference on Intelligent Technologies, Bangkok, Thailand, 13–15 December 2000; pp. 307–312.
8. Kim, C.S.; Kang, J.G.; Kang, J.M. Ideal theory of semigroups based on the bipolar valued fuzzy set theory. *Ann. Fuzzy Math. Inform.* **2011**, *2*, 193–206.
9. Mordeson, J.N.; Malik, D.S.; Kuroki, N. *Fuzzy Semigroups*; Springer: Berlin, Germany, 2003.

mathematics

MDPI

Article

Hyperfuzzy Ideals in BCK/BCI-Algebras †

Seok-Zun Song [1], Seon Jeong Kim [2] and Young Bae Jun [3,*]

1 Department of Mathematics, Jeju National University, Jeju 63243, Korea; szsong@jejunu.ac.kr
2 Department of Mathematics, Natural Science of College, Gyeongsang National University, Jinju 52828, Korea; skim@gnu.ac.kr
3 Department of Mathematics Education, Gyeongsang National University, Jinju 52828, Korea
* Correspondence: skywine@gmail.com
† To the memory of Professor Lotfi A. Zadeh.

Received: 17 November 2017; Accepted: 10 December 2017; Published: 14 December 2017

Abstract: The notions of hyperfuzzy ideals in BCK/BCI-algebras are introduced, and related properties are investigated. Characterizations of hyperfuzzy ideals are established. Relations between hyperfuzzy ideals and hyperfuzzy subalgebras are discussed. Conditions for hyperfuzzy subalgebras to be hyperfuzzy ideals are provided.

Keywords: hyperfuzzy set; hyperfuzzy subalgebra; hyperfuzzy ideal

1. Introduction

After Zadeh [1] has introduced the fundamental concept of fuzzy sets, several generalizations of fuzzy sets are achieved. As a generalization of fuzzy sets and interval-valued fuzzy sets, Ghosh and Samanta [2] introduced the notion of hyperfuzzy sets, and then they applied it to group theory. They defined hyperfuzzy (normal) subgroups and hyperfuzzy cosets, and investigated their properties. The hyperfuzzy set has a subset of the interval $[0,1]$ as its image. Hence, it is a generalization of an interval-valued fuzzy set. In mathematics, BCK and BCI-algebras are algebraic structures, introduced by Imai, Iseki and Tanaka, that describe fragments of the propositional calculus involving implication known as BCK and BCI logics (see [3–5]). Jun et al. [6] applied hyperfuzzy sets to BCK/BCI-algebras by using the infimum and supremum of the image of hyperfuzzy sets. They introduced the notion of k-fuzzy substructure for $k \in \{1,2,3,4\}$ and then they introduced the concepts of hyperfuzzy substructures of several types by using k-fuzzy substructures, and investigated their basic properties. They also introduced the notion of hyperfuzzy subalgebras of type (i,j) for $i,j \in \{1,2,3,4\}$, and discussed relations between hyperfuzzy substructure/subalgebra and its length. They investigated the properties of hyperfuzzy subalgebras related to upper and lower level subsets.

The aim of this paper is to study BCK/BCI-algebraic structures based on hyperfuzzy structures. So, the notions and results in this manuscript are a generalization of BCK/BCI-algebraic structures based on fuzzy and interval-valued fuzzy structures. We introduce the notion of hyperfuzzy ideals in BCK/BCI-algebras, and investigate several properties. We consider characterizations of hyperfuzzy ideals, and discuss relations between hyperfuzzy subalgebras and hyperfuzzy ideals. We provide conditions for hyperfuzzy subalgebras to be hyperfuzzy ideals.

2. Preliminaries

By a *BCI-algebra* (see [7,8]) we mean a system $X := (X, *, 0)$ in which the following axioms hold:

(I) $((x*y)*(x*z))*(z*y) = 0,$
(II) $(x*(x*y))*y = 0,$
(III) $x*x = 0,$
(IV) $x*y = y*x = 0 \Rightarrow x = y$

for all $x, y, z \in X$. If a BCI-algebra X satisfies $0 * x = 0$ for all $x \in X$, then we say that X is a *BCK-algebra* (see [7,8]). We can define a partial ordering $\leq$ by

$$(\forall x, y \in X)\ (x \leq y \iff x * y = 0).$$

In a BCK/BCI-algebra X, the following hold (see [7,8]):

$$(\forall x \in X)\ (x * 0 = x), \tag{1}$$

$$(\forall x, y, z \in X)\ ((x * y) * z = (x * z) * y). \tag{2}$$

A non-empty subset S of a BCK/BCI-algebra X is called a *subalgebra* of X (see [7,8]) if $x * y \in S$ for all $x, y \in S$.

We refer the reader to the books [7,8] for further information regarding BCK/BCI-algebras.

By a *fuzzy structure* over a nonempty set X we mean an ordered pair (X, ρ) of X and a fuzzy set ρ on X.

Let X be a nonempty set. A mapping $\tilde{\mu} : X \to \tilde{\mathcal{P}}([0,1])$ is called a *hyperfuzzy set* over X (see [2]), where $\tilde{\mathcal{P}}([0,1])$ is the family of all nonempty subsets of $[0,1]$. An ordered pair $(X, \tilde{\mu})$ is called a *hyper structure* over X.

Given a hyper structure $(X, \tilde{\mu})$ over a nonempty set X, we consider two fuzzy structures $(X, \tilde{\mu}_{\inf})$ and $(X, \tilde{\mu}_{\sup})$ over X (see [6]) in which

$$\tilde{\mu}_{\inf} : X \to [0,1],\ \ x \mapsto \inf\{\tilde{\mu}(x)\},$$
$$\tilde{\mu}_{\sup} : X \to [0,1],\ \ x \mapsto \sup\{\tilde{\mu}(x)\}.$$

Given a nonempty set X, let $\mathcal{B}_K(X)$ and $\mathcal{B}_I(X)$ denote the collection of all BCK-algebras and all BCI-algebras, respectively. Also $\mathcal{B}(X) := \mathcal{B}_K(X) \cup \mathcal{B}_I(X)$.

Definition 1 ([6]). *For any $(X, *, 0) \in \mathcal{B}(X)$, a fuzzy structure (X, μ) over $(X, *, 0)$ is called a*

- *fuzzy subalgebra of $(X, *, 0)$ with type 1 (briefly, 1-fuzzy subalgebra of $(X, *, 0)$) if*

$$(\forall x, y \in X)\ (\mu(x * y) \geq \min\{\mu(x), \mu(y)\}), \tag{3}$$

- *fuzzy subalgebra of $(X, *, 0)$ with type 2 (briefly, 2-fuzzy subalgebra of $(X, *, 0)$) if*

$$(\forall x, y \in X)\ (\mu(x * y) \leq \min\{\mu(x), \mu(y)\}), \tag{4}$$

- *fuzzy subalgebra of $(X, *, 0)$ with type 3 (briefly, 3-fuzzy subalgebra of $(X, *, 0)$) if*

$$(\forall x, y \in X)\ (\mu(x * y) \geq \max\{\mu(x), \mu(y)\}), \tag{5}$$

- *fuzzy subalgebra of $(X, *, 0)$ with type 4 (briefly, 4-fuzzy subalgebra of $(X, *, 0)$) if*

$$(\forall x, y \in X)\ (\mu(x * y) \leq \max\{\mu(x), \mu(y)\}). \tag{6}$$

It is clear that every 3-fuzzy subalgebra is a 1-fuzzy subalgebra and every 2-fuzzy subalgebra is a 4-fuzzy subalgebra.

Definition 2 ([6]). *For any $(X, *, 0) \in \mathcal{B}(X)$ and $i, j \in \{1, 2, 3, 4\}$, a hyper structure $(X, \tilde{\mu})$ over $(X, *, 0)$ is called an (i, j)-hyperfuzzy subalgebra of $(X, *, 0)$ if $(X, \tilde{\mu}_{\inf})$ is an i-fuzzy subalgebra of $(X, *, 0)$ and $(X, \tilde{\mu}_{\sup})$ is a j-fuzzy subalgebra of $(X, *, 0)$.*

3. Hyperfuzzy Ideals

In what follows, let $(X, *, 0) \in \mathcal{B}(X)$ unless otherwise specified.

Definition 3. *A fuzzy structure (X, μ) over $(X, *, 0)$ is called a*

- *fuzzy ideal of $(X, *, 0)$ with type 1 (briefly, 1-fuzzy ideal of $(X, *, 0)$) if*

$$(\forall x \in X)\ (\mu(0) \geq \mu(x)), \tag{7}$$
$$(\forall x, y \in X)\ (\mu(x) \geq \min\{\mu(x * y), \mu(y)\}), \tag{8}$$

- *fuzzy ideal of $(X, *, 0)$ with type 2 (briefly, 2-fuzzy ideal of $(X, *, 0)$) if*

$$(\forall x \in X)\ (\mu(0) \leq \mu(x)), \tag{9}$$
$$(\forall x, y \in X)\ (\mu(x) \leq \min\{\mu(x * y), \mu(y)\}), \tag{10}$$

- *fuzzy ideal of $(X, *, 0)$ with type 3 (briefly, 3-fuzzy ideal of $(X, *, 0)$) if it satisfies (7) and*

$$(\forall x, y \in X)\ (\mu(x) \geq \max\{\mu(x * y), \mu(y)\}), \tag{11}$$

- *fuzzy ideal of $(X, *, 0)$ with type 4 (briefly, 4-fuzzy ideal of $(X, *, 0)$) if it satisfies (9) and*

$$(\forall x, y \in X)\ (\mu(x) \leq \max\{\mu(x * y), \mu(y)\}). \tag{12}$$

It is clear that every 3-fuzzy ideal is a 1-fuzzy ideal and every 2-fuzzy ideal is a 4-fuzzy ideal.

Definition 4. *For any $i, j \in \{1, 2, 3, 4\}$, a hyper structure $(X, \tilde{\mu})$ over $(X, *, 0)$ is called an (i, j)-hyperfuzzy ideal of $(X, *, 0)$ if $(X, \tilde{\mu}_{\inf})$ is an i-fuzzy ideal of $(X, *, 0)$ and $(X, \tilde{\mu}_{\sup})$ is a j-fuzzy ideal of $(X, *, 0)$.*

Example 1. *Consider a BCK-algebra $X = \{0, 1, 2, 3, 4\}$ with the binary operation $*$ which is given in Table 1 (see [8]).*

Table 1. Cayley table for the binary operation "$*$".

$*$	**0**	**1**	**2**	**3**	**4**
0	0	0	0	0	0
1	1	0	1	0	0
2	2	2	0	0	0
3	3	3	3	0	0
4	4	3	4	1	0

(1) Let $(X, \tilde{\mu})$ be a hyper structure over $(X, *, 0)$ in which $\tilde{\mu}$ is given as follows:

$$\tilde{\mu} : X \to \tilde{\mathcal{P}}([0,1]),\ x \mapsto \begin{cases} [0.5, 0.6) & \text{if } x = 0, \\ (0.4, 0.8] & \text{if } x = 1, \\ [0.3, 0.7] & \text{if } x = 2, \\ [0.2, 0.9] & \text{if } x \in \{3, 4\}. \end{cases}$$

It is routine to verify that $(X, \tilde{\mu})$ is a $(1, 4)$-hyperfuzzy ideal of $(X, *, 0)$.

(2) Let $(X, \tilde{\mu})$ be a hyper structure over $(X, *, 0)$ in which $\tilde{\mu}$ is given as follows:

$$\tilde{\mu} : X \to \tilde{\mathcal{P}}([0,1]), \ x \mapsto \begin{cases} [0.5, 0.9) & \text{if } x = 0, \\ (0.4, 0.9] & \text{if } x = 1, \\ [0.3, 0.5] & \text{if } x = 2, \\ [0.2, 0.4] & \text{if } x \in \{3, 4\}. \end{cases}$$

It is routine to verify that $(X, \tilde{\mu})$ is a $(1,1)$-hyperfuzzy ideal of $(X, *, 0)$.

Example 2. *Consider a BCI-algebra $X = \{0, 1, 2, a, b\}$ with the binary operation $*$ which is given in Table 2 (see [8]).*

Table 2. Cayley table for the binary operation "$*$".

$*$	**0**	**1**	**2**	***a***	***b***
0	0	0	0	a	a
1	1	0	1	b	a
2	2	2	0	a	a
a	a	a	a	0	0
b	b	a	b	1	0

(1) Let $(X, \tilde{\mu})$ be a hyper structure over $(X, *, 0)$ in which $\tilde{\mu}$ is given as follows:

$$\tilde{\mu} : X \to \tilde{\mathcal{P}}([0,1]), \ x \mapsto \begin{cases} [0.33, 0.75) & \text{if } x = 0, \\ (0.63, 0.75] & \text{if } x = 1, \\ [0.43, 0.70] & \text{if } x = 2, \\ [0.53, 0.65] & \text{if } x = a, \\ [0.63, 0.65] & \text{if } x = b. \end{cases}$$

By routine calculations, we know that $(X, \tilde{\mu})$ is a $(4,1)$-hyperfuzzy ideal of $(X, *, 0)$.
(2) Let $(X, \tilde{\mu})$ be a hyper structure over $(X, *, 0)$ in which $\tilde{\mu}$ is given as follows:

$$\tilde{\mu} : X \to \tilde{\mathcal{P}}([0,1]), \ x \mapsto \begin{cases} [0.33, 0.39) & \text{if } x = 0, \\ (0.63, 0.69] & \text{if } x = 1, \\ [0.43, 0.49] & \text{if } x = 2, \\ [0.53, 0.59] & \text{if } x = a, \\ [0.63, 0.69] & \text{if } x = b. \end{cases}$$

By routine calculations, we know that $(X, \tilde{\mu})$ is a $(4,4)$-hyperfuzzy ideal of $(X, *, 0)$.

Proposition 1. *Given a hyper structure $(X, \tilde{\mu})$ over $(X, *, 0)$, we have the following assertions.*

(1) *If $(X, \tilde{\mu})$ is a $(1,4)$-hyperfuzzy ideal of $(X, *, 0)$, then*

$$(\forall x, y \in X) \left(x \leq y \ \Rightarrow \ \tilde{\mu}_{\inf}(x) \geq \tilde{\mu}_{\inf}(y), \ \tilde{\mu}_{\sup}(x) \leq \tilde{\mu}_{\sup}(y) \right). \tag{13}$$

(2) *If $(X, \tilde{\mu})$ is a $(1,1)$-hyperfuzzy ideal of $(X, *, 0)$, then*

$$(\forall x, y \in X) \left(x \leq y \ \Rightarrow \ \tilde{\mu}_{\inf}(x) \geq \tilde{\mu}_{\inf}(y), \ \tilde{\mu}_{\sup}(x) \geq \tilde{\mu}_{\sup}(y) \right). \tag{14}$$

(3) *If $(X, \tilde{\mu})$ is a $(4,1)$-hyperfuzzy ideal of $(X, *, 0)$, then*

$$(\forall x, y \in X) \left(x \leq y \ \Rightarrow \ \tilde{\mu}_{\inf}(x) \leq \tilde{\mu}_{\inf}(y), \ \tilde{\mu}_{\sup}(x) \geq \tilde{\mu}_{\sup}(y) \right). \tag{15}$$

(4) *If* $(X, \tilde{\mu})$ *is a* $(4,4)$*-hyperfuzzy ideal of* $(X, *, 0)$*, then*

$$(\forall x, y \in X)\ (x \leq y \Rightarrow \tilde{\mu}_{\inf}(x) \leq \tilde{\mu}_{\inf}(y),\ \tilde{\mu}_{\sup}(x) \leq \tilde{\mu}_{\sup}(y)). \quad (16)$$

Proof. If $(X, \tilde{\mu})$ is a $(1,4)$-hyperfuzzy ideal of $(X, *, 0)$, then $(X, \tilde{\mu}_{\inf})$ is a 1-fuzzy ideal of $(X, *, 0)$ and $(X, \tilde{\mu}_{\sup})$ is a 4-fuzzy ideal of $(X, *, 0)$. Let $x, y \in X$ be such that $x \leq y$. Then $x * y = 0$, and so

$$\tilde{\mu}_{\inf}(x) \geq \min\{\tilde{\mu}_{\inf}(x * y), \tilde{\mu}_{\inf}(y)\} = \min\{\tilde{\mu}_{\inf}(0), \tilde{\mu}_{\inf}(y)\} = \tilde{\mu}_{\inf}(y)$$

by (8) and (7), and

$$\tilde{\mu}_{\sup}(x) \leq \max\{\tilde{\mu}_{\sup}(x * y), \tilde{\mu}_{\sup}(y)\} = \max\{\tilde{\mu}_{\sup}(0), \tilde{\mu}_{\sup}(y)\} = \tilde{\mu}_{\sup}(y)$$

by (12) and (9). Similarly, we can prove (2), (3) and (4). □

Proposition 2. *Given a hyper structure* $(X, \tilde{\mu})$ *over* $(X, *, 0)$*, we have the following assertions.*

(1) *If* $(X, \tilde{\mu})$ *is an* (i, j)*-hyperfuzzy ideal of* $(X, *, 0)$ *for* $(i, j) \in \{(2,2), (2,3), (3,2), (3,3)\}$*, then*

$$(\forall x, y \in X)\ (x \leq y \Rightarrow \tilde{\mu}_{\inf}(x) = \tilde{\mu}_{\inf}(0),\ \tilde{\mu}_{\sup}(x) = \tilde{\mu}_{\sup}(0)). \quad (17)$$

(2) *If* $(X, \tilde{\mu})$ *is either a* $(1,2)$*-hyperfuzzy ideal or a* $(1,3)$*-hyperfuzzy ideal of* $(X, *, 0)$*, then the following assertion is valid.*

$$(\forall x, y \in X)\ (x \leq y \Rightarrow \tilde{\mu}_{\inf}(x) \geq \tilde{\mu}_{\inf}(y),\ \tilde{\mu}_{\sup}(x) = \tilde{\mu}_{\sup}(0)). \quad (18)$$

(3) *If* $(X, \tilde{\mu})$ *is either a* $(2,1)$*-hyperfuzzy ideal or a* $(3,1)$*-hyperfuzzy ideal of* $(X, *, 0)$*, then the following assertion is valid.*

$$(\forall x, y \in X)\ (x \leq y \Rightarrow \tilde{\mu}_{\inf}(x) = \tilde{\mu}_{\inf}(0),\ \tilde{\mu}_{\sup}(x) \geq \tilde{\mu}_{\sup}(y)). \quad (19)$$

(4) *If* $(X, \tilde{\mu})$ *is either a* $(2,4)$*-hyperfuzzy ideal or a* $(3,4)$*-hyperfuzzy ideal of* $(X, *, 0)$*, then the following assertion is valid.*

$$(\forall x, y \in X)\ (x \leq y \Rightarrow \tilde{\mu}_{\inf}(x) = \tilde{\mu}_{\inf}(0),\ \tilde{\mu}_{\sup}(x) \leq \tilde{\mu}_{\sup}(y)). \quad (20)$$

(5) *If* $(X, \tilde{\mu})$ *is either a* $(4,2)$*-hyperfuzzy ideal or a* $(4,3)$*-hyperfuzzy ideal of* $(X, *, 0)$*, then the following assertion is valid.*

$$(\forall x, y \in X)\ (x \leq y \Rightarrow \tilde{\mu}_{\inf}(x) \leq \tilde{\mu}_{\inf}(y),\ \tilde{\mu}_{\sup}(x) = \tilde{\mu}_{\sup}(0)). \quad (21)$$

Proof. We prove (1) only, Others can be verified by the similar way. If $(X, \tilde{\mu})$ is a $(2,3)$-hyperfuzzy ideal of $(X, *, 0)$, then $(X, \tilde{\mu}_{\inf})$ is a 2-fuzzy ideal of $(X, *, 0)$ and $(X, \tilde{\mu}_{\sup})$ is a 3-fuzzy ideal of $(X, *, 0)$. Let $x, y \in X$ be such that $x \leq y$. Then $x * y = 0$, and thus

$$\begin{aligned}\tilde{\mu}_{\inf}(x) &\leq \min\{\tilde{\mu}_{\inf}(x * y), \tilde{\mu}_{\inf}(y)\} = \min\{\tilde{\mu}_{\inf}(0), \tilde{\mu}_{\inf}(y)\} = \tilde{\mu}_{\inf}(0),\\ \tilde{\mu}_{\sup}(x) &\geq \max\{\tilde{\mu}_{\sup}(x * y), \tilde{\mu}_{\sup}(y)\} = \max\{\tilde{\mu}_{\sup}(0), \tilde{\mu}_{\sup}(y)\} = \tilde{\mu}_{\sup}(0)\end{aligned}$$

by (10), (9), (11) and (7). Since $\tilde{\mu}_{\inf}(0) \leq \tilde{\mu}_{\inf}(x)$ and $\tilde{\mu}_{\sup}(0) \geq \tilde{\mu}_{\sup}(x)$ for all $x \in X$, it follows that $\tilde{\mu}_{\inf}(0) = \tilde{\mu}_{\inf}(x)$ and $\tilde{\mu}_{\sup}(0) = \tilde{\mu}_{\sup}(x)$ for all $x \in X$. Similarly, we can verify that (17) is true for $(i, j) \in \{(2,2), (3,2), (3,3)\}$. □

Proposition 3. *Given a hyper structure* $(X, \tilde{\mu})$ *over* $(X, *, 0)$*, we have the following assertions.*

(1) *If* $(X, \tilde{\mu})$ *is a* $(1,4)$*-hyperfuzzy ideal of* $(X, *, 0)$*, then*

$$(\forall x, y, z \in X) \left(x * y \leq z \Rightarrow \begin{cases} \tilde{\mu}_{\inf}(x) \geq \min\{\tilde{\mu}_{\inf}(y), \tilde{\mu}_{\inf}(z)\} \\ \tilde{\mu}_{\sup}(x) \leq \max\{\tilde{\mu}_{\sup}(y), \tilde{\mu}_{\sup}(z)\} \end{cases} \right). \tag{22}$$

(2) *If* $(X, \tilde{\mu})$ *is a* $(1,1)$*-hyperfuzzy ideal of* $(X, *, 0)$*, then*

$$(\forall x, y, z \in X) \left(x * y \leq z \Rightarrow \begin{cases} \tilde{\mu}_{\inf}(x) \geq \min\{\tilde{\mu}_{\inf}(y), \tilde{\mu}_{\inf}(z)\} \\ \tilde{\mu}_{\sup}(x) \geq \min\{\tilde{\mu}_{\sup}(y), \tilde{\mu}_{\sup}(z)\} \end{cases} \right). \tag{23}$$

(3) *If* $(X, \tilde{\mu})$ *is a* $(4,1)$*-hyperfuzzy ideal of* $(X, *, 0)$*, then*

$$(\forall x, y, z \in X) \left(x * y \leq z \Rightarrow \begin{cases} \tilde{\mu}_{\inf}(x) \leq \max\{\tilde{\mu}_{\inf}(y), \tilde{\mu}_{\inf}(z)\} \\ \tilde{\mu}_{\sup}(x) \geq \min\{\tilde{\mu}_{\sup}(y), \tilde{\mu}_{\sup}(z)\} \end{cases} \right). \tag{24}$$

(4) *If* $(X, \tilde{\mu})$ *is a* $(4,4)$*-hyperfuzzy ideal of* $(X, *, 0)$*, then*

$$(\forall x, y, z \in X) \left(x * y \leq z \Rightarrow \begin{cases} \tilde{\mu}_{\inf}(x) \leq \max\{\tilde{\mu}_{\inf}(y), \tilde{\mu}_{\inf}(z)\} \\ \tilde{\mu}_{\sup}(x) \leq \max\{\tilde{\mu}_{\sup}(y), \tilde{\mu}_{\sup}(z)\} \end{cases} \right). \tag{25}$$

Proof. Assume that $(X, \tilde{\mu})$ is a $(1,4)$-hyperfuzzy ideal of $(X, *, 0)$. Let $x, y, z \in X$ be such that $x * y \leq z$. Then $(x * y) * z = 0$, and so

$$\begin{aligned} \tilde{\mu}_{\inf}(x) &\geq \min\{\tilde{\mu}_{\inf}(x * y), \tilde{\mu}_{\inf}(y)\} \\ &\geq \min\{\min\{\tilde{\mu}_{\inf}((x * y) * z), \tilde{\mu}_{\inf}(z)\}, \tilde{\mu}_{\inf}(y)\} \\ &= \min\{\min\{\tilde{\mu}_{\inf}(0), \tilde{\mu}_{\inf}(z)\}, \tilde{\mu}_{\inf}(y)\} \\ &= \min\{\tilde{\mu}_{\inf}(y), \tilde{\mu}_{\inf}(z)\} \end{aligned}$$

by (8) and (7), and

$$\begin{aligned} \tilde{\mu}_{\sup}(x) &\leq \max\{\tilde{\mu}_{\sup}(x * y), \tilde{\mu}_{\sup}(y)\} \\ &\leq \max\{\max\{\tilde{\mu}_{\sup}((x * y) * z), \tilde{\mu}_{\sup}(z)\}, \tilde{\mu}_{\sup}(y)\} \\ &= \max\{\max\{\tilde{\mu}_{\sup}(0), \tilde{\mu}_{\sup}(z)\}, \tilde{\mu}_{\sup}(y)\} \\ &= \max\{\tilde{\mu}_{\sup}(y), \tilde{\mu}_{\sup}(z)\} \end{aligned}$$

by (12) and (9). Similarly, we can check that (2), (3) and (4) hold. □

Proposition 4. *Given a hyper structure* $(X, \tilde{\mu})$ *over* $(X, *, 0)$*, we have the following assertions.*

(1) *If* $(X, \tilde{\mu})$ *is an* (i, j)*-hyperfuzzy ideal of* $(X, *, 0)$ *for* $(i, j) \in \{(2,2), (2,3), (3,2), (3,3)\}$*, then*

$$(\forall x, y, z \in X) \left(x * y \leq z \Rightarrow \tilde{\mu}_{\inf}(x) = \tilde{\mu}_{\inf}(0),\ \tilde{\mu}_{\sup}(x) = \tilde{\mu}_{\sup}(0)\right). \tag{26}$$

(2) *If* $(X, \tilde{\mu})$ *is either a* $(1,2)$*-hyperfuzzy ideal or a* $(1,3)$*-hyperfuzzy ideal of* $(X, *, 0)$*, then the following assertion is valid.*

$$(\forall x, y, z \in X) \left(x * y \leq z \Rightarrow \begin{cases} \tilde{\mu}_{\inf}(x) \geq \min\{\tilde{\mu}_{\inf}(y), \tilde{\mu}_{\inf}(z)\} \\ \tilde{\mu}_{\sup}(x) = \tilde{\mu}_{\sup}(0) \end{cases} \right). \tag{27}$$

(3) *If* $(X,\tilde{\mu})$ *is either a* $(2,1)$*-hyperfuzzy ideal or a* $(3,1)$*-hyperfuzzy ideal of* $(X,*,0)$*, then the following assertion is valid.*

$$(\forall x,y,z\in X)\left(\ x*y\le z\ \Rightarrow\begin{cases}\tilde{\mu}_{\inf}(x)=\tilde{\mu}_{\inf}(0)\\ \tilde{\mu}_{\sup}(x)\ge\min\{\tilde{\mu}_{\sup}(y),\tilde{\mu}_{\sup}(z)\}\end{cases}\right). \tag{28}$$

(4) *If* $(X,\tilde{\mu})$ *is either a* $(2,4)$*-hyperfuzzy ideal or a* $(3,4)$*-hyperfuzzy ideal of* $(X,*,0)$*, then the following assertion is valid.*

$$(\forall x,y,z\in X)\left(\ x*y\le z\ \Rightarrow\begin{cases}\tilde{\mu}_{\inf}(x)=\tilde{\mu}_{\inf}(0)\\ \tilde{\mu}_{\sup}(x)\le\max\{\tilde{\mu}_{\sup}(y),\tilde{\mu}_{\sup}(z)\}\end{cases}\right). \tag{29}$$

(5) *If* $(X,\tilde{\mu})$ *is either a* $(4,2)$*-hyperfuzzy ideal or a* $(4,3)$*-hyperfuzzy ideal of* $(X,*,0)$*, then the following assertion is valid.*

$$(\forall x,y,z\in X)\left(\ x*y\le z\ \Rightarrow\begin{cases}\tilde{\mu}_{\inf}(x)\le\max\{\tilde{\mu}_{\inf}(y),\tilde{\mu}_{\inf}(z)\}\\ \tilde{\mu}_{\sup}(x)=\tilde{\mu}_{\sup}(0)\end{cases}\right). \tag{30}$$

Proof. If $(X,\tilde{\mu})$ is a $(2,3)$-hyperfuzzy ideal of $(X,*,0)$, then $(X,\tilde{\mu}_{\inf})$ is a 2-fuzzy ideal of $(X,*,0)$ and $(X,\tilde{\mu}_{\sup})$ is a 3-fuzzy ideal of $(X,*,0)$. Let $x,y,z\in X$ be such that $x*y\le z$. Then $(x*y)*z=0$, and so

$$\begin{aligned}\tilde{\mu}_{\inf}(x)&\le\min\{\tilde{\mu}_{\inf}(x*y),\tilde{\mu}_{\inf}(y)\}\\&\le\min\{\min\{\tilde{\mu}_{\inf}((x*y)*z),\tilde{\mu}_{\inf}(z)\},\tilde{\mu}_{\inf}(y)\}\\&=\min\{\min\{\tilde{\mu}_{\inf}(0),\tilde{\mu}_{\inf}(z)\},\tilde{\mu}_{\inf}(y)\}\\&=\min\{\tilde{\mu}_{\inf}(0),\tilde{\mu}_{\inf}(y)\}\\&=\tilde{\mu}_{\inf}(0)\end{aligned}$$

by (10) and (9), and

$$\begin{aligned}\tilde{\mu}_{\sup}(x)&\ge\max\{\tilde{\mu}_{\sup}(x*y),\tilde{\mu}_{\sup}(y)\}\\&\ge\max\{\max\{\tilde{\mu}_{\sup}((x*y)*z),\tilde{\mu}_{\sup}(z)\},\tilde{\mu}_{\sup}(y)\}\\&=\max\{\max\{\tilde{\mu}_{\sup}(0),\tilde{\mu}_{\sup}(z)\},\tilde{\mu}_{\sup}(y)\}\\&=\max\{\tilde{\mu}_{\sup}(0),\tilde{\mu}_{\sup}(y)\}\\&=\tilde{\mu}_{\sup}(0)\end{aligned}$$

by (11) and (7). Since $\tilde{\mu}_{\inf}(0)\le\tilde{\mu}_{\inf}(x)$ and $\tilde{\mu}_{\sup}(0)\ge\tilde{\mu}_{\sup}(x)$ for all $x\in X$, it follows that $\tilde{\mu}_{\inf}(x)=\tilde{\mu}_{\inf}(0)$ and $\tilde{\mu}_{\sup}(x)=\tilde{\mu}_{\sup}(0)$. Similarly, we can verify that (26) is true for $(i,j)\in\{(2,2),(3,2),(3,3)\}$. Similarly, we can show that (2), (3), (4) and (5) are true. □

Given a hyper structure $(X,\tilde{\mu})$ over X and $\alpha,\beta\in[0,1]$, we consider the following sets:

$$\begin{aligned}U(\tilde{\mu}_{\inf};\alpha)&:=\{x\in X\mid\tilde{\mu}_{\inf}(x)\ge\alpha\},\\L(\tilde{\mu}_{\inf};\alpha)&:=\{x\in X\mid\tilde{\mu}_{\inf}(x)\le\alpha\},\\U(\tilde{\mu}_{\sup};\beta)&:=\{x\in X\mid\tilde{\mu}_{\sup}(x)\ge\beta\},\\L(\tilde{\mu}_{\sup};\beta)&:=\{x\in X\mid\tilde{\mu}_{\sup}(x)\le\beta\}.\end{aligned}$$

Theorem 1. (1) *A hyper structure* $(X,\tilde{\mu})$ *over* $(X,*,0)$ *is a* $(1,4)$*-hyperfuzzy ideal of* $(X,*,0)$ *if and only if the sets* $U(\tilde{\mu}_{\inf};\alpha)$ *and* $L(\tilde{\mu}_{\sup};\beta)$ *are either empty or ideals of* X *for all* $\alpha,\beta\in[0,1]$.

(2) *A hyper structure* $(X, \tilde{\mu})$ *over* $(X, *, 0)$ *is a* $(1,1)$*-hyperfuzzy ideal of* $(X, *, 0)$ *if and only if the sets* $U(\tilde{\mu}_{\inf}; \alpha)$ *and* $U(\tilde{\mu}_{\sup}; \beta)$ *are either empty or ideals of* X *for all* $\alpha, \beta \in [0,1]$.

(3) *A hyper structure* $(X, \tilde{\mu})$ *over* $(X, *, 0)$ *is a* $(4,1)$*-hyperfuzzy ideal of* $(X, *, 0)$ *if and only if the sets* $L(\tilde{\mu}_{\inf}; \alpha)$ *and* $U(\tilde{\mu}_{\sup}; \beta)$ *are either empty or ideals of* X *for all* $\alpha, \beta \in [0,1]$.

(4) *A hyper structure* $(X, \tilde{\mu})$ *over* $(X, *, 0)$ *is a* $(4,4)$*-hyperfuzzy ideal of* $(X, *, 0)$ *if and only if the sets* $L(\tilde{\mu}_{\inf}; \alpha)$ *and* $L(\tilde{\mu}_{\sup}; \beta)$ *are either empty or ideals of* X *for all* $\alpha, \beta \in [0,1]$.

Proof. Assume that $(X, \tilde{\mu})$ is a $(1,4)$-hyperfuzzy ideal of $(X, *, 0)$ and $U(\tilde{\mu}_{\inf}; \alpha) \neq \emptyset \neq L(\tilde{\mu}_{\sup}; \beta)$ for $\alpha, \beta \in [0,1]$. Then there exist $a \in U(\tilde{\mu}_{\inf}; \alpha)$ and $b \in L(\tilde{\mu}_{\sup}; \beta)$. Hence $\tilde{\mu}_{\inf}(0) \geq \tilde{\mu}_{\inf}(a) \geq \alpha$ and $\tilde{\mu}_{\sup}(0) \leq \tilde{\mu}_{\sup}(b) \leq \beta$, that is, $0 \in U(\tilde{\mu}_{\inf}; \alpha) \cap L(\tilde{\mu}_{\sup}; \beta)$. Let $x, y \in X$ be such that $x * y \in U(\tilde{\mu}_{\inf}; \alpha)$ and $y \in U(\tilde{\mu}_{\inf}; \alpha)$. Then $\tilde{\mu}_{\inf}(x * y) \geq \alpha$ and $\tilde{\mu}_{\inf}(y) \geq \alpha$. It follows that

$$\tilde{\mu}_{\inf}(x) \geq \min\{\tilde{\mu}_{\inf}(x * y), \tilde{\mu}_{\inf}(y)\} \geq \alpha$$

and so that $x \in U(\tilde{\mu}_{\inf}; \alpha)$. Thus $U(\tilde{\mu}_{\inf}; \alpha)$ is an ideal of $(X, *, 0)$. Now let $a, b \in X$ be such that $a * b \in L(\tilde{\mu}_{\sup}; \beta)$ and $b \in L(\tilde{\mu}_{\sup}; \beta)$. Then $\tilde{\mu}_{\sup}(a * b) \leq \beta$ and $\tilde{\mu}_{\sup}(b) \leq \beta$, which imply that

$$\tilde{\mu}_{\sup}(a) \leq \max\{\tilde{\mu}_{\sup}(a * b), \tilde{\mu}_{\sup}(b)\} \leq \beta.$$

Thus $a \in L(\tilde{\mu}_{\sup}; \beta)$, and therefore $L(\tilde{\mu}_{\sup}; \beta)$ is an ideal of $(X, *, 0)$.

Conversely, suppose that the sets $U(\tilde{\mu}_{\inf}; \alpha)$ and $L(\tilde{\mu}_{\sup}; \beta)$ are either empty or ideals of X for all $\alpha, \beta \in [0,1]$. For any $x \in X$, let $\tilde{\mu}_{\inf}(x) = \alpha$ and $\tilde{\mu}_{\sup}(x) = \beta$. Then $x \in U(\tilde{\mu}_{\inf}; \alpha) \cap L(\tilde{\mu}_{\sup}; \beta)$, and so $U(\tilde{\mu}_{\inf}; \alpha)$ and $L(\tilde{\mu}_{\sup}; \beta)$ are nonempty. Hence $U(\tilde{\mu}_{\inf}; \alpha)$ and $L(\tilde{\mu}_{\sup}; \beta)$ are ideals of $(X, *, 0)$, and thus $0 \in U(\tilde{\mu}_{\inf}; \alpha) \cap L(\tilde{\mu}_{\sup}; \beta)$. It follows that $\tilde{\mu}_{\inf}(0) \geq \alpha = \tilde{\mu}_{\inf}(x)$ and $\tilde{\mu}_{\sup}(0) \leq \beta = \tilde{\mu}_{\sup}(x)$ for all $x \in X$. Assume that there exist $a, b \in X$ such that

$$\tilde{\mu}_{\inf}(a) < \min\{\tilde{\mu}_{\inf}(a * b), \tilde{\mu}_{\inf}(b)\}.$$

If we take $\gamma := \min\{\tilde{\mu}_{\inf}(a * b), \tilde{\mu}_{\inf}(b)\}$, then $\gamma \in [0,1]$, $a * b \in U(\tilde{\mu}_{\inf}; \gamma)$ and $b \in U(\tilde{\mu}_{\inf}; \gamma)$. Since $U(\tilde{\mu}_{\inf}; \gamma)$ is an ideal of X, we have $a \in U(\tilde{\mu}_{\inf}; \gamma)$, that is, $\tilde{\mu}_{\inf}(a) \geq \gamma$. This is a contradiction, and so

$$\tilde{\mu}_{\inf}(x) \geq \min\{\tilde{\mu}_{\inf}(x * y), \tilde{\mu}_{\inf}(y)\}$$

for all $x, y \in X$. Now, suppose that

$$\tilde{\mu}_{\sup}(x) > \max\{\tilde{\mu}_{\sup}(x * y), \tilde{\mu}_{\sup}(y)\}$$

for some $x, y \in X$, and take

$$\beta := \tfrac{1}{2}\left(\tilde{\mu}_{\sup}(x) + \max\{\tilde{\mu}_{\sup}(x * y), \tilde{\mu}_{\sup}(y)\}\right).$$

Then $x * y \in L(\tilde{\mu}_{\sup}; \beta)$ and $y \in L(\tilde{\mu}_{\sup}; \beta)$, which imply that $x \in L(\tilde{\mu}_{\sup}; \beta)$ since $L(\tilde{\mu}_{\sup}; \beta)$ is an ideal of X. Hence $\tilde{\mu}_{\sup}(x) \leq \beta$, which is a contradiction, and so

$$\tilde{\mu}_{\sup}(x) \leq \max\{\tilde{\mu}_{\sup}(x * y), \tilde{\mu}_{\sup}(y)\}$$

for all $x, y \in X$. Therefore $(X, \tilde{\mu})$ is a $(1,4)$-hyperfuzzy ideal of $(X, *, 0)$. Similarly, we can verify that (2), (3), and (4) hold. □

Theorem 2. *If a hyper structure* $(X, \tilde{\mu})$ *over* $(X, *, 0)$ *is a* $(2,3)$*-hyperfuzzy ideal of* $(X, *, 0)$*, then the sets* $U(\tilde{\mu}_{\inf}; \alpha)^c$ *and* $L(\tilde{\mu}_{\sup}; \beta)^c$ *are either empty or ideals of* $(X, *, 0)$ *for all* $\alpha, \beta \in [0,1]$.

Proof. If a hyper structure $(X, \tilde{\mu})$ over $(X, *, 0)$ is a $(2,3)$-hyperfuzzy ideal of $(X, *, 0)$, then $(X, \tilde{\mu}_{\inf})$ is a 2-fuzzy ideal of X and $(X, \tilde{\mu}_{\sup})$ is a 3-fuzzy ideal of $(X, *, 0)$. Let $\alpha, \beta \in [0,1]$ be such that $U(\tilde{\mu}_{\inf}; \alpha)^c$ and $L(\tilde{\mu}_{\sup}; \beta)^c$ are nonempty. Then there exist $x, a \in X$ such that $x \in U(\tilde{\mu}_{\inf}; \alpha)^c$ and $a \in L(\tilde{\mu}_{\sup}; \beta)^c$. Hence $\tilde{\mu}_{\inf}(0) \leq \tilde{\mu}_{\inf}(x) < \alpha$ and $\tilde{\mu}_{\sup}(0) \geq \tilde{\mu}_{\sup}(a) > \beta$, which imply that $0 \in U(\tilde{\mu}_{\inf}; \alpha)^c \cap L(\tilde{\mu}_{\sup}; \beta)^c$. Let $x, y \in X$ be such that $x * y \in U(\tilde{\mu}_{\inf}; \alpha)^c$ and $y \in U(\tilde{\mu}_{\inf}; \alpha)^c$. Then $\tilde{\mu}_{\inf}(x * y) < \alpha$ and $\tilde{\mu}_{\inf}(y) < \alpha$. It follows from (10) that

$$\tilde{\mu}_{\inf}(x) \leq \min\{\tilde{\mu}_{\inf}(x * y), \tilde{\mu}_{\inf}(y)\} < \alpha$$

and so that $x \in U(\tilde{\mu}_{\inf}; \alpha)^c$. Hence $U(\tilde{\mu}_{\inf}; \alpha)^c$ is an ideal of $(X, *, 0)$. Now let $a, b \in X$ be such that $a * b \in L(\tilde{\mu}_{\sup}; \beta)^c$ and $b \in L(\tilde{\mu}_{\sup}; \beta)^c$. Then $\tilde{\mu}_{\sup}(a * b) > \beta$ and $\tilde{\mu}_{\sup}(b) > \beta$, which imply from (11) that

$$\tilde{\mu}_{\sup}(a) \geq \max\{\tilde{\mu}_{\sup}(a * b), \tilde{\mu}_{\sup}(b)\} > \beta$$

Thus $a \in L(\tilde{\mu}_{\sup}; \beta)^c$, and therefore $L(\tilde{\mu}_{\sup}; \beta)^c$ is an ideal of X. □

The following example shows that the converse of Theorem 2 is not true, that is, there exists a hyper structure $(X, \tilde{\mu})$ over $(X, *, 0)$ such that

(1) $(X, \tilde{\mu})$ is not a $(2,3)$-hyperfuzzy ideal of $(X, *, 0)$,
(2) The nonempty sets $U(\tilde{\mu}_{\inf}; \alpha)^c$ and $L(\tilde{\mu}_{\sup}; \beta)^c$ are ideals of $(X, *, 0)$ for all $\alpha, \beta \in [0,1]$.

Example 3. *Consider a BCI-algebra $X = \{0, 1, a, b, c\}$ with the binary operation $*$ which is given in Table 3 (see [8]).*

Table 3. Cayley table for the binary operation "$*$".

$*$	**0**	**1**	***a***	***b***	***c***
0	0	0	a	b	c
1	1	0	a	b	c
a	a	a	0	c	b
b	b	b	c	0	a
c	c	c	b	a	0

Let $(X, \tilde{\mu})$ be a hyper structure over $(X, *, 0)$ in which $\tilde{\mu}$ is given as follows:

$$\tilde{\mu} : X \to \tilde{\mathcal{P}}([0,1]),\ x \mapsto \begin{cases} [0.23, 0.85) & \text{if } x = 0, \\ (0.43, 0.83] & \text{if } x = 1, \\ [0.53, 0.73] & \text{if } x = a, \\ (0.63, 0.73] & \text{if } x = b, \\ [0.63, 0.75) & \text{if } x = c. \end{cases}$$

Then

$$U(\tilde{\mu}_{\inf}; \alpha)^c = \begin{cases} \varnothing & \text{if } \alpha \in [0, 0.23], \\ \{0\} & \text{if } \alpha \in (0.23, 0.43], \\ \{0, 1\} & \text{if } \alpha \in (0.43, 0.53], \\ \{0, 1, a\} & \text{if } \alpha \in (0.53, 0.63], \\ X & \text{if } \alpha \in (0.63, 1.0], \end{cases}$$

and

$$L(\tilde{\mu}_{\sup};\beta)^c = \begin{cases} \emptyset & \text{if } \beta \in (0.85, 1.0], \\ \{0\} & \text{if } \beta \in (0.83, 0.85], \\ \{0,1\} & \text{if } \beta \in (0.75, 0.83], \\ \{0,1,c\} & \text{if } \beta \in (0.73, 0.75], \\ X & \text{if } \beta \in (0, 0.73]. \end{cases}$$

Hence the nonempty sets $U(\tilde{\mu}_{\inf};\alpha)^c$ and $L(\tilde{\mu}_{\sup};\beta)^c$ are ideals of $(X,*,0)$ for all $\alpha,\beta \in [0,1]$. But $(X,\tilde{\mu})$ is not a $(2,3)$-hyperfuzzy ideal of $(X,*,0)$ since

$$\tilde{\mu}_{\inf}(c) = 0.63 > 0.53 = \min\{\tilde{\mu}_{\inf}(c*a), \tilde{\mu}_{\inf}(a)\}$$

and/or

$$\tilde{\mu}_{\sup}(a) = 0.73 < 0.75 = \max\{\tilde{\mu}_{\sup}(a*c), \tilde{\mu}_{\inf}(c)\}.$$

Using the similar way to the proof of Theorem 2, we have the following theorem.

Theorem 3. (1) *If a hyper structure $(X,\tilde{\mu})$ over $(X,*,0)$ is a $(2,2)$-hyperfuzzy ideal of $(X,*,0)$, then the sets $U(\tilde{\mu}_{\inf};\alpha)^c$ and $U(\tilde{\mu}_{\sup};\beta)^c$ are either empty or ideals of $(X,*,0)$ for all $\alpha,\beta \in [0,1]$.*

(2) *If a hyper structure $(X,\tilde{\mu})$ over $(X,*,0)$ is a $(3,2)$-hyperfuzzy ideal of $(X,*,0)$, then the sets $L(\tilde{\mu}_{\inf};\alpha)^c$ and $U(\tilde{\mu}_{\sup};\beta)^c$ are either empty or ideals of $(X,*,0)$ for all $\alpha,\beta \in [0,1]$.*

(3) *If a hyper structure $(X,\tilde{\mu})$ over $(X,*,0)$ is a $(3,3)$-hyperfuzzy ideal of $(X,*,0)$, then the sets $L(\tilde{\mu}_{\inf};\alpha)^c$ and $L(\tilde{\mu}_{\sup};\beta)^c$ are either empty or ideals of $(X,*,0)$ for all $\alpha,\beta \in [0,1]$.*

Using the similar way to the proof of Theorems 1 and 2, we have the following theorem.

Theorem 4. (1) *If a hyper structure $(X,\tilde{\mu})$ over $(X,*,0)$ is a $(1,2)$-hyperfuzzy ideal of $(X,*,0)$, then the sets $U(\tilde{\mu}_{\inf};\alpha)$ and $U(\tilde{\mu}_{\sup};\beta)^c$ are either empty or ideals of $(X,*,0)$ for all $\alpha,\beta \in [0,1]$.*

(2) *If a hyper structure $(X,\tilde{\mu})$ over $(X,*,0)$ is a $(1,3)$-hyperfuzzy ideal of $(X,*,0)$, then the sets $U(\tilde{\mu}_{\inf};\alpha)$ and $L(\tilde{\mu}_{\sup};\beta)^c$ are either empty or ideals of $(X,*,0)$ for all $\alpha,\beta \in [0,1]$.*

(3) *If a hyper structure $(X,\tilde{\mu})$ over $(X,*,0)$ is a $(2,1)$-hyperfuzzy ideal of $(X,*,0)$, then the sets $U(\tilde{\mu}_{\inf};\alpha)^c$ and $U(\tilde{\mu}_{\sup};\beta)$ are either empty or ideals of $(X,*,0)$ for all $\alpha,\beta \in [0,1]$.*

(4) *If a hyper structure $(X,\tilde{\mu})$ over $(X,*,0)$ is a $(3,1)$-hyperfuzzy ideal of $(X,*,0)$, then the sets $L(\tilde{\mu}_{\inf};\alpha)^c$ and $U(\tilde{\mu}_{\sup};\beta)$ are either empty or ideals of $(X,*,0)$ for all $\alpha,\beta \in [0,1]$.*

(5) *If a hyper structure $(X,\tilde{\mu})$ over $(X,*,0)$ is a $(2,4)$-hyperfuzzy ideal of $(X,*,0)$, then the sets $U(\tilde{\mu}_{\inf};\alpha)^c$ and $L(\tilde{\mu}_{\sup};\beta)$ are either empty or ideals of $(X,*,0)$ for all $\alpha,\beta \in [0,1]$.*

(6) *If a hyper structure $(X,\tilde{\mu})$ over $(X,*,0)$ is a $(3,4)$-hyperfuzzy ideal of $(X,*,0)$, then the sets $L(\tilde{\mu}_{\inf};\alpha)^c$ and $L(\tilde{\mu}_{\sup};\beta)$ are either empty or ideals of $(X,*,0)$ for all $\alpha,\beta \in [0,1]$.*

(7) *If a hyper structure $(X,\tilde{\mu})$ over $(X,*,0)$ is a $(4,2)$-hyperfuzzy ideal of $(X,*,0)$, then the sets $L(\tilde{\mu}_{\inf};\alpha)$ and $U(\tilde{\mu}_{\sup};\beta)^c$ are either empty or ideals of $(X,*,0)$ for all $\alpha,\beta \in [0,1]$.*

(8) *If a hyper structure $(X,\tilde{\mu})$ over $(X,*,0)$ is a $(4,3)$-hyperfuzzy ideal of $(X,*,0)$, then the sets $L(\tilde{\mu}_{\inf};\alpha)$ and $L(\tilde{\mu}_{\sup};\beta)^c$ are either empty or ideals of $(X,*,0)$ for all $\alpha,\beta \in [0,1]$.*

4. Relations between Hyperfuzzy Ideals and Hyperfuzzy Subalgebras

Theorem 5. *Let $(X,*,0) \in \mathcal{B}_K(X)$. For any $i,j \in \{1,4\}$, every (i,j)-hyperfuzzy ideal is an (i,j)-hyperfuzzy subalgebra.*

Proof. Let $(X,*,0) \in \mathcal{B}_K(X)$ and let $(X,\tilde{\mu})$ be a $(1,4)$-hyperfuzzy ideal of $(X,*,0)$. Then $(X,\tilde{\mu}_{\inf})$ is a 1-fuzzy ideal of $(X,*,0)$ and $(X,\tilde{\mu}_{\sup})$ is a 4-fuzzy ideal of $(X,*,0)$. Since $x*y \leq x$ for all $x,y \in X$, it follows from Proposition 1, (8) and (12) that

$$\tilde{\mu}_{\inf}(x*y) \geq \tilde{\mu}_{\inf}(x) \geq \min\{\tilde{\mu}_{\inf}(x*y), \tilde{\mu}_{\inf}(y)\} \geq \min\{\tilde{\mu}_{\inf}(x), \tilde{\mu}_{\inf}(y)\}$$
$$\tilde{\mu}_{\sup}(x*y) \leq \tilde{\mu}_{\sup}(x) \leq \max\{\tilde{\mu}_{\sup}(x*y), \tilde{\mu}_{\sup}(y)\} \leq \max\{\tilde{\mu}_{\sup}(x), \tilde{\mu}_{\sup}(y)\}$$

for all $x,y \in X$. Therefore $(X,\tilde{\mu})$ is a $(1,4)$-hyperfuzzy subalgebra of $(X,*,0)$. Similarly, we can prove the result for $(i,j) \in \{(1,1),(4,1),(4,4)\}$. □

The converse of Theorem 5 is not true for $(i,j) = (1,4)$ as seen in the following example.

Example 4. *Consider a BCK-algebra $X = \{0,a,b,c\}$ with the binary operation $*$ which is given in Table 4 (see [8]).*

Table 4. Cayley table for the binary operation "$*$".

$*$	0	a	b	c
0	0	0	0	0
a	a	0	0	a
b	b	a	0	b
c	c	c	c	0

Let $(X,\tilde{\mu})$ be a hyper structure over $(X,*,0)$ in which $\tilde{\mu}$ is given as follows:

$$\tilde{\mu} : X \to \tilde{\mathcal{P}}([0,1]),\ x \mapsto \begin{cases} [0.4,0.7] & \text{if } x=0, \\ (0.4,0.7] & \text{if } x=a, \\ [0.2,0.9] & \text{if } x=b, \\ [0.4,0.5) \cup (0.5,0.7) & \text{if } x=c. \end{cases}$$

It is routine to verify that $(X,\tilde{\mu})$ is a $(1,4)$-hyperfuzzy subalgebra of $(X,*,0)$. But it is not a $(1,4)$-hyperfuzzy ideal of $(X,*,0)$ since $\tilde{\mu}_{\sup}(b) = 0.9 > 0.7 = \max\{\tilde{\mu}_{\sup}(b*a), \tilde{\mu}_{\sup}(a)\}$.

Example 5. *Let $X = \{0,1,2,3,4\}$ be the BCK-algebra in Example 1. Let $(X,\tilde{\mu})$ be a hyper structure over $(X,*,0)$ in which $\tilde{\mu}$ is given as follows:*

$$\tilde{\mu} : X \to \tilde{\mathcal{P}}([0,1]),\ x \mapsto \begin{cases} [0.5,0.9) & \text{if } x=0, \\ (0.4,0.6) \cup (0.6,0.8] & \text{if } x=1, \\ [0.3,0.5] & \text{if } x=2, \\ [0.2,0.4) \cup (0.5,0.6] & \text{if } x=3, \\ [0.2,0.5] & \text{if } x=4. \end{cases}$$

Then $(X,\tilde{\mu})$ is a $(1,1)$-hyperfuzzy subalgebra of $(X,*,0)$. Since

$$\tilde{\mu}_{\sup}(4) = 0.5 < 0.6 = \min\{\tilde{\mu}_{\sup}(4*3), \tilde{\mu}_{\sup}(3)\},$$

$(X,\tilde{\mu}_{\sup})$ is not a 1-fuzzy ideal of X. Hence $(X,\tilde{\mu})$ is not a $(1,1)$-hyperfuzzy ideal of $(X,*,0)$.

Example 6. *Consider a BCK-algebra $X = \{0,a,b,c\}$ with the binary operation $*$ which is given in Table 5 (see [8]).*

Table 5. Cayley table for the binary operation "$*$".

$*$	**0**	***a***	***b***	***c***
0	0	0	0	0
a	a	0	0	0
b	b	a	0	a
c	c	c	c	0

(1) Let $(X, \tilde{\mu})$ be a hyper structure over $(X, *, 0)$ in which $\tilde{\mu}$ is given as follows:

$$\tilde{\mu} : X \to \tilde{\mathcal{P}}([0,1]), \quad x \mapsto \begin{cases} [0.3, 0.9] & \text{if } x = 0, \\ (0.5, 0.8] & \text{if } x = a, \\ [0.4, 0.5] & \text{if } x = b, \\ [0.6, 0.7) & \text{if } x = c. \end{cases}$$

Then $(X, \tilde{\mu})$ is a $(4,1)$-hyperfuzzy subalgebra of $(X, *, 0)$. Since

$$\tilde{\mu}_{\inf}(1) = 0.5 > 0.4 = \max\{\tilde{\mu}_{\inf}(1 * 2), \tilde{\mu}_{\inf}(2)\}$$

and/or

$$\tilde{\mu}_{\sup}(2) = 0.5 < 0.8 = \min\{\tilde{\mu}_{\sup}(2 * 1), \tilde{\mu}_{\sup}(1)\},$$

$(X, \tilde{\mu}_{\inf})$ is not a 4-fuzzy ideal of $(X, *, 0)$ and/or $(X, \tilde{\mu}_{\sup})$ is not a 1-fuzzy ideal of $(X, *, 0)$. Therefore $(X, \tilde{\mu})$ is not a $(4,1)$-hyperfuzzy ideal of $(X, *, 0)$.

(2) Let $(X, \tilde{\mu})$ be a hyper structure over $(X, *, 0)$ in which $\tilde{\mu}$ is given as follows:

$$\tilde{\mu} : X \to \tilde{\mathcal{P}}([0,1]), \quad x \mapsto \begin{cases} [0.3, 0.5] & \text{if } x = 0, \\ (0.5, 0.7] & \text{if } x = a, \\ [0.4, 0.7) & \text{if } x = b, \\ [0.6, 0.8] & \text{if } x = c. \end{cases}$$

Then $(X, \tilde{\mu})$ is a $(4,4)$-hyperfuzzy subalgebra of $(X, *, 0)$ and $(X, \tilde{\mu}_{\sup})$ is a 4-fuzzy ideal of $(X, *, 0)$. But $(X, \tilde{\mu}_{\inf})$ is not a 4-fuzzy ideal of $(X, *, 0)$ since

$$\tilde{\mu}_{\inf}(1) = 0.5 > 0.4 = \max\{\tilde{\mu}_{\inf}(1 * 2), \tilde{\mu}_{\inf}(2)\}.$$

Hence $(X, \tilde{\mu})$ is not a $(4,4)$-hyperfuzzy ideal of $(X, *, 0)$.

We provide conditions for a $(1,4)$-hyperfuzzy subalgebra to be a $(1,4)$-hyperfuzzy ideal.

Theorem 6. *For any $(X, *, 0) \in \mathcal{B}_K(X)$, if a $(1,4)$-hyperfuzzy subalgebra $(X, \tilde{\mu})$ of $(X, *, 0)$ satisfies the condition (22), then $(X, \tilde{\mu})$ is a $(1,4)$-hyperfuzzy ideal of $(X, *, 0)$.*

Proof. Let $(X, *, 0) \in \mathcal{B}_K(X)$ and let $(X, \tilde{\mu})$ be a $(1,4)$-hyperfuzzy subalgebra of $(X, *, 0)$ that satisfies (22). Since $0 * x \leq x$ for all $x \in X$, we have $\tilde{\mu}_{\inf}(0) \geq \tilde{\mu}_{\inf}(x)$ and $\tilde{\mu}_{\sup}(0) \leq \tilde{\mu}_{\sup}(x)$ for all $x \in X$ by (22). Since $x * (x * y) \leq y$ for all $x, y \in X$, it follows from (22) that

$$\tilde{\mu}_{\inf}(x) \geq \min\{\tilde{\mu}_{\inf}(x * y), \tilde{\mu}_{\inf}(y)\}$$

and

$$\tilde{\mu}_{\sup}(x) \leq \max\{\tilde{\mu}_{\sup}(x * y), \tilde{\mu}_{\sup}(y)\}$$

for all $x, y \in X$. Hence $(X, \tilde{\mu})$ is a $(1,4)$-hyperfuzzy ideal of $(X, *, 0)$. □

Using the similar way to the proof of Theorem 6, we have the following theorem.

Theorem 7. *For any $(X, *, 0) \in \mathcal{B}_K(X)$, we have the following assertions.*

(1) *If $(X, \tilde{\mu})$ is a $(1,1)$-hyperfuzzy subalgebra of $(X, *, 0)$ which satisfies the condition (23), then $(X, \tilde{\mu})$ is a $(1,1)$-hyperfuzzy ideal of $(X, *, 0)$.*
(2) *If $(X, \tilde{\mu})$ is a $(4,1)$-hyperfuzzy subalgebra of $(X, *, 0)$ which satisfies the condition (24), then $(X, \tilde{\mu})$ is a $(4,1)$-hyperfuzzy ideal of $(X, *, 0)$.*
(3) *If $(X, \tilde{\mu})$ is a $(4,4)$-hyperfuzzy subalgebra of $(X, *, 0)$ which satisfies the condition (25), then $(X, \tilde{\mu})$ is a $(4,4)$-hyperfuzzy ideal of $(X, *, 0)$.*

Theorem 8. *For any $(X, *, 0) \in \mathcal{B}_K(X)$ and $i, j \in \{2, 3\}$, every (i, j)-hyperfuzzy ideal is an (i, j)-hyperfuzzy subalgebra.*

Proof. Let $(X, *, 0) \in \mathcal{B}_K(X)$ and let $(X, \tilde{\mu})$ be a $(2,3)$-hyperfuzzy ideal of $(X, *, 0)$. Then $(X, \tilde{\mu}_{\inf})$ is a 2-fuzzy ideal of $(X, *, 0)$ and $(X, \tilde{\mu}_{\sup})$ is a 3-fuzzy ideal of $(X, *, 0)$. Since $x * y \leq x$ for all $x, y \in X$, we have

$$\tilde{\mu}_{\inf}(x * y) = \tilde{\mu}_{\inf}(0) \leq \min\{\tilde{\mu}_{\inf}(x), \tilde{\mu}_{\inf}(y)\}$$
$$\tilde{\mu}_{\sup}(x * y) = \tilde{\mu}_{\sup}(0) \geq \max\{\tilde{\mu}_{\sup}(x), \tilde{\mu}_{\sup}(y)\}$$

for all $x, y \in X$ by Proposition 2, (9) and (7). Hence $(X, \tilde{\mu})$ is a $(2,3)$-hyperfuzzy subalgebra of $(X, *, 0)$. Similarly, we can prove it for $(i, j) \in \{(2,2), (3,2), (3,3)\}$. □

Using the similar way to the proof of Theorems 5 and 8, we have the following theorem.

Theorem 9. *For any $(X, *, 0) \in \mathcal{B}_K(X)$, every (i, j)-hyperfuzzy ideal is an (i, j)-hyperfuzzy subalgebra for $(i, j) \in \{(1,2), (1,3), (2,1), (2,4), (3,1), (3,4), (4,2), (4,3)\}$.*

5. Conclusions

In the paper [2], Ghosh and Samanta have introduced the concept of hyperfuzzy sets as a generalization of fuzzy sets and interval-valued fuzzy sets, and have presented an application of hyperfuzzy sets in group theory. Jun et al. [6] have applied the hyperfuzzy sets to BCK/BCI-algebras. In this article, we have discussed ideal theory in BCK/BCI-algebras by using the hyperfuzzy sets, and have introduced the notion of hyperfuzzy ideals in BCK/BCI-algebras, and have investigate several properties. We have considered characterizations of hyperfuzzy ideals, and have discussed relations between hyperfuzzy subalgebras and hyperfuzzy ideals. We have provided conditions for hyperfuzzy subalgebras to be hyperfuzzy ideals. Recently, many kinds of fuzzy sets have several applications to deal with uncertainties from our different kinds of daily life problems, in particular, for solving decision making problems (see [9–13]). In the future, we shall extend our proposed approach to some decision making problem under the field of fuzzy cluster analysis, decision-making, uncertain programming and mathematical programming [9]. Moreover, we will apply the notions and results in this manuscript to related algebraic structures, for example, MV-algebras, BL-algebras, MTL-algebras, EQ-algebras, effect algebras, and so on.

Acknowledgments: The authors wish to thank the anonymous reviewers for their valuable suggestions. The first author, S. Z. Song, was supported by Basic Science Research Program through the National Research Foundation of Korea (NRF) funded by the Ministry of Education (No. 2016R1D1A1B02006812).

Author Contributions: All authors contributed equally and significantly to the study and preparation of the article. They have read and approved the final manuscript.

Conflicts of Interest: The authors declare no conflict of interest.

References

1. Zadeh, L.A. Fuzzy sets. *Inf. Control.* **1965**, *8*, 338–353.
2. Ghosh, J.; Samanta, T.K. Hyperfuzzy sets and hyperfuzzy group. *Int. J. Adv. Sci. Technol.* **2012**, 41, 27–37.
3. Imai, Y.; Iséki, K. On axiom systems of propositional calculi. *Proc. Jpn. Acad. Ser. A Math. Sci.* **1966**, *42*, 19–21.
4. Iséki, K. An algebra related with a propositional calculus. *Proc. Jpn. Acad. Ser. A Math. Sci.* **1966**, *42*, 26–29.
5. Iseki, K.; Tanaka, S. An introduction to the theory of BCK-algebras. *Math. Jpn.* **1978**, 23, 1–26
6. Jun, Y.B.; Hur, K.; Lee, K.J. Hyperfuzzy subalgebras of BCK/BCI-algebras. *Ann. Fuzzy Math. Inf.* **2017**, in press.
7. Huang, Y.S. *BCI-Algebra*; Science Press: Beijing, China, 2006.
8. Meng, J.; Jun, Y.B. *BCK-Algebras*; Kyungmoon Sa Co.: Seoul, Korea, 1994.
9. Garg, H. A robust ranking method for intuitionistic multiplicative sets under crisp, interval environments and its applications. *IEEE Trans. Emerg. Top. Comput. Intell.* **2017**, *1*, 366–374.
10. Feng, F.; Jun, Y.B.; Liu, X.; Li, L. An adjustable approach to fuzzy soft set based decision making. *J. Comput. Appl. Math.* **2010**, *234*, 10–20.
11. Xia, M.; Xu, Z. Hesitant fuzzy information aggregation in decision making. *Int. J. Approx. Reason.* **2011**, *52*, 395–407.
12. Tang, H. Decision making based on interval-valued intuitionistic fuzzy soft sets and its algorithm. *J. Comput. Anal. Appl.* **2017**, *23*, 119–131.
13. Wei, G.; Alsaadi, F.E.; Hayat, T.; Alsaedi, A. Hesitant bipolar fuzzy aggregation operators in multiple attribute decision making. *J. Intell. Fuzzy Syst.* **2017**, *33*, 1119–1128.

Σ mathematics

MDPI

Article

Length-Fuzzy Subalgebras in *BCK*/*BCI*-Algebras

Young Bae Jun [1], Seok-Zun Song [2,*] and Seon Jeong Kim [3]

1 Department of Mathematics Education, Gyeongsang National University, Jinju 52828, Korea; skywine@gmail.com
2 Department of Mathematics, Jeju National University, Jeju 63243, Korea
3 Department of Mathematics, Natural Science of College, Gyeongsang National University, Jinju 52828, Korea; skim@gnu.ac.kr
* Correspondence: szsong@jejunu.ac.kr

Received: 1 December 2017; Accepted: 5 January 2018; Published: 12 January 2018

Abstract: As a generalization of interval-valued fuzzy sets and fuzzy sets, the concept of hyperfuzzy sets was introduced by Ghosh and Samanta in the paper [J. Ghosh and T.K. Samanta, Hyperfuzzy sets and hyperfuzzy group, Int. J. Advanced Sci Tech. 41 (2012), 27–37]. The aim of this manuscript is to introduce the length-fuzzy set and apply it to BCK/BCI-algebras. The notion of length-fuzzy subalgebras in BCK/BCI-algebras is introduced, and related properties are investigated. Characterizations of a length-fuzzy subalgebra are discussed. Relations between length-fuzzy subalgebras and hyperfuzzy subalgebras are established.

Keywords: hyperfuzzy set; hyperfuzzy subalgebra; length of hyperfuzzy set; length-fuzzy subalgebra

MSC: 06F35; 03G25; 03B52

1. Introduction

Fuzzy set theory was firstly introduced by Zadeh [1] and opened a new path of thinking to mathematicians, physicists, chemists, engineers and many others due to its diverse applications in various fields. Algebraic hyperstructure, which was introduced by the French mathematician Marty [2], represents a natural extension of classical algebraic structures. Since then, many papers and several books have been written in this area. Nowadays, hyperstructures have a lot of applications in several domains of mathematics and computer science. In a classical algebraic structure, the composition of two elements is an element, while in an algebraic hyperstructure, the composition of two elements is a set. The study of fuzzy hyperstructures is an interesting research area of fuzzy sets. As a generalization of fuzzy sets and interval-valued fuzzy sets, Ghosh and Samanta [3] introduced the notion of hyperfuzzy sets, and applied it to group theory. Jun et al. [4] applied the hyperfuzzy sets to BCK/BCI-algebras, and introduced the notion of k-fuzzy substructures for $k \in \{1,2,3,4\}$. They introduced the concepts of hyperfuzzy substructures of several types by using k-fuzzy substructures, and investigated their basic properties. They also defined hyperfuzzy subalgebras of type (i,j) for $i,j \in \{1,2,3,4\}$, and discussed relations between the hyperfuzzy substructure/subalgebra and its length. They investigated the properties of hyperfuzzy subalgebras related to upper- and lower-level subsets.

In this paper, we introduce the length-fuzzy subalgebra in BCK/BCI-algebras based on hyperfuzzy structures, and investigate several properties.

2. Preliminaries

By a *BCI-algebra* we mean a system $X := (X, *, 0) \in K(\tau)$ in which the following axioms hold:

(I) $((x*y)*(x*z))*(z*y) = 0$,
(II) $(x*(x*y))*y = 0$,

(III) $x * x = 0$,
(IV) $x * y = y * x = 0 \Rightarrow x = y$,

for all $x, y, z \in X$. If a BCI-algebra X satisfies $0 * x = 0$ for all $x \in X$, then we say that X is a *BCK-algebra*. We can define a partial ordering $\leq$ by

$$(\forall x, y \in X)\ (x \leq y \iff x * y = 0).$$

In a BCK/BCI-algebra X, the following hold:

$$(\forall x \in X)\ (x * 0 = x), \tag{1}$$
$$(\forall x, y, z \in X)\ ((x * y) * z = (x * z) * y). \tag{2}$$

A non-empty subset S of a BCK/BCI-algebra X is called a *subalgebra* of X if $x * y \in S$ for all $x, y \in S$.

We refer the reader to the books [5,6] for further information regarding BCK/BCI-algebras.

An ordered pair (X, ρ) of a nonempty set X and a fuzzy set ρ on X is called a *fuzzy structure* over X.

Let X be a nonempty set. A mapping $\tilde{\mu} : X \to \tilde{\mathcal{P}}([0,1])$ is called a *hyperfuzzy set* over X (see [3]), where $\tilde{\mathcal{P}}([0,1])$ is the family of all nonempty subsets of $[0,1]$. An ordered pair $(X, \tilde{\mu})$ is called a *hyper structure* over X.

Given a hyper structure $(X, \tilde{\mu})$ over a nonempty set X, we consider two fuzzy structures $(X, \tilde{\mu}_{\inf})$ and $(X, \tilde{\mu}_{\sup})$ over X in which

$$\tilde{\mu}_{\inf} : X \to [0,1],\ \ x \mapsto \inf\{\tilde{\mu}(x)\},$$
$$\tilde{\mu}_{\sup} : X \to [0,1],\ \ x \mapsto \sup\{\tilde{\mu}(x)\}.$$

Given a nonempty set X, let $\mathcal{B}_K(X)$ and $\mathcal{B}_I(X)$ denote the collection of all BCK-algebras and all BCI-algebras, respectively. Also, $\mathcal{B}(X) := \mathcal{B}_K(X) \cup \mathcal{B}_I(X)$.

Definition 1. *[4] For any $(X, *, 0) \in \mathcal{B}(X)$, a fuzzy structure (X, μ) over $(X, *, 0)$ is called a*

- *fuzzy subalgebra of $(X, *, 0)$ with type 1 (briefly, 1-fuzzy subalgebra of $(X, *, 0)$) if*

$$(\forall x, y \in X)\ (\mu(x * y) \geq \min\{\mu(x), \mu(y)\}), \tag{3}$$

- *fuzzy subalgebra of $(X, *, 0)$ with type 2 (briefly, 2-fuzzy subalgebra of $(X, *, 0)$) if*

$$(\forall x, y \in X)\ (\mu(x * y) \leq \min\{\mu(x), \mu(y)\}), \tag{4}$$

- *fuzzy subalgebra of $(X, *, 0)$ with type 3 (briefly, 3-fuzzy subalgebra of $(X, *, 0)$) if*

$$(\forall x, y \in X)\ (\mu(x * y) \geq \max\{\mu(x), \mu(y)\}), \tag{5}$$

- *fuzzy subalgebra of $(X, *, 0)$ with type 4 (briefly, 4-fuzzy subalgebra of $(X, *, 0)$) if*

$$(\forall x, y \in X)\ (\mu(x * y) \leq \max\{\mu(x), \mu(y)\}). \tag{6}$$

It is clear that every 3-fuzzy subalgebra is a 1-fuzzy subalgebra and every 2-fuzzy subalgebra is a 4-fuzzy subalgebra.

Definition 2. *[4] For any* $(X, *, 0) \in \mathcal{B}(X)$ *and* $i, j \in \{1, 2, 3, 4\}$, *a hyper structure* $(X, \tilde{\mu})$ *over* $(X, *, 0)$ *is called an* (i, j)*-hyperfuzzy subalgebra of* $(X, *, 0)$ *if* $(X, \tilde{\mu}_{\inf})$ *is an i-fuzzy subalgebra of* $(X, *, 0)$ *and* $(X, \tilde{\mu}_{\sup})$ *is a j-fuzzy subalgebra of* $(X, *, 0)$.

3. Length-Fuzzy Subalgebras

In what follows, let $(X, *, 0) \in \mathcal{B}(X)$ unless otherwise specified.

Definition 3. *[4] Given a hyper structure* $(X, \tilde{\mu})$ *over* $(X, *, 0)$, *we define*

$$\tilde{\mu}_\ell : X \to [0,1], \; x \mapsto \tilde{\mu}_{\sup}(x) - \tilde{\mu}_{\inf}(x), \tag{7}$$

which is called the length of $\tilde{\mu}$.

Definition 4. *A hyper structure* $(X, \tilde{\mu})$ *over* $(X, *, 0)$ *is called a length* 1*-fuzzy (resp.* 2*-fuzzy,* 3*-fuzzy and* 4*-fuzzy) subalgebra of* $(X, *, 0)$ *if* $\tilde{\mu}_\ell$ *satisfies the condition* (3) *(resp.* (4)–(6)*).*

Example 1. *Consider a BCK-algebra* $X = \{0, 1, 2, 3, 4\}$ *with the binary operation* $*$ *which is given in Table 1 (see [6]).*

Table 1. Cayley table for the binary operation "$*$".

$*$	0	1	2	3	4
0	0	0	0	0	0
1	1	0	1	0	0
2	2	2	0	0	0
3	3	3	3	0	0
4	4	3	4	1	0

Let $(X, \tilde{\mu})$ *be a hyper structure over* $(X, *, 0)$ *in which* $\tilde{\mu}$ *is given as follows:*

$$\tilde{\mu} : X \to \tilde{\mathcal{P}}([0,1]), \; x \mapsto \begin{cases} [0.2, 0.4) \cup [0.5, 0.8) & \text{if } x = 0, \\ (0.5, 0.9] & \text{if } x = 1, \\ [0.1, 0.3] \cup (0.4, 0.6] & \text{if } x = 2, \\ [0.6, 0.9] & \text{if } x = 3, \\ [0.3, 0.5] & \text{if } x = 4. \end{cases}$$

Then, the length of $\tilde{\mu}$ *is given by Table 2.*

Table 2. The length of $(X, \tilde{\mu})$.

X	0	1	2	3	4
$\tilde{\mu}_\ell$	0.6	0.4	0.5	0.3	0.2

It is routine to verify that $(X, \tilde{\mu})$ *is a length* 1*-fuzzy subalgebra of* $(X, *, 0)$.

Proposition 1. *If* $(X, \tilde{\mu})$ *is a length k-fuzzy subalgebra of* $(X, *, 0)$ *for* $k = 1, 3$, *then* $\tilde{\mu}_\ell(0) \geq \tilde{\mu}_\ell(x)$ *for all* $x \in X$.

Proof. Let $(X, \tilde{\mu})$ be a length 1-fuzzy subalgebra of $(X, *, 0)$. Then,

$$\tilde{\mu}_\ell(0) = \tilde{\mu}_\ell(x * x) \geq \min\{\tilde{\mu}_\ell(x), \tilde{\mu}_\ell(x)\} = \tilde{\mu}_\ell(x) \tag{8}$$

for all $x \in X$. If $(X, \tilde{\mu})$ is a length 3-fuzzy subalgebra of $(X, *, 0)$, then

$$\tilde{\mu}_\ell(0) = \tilde{\mu}_\ell(x * x) \geq \max\{\tilde{\mu}_\ell(x), \tilde{\mu}_\ell(x)\} = \tilde{\mu}_\ell(x) \tag{9}$$

for all $x \in X$. □

Proposition 2. *If $(X, \tilde{\mu})$ is a length k-fuzzy subalgebra of $(X, *, 0)$ for $k = 2, 4$, then $\tilde{\mu}_\ell(0) \leq \tilde{\mu}_\ell(x)$ for all $x \in X$.*

Proof. It is similar to the proof of Proposition 1. □

Theorem 1. *Given a subalgebra A of $(X, *, 0)$ and $B_1, B_2 \in \tilde{\mathcal{P}}([0,1])$, let $(X, \tilde{\mu})$ be a hyper structure over $(X, *, 0)$ given by*

$$\tilde{\mu} : X \to \tilde{\mathcal{P}}([0,1]),\ x \mapsto \begin{cases} B_2 & \text{if } x \in A, \\ B_1 & \text{otherwise.} \end{cases} \tag{10}$$

If $B_1 \subsetneq B_2$, then $(X, \tilde{\mu})$ is a length 1-fuzzy subalgebra of $(X, *, 0)$. Also, if $B_1 \supsetneq B_2$, then $(X, \tilde{\mu})$ is a length 4-fuzzy subalgebra of $(X, *, 0)$.

Proof. If $x \in A$, then $\tilde{\mu}(x) = B_2$ and so

$$\tilde{\mu}_\ell(x) = \tilde{\mu}_{\sup}(x) - \tilde{\mu}_{\inf}(x) = \sup\{\tilde{\mu}(x)\} - \inf\{\tilde{\mu}(x)\} = \sup\{B_2\} - \inf\{B_2\}.$$

If $x \notin A$, then $\tilde{\mu}(x) = B_1$ and so

$$\tilde{\mu}_\ell(x) = \tilde{\mu}_{\sup}(x) - \tilde{\mu}_{\inf}(x) = \sup\{\tilde{\mu}(x)\} - \inf\{\tilde{\mu}(x)\} = \sup\{B_1\} - \inf\{B_1\}.$$

Assume that $B_1 \subsetneq B_2$. Then, $\sup\{B_2\} - \inf\{B_2\} \geq \sup\{B_1\} - \inf\{B_1\}$. Let $x, y \in X$. If $x, y \in A$, then $x * y \in A$ and so

$$\tilde{\mu}_\ell(x * y) = \sup\{B_2\} - \inf\{B_2\} = \min\{\tilde{\mu}_\ell(x), \tilde{\mu}_\ell(y)\}.$$

If $x, y \notin A$, then $\tilde{\mu}_\ell(x * y) \geq \sup\{B_1\} - \inf\{B_1\} = \min\{\tilde{\mu}_\ell(x), \tilde{\mu}_\ell(y)\}$. Suppose that $x \in A$ and $y \notin A$ (or, $x \notin A$ and $y \in A$). Then,

$$\tilde{\mu}_\ell(x * y) \geq \sup B_1 - \inf B_1 = \min\{\tilde{\mu}_\ell(x), \tilde{\mu}_\ell(y)\}.$$

Therefore, $(X, \tilde{\mu})$ is a length 1-fuzzy subalgebra of $(X, *, 0)$.

Assume that $B_1 \supsetneq B_2$. Then,

$$\sup\{B_2\} - \inf\{B_2\} \leq \sup\{B_1\} - \inf\{B_1\},$$

and so

$$\tilde{\mu}_\ell(x * y) = \sup\{B_2\} - \inf\{B_2\} = \max\{\tilde{\mu}_\ell(x), \tilde{\mu}_\ell(y)\}$$

for all $x, y \in A$. If $x \notin A$ or $y \notin A$, then $\tilde{\mu}_\ell(x * y) \leq \max\{\tilde{\mu}_\ell(x), \tilde{\mu}_\ell(y)\}$. Hence, $(X, \tilde{\mu})$ is a length 4-fuzzy subalgebra of $(X, *, 0)$. □

It is clear that every length 3-fuzzy subalgebra is a length 1-fuzzy subalgebra and every length 2-fuzzy subalgebra is a length 4-fuzzy subalgebra. However, the converse is not true, as seen in the following example.

Example 2. *Consider the BCK-algebra $(X, *, 0)$ in Example 1. Given a subalgebra $A = \{0,1,2\}$ of $(X, *, 0)$, let $(X, \tilde{\mu})$ be a hyper structure over $(X, *, 0)$ given by*

$$\tilde{\mu} : X \to \tilde{\mathcal{P}}([0,1]),\ x \mapsto \begin{cases} \{0.2n \mid n \in [0.2, 0.9)\} & \text{if } x \in A, \\ \{0.2n \mid n \in (0.3, 0.7]\} & \text{otherwise.} \end{cases}$$

Then, $(X, \tilde{\mu})$ is a length 1-fuzzy subalgebra of $(X, *, 0)$ by Theorem 1. Since

$$\begin{aligned} \tilde{\mu}_\ell(2) &= \tilde{\mu}_{\sup}(2) - \tilde{\mu}_{\inf}(2) \\ &= \sup\{0.2n \mid n \in [0.2, 0.9)\} - \inf\{0.2n \mid n \in [0.2, 0.9)\} \\ &= 0.18 - 0.04 = 0.14 \end{aligned}$$

and

$$\begin{aligned} \tilde{\mu}_\ell(3 * 2) = \tilde{\mu}_\ell(3) &= \tilde{\mu}_{\sup}(3) - \tilde{\mu}_{\inf}(3) \\ &= \sup\{0.2n \mid n \in (0.3, 0.7]\} - \inf\{0.2n \mid n \in (0.3, 0.7]\} \\ &= 0.14 - 0.06 = 0.08, \end{aligned}$$

we have $\tilde{\mu}_\ell(3 * 2) = 0.08 < 0.14 = \max\{0.08, 0.14\} = \max\{\tilde{\mu}_\ell(3), \tilde{\mu}_\ell(2)\}$. Therefore, $(X, \tilde{\mu})$ is not a length 3-fuzzy subalgebra of $(X, *, 0)$.

Give a subalgebra $A = \{0,1,2,3\}$ of $(X, *, 0)$, let $(X, \tilde{\mu})$ be a hyper structure over $(X, *, 0)$ given by

$$\tilde{\mu} : X \to \tilde{\mathcal{P}}([0,1]),\ x \mapsto \begin{cases} (0.4, 0.7) & \text{if } x \in A, \\ [0.3, 0.9) & \text{otherwise.} \end{cases}$$

Then, $(X, \tilde{\mu})$ is a length 4-fuzzy subalgebra of $(X, *, 0)$ by Theorem 1. However, it is not a length 2-fuzzy subalgebra of $(X, *, 0)$, since

$$\tilde{\mu}_\ell(4 * 2) = \tilde{\mu}_\ell(4) = 0.6 > 0.3 = \min\{\tilde{\mu}_\ell(4), \tilde{\mu}_\ell(2)\}.$$

Theorem 2. *A hyper structure $(X, \tilde{\mu})$ over $(X, *, 0)$ is a length 1-fuzzy subalgebra of $(X, *, 0)$ if and only if the set*

$$U_\ell(\tilde{\mu}; t) := \{x \in X \mid \tilde{\mu}_\ell(x) \geq t\} \tag{11}$$

is a subalgebra of $(X, *, 0)$ for all $t \in [0,1]$ with $U_\ell(\tilde{\mu}; t) \neq \emptyset$.

Proof. Assume that $(X, \tilde{\mu})$ is a length 1-fuzzy subalgebra of $(X, *, 0)$ and let $t \in [0,1]$ be such that $U_\ell(\tilde{\mu}; t)$ is nonempty. If $x, y \in U_\ell(\tilde{\mu}; t)$, then $\tilde{\mu}_\ell(x) \geq t$ and $\tilde{\mu}_\ell(y) \geq t$. It follows from (3) that

$$\tilde{\mu}_\ell(x * y) \geq \min\{\tilde{\mu}_\ell(x), \tilde{\mu}_\ell(y)\} \geq t,$$

and so $x * y \in U_\ell(\tilde{\mu}; t)$. Hence, $U_\ell(\tilde{\mu}; t)$ is a subalgebra of $(X, *, 0)$.

Conversely, suppose that $U_\ell(\tilde{\mu}; t)$ is a subalgebra of $(X, *, 0)$ for all $t \in [0,1]$ with $U_\ell(\tilde{\mu}; t) \neq \emptyset$. Assume that there exist $a, b \in X$ such that

$$\tilde{\mu}_\ell(a * b) < \min\{\tilde{\mu}_\ell(a), \tilde{\mu}_\ell(b)\}.$$

If we take $t := \min\{\tilde{\mu}_\ell(a), \tilde{\mu}_\ell(b)\}$, then $a, b \in U_\ell(\tilde{\mu}; t)$ and so $a * b \in U_\ell(\tilde{\mu}; t)$. Thus, $\tilde{\mu}_\ell(a * b) \geq t$, which is a contradiction. Hence,

$$\tilde{\mu}_\ell(x * y) \geq \min\{\tilde{\mu}_\ell(x), \tilde{\mu}_\ell(y)\}$$

for all $x, y \in X$. Therefore, $(X, \tilde{\mu})$ is a length 1-fuzzy subalgebra of $(X, *, 0)$. □

Corollary 1. *If $(X, \tilde{\mu})$ is a length 3-fuzzy subalgebra of $(X, *, 0)$, then the set $U_\ell(\tilde{\mu}; t)$ is a subalgebra of $(X, *, 0)$ for all $t \in [0, 1]$ with $U_\ell(\tilde{\mu}; t) \neq \emptyset$.*

The converse of Corollary 1 is not true, as seen in the following example.

Example 3. *Consider a BCI-algebra $X = \{0, 1, 2, a, b\}$ with the binary operation $*$, which is given in Table 3 (see [6]).*

Table 3. Cayley table for the binary operation "$*$".

$*$	**0**	**1**	**2**	a	b
0	0	0	0	a	a
1	1	0	1	b	a
2	2	2	0	a	a
a	a	a	a	0	0
b	b	a	b	1	0

Let $(X, \tilde{\mu})$ be a hyper structure over $(X, *, 0)$ in which $\tilde{\mu}$ is given as follows:

$$\tilde{\mu} : X \to \tilde{\mathcal{P}}([0,1]), \ x \mapsto \begin{cases} [0.3, 0.4) \cup [0.6, 0.9) & \text{if } x = 0, \\ (0.5, 0.7] & \text{if } x = 1, \\ [0.1, 0.3] \cup (0.5, 0.6] & \text{if } x = 2, \\ [0.4, 0.7] & \text{if } x = a, \\ (0.3, 0.5] & \text{if } x = b. \end{cases}$$

Then, the length of $\tilde{\mu}$ is given by Table 4.

Table 4. The length of $(X, \tilde{\mu})$.

X	**0**	**1**	**2**	a	b
$\tilde{\mu}_\ell$	0.6	0.2	0.5	0.3	0.2

Hence, we have

$$U_\ell(\tilde{\mu}; t) = \begin{cases} \emptyset & \text{if } t \in (0.6, 1], \\ \{0\} & \text{if } t \in (0.5, 0.6], \\ \{0, 2\} & \text{if } t \in (0.3, 0.5], \\ \{0, 2, a\} & \text{if } t \in (0.2, 0.3], \\ X & \text{if } t \in [0, 0.2], \end{cases}$$

and so $U_\ell(\tilde{\mu}; t)$ is a subalgebra of $(X, *, 0)$ for all $t \in [0, 1]$ with $U_\ell(\tilde{\mu}; t) \neq \emptyset$. Since

$$\tilde{\mu}_\ell(b * 2) = \tilde{\mu}_\ell(b) = 0.2 \ngeq 0.5 = \max\{\tilde{\mu}_\ell(b), \tilde{\mu}_\ell(2)\},$$

$(X, \tilde{\mu})$ is not a length 3-fuzzy subalgebra of $(X, *, 0)$.

Theorem 3. *A hyper structure $(X, \tilde{\mu})$ over $(X, *, 0)$ is a length 4-fuzzy subalgebra of $(X, *, 0)$ if and only if the set*

$$L_\ell(\tilde{\mu}; t) := \{x \in X \mid \tilde{\mu}_\ell(x) \leq t\} \tag{12}$$

is a subalgebra of $(X, *, 0)$ *for all* $t \in [0, 1]$ *with* $L_\ell(\tilde{\mu}; t) \neq \emptyset$.

Proof. Suppose that $(X, \tilde{\mu})$ is a length 4-fuzzy subalgebra of $(X, *, 0)$ and $L_\ell(\tilde{\mu}; t) \neq \emptyset$ for all $t \in [0, 1]$. Let $x, y \in L_\ell(\tilde{\mu}; t)$. Then, $\tilde{\mu}_\ell(x) \leq t$ and $\tilde{\mu}_\ell(y) \leq t$, which implies from (6) that

$$\tilde{\mu}_\ell(x * y) \leq \max\{\tilde{\mu}_\ell(x), \tilde{\mu}_\ell(y)\} \leq t.$$

Hence, $x * y \in L_\ell(\tilde{\mu}; t)$, and so $L_\ell(\tilde{\mu}; t)$ is a subalgebra of $(X, *, 0)$.

Conversely, assume that $L_\ell(\tilde{\mu}; t)$ is a subalgebra of $(X, *, 0)$ for all $t \in [0, 1]$ with $L_\ell(\tilde{\mu}; t) \neq \emptyset$. If there exist $a, b \in X$ such that

$$\tilde{\mu}_\ell(a * b) > \max\{\tilde{\mu}_\ell(a), \tilde{\mu}_\ell(b)\},$$

then $a, b \in L_\ell(\tilde{\mu}; t)$ by taking $t = \max\{\tilde{\mu}_\ell(a), \tilde{\mu}_\ell(b)\}$. It follows that $a * b \in L_\ell(\tilde{\mu}; t)$, and so $\tilde{\mu}_\ell(a * b) \leq t$, which is a contradiction. Hence,

$$\tilde{\mu}_\ell(x * y) \leq \max\{\tilde{\mu}_\ell(x), \tilde{\mu}_\ell(y)\}$$

for all $x, y \in X$, and therefore $(X, \tilde{\mu})$ is a length 4-fuzzy subalgebra of $(X, *, 0)$. □

Corollary 2. *If* $(X, \tilde{\mu})$ *is a length 2-fuzzy subalgebra of* $(X, *, 0)$, *then the set* $L_\ell(\tilde{\mu}; t)$ *is a subalgebra of* $(X, *, 0)$ *for all* $t \in [0, 1]$ *with* $L_\ell(\tilde{\mu}; t) \neq \emptyset$.

The converse of Corollary 2 is not true, as seen in the following example.

Example 4. *Consider the BCI-algebra* $X = \{0, 1, 2, a, b\}$ *in Example 3 and let* $(X, \tilde{\mu})$ *be a hyper structure over* $(X, *, 0)$ *in which* $\tilde{\mu}$ *is given as follows:*

$$\tilde{\mu} : X \to \tilde{\mathcal{P}}([0,1]),\ x \mapsto \begin{cases} [0.6, 0.8) & \text{if } x = 0, \\ (0.3, 0.7] & \text{if } x = 1, \\ [0.4, 0.6) \cup (0.6, 0.7] & \text{if } x = 2, \\ [0.1, 0.7] & \text{if } x = a, \\ (0.2, 0.8] & \text{if } x = b. \end{cases}$$

Then, the length of $\tilde{\mu}$ *is given by Table 5.*

Table 5. The length of $(X, \tilde{\mu})$.

X	0	1	2	a	b
$\tilde{\mu}_\ell$	0.2	0.4	0.3	0.6	0.6

Hence, we have

$$L_\ell(\tilde{\mu}; t) = \begin{cases} X & \text{if } t \in [0.6, 1], \\ \{0, 1, 2\} & \text{if } t \in [0.4, 0.6), \\ \{0, 2\} & \text{if } t \in [0.3, 0.4), \\ \{0\} & \text{if } t \in [0.2, 0.3), \\ \emptyset & \text{if } t \in [0, 0.2), \end{cases}$$

and so $L_\ell(\tilde{\mu}; t)$ *is a subalgebra of* $(X, *, 0)$ *for all* $t \in [0, 1]$ *with* $L_\ell(\tilde{\mu}; t) \neq \emptyset$. *However,* $(X, \tilde{\mu})$ *is not a length 2-fuzzy subalgebra of* $(X, *, 0)$ *since*

$$\tilde{\mu}_\ell(a * 1) = 0.6 \not\leq 0.4 = \min\{\tilde{\mu}_\ell(a), \tilde{\mu}_\ell(1)\}.$$

Theorem 4. *Let $(X, \tilde{\mu})$ be a hyper structure over $(X, *, 0)$ in which $(X, \tilde{\mu}_{\inf})$ satisfies the Condition (4). If $(X, \tilde{\mu})$ is a $(k, 1)$-hyperfuzzy subalgebra of $(X, *, 0)$ for $k \in \{1, 2, 3, 4\}$, then it is a length 1-fuzzy subalgebra of $(X, *, 0)$.*

Proof. Assume that $(X, \tilde{\mu})$ is a $(k, 1)$-hyperfuzzy subalgebra of $(X, *, 0)$ for $k \in \{1, 2, 3, 4\}$ in which $(X, \tilde{\mu}_{\inf})$ satisfies the Condition (4). Then, $\tilde{\mu}_{\inf}(x * y) \leq \tilde{\mu}_{\inf}(x)$ and $\tilde{\mu}_{\inf}(x * y) \leq \tilde{\mu}_{\inf}(y)$ for all $x, y \in X$, and $(X, \tilde{\mu}_{\sup})$ is a 1-fuzzy subalgebra of X. It follows from (3) that

$$\begin{aligned}
\tilde{\mu}_{\ell}(x * y) &= \tilde{\mu}_{\sup}(x * y) - \tilde{\mu}_{\inf}(x * y) \\
&\geq \min\{\tilde{\mu}_{\sup}(x), \tilde{\mu}_{\sup}(y)\} - \tilde{\mu}_{\inf}(x * y) \\
&= \min\{\tilde{\mu}_{\sup}(x) - \tilde{\mu}_{\inf}(x * y),\ \tilde{\mu}_{\sup}(y) - \tilde{\mu}_{\inf}(x * y)\} \\
&\geq \min\{\tilde{\mu}_{\sup}(x) - \tilde{\mu}_{\inf}(x),\ \tilde{\mu}_{\sup}(y) - \tilde{\mu}_{\inf}(y)\} \\
&= \min\{\tilde{\mu}_{\ell}(x), \tilde{\mu}_{\ell}(y)\}
\end{aligned}$$

for all $x, y \in X$. Therefore $(X, \tilde{\mu})$ is a length 1-fuzzy subalgebra of $(X, *, 0)$. □

Corollary 3. *Let $(X, \tilde{\mu})$ be a hyper structure over $(X, *, 0)$ in which $(X, \tilde{\mu}_{\inf})$ satisfies the Condition (4). If $(X, \tilde{\mu})$ is a $(k, 3)$-hyperfuzzy subalgebra of $(X, *, 0)$ for $k \in \{1, 2, 3, 4\}$, then it is a length 1-fuzzy subalgebra of $(X, *, 0)$.*

Corollary 4. *For $j \in \{1, 3\}$, every $(2, j)$-hyperfuzzy subalgebra is a length 1-fuzzy subalgebra.*

In general, any length 1-fuzzy subalgebra may not be a $(k, 1)$-hyperfuzzy subalgebra for $k \in \{1, 2, 3, 4\}$, as seen in the following example.

Example 5. *Consider a BCI-algebra $X = \{0, 1, a, b, c\}$ with the binary operation $*$, which is given in Table 6 (see [6]).*

Table 6. Cayley table for the binary operation "$*$".

$*$	0	1	a	b	c
0	0	0	a	b	c
1	1	0	a	b	c
a	a	a	0	c	b
b	b	b	c	0	a
c	c	c	b	a	0

Let $(X, \tilde{\mu})$ be a hyper structure over $(X, *, 0)$ in which $\tilde{\mu}$ is given as follows:

$$\tilde{\mu} : X \to \tilde{\mathcal{P}}([0,1]),\ x \mapsto \begin{cases} [0.1, 0.9) & \text{if } x = 0, \\ (0.1, 0.8] & \text{if } x = 1, \\ [0.4, 0.9] & \text{if } x = a, \\ [0.3, 0.6) & \text{if } x \in \{b, c\}. \end{cases}$$

The length of $\tilde{\mu}$ is given by Table 7 and it is routine to verify that $(X, \tilde{\mu})$ is a length 1-fuzzy subalgebra of $(X, *, 0)$.

Table 7. The length of $(X, \tilde{\mu})$.

X	0	1	a	b	c
$\tilde{\mu}_{\ell}$	0.8	0.7	0.5	0.3	0.3

However, it is not a $(k,1)$*-hyperfuzzy subalgebra of* X *since*

$$\begin{aligned}\tilde{\mu}_{\inf}(a*a)&=\tilde{\mu}_{\inf}(0)=0.1<0.4=\min\{\tilde{\mu}_{\inf}(a),\tilde{\mu}_{\inf}(a)\},\\ \tilde{\mu}_{\inf}(b*c)&=\tilde{\mu}_{\inf}(a)=0.4>0.3=\min\{\tilde{\mu}_{\inf}(b),\tilde{\mu}_{\inf}(c)\},\\ \tilde{\mu}_{\inf}(b*b)&=\tilde{\mu}_{\inf}(0)=0.1<0.3=\max\{\tilde{\mu}_{\inf}(b),\tilde{\mu}_{\inf}(b)\},\\ \tilde{\mu}_{\inf}(b*c)&=\tilde{\mu}_{\inf}(a)=0.4>0.3=\max\{\tilde{\mu}_{\inf}(b),\tilde{\mu}_{\inf}(c)\}.\end{aligned}$$

We provide a condition for a length 1-fuzzy subalgebra to be a $(k,1)$-hyperfuzzy subalgebra for $k\in\{1,2,3,4\}$.

Theorem 5. *If* $(X,\tilde{\mu})$ *is a length* 1*-fuzzy subalgebra of* $(X,*,0)$ *in which* $\tilde{\mu}_{\inf}$ *is constant on* X*, then it is a* $(k,1)$*-hyperfuzzy subalgebra of* $(X,*,0)$ *for* $k\in\{1,2,3,4\}$.

Proof. Assume that $(X,\tilde{\mu})$ is a length 1-fuzzy subalgebra of $(X,*,0)$ in which $\tilde{\mu}_{\inf}$ is constant on X. It is clear that $(X,\tilde{\mu}_{\inf})$ is a k-fuzzy subalgebra of $(X,*,0)$ for $k\in\{1,2,3,4\}$. Let $\tilde{\mu}_{\inf}(x)=k$ for all $x\in X$. Then,

$$\begin{aligned}\tilde{\mu}_{\sup}(x*y)&=\tilde{\mu}_{\ell}(x*y)+\tilde{\mu}_{\inf}(x*y)\\ &=\tilde{\mu}_{\ell}(x*y)+k\\ &\geq\min\{\tilde{\mu}_{\ell}(x),\tilde{\mu}_{\ell}(y)\}+k\\ &=\min\{\tilde{\mu}_{\ell}(x)+k,\tilde{\mu}_{\ell}(y)+k\}\\ &=\min\{\tilde{\mu}_{\sup}(x),\tilde{\mu}_{\sup}(y)\}\end{aligned}$$

for all $x,y\in X$. Thus, $(X,\tilde{\mu}_{\sup})$ is a 1-fuzzy subalgebra of X. Therefore, $(X,\tilde{\mu})$ is a $(k,1)$-hyperfuzzy subalgebra of $(X,*,0)$ for $k\in\{1,2,3,4\}$. □

Corollary 5. *If* $(X,\tilde{\mu})$ *is a length* 3*-fuzzy subalgebra of* $(X,*,0)$ *in which* $\tilde{\mu}_{\inf}$ *is constant on* X*, then it is a* $(k,1)$*-hyperfuzzy subalgebra of* $(X,*,0)$ *for* $k\in\{1,2,3,4\}$.

Corollary 6. *Let* $(X,\tilde{\mu})$ *be a hyper structure over* $(X,*,0)$ *in which* $\tilde{\mu}_{\inf}$ *is constant on* X*. Then,* $(X,\tilde{\mu})$ *is a* $(k,1)$*-hyperfuzzy subalgebra of* $(X,*,0)$ *for* $k\in\{1,2,3,4\}$ *if and only if* $(X,\tilde{\mu})$ *is a length* 1*-fuzzy subalgebra of* $(X,*,0)$.

Theorem 6. *If* $(X,\tilde{\mu})$ *is a length* 1*-fuzzy subalgebra of* $(X,*,0)$ *in which* $\tilde{\mu}_{\sup}$ *is constant on* X*, then it is a* $(4,k)$*-hyperfuzzy subalgebra of* $(X,*,0)$ *for* $k\in\{1,2,3,4\}$.

Proof. Let $(X,\tilde{\mu})$ be a length 1-fuzzy subalgebra of $(X,*,0)$ in which $\tilde{\mu}_{\sup}$ is constant on X. Clearly, $(X,\tilde{\mu}_{\sup})$ is a k-fuzzy subalgebra of $(X,*,0)$ for $k\in\{1,2,3,4\}$. Let $\tilde{\mu}_{\sup}(x)=t$ for all $x\in X$. Then,

$$\begin{aligned}\tilde{\mu}_{\inf}(x*y)&=\tilde{\mu}_{\sup}(x*y)-\tilde{\mu}_{\ell}(x*y)\\ &=t-\tilde{\mu}_{\ell}(x*y)\\ &\leq t-\min\{\tilde{\mu}_{\ell}(x),\tilde{\mu}_{\ell}(y)\}\\ &=t+\max\{-\tilde{\mu}_{\ell}(x),-\tilde{\mu}_{\ell}(y)\}\\ &=\max\{t-\tilde{\mu}_{\ell}(x),t-\tilde{\mu}_{\ell}(y)\}\\ &=\max\{\tilde{\mu}_{\inf}(x),\tilde{\mu}_{\inf}(y)\}\end{aligned}$$

for all $x,y\in X$, and so $(X,\tilde{\mu}_{\inf})$ is a 4-fuzzy subalgebra of $(X,*,0)$. Therefore, $(X,\tilde{\mu})$ is a $(4,k)$-hyperfuzzy subalgebra of $(X,*,0)$ for $k\in\{1,2,3,4\}$. □

Theorem 7. *Let $(X,\tilde{\mu})$ be a hyper structure over $(X,*,0)$ in which $(X,\tilde{\mu}_{\sup})$ satisfies the Condition (5). For any $k \in \{1,2,3,4\}$, if $(X,\tilde{\mu})$ is a $(4,k)$-hyperfuzzy subalgebra of $(X,*,0)$, then it is a length 1-fuzzy subalgebra of $(X,*,0)$.*

Proof. Let $(X,\tilde{\mu})$ be a $(4,k)$-hyperfuzzy subalgebra of $(X,*,0)$ for $k \in \{1,2,3,4\}$ in which $(X,\tilde{\mu}_{\sup})$ satisfies the Condition (5). Then, $\tilde{\mu}_{\sup}(x*y) \geq \tilde{\mu}_{\sup}(x)$ and $\tilde{\mu}_{\sup}(x*y) \geq \tilde{\mu}_{\sup}(y)$ for all $x,y \in X$, and $(X,\tilde{\mu}_{\inf})$ is a 4-fuzzy subalgebra of $(X,*,0)$. It follows from (6) that

$$\begin{aligned}
\tilde{\mu}_{\ell}(x*y) &= \tilde{\mu}_{\sup}(x*y) - \tilde{\mu}_{\inf}(x*y) \\
&\geq \tilde{\mu}_{\sup}(x*y) - \max\{\tilde{\mu}_{\inf}(x), \tilde{\mu}_{\inf}(y)\} \\
&= \min\{\tilde{\mu}_{\sup}(x*y) - \tilde{\mu}_{\inf}(x),\ \tilde{\mu}_{\sup}(x*y) - \tilde{\mu}_{\inf}(y)\} \\
&\geq \min\{\tilde{\mu}_{\sup}(x) - \tilde{\mu}_{\inf}(x),\ \tilde{\mu}_{\sup}(y) - \tilde{\mu}_{\inf}(y)\} \\
&= \min\{\tilde{\mu}_{\ell}(x), \tilde{\mu}_{\ell}(y)\}
\end{aligned}$$

for all $x,y \in X$. Hence, $(X,\tilde{\mu})$ is a length 1-fuzzy subalgebra of $(X,*,0)$. □

Corollary 7. *Let $(X,\tilde{\mu})$ be a hyper structure over $(X,*,0)$ in which $(X,\tilde{\mu}_{\sup})$ satisfies the Condition (5). For any $k \in \{1,2,3,4\}$, every $(2,k)$-hyperfuzzy subalgebra is a length 1-fuzzy subalgebra.*

Theorem 8. *Let $(X,\tilde{\mu})$ be a hyper structure over $(X,*,0)$ in which $(X,\tilde{\mu}_{\inf})$ satisfies the Condition (5). If $(X,\tilde{\mu})$ is a $(k,4)$-hyperfuzzy subalgebra of $(X,*,0)$ for $k \in \{1,2,3,4\}$, then it is a length 4-fuzzy subalgebra of $(X,*,0)$.*

Proof. Assume that $(X,\tilde{\mu})$ is a $(k,4)$-hyperfuzzy subalgebra of $(X,*,0)$ for $k \in \{1,2,3,4\}$ in which $(X,\tilde{\mu}_{\inf})$ satisfies the Condition (5). Then, $\tilde{\mu}_{\inf}(x*y) \geq \tilde{\mu}_{\inf}(x)$ and $\tilde{\mu}_{\inf}(x*y) \geq \tilde{\mu}_{\inf}(y)$ for all $x,y \in X$, and $(X,\tilde{\mu}_{\sup})$ is a 4-fuzzy subalgebra of X. Hence,

$$\begin{aligned}
\tilde{\mu}_{\ell}(x*y) &= \tilde{\mu}_{\sup}(x*y) - \tilde{\mu}_{\inf}(x*y) \\
&\leq \max\{\tilde{\mu}_{\sup}(x), \tilde{\mu}_{\sup}(y)\} - \tilde{\mu}_{\inf}(x*y) \\
&= \max\{\tilde{\mu}_{\sup}(x) - \tilde{\mu}_{\inf}(x*y),\ \tilde{\mu}_{\sup}(y) - \tilde{\mu}_{\inf}(x*y)\} \\
&\leq \max\{\tilde{\mu}_{\sup}(x) - \tilde{\mu}_{\inf}(x),\ \tilde{\mu}_{\sup}(y) - \tilde{\mu}_{\inf}(y)\} \\
&= \max\{\tilde{\mu}_{\ell}(x), \tilde{\mu}_{\ell}(y)\}
\end{aligned}$$

for all $x,y \in X$, and so $(X,\tilde{\mu})$ is a length 4-fuzzy subalgebra of $(X,*,0)$. □

Corollary 8. *Let $(X,\tilde{\mu})$ be a hyper structure over $(X,*,0)$ in which $(X,\tilde{\mu}_{\inf})$ satisfies the Condition (5). If $(X,\tilde{\mu})$ is a $(k,2)$-hyperfuzzy subalgebra of $(X,*,0)$ for $k \in \{1,2,3,4\}$, then it is a length 4-fuzzy subalgebra of $(X,*,0)$.*

Corollary 9. *For $j \in \{2,4\}$, every $(3,j)$-hyperfuzzy subalgebra is a length 4-fuzzy subalgebra.*

Theorem 9. *Let $(X,\tilde{\mu})$ be a hyper structure over $(X,*,0)$ in which $\tilde{\mu}_{\inf}$ is constant. Then, every length 4-fuzzy subalgebra is a $(k,4)$-hyperfuzzy subalgebra for $k \in \{1,2,3,4\}$.*

Proof. Let $(X,\tilde{\mu})$ be a length 4-fuzzy subalgebra of $(X,*,0)$ in which $\tilde{\mu}_{\inf}$ is constant. It is obvious that $(X,\tilde{\mu}_{\inf})$ is a k-fuzzy subalgebra of $(X,*,0)$ for $k \in \{1,2,3,4\}$. Let $\tilde{\mu}_{\inf}(x) = t$ for all $x \in X$. Then,

$$
\begin{aligned}
\tilde{\mu}_{\sup}(x*y) &= \tilde{\mu}_{\ell}(x*y) + \tilde{\mu}_{\inf}(x*y) = \tilde{\mu}_{\ell}(x*y) + t \\
&\leq \max\{\tilde{\mu}_{\ell}(x), \tilde{\mu}_{\ell}(y)\} + t \\
&= \max\{\tilde{\mu}_{\ell}(x) + t, \tilde{\mu}_{\ell}(y) + t\} \\
&= \max\{\tilde{\mu}_{\ell}(x), \tilde{\mu}_{\ell}(y)\}
\end{aligned}
$$

for all $x, y \in X$, and hence $(X, \tilde{\mu}_{\sup})$ is a 4-fuzzy subalgebra of $(X, *, 0)$. Therefore, $(X, \tilde{\mu})$ is a $(k, 4)$-hyperfuzzy subalgebra of $(X, *, 0)$ for $k \in \{1, 2, 3, 4\}$. □

Corollary 10. *Let $(X, \tilde{\mu})$ be a hyper structure over $(X, *, 0)$ in which $\tilde{\mu}_{\inf}$ is constant. Then, every length 2-fuzzy subalgebra is a $(k, 4)$-hyperfuzzy subalgebra for $k \in \{1, 2, 3, 4\}$.*

Theorem 10. *Let $(X, \tilde{\mu})$ be a hyper structure over $(X, *, 0)$ in which $(X, \tilde{\mu}_{\sup})$ satisfies the Condition (4). For every $k \in \{1, 2, 3, 4\}$, every $(1, k)$-hyperfuzzy subalgebra is a length 4-fuzzy subalgebra.*

Proof. For every $k \in \{1, 2, 3, 4\}$, let $(X, \tilde{\mu})$ be a $(1, k)$-hyperfuzzy subalgebra of $(X, *, 0)$ in which $(X, \tilde{\mu}_{\sup})$ satisfies the Condition (4). Then, $\tilde{\mu}_{\sup}(x*y) \leq \tilde{\mu}_{\sup}(x)$ and $\tilde{\mu}_{\sup}(x*y) \leq \tilde{\mu}_{\sup}(y)$ for all $x, y \in X$. Since $(X, \tilde{\mu}_{\inf})$ is a 1-fuzzy subalgebra of $(X, *, 0)$, we have

$$
\begin{aligned}
\tilde{\mu}_{\ell}(x*y) &= \tilde{\mu}_{\sup}(x*y) - \tilde{\mu}_{\inf}(x*y) \\
&\leq \tilde{\mu}_{\sup}(x*y) - \min\{\tilde{\mu}_{\inf}(x), \tilde{\mu}_{\inf}(y)\} \\
&= \max\{\tilde{\mu}_{\sup}(x*y) - \tilde{\mu}_{\inf}(x),\ \tilde{\mu}_{\sup}(x*y) - \tilde{\mu}_{\inf}(y)\} \\
&\leq \max\{\tilde{\mu}_{\sup}(x) - \tilde{\mu}_{\inf}(x),\ \tilde{\mu}_{\sup}(y) - \tilde{\mu}_{\inf}(y)\} \\
&= \max\{\tilde{\mu}_{\ell}(x), \tilde{\mu}_{\ell}(y)\}
\end{aligned}
$$

for all $x, y \in X$. Thus, $(X, \tilde{\mu})$ is a length 4-fuzzy subalgebra of $(X, *, 0)$. □

Corollary 11. *Let $(X, \tilde{\mu})$ be a hyper structure over $(X, *, 0)$ in which $(X, \tilde{\mu}_{\sup})$ satisfies the Condition (4). For every $k \in \{1, 2, 3, 4\}$, every $(3, k)$-hyperfuzzy subalgebra is a length 4-fuzzy subalgebra.*

Theorem 11. *Let $(X, \tilde{\mu})$ be a length 4-fuzzy subalgebra of $(X, *, 0)$. If $\tilde{\mu}_{\sup}$ is constant on X, then $(X, \tilde{\mu})$ is a $(1, k)$-hyperfuzzy subalgebra of $(X, *, 0)$ for $k \in \{1, 2, 3, 4\}$.*

Proof. Assume that $\tilde{\mu}_{\sup}$ is constant on X in a length 4-fuzzy subalgebra $(X, \tilde{\mu})$ of $(X, *, 0)$. Obviously, $(X, \tilde{\mu}_{\sup})$ is a k-fuzzy subalgebra of $(X, *, 0)$ for $k \in \{1, 2, 3, 4\}$. Let $\tilde{\mu}_{\sup}(x) = t$ for all $x \in X$. Then,

$$
\begin{aligned}
\tilde{\mu}_{\inf}(x*y) &= \tilde{\mu}_{\sup}(x*y) - \tilde{\mu}_{\ell}(x*y) \\
&= t - \tilde{\mu}_{\ell}(x*y) \\
&\geq t - \max\{\tilde{\mu}_{\ell}(x), \tilde{\mu}_{\ell}(y)\} \\
&= \min\{t - \tilde{\mu}_{\ell}(x), t - \tilde{\mu}_{\ell}(y)\} \\
&= \min\{\tilde{\mu}_{\inf}(x), \tilde{\mu}_{\inf}(y)\}
\end{aligned}
$$

for all $x, y \in X$, and so $(X, \tilde{\mu}_{\inf})$ is a 1-fuzzy subalgebra of $(X, *, 0)$. Therefore, $(X, \tilde{\mu}_{\inf})$ is a $(1, k)$-hyperfuzzy subalgebra of $(X, *, 0)$ for $k \in \{1, 2, 3, 4\}$. □

Corollary 12. *Let $(X, \tilde{\mu})$ be a length 2-fuzzy subalgebra of $(X, *, 0)$. If $\tilde{\mu}_{\sup}$ is constant on X, then $(X, \tilde{\mu})$ is a $(1, k)$-hyperfuzzy subalgebra of $(X, *, 0)$ for $k \in \{1, 2, 3, 4\}$.*

4. Conclusions

In order to consider a generalization of fuzzy sets and interval-valued fuzzy sets, the notion of hyperfuzzy sets was introduced by Ghosh and Samanta (see [3]). Jun et al. [4] and Song et al. [7] have applied the hyperfuzzy sets to BCK/BCI-algebras. In this article, we have introduced the concept of length-fuzzy sets based on hyperfuzzy sets, and have presented an application in BCK/BCI-algebras. We have introduced the notion of length fuzzy subalgebras in BCK/BCI-algebras, and have investigated related properties. We have discussed characterizations of a length fuzzy subalgebra, and have established relations between length fuzzy subalgebras and hyperfuzzy subalgebras. Recently, many kinds of fuzzy sets have several applications to deal with uncertainties from our different kinds of daily life problems, in particular, for solving decision-making problems (see [8–12]). In the future, from a purely mathematical standpoint, we will apply the notions and results in this manuscript to related algebraic structures, for example, MV-algebras, BL-algebras, MTL-algebras, EQ-algebras, effect algebras and so on. From an applicable standpoint, we shall extend our proposed approach to some decision-making problems under the field of fuzzy cluster analysis, uncertain programming, mathematical programming, decision-making problems and so on.

Acknowledgments: The authors wish to thank the anonymous reviewers for their valuable suggestions. To the memory of Lotfi A. Zadeh.

Author Contributions: All authors contributed equally and significantly to the study and preparation of the article. They have read and approved the final manuscript.

Conflicts of Interest: The authors declare no conflict of interest.

References

1. Zadeh, L.A. Fuzzy sets. *Inf. Control* **1965**, *8*, 338–353.
2. Marty, F. Sur une generalization de la notion de groupe. In Proceedings of the 8th Congress Math Scandenaves, Stockholm, Sweden, 1934; pp. 45–49.
3. Ghosh, J.; Samanta, T.K. Hyperfuzzy sets and hyperfuzzy group. *Int. J. Adv. Sci. Technol.* **2012**, *41*, 27–37.
4. Jun, Y.B.; Hur, K.; Lee, K.J. Hyperfuzzy subalgebras of BCK/BCI-algebras. *Ann. Fuzzy Math. Inform.* **2017**, in press.
5. Huang, Y.S. *BCI-Algebra*; Science Press: Beijing, China, 2006.
6. Meng, J.; Jun, Y.B. *BCK-Algebras*; Kyungmoon Sa Co.: Seoul, Korea, 1994.
7. Song, S.Z.; Kim, S.J.; Jun, Y.B. Hyperfuzzy ideals in BCK/BCI-algebras. *Mathematics* **2017**, *5*, 81, doi:10.3390/math5040081.
8. Garg, H. A robust ranking method for intuitionistic multiplicative sets under crisp, interval environments and its applications. *IEEE Trans. Emerg. Top. Comput. Intell.* **2017**, *1*, 366–374.
9. Feng, F.; Jun, Y.B.; Liu, X.; Li, L. An adjustable approach to fuzzy soft set based decision making. *J. Comput. Appl. Math.* **2010**, *234*, 10–20.
10. Xia, M.; Xu, Z. Hesitant fuzzy information aggregation in decision making. *Int. J. Approx. Reason.* **2011**, *52*, 395–407.
11. Tang, H. Decision making based on interval-valued intuitionistic fuzzy soft sets and its algorithm. *J. Comput. Anal. Appl.* **2017**, *23*, 119–131.
12. Wei, G.; Alsaadi, F.E.; Hayat, T.; Alsaedi, A. Hesitant bipolar fuzzy aggregation operators in multiple attribute decision making. *J. Intell. Fuzzy Syst.* **2017**, *33*, 1119–1128.

mathematics

MDPI

Article

Neutrosophic Permeable Values and Energetic Subsets with Applications in BCK/BCI-Algebras

Young Bae Jun [1,*], Florentin Smarandache [2], Seok-Zun Song [3] and Hashem Bordbar [4]

1 Department of Mathematics Education, Gyeongsang National University, Jinju 52828, Korea
2 Mathematics & Science Department, University of New Mexico, 705 Gurley Ave., Gallup, NM 87301, USA; fsmarandache@gmail.com
3 Department of Mathematics, Jeju National University, Jeju 63243, Korea; szsong@jejunu.ac.kr
4 Postdoctoral Research Fellow, Shahid Beheshti University, Tehran, District 1, Daneshjou Boulevard 1983969411, Iran; bordbar.amirh@gmail.com
* Correspondence: skywine@gmail.com

Received: 14 March 2018; Accepted: 2 May 2018; Published: 7 May 2018

Abstract: The concept of a $(\in, \in)$-neutrosophic ideal is introduced, and its characterizations are established. The notions of neutrosophic permeable values are introduced, and related properties are investigated. Conditions for the neutrosophic level sets to be energetic, right stable, and right vanished are discussed. Relations between neutrosophic permeable S- and I-values are considered.

Keywords: $(\in, \in)$-neutrosophic subalgebra; $(\in, \in)$-neutrosophic ideal; neutrosophic (anti-)permeable S-value; neutrosophic (anti-)permeable I-value; S-energetic set; I-energetic set

MSC: 06F35; 03G25; 08A72

1. Introduction

The notion of neutrosophic set (NS) theory developed by Smarandache (see [1,2]) is a more general platform that extends the concepts of classic and fuzzy sets, intuitionistic fuzzy sets, and interval-valued (intuitionistic) fuzzy sets and that is applied to various parts: pattern recognition, medical diagnosis, decision-making problems, and so on (see [3–6]). Smarandache [2] mentioned that a cloud is a NS because its borders are ambiguous and because each element (water drop) belongs with a neutrosophic probability to the set (e.g., there are types of separated water drops around a compact mass of water drops, such that we do not know how to consider them: in or out of the cloud). Additionally, we are not sure where the cloud ends nor where it begins, and neither whether some elements are or are not in the set. This is why the percentage of indeterminacy is required and the neutrosophic probability (using subsets—not numbers—as components) should be used for better modeling: it is a more organic, smooth, and particularly accurate estimation. Indeterminacy is the zone of ignorance of a proposition's value, between truth and falsehood.

Algebraic structures play an important role in mathematics with wide-ranging applications in several disciplines such as coding theory, information sciences, computer sciences, control engineering, theoretical physics, and so on. NS theory is also applied to several algebraic structures. In particular, Jun et al. applied it to BCK/BCI-algebras (see [7–12]). Jun et al. [8] introduced the notions of energetic subsets, right vanished subsets, right stable subsets, and (anti-)permeable values in BCK/BCI-algebras and investigated relations between these sets.

In this paper, we introduce the notions of neutrosophic permeable S-values, neutrosophic permeable I-values, $(\in, \in)$-neutrosophic ideals, neutrosophic anti-permeable S-values, and neutrosophic anti-permeable I-values, which are motivated by the idea of subalgebras (i.e., S-values) and ideals (i.e., I-values), and investigate their properties. We consider characterizations of $(\in, \in)$-neutrosophic ideals. We discuss conditions for the lower (upper) neutrosophic $\in_\Phi$-subsets to be S- and I-energetic. We provide conditions for a triple (α, β, γ) of numbers to be a neutrosophic (anti-)permeable S- or I-value. We consider conditions for the upper (lower) neutrosophic $\in_\Phi$-subsets to be right stable (right vanished) subsets. We establish relations between neutrosophic (anti-)permeable S- and I-values.

2. Preliminaries

An algebra $(X; *, 0)$ of type $(2, 0)$ is called a *BCI-algebra* if it satisfies the following conditions:

(I) $(\forall x, y, z \in X)\ (((x * y) * (x * z)) * (z * y) = 0)$;
(II) $(\forall x, y \in X)\ ((x * (x * y)) * y = 0)$;
(III) $(\forall x \in X)\ (x * x = 0)$;
(IV) $(\forall x, y \in X)\ (x * y = 0, y * x = 0 \Rightarrow x = y)$.

If a BCI-algebra X satisfies the following identity:

(V) $(\forall x \in X)\ (0 * x = 0)$,

then X is called a *BCK-algebra*. Any BCK/BCI-algebra X satisfies the following conditions:

$$(\forall x \in X)\ (x * 0 = x), \tag{1}$$
$$(\forall x, y, z \in X)\ (x \leq y \Rightarrow x * z \leq y * z,\ z * y \leq z * x), \tag{2}$$
$$(\forall x, y, z \in X)\ ((x * y) * z = (x * z) * y), \tag{3}$$
$$(\forall x, y, z \in X)\ ((x * z) * (y * z) \leq x * y), \tag{4}$$

where $x \leq y$ if and only if $x * y = 0$. A nonempty subset S of a BCK/BCI-algebra X is called a *subalgebra* of X if $x * y \in S$ for all $x, y \in S$. A subset I of a BCK/BCI-algebra X is called an *ideal* of X if it satisfies the following:

$$0 \in I, \tag{5}$$
$$(\forall x, y \in X)\ (x * y \in I,\ y \in I \rightarrow x \in I). \tag{6}$$

We refer the reader to the books [13] and [14] for further information regarding BCK/BCI-algebras.

For any family $\{a_i \mid i \in \Lambda\}$ of real numbers, we define

$$\bigvee\{a_i \mid i \in \Lambda\} = \sup\{a_i \mid i \in \Lambda\}$$

and

$$\bigwedge\{a_i \mid i \in \Lambda\} = \inf\{a_i \mid i \in \Lambda\}.$$

If $\Lambda = \{1, 2\}$, we also use $a_1 \vee a_2$ and $a_1 \wedge a_2$ instead of $\bigvee\{a_i \mid i \in \{1, 2\}\}$ and $\bigwedge\{a_i \mid i \in \{1, 2\}\}$, respectively.

We let X be a nonempty set. A NS in X (see [1]) is a structure of the form

$$A := \{\langle x; A_T(x), A_I(x), A_F(x)\rangle \mid x \in X\},$$

where $A_T : X \rightarrow [0, 1]$ is a truth membership function, $A_I : X \rightarrow [0, 1]$ is an indeterminate membership function, and $A_F : X \rightarrow [0, 1]$ is a false membership function. For the sake of simplicity, we use the symbol $A = (A_T, A_I, A_F)$ for the NS

$$A := \{\langle x; A_T(x), A_I(x), A_F(x)\rangle \mid x \in X\}.$$

A subset A of a BCK/BCI-algebra X is said to be S-energetic (see [8]) if it satisfies

$$(\forall x, y \in X)\,(x * y \in A \Rightarrow \{x, y\} \cap A \neq \emptyset). \tag{7}$$

A subset A of a BCK/BCI-algebra X is said to be *I-energetic* (see [8]) if it satisfies

$$(\forall x, y \in X)\,(y \in A \Rightarrow \{x, y * x\} \cap A \neq \emptyset). \tag{8}$$

A subset A of a BCK/BCI-algebra X is said to be *right vanished* (see [8]) if it satisfies

$$(\forall x, y \in X)\,(x * y \in A \Rightarrow x \in A). \tag{9}$$

A subset A of a BCK/BCI-algebra X is said to be *right stable* (see [8]) if $A * X := \{a * x \mid a \in A,\ x \in X\} \subseteq A$.

3. Neutrosophic Permeable Values

Given a NS $A = (A_T, A_I, A_F)$ in a set X, $\alpha, \beta \in (0,1]$ and $\gamma \in [0,1)$, we consider the following sets:

$$\begin{aligned}
U_T^{\in}(A;\alpha) &= \{x \in X \mid A_T(x) \geq \alpha\},\ U_T^{\in}(A;\alpha)^* = \{x \in X \mid A_T(x) > \alpha\},\\
U_I^{\in}(A;\beta) &= \{x \in X \mid A_I(x) \geq \beta\},\ U_I^{\in}(A;\beta)^* = \{x \in X \mid A_I(x) > \beta\},\\
U_F^{\in}(A;\gamma) &= \{x \in X \mid A_F(x) \leq \gamma\},\ U_F^{\in}(A;\gamma)^* = \{x \in X \mid A_F(x) < \gamma\},\\
L_T^{\in}(A;\alpha) &= \{x \in X \mid A_T(x) \leq \alpha\},\ L_T^{\in}(A;\alpha)^* = \{x \in X \mid A_T(x) < \alpha\},\\
L_I^{\in}(A;\beta) &= \{x \in X \mid A_I(x) \leq \beta\},\ L_I^{\in}(A;\beta)^* = \{x \in X \mid A_I(x) < \beta\},\\
L_F^{\in}(A;\gamma) &= \{x \in X \mid A_F(x) \geq \gamma\},\ L_F^{\in}(A;\gamma)^* = \{x \in X \mid A_F(x) > \gamma\}.
\end{aligned}$$

We say $U_T^{\in}(A;\alpha)$, $U_I^{\in}(A;\beta)$, and $U_F^{\in}(A;\gamma)$ are *upper neutrosophic $\in_\Phi$-subsets* of X, and $L_T^{\in}(A;\alpha)$, $L_I^{\in}(A;\beta)$, and $L_F^{\in}(A;\gamma)$ are *lower neutrosophic $\in_\Phi$-subsets* of X, where $\Phi \in \{T, I, F\}$. We say $U_T^{\in}(A;\alpha)^*$, $U_I^{\in}(A;\beta)^*$, and $U_F^{\in}(A;\gamma)^*$ are *strong upper neutrosophic $\in_\Phi$-subsets* of X, and $L_T^{\in}(A;\alpha)^*$, $L_I^{\in}(A;\beta)^*$, and $L_F^{\in}(A;\gamma)^*$ are *strong lower neutrosophic $\in_\Phi$-subsets* of X, where $\Phi \in \{T, I, F\}$.

Definition 1 ([7]). *A NS $A = (A_T, A_I, A_F)$ in a BCK/BCI-algebra X is called an $(\in, \in)$-neutrosophic subalgebra of X if the following assertions are valid:*

$$\begin{aligned}
x \in U_T^{\in}(A;\alpha_x),\ y \in U_T^{\in}(A;\alpha_y) &\Rightarrow x * y \in U_T^{\in}(A;\alpha_x \wedge \alpha_y),\\
x \in U_I^{\in}(A;\beta_x),\ y \in U_I^{\in}(A;\beta_y) &\Rightarrow x * y \in U_I^{\in}(A;\beta_x \wedge \beta_y),\\
x \in U_F^{\in}(A;\gamma_x),\ y \in U_F^{\in}(A;\gamma_y) &\Rightarrow x * y \in U_F^{\in}(A;\gamma_x \vee \gamma_y),
\end{aligned} \tag{10}$$

for all $x, y \in X$, $\alpha_x, \alpha_y, \beta_x, \beta_y \in (0,1]$ and $\gamma_x, \gamma_y \in [0,1)$.

Lemma 1 ([7]). *A NS $A = (A_T, A_I, A_F)$ in a BCK/BCI-algebra X is an $(\in, \in)$-neutrosophic subalgebra of X if and only if $A = (A_T, A_I, A_F)$ satisfies*

$$(\forall x, y \in X) \begin{pmatrix} A_T(x * y) \geq A_T(x) \wedge A_T(y) \\ A_I(x * y) \geq A_I(x) \wedge A_I(y) \\ A_F(x * y) \leq A_F(x) \vee A_F(y) \end{pmatrix}. \tag{11}$$

Proposition 1. *Every $(\in, \in)$-neutrosophic subalgebra $A = (A_T, A_I, A_F)$ of a BCK/BCI-algebra X satisfies*

$$(\forall x \in X)\,(A_T(0) \geq A_T(x),\ A_I(0) \geq A_I(x),\ A_F(0) \leq A_F(x)). \tag{12}$$

Proof. Straightforward. □

Theorem 1. *If $A = (A_T, A_I, A_F)$ is an $(\in, \in)$-neutrosophic subalgebra of a BCK/BCI-algebra X, then the lower neutrosophic $\in_\Phi$-subsets of X are S-energetic subsets of X, where $\Phi \in \{T, I, F\}$.*

Proof. Let $x, y \in X$ and $\alpha \in (0, 1]$ be such that $x * y \in L_T^{\in}(A; \alpha)$. Then

$$\alpha \geq A_T(x * y) \geq A_T(x) \wedge A_T(y),$$

and thus $A_T(x) \leq \alpha$ or $A_T(y) \leq \alpha$; that is, $x \in L_T^{\in}(A; \alpha)$ or $y \in L_T^{\in}(A; \alpha)$. Thus $\{x, y\} \cap L_T^{\in}(A; \alpha) \neq \emptyset$. Therefore $L_T^{\in}(A; \alpha)$ is an S-energetic subset of X. Similarly, we can verify that $L_I^{\in}(A; \beta)$ is an S-energetic subset of X. We let $x, y \in X$ and $\gamma \in [0, 1)$ be such that $x * y \in L_F^{\in}(A; \gamma)$. Then

$$\gamma \leq A_F(x * y) \leq A_F(x) \vee A_F(y).$$

It follows that $A_F(x) \geq \gamma$ or $A_F(y) \geq \gamma$; that is, $x \in L_F^{\in}(A; \gamma)$ or $y \in L_F^{\in}(A; \gamma)$. Hence $\{x, y\} \cap L_F^{\in}(A; \gamma) \neq \emptyset$, and therefore $L_F^{\in}(A; \gamma)$ is an S-energetic subset of X. □

Corollary 1. *If $A = (A_T, A_I, A_F)$ is an $(\in, \in)$-neutrosophic subalgebra of a BCK/BCI-algebra X, then the strong lower neutrosophic $\in_\Phi$-subsets of X are S-energetic subsets of X, where $\Phi \in \{T, I, F\}$.*

Proof. Straightforward. □

The converse of Theorem 1 is not true, as seen in the following example.

Example 1. *Consider a BCK-algebra $X = \{0, 1, 2, 3, 4\}$ with the binary operation $*$ that is given in Table 1 (see [14]).*

Table 1. Cayley table for the binary operation "$*$".

*	0	1	2	3	4
0	0	0	0	0	0
1	1	0	0	0	0
2	2	1	0	0	1
3	3	2	1	0	2
4	4	1	1	1	0

Let $A = (A_T, A_I, A_F)$ be a NS in X that is given in Table 2.

Table 2. Tabulation representation of $A = (A_T, A_I, A_F)$.

X	$A_T(x)$	$A_I(x)$	$A_F(x)$
0	0.6	0.8	0.2
1	0.4	0.5	0.7
2	0.4	0.5	0.6
3	0.4	0.5	0.5
4	0.7	0.8	0.2

If $\alpha \in [0.4, 0.6)$, $\beta \in [0.5, 0.8)$, and $\gamma \in (0.2, 0.5]$, then $L_T^{\in}(A; \alpha) = \{1, 2, 3\}$, $L_I^{\in}(A; \beta) = \{1, 2, 3\}$, and $L_F^{\in}(A; \gamma) = \{1, 2, 3\}$ are S-energetic subsets of X. Because

$$A_T(4 * 4) = A_T(0) = 0.6 \ngeq 0.7 = A_T(4) \wedge A_T(4)$$

and/or

$$A_F(3*2) = A_F(1) = 0.7 \nleq 0.6 = A_F(3) \vee A_F(2),$$

it follows from Lemma 1 that $A = (A_T, A_I, A_F)$ is not an $(\in, \in)$-neutrosophic subalgebra of X.

Definition 2. *Let $A = (A_T, A_I, A_F)$ be a NS in a BCK/BCI-algebra X and $(\alpha, \beta, \gamma) \in \Lambda_T \times \Lambda_I \times \Lambda_F$, where Λ_T, Λ_I, and Λ_F are subsets of $[0,1]$. Then (α, β, γ) is called a neutrosophic permeable S-value for $A = (A_T, A_I, A_F)$ if the following assertion is valid:*

$$(\forall x, y \in X) \left(\begin{array}{l} x*y \in U_T^{\in}(A;\alpha) \Rightarrow A_T(x) \vee A_T(y) \geq \alpha, \\ x*y \in U_I^{\in}(A;\beta) \Rightarrow A_I(x) \vee A_I(y) \geq \beta, \\ x*y \in U_F^{\in}(A;\gamma) \Rightarrow A_F(x) \wedge A_F(y) \leq \gamma \end{array} \right) \tag{13}$$

Example 2. *Let $X = \{0, 1, 2, 3, 4\}$ be a set with the binary operation $*$ that is given in Table 3.*

Table 3. Cayley table for the binary operation "$*$".

*	0	1	2	3	4
0	0	0	0	0	0
1	1	0	1	1	0
2	2	2	0	2	0
3	3	3	3	0	3
4	4	4	4	4	0

Then $(X, *, 0)$ is a BCK-algebra (see [14]). Let $A = (A_T, A_I, A_F)$ be a NS in X that is given in Table 4.

Table 4. Tabulation representation of $A = (A_T, A_I, A_F)$.

X	$A_T(x)$	$A_I(x)$	$A_F(x)$
0	0.2	0.3	0.7
1	0.6	0.4	0.6
2	0.5	0.3	0.4
3	0.4	0.8	0.5
4	0.7	0.6	0.2

It is routine to verify that $(\alpha, \beta, \gamma) \in (0, 2, 1] \times (0.3, 1] \times [0, 0.7)$ is a neutrosophic permeable S-value for $A = (A_T, A_I, A_F)$.

Theorem 2. *Let $A = (A_T, A_I, A_F)$ be a NS in a BCK/BCI-algebra X and $(\alpha, \beta, \gamma) \in \Lambda_T \times \Lambda_I \times \Lambda_F$, where Λ_T, Λ_I, and Λ_F are subsets of $[0,1]$. If $A = (A_T, A_I, A_F)$ satisfies the following condition:*

$$(\forall x, y \in X) \left(\begin{array}{l} A_T(x*y) \leq A_T(x) \vee A_T(y) \\ A_I(x*y) \leq A_I(x) \vee A_I(y) \\ A_F(x*y) \geq A_F(x) \wedge A_F(y) \end{array} \right), \tag{14}$$

then (α, β, γ) is a neutrosophic permeable S-value for $A = (A_T, A_I, A_F)$.

Proof. Let $x, y \in X$ be such that $x * y \in U_T^{\in}(A;\alpha)$. Then

$$\alpha \leq A_T(x*y) \leq A_T(x) \vee A_T(y).$$

Similarly, if $x * y \in U_I^{\in}(A;\beta)$ for $x, y \in X$, then $A_I(x) \vee A_I(y) \geq \beta$. Now, let $a, b \in X$ be such that $a * b \in U_F^{\in}(A;\gamma)$. Then

$$\gamma \geq A_F(a * b) \geq A_F(a) \wedge A_F(b).$$

Therefore (α, β, γ) is a neutrosophic permeable S-value for $A = (A_T, A_I, A_F)$. □

Theorem 3. *Let $A = (A_T, A_I, A_F)$ be a NS in a BCK-algebra X and $(\alpha, \beta, \gamma) \in \Lambda_T \times \Lambda_I \times \Lambda_F$, where Λ_T, Λ_I, and Λ_F are subsets of $[0,1]$. If $A = (A_T, A_I, A_F)$ satisfies the following conditions:*

$$(\forall x \in X)\ (A_T(0) \leq A_T(x), A_I(0) \leq A_I(x), A_F(0) \geq A_F(x)) \tag{15}$$

and

$$(\forall x, y \in X) \begin{pmatrix} A_T(x) \leq A_T(x * y) \vee A_T(y) \\ A_I(x) \leq A_I(x * y) \vee A_I(y) \\ A_F(x) \geq A_F(x * y) \wedge A_F(y) \end{pmatrix}, \tag{16}$$

then (α, β, γ) is a neutrosophic permeable S-value for $A = (A_T, A_I, A_F)$.

Proof. Let $x, y, a, b, u, v \in X$ be such that $x * y \in U_T^{\in}(A;\alpha)$, $a * b \in U_I^{\in}(A;\beta)$, and $u * v \in U_F^{\in}(A;\gamma)$. Then

$$\begin{aligned} \alpha &\leq A_T(x * y) \leq A_T((x * y) * x) \vee A_T(x) \\ &= A_T((x * x) * y) \vee A_T(x) = A_T(0 * y) \vee A_T(x) \\ &= A_T(0) \vee A_T(x) = A_T(x), \end{aligned}$$

$$\begin{aligned} \beta &\leq A_I(a * b) \leq A_I((a * b) * a) \vee A_I(a) \\ &= A_I((a * a) * b) \vee A_I(a) = A_I(0 * b) \vee A_I(a) \\ &= A_I(0) \vee A_I(a) = A_I(a), \end{aligned}$$

and

$$\begin{aligned} \gamma &\geq A_F(u * v) \geq A_F((u * v) * u) \wedge A_F(u) \\ &= A_F((u * u) * v) \wedge A_F(u) = A_F(0 * v) \wedge A_F(v) \\ &= A_F(0) \wedge A_F(v) = A_F(v) \end{aligned}$$

by Equations (3), (V), (15), and (16). It follows that

$$\begin{aligned} A_T(x) \vee A_T(y) &\geq A_T(x) \geq \alpha, \\ A_I(a) \vee A_I(b) &\geq A_I(a) \geq \beta, \\ A_F(u) \wedge A_F(v) &\leq A_F(u) \leq \gamma. \end{aligned}$$

Therefore (α, β, γ) is a neutrosophic permeable S-value for $A = (A_T, A_I, A_F)$. □

Theorem 4. *Let $A = (A_T, A_I, A_F)$ be a NS in a BCK/BCI-algebra X and $(\alpha, \beta, \gamma) \in \Lambda_T \times \Lambda_I \times \Lambda_F$, where Λ_T, Λ_I, and Λ_F are subsets of $[0,1]$. If (α, β, γ) is a neutrosophic permeable S-value for $A = (A_T, A_I, A_F)$, then upper neutrosophic $\in_\Phi$-subsets of X are S-energetic where $\Phi \in \{T, I, F\}$.*

Proof. Let $x, y, a, b, u, v \in X$ be such that $x * y \in U_T^{\in}(A;\alpha)$, $a * b \in U_I^{\in}(A;\beta)$, and $u * v \in U_F^{\in}(A;\gamma)$. Using Equation (13), we have $A_T(x) \vee A_T(y) \geq \alpha$, $A_I(a) \vee A_I(b) \geq \beta$, and $A_F(u) \wedge A_F(v) \leq \gamma$. It follows that

$$A_T(x) \geq \alpha \text{ or } A_T(y) \geq \alpha, \text{ that is, } x \in U_T^{\in}(A;\alpha) \text{ or } y \in U_T^{\in}(A;\alpha);$$

$$A_I(a) \geq \beta \text{ or } A_I(b) \geq \beta, \text{ that is, } a \in U_I^{\in}(A;\beta) \text{ or } b \in U_I^{\in}(A;\beta);$$

and

$$A_F(u) \leq \gamma \text{ or } A_F(v) \leq \gamma, \text{ that is, } u \in U_F^{\in}(A;\gamma) \text{ or } v \in U_F^{\in}(A;\gamma).$$

Hence $\{x, y\} \cap U_T^{\in}(A;\alpha) \neq \emptyset$, $\{a, b\} \cap U_I^{\in}(A;\beta) \neq \emptyset$, and $\{u, v\} \cap U_F^{\in}(A;\gamma) \neq \emptyset$. Therefore $U_T^{\in}(A;\alpha)$, $U_I^{\in}(A;\beta)$, and $U_F^{\in}(A;\gamma)$ are S-energetic subsets of X. □

Definition 3. *Let $A = (A_T, A_I, A_F)$ be a NS in a BCK/BCI-algebra X and $(\alpha, \beta, \gamma) \in \Lambda_T \times \Lambda_I \times \Lambda_F$, where Λ_T, Λ_I, and Λ_F are subsets of $[0, 1]$. Then (α, β, γ) is called a neutrosophic anti-permeable S-value for $A = (A_T, A_I, A_F)$ if the following assertion is valid:*

$$(\forall x, y \in X)\left(\begin{array}{l} x * y \in L_T^{\in}(A;\alpha) \Rightarrow A_T(x) \wedge A_T(y) \leq \alpha, \\ x * y \in L_I^{\in}(A;\beta) \Rightarrow A_I(x) \wedge A_I(y) \leq \beta, \\ x * y \in L_F^{\in}(A;\gamma) \Rightarrow A_F(x) \vee A_F(y) \geq \gamma \end{array}\right). \tag{17}$$

Example 3. *Let $X = \{0, 1, 2, 3, 4\}$ be a set with the binary operation $*$ that is given in Table 5.*

Table 5. Cayley table for the binary operation "$*$".

*	0	1	2	3	4
0	0	0	0	0	0
1	1	0	0	1	0
2	2	1	0	2	0
3	3	3	3	0	3
4	4	4	4	4	0

Then $(X, *, 0)$ is a BCK-algebra (see [14]). Let $A = (A_T, A_I, A_F)$ be a NS in X that is given in Table 6.

Table 6. Tabulation representation of $A = (A_T, A_I, A_F)$.

X	$A_T(x)$	$A_I(x)$	$A_F(x)$
0	0.7	0.6	0.4
1	0.4	0.5	0.6
2	0.4	0.5	0.6
3	0.5	0.2	0.7
4	0.3	0.3	0.9

It is routine to verify that $(\alpha, \beta, \gamma) \in (0.3, 1] \times (0.2, 1] \times [0, 0.9)$ is a neutrosophic anti-permeable S-value for $A = (A_T, A_I, A_F)$.

Theorem 5. *Let $A = (A_T, A_I, A_F)$ be a NS in a BCK/BCI-algebra X and $(\alpha, \beta, \gamma) \in \Lambda_T \times \Lambda_I \times \Lambda_F$, where Λ_T, Λ_I, and Λ_F are subsets of $[0, 1]$. If $A = (A_T, A_I, A_F)$ is an $(\in, \in)$-neutrosophic subalgebra of X, then (α, β, γ) is a neutrosophic anti-permeable S-value for $A = (A_T, A_I, A_F)$.*

Proof. Let $x, y, a, b, u, v \in X$ be such that $x * y \in L_T^{\in}(A;\alpha)$, $a * b \in L_I^{\in}(A;\beta)$, and $u * v \in L_F^{\in}(A;\gamma)$. Using Lemma 1, we have

$$\begin{aligned} A_T(x) \wedge A_T(y) &\leq A_T(x * y) \leq \alpha, \\ A_I(a) \wedge A_I(b) &\leq A_I(a * b) \leq \beta, \\ A_F(u) \vee A_F(v) &\geq A_F(u * v) \geq \gamma, \end{aligned}$$

and thus (α, β, γ) is a neutrosophic anti-permeable S-value for $A = (A_T, A_I, A_F)$. □

Theorem 6. *Let $A = (A_T, A_I, A_F)$ be a NS in a BCK/BCI-algebra X and $(\alpha, \beta, \gamma) \in \Lambda_T \times \Lambda_I \times \Lambda_F$, where Λ_T, Λ_I, and Λ_F are subsets of $[0,1]$. If (α, β, γ) is a neutrosophic anti-permeable S-value for $A = (A_T, A_I, A_F)$, then lower neutrosophic $\in_\Phi$-subsets of X are S-energetic where $\Phi \in \{T, I, F\}$.*

Proof. Let $x, y, a, b, u, v \in X$ be such that $x * y \in L_T^{\in}(A;\alpha)$, $a * b \in L_I^{\in}(A;\beta)$, and $u * v \in L_F^{\in}(A;\gamma)$. Using Equation (17), we have $A_T(x) \wedge A_T(y) \leq \alpha$, $A_I(a) \wedge A_I(b) \leq \beta$, and $A_F(u) \vee A_F(v) \geq \gamma$, which imply that

$$A_T(x) \leq \alpha \text{ or } A_T(y) \leq \alpha, \text{ that is, } x \in L_T^{\in}(A;\alpha) \text{ or } y \in L_T^{\in}(A;\alpha);$$

$$A_I(a) \leq \beta \text{ or } A_I(b) \leq \beta, \text{ that is, } a \in L_I^{\in}(A;\beta) \text{ or } b \in L_I^{\in}(A;\beta);$$

and

$$A_F(u) \geq \gamma \text{ or } A_F(v) \geq \gamma, \text{ that is, } u \in L_F^{\in}(A;\gamma) \text{ or } v \in L_F^{\in}(A;\gamma).$$

Hence $\{x, y\} \cap L_T^{\in}(A;\alpha) \neq \emptyset$, $\{a, b\} \cap L_I^{\in}(A;\beta) \neq \emptyset$, and $\{u, v\} \cap L_F^{\in}(A;\gamma) \neq \emptyset$. Therefore $L_T^{\in}(A;\alpha)$, $L_I^{\in}(A;\beta)$, and $L_F^{\in}(A;\gamma)$ are S-energetic subsets of X. □

Definition 4. *A NS $A = (A_T, A_I, A_F)$ in a BCK/BCI-algebra X is called an $(\in, \in)$-neutrosophic ideal of X if the following assertions are valid:*

$$(\forall x \in X) \begin{pmatrix} x \in U_T^{\in}(A;\alpha) \Rightarrow 0 \in U_T^{\in}(A;\alpha) \\ x \in U_I^{\in}(A;\beta) \Rightarrow 0 \in U_I^{\in}(A;\beta) \\ x \in U_F^{\in}(A;\gamma) \Rightarrow 0 \in U_F^{\in}(A;\gamma) \end{pmatrix}, \tag{18}$$

$$(\forall x, y \in X) \begin{pmatrix} x * y \in U_T^{\in}(A;\alpha_x), y \in U_T^{\in}(A;\alpha_y) \Rightarrow x \in U_T^{\in}(A;\alpha_x \wedge \alpha_y) \\ x * y \in U_I^{\in}(A;\beta_x), y \in U_I^{\in}(A;\beta_y) \Rightarrow x \in U_I^{\in}(A;\beta_x \wedge \beta_y) \\ x * y \in U_F^{\in}(A;\gamma_x), y \in U_F^{\in}(A;\gamma_y) \Rightarrow x \in U_F^{\in}(A;\gamma_x \vee \gamma_y) \end{pmatrix}, \tag{19}$$

for all $\alpha, \beta, \alpha_x, \alpha_y, \beta_x, \beta_y \in (0,1]$ and $\gamma, \gamma_x, \gamma_y \in [0,1)$.

Theorem 7. *A NS $A = (A_T, A_I, A_F)$ in a BCK/BCI-algebra X is an $(\in, \in)$-neutrosophic ideal of X if and only if $A = (A_T, A_I, A_F)$ satisfies*

$$(\forall x, y \in X) \begin{pmatrix} A_T(0) \geq A_T(x) \geq A_T(x * y) \wedge A_T(y) \\ A_I(0) \geq A_I(x) \geq A_I(x * y) \wedge A_I(y) \\ A_F(0) \leq A_F(x) \leq A_F(x * y) \vee A_F(y) \end{pmatrix}. \tag{20}$$

Proof. Assume that Equation (20) is valid, and let $x \in U_T^{\in}(A;\alpha)$, $a \in U_I^{\in}(A;\beta)$, and $u \in U_F^{\in}(A;\gamma)$ for any $x, a, u \in X$, $\alpha, \beta \in (0,1]$ and $\gamma \in [0,1)$. Then $A_T(0) \geq A_T(x) \geq \alpha$, $A_I(0) \geq A_I(a) \geq \beta$, and $A_F(0) \leq A_F(u) \leq \gamma$. Hence $0 \in U_T^{\in}(A;\alpha)$, $0 \in U_I^{\in}(A;\beta)$, and $0 \in U_F^{\in}(A;\gamma)$, and thus Equation (18) is valid. Let $x, y, a, b, u, v \in X$ be such that $x * y \in U_T^{\in}(A;\alpha_x)$, $y \in U_T^{\in}(A;\alpha_y)$, $a * b \in U_I^{\in}(A;\beta_a)$, $b \in U_I^{\in}(A;\beta_b)$, $u * v \in U_F^{\in}(A;\gamma_u)$, and $v \in U_F^{\in}(A;\gamma_v)$ for all $\alpha_x, \alpha_y, \beta_a, \beta_b \in (0,1]$

and $\gamma_u, \gamma_v \in [0,1)$. Then $A_T(x*y) \geq \alpha_x$, $A_T(y) \geq \alpha_y$, $A_I(a*b) \geq \beta_a$, $A_I(b) \geq \beta_b$, $A_F(u*v) \leq \gamma_u$, and $A_F(v) \leq \gamma_v$. It follows from Equation (20) that

$$\begin{aligned} A_T(x) &\geq A_T(x*y) \wedge A_T(y) \geq \alpha_x \wedge \alpha_y, \\ A_I(a) &\geq A_I(a*b) \wedge A_I(b) \geq \beta_a \wedge \beta_b, \\ A_F(u) &\leq A_F(u*v) \vee A_F(v) \leq \gamma_u \vee \gamma_v. \end{aligned}$$

Hence $x \in U_T^{\in}(A;\alpha_x \wedge \alpha_y)$, $a \in U_I^{\in}(A;\beta_a \wedge \beta_b)$, and $u \in U_F^{\in}(A;\gamma_u \vee \gamma_v)$. Therefore $A = (A_T, A_I, A_F)$ is an $(\in, \in)$-neutrosophic ideal of X.

Conversely, let $A = (A_T, A_I, A_F)$ be an $(\in, \in)$-neutrosophic ideal of X. If there exists $x_0 \in X$ such that $A_T(0) < A_T(x_0)$, then $x_0 \in U_T^{\in}(A;\alpha)$ and $0 \notin U_T^{\in}(A;\alpha)$, where $\alpha = A_T(x_0)$. This is a contradiction, and thus $A_T(0) \geq A_T(x)$ for all $x \in X$. Assume that $A_T(x_0) < A_T(x_0 * y_0) \wedge A_T(y_0)$ for some $x_0, y_0 \in X$. Taking $\alpha := A_T(x_0 * y_0) \wedge A_T(y_0)$ implies that $x_0 * y_0 \in U_T^{\in}(A;\alpha)$ and $y_0 \in U_T^{\in}(A;\alpha)$; but $x_0 \notin U_T^{\in}(A;\alpha)$. This is a contradiction, and thus $A_T(x) \geq A_T(x*y) \wedge A_T(y)$ for all $x, y \in X$. Similarly, we can verify that $A_I(0) \geq A_I(x) \geq A_I(x*y) \wedge A_I(y)$ for all $x, y \in X$. Now, suppose that $A_F(0) > A_F(a)$ for some $a \in X$. Then $a \in U_F^{\in}(A;\gamma)$ and $0 \notin U_F^{\in}(A;\gamma)$ by taking $\gamma = A_F(a)$. This is impossible, and thus $A_F(0) \leq A_F(x)$ for all $x \in X$. Suppose there exist $a_0, b_0 \in X$ such that $A_F(a_0) > A_F(a_0 * b_0) \vee A_F(b_0)$, and take $\gamma := A_F(a_0 * b_0) \vee A_F(b_0)$. Then $a_0 * b_0 \in U_F^{\in}(A;\gamma)$, $b_0 \in U_F^{\in}(A;\gamma)$, and $a_0 \notin U_F^{\in}(A;\gamma)$, which is a contradiction. Thus $A_F(x) \leq A_F(x*y) \vee A_F(y)$ for all $x, y \in X$. Therefore $A = (A_T, A_I, A_F)$ satisfies Equation (20). □

Lemma 2. *Every $(\in, \in)$-neutrosophic ideal $A = (A_T, A_I, A_F)$ of a BCK/BCI-algebra X satisfies*

$$(\forall x, y \in X)\,(x \leq y \Rightarrow A_T(x) \geq A_T(y),\, A_I(x) \geq A_I(y),\, A_F(x) \leq A_F(y)). \tag{21}$$

Proof. Let $x, y \in X$ be such that $x \leq y$. Then $x * y = 0$, and thus

$$\begin{aligned} A_T(x) &\geq A_T(x*y) \wedge A_T(y) = A_T(0) \wedge A_T(y) = A_T(y), \\ A_I(x) &\geq A_I(x*y) \wedge A_I(y) = A_I(0) \wedge A_I(y) = A_I(y), \\ A_F(x) &\leq A_F(x*y) \vee A_F(y) = A_F(0) \vee A_F(y) = A_F(y), \end{aligned}$$

by Equation (20). This completes the proof. □

Theorem 8. *A NS $A = (A_T, A_I, A_F)$ in a BCK-algebra X is an $(\in, \in)$-neutrosophic ideal of X if and only if $A = (A_T, A_I, A_F)$ satisfies*

$$(\forall x, y, z \in X) \left(x * y \leq z \Rightarrow \begin{cases} A_T(x) \geq A_T(y) \wedge A_T(z) \\ A_I(x) \geq A_I(y) \wedge A_I(z) \\ A_F(x) \leq A_F(y) \vee A_F(z) \end{cases} \right) \tag{22}$$

Proof. Let $A = (A_T, A_I, A_F)$ be an $(\in, \in)$-neutrosophic ideal of X, and let $x, y, z \in X$ be such that $x * y \leq z$. Using Theorem 7 and Lemma 2, we have

$$\begin{aligned} A_T(x) &\geq A_T(x*y) \wedge A_T(y) \geq A_T(y) \wedge A_T(z), \\ A_I(x) &\geq A_I(x*y) \wedge A_I(y) \geq A_I(y) \wedge A_I(z), \\ A_F(x) &\leq A_F(x*y) \vee A_F(y) \leq A_F(y) \vee A_F(z). \end{aligned}$$

Conversely, assume that $A = (A_T, A_I, A_F)$ satisfies Equation (22). Because $0 * x \leq x$ for all $x \in X$, it follows from Equation (22) that

$$A_T(0) \geq A_T(x) \wedge A_T(x) = A_T(x),$$
$$A_I(0) \geq A_I(x) \wedge A_I(x) = A_I(x),$$
$$A_F(0) \leq A_F(x) \vee A_F(x) = A_F(x),$$

for all $x \in X$. Because $x * (x * y) \leq y$ for all $x, y \in X$, we have

$$A_T(x) \geq A_T(x * y) \wedge A_T(y),$$
$$A_I(x) \geq A_I(x * y) \wedge A_I(y),$$
$$A_F(x) \leq A_F(x * y) \vee A_F(y),$$

for all $x, y \in X$ by Equation (22). It follows from Theorem 7 that $A = (A_T, A_I, A_F)$ is an $(\in, \in)$-neutrosophic ideal of X. □

Theorem 9. *If $A = (A_T, A_I, A_F)$ is an $(\in, \in)$-neutrosophic ideal of a BCK/BCI-algebra X, then the lower neutrosophic $\in_\Phi$-subsets of X are I-energetic subsets of X where $\Phi \in \{T, I, F\}$.*

Proof. Let $x, a, u \in X$, $\alpha, \beta \in (0, 1]$, and $\gamma \in [0, 1)$ be such that $x \in L_T^{\in}(A; \alpha)$, $a \in L_I^{\in}(A; \beta)$, and $u \in L_F^{\in}(A; \gamma)$. Using Theorem 7, we have

$$\alpha \geq A_T(x) \geq A_T(x * y) \wedge A_T(y),$$
$$\beta \geq A_I(a) \geq A_I(a * b) \wedge A_I(b),$$
$$\gamma \leq A_F(u) \leq A_F(u * v) \vee A_F(v),$$

for all $y, b, v \in X$. It follows that

$$A_T(x * y) \leq \alpha \text{ or } A_T(y) \leq \alpha, \text{ that is, } x * y \in L_T^{\in}(A; \alpha) \text{ or } y \in L_T^{\in}(A; \alpha);$$
$$A_I(a * b) \leq \beta \text{ or } A_I(b) \leq \beta, \text{ that is, } a * b \in L_T^{\in}(A; \beta) \text{ or } b \in L_T^{\in}(A; \beta);$$

and

$$A_F(u * v) \geq \gamma \text{ or } A_F(v) \geq \gamma, \text{ that is, } u * v \in L_T^{\in}(A; \gamma) \text{ or } v \in L_T^{\in}(A; \gamma).$$

Hence $\{y, x * y\} \cap L_T^{\in}(A; \alpha)$, $\{b, a * b\} \cap L_I^{\in}(A; \beta)$, and $\{v, u * v\} \cap L_F^{\in}(A; \gamma)$ are nonempty, and therefore $L_T^{\in}(A; \alpha)$, $L_I^{\in}(A; \beta)$ and $L_F^{\in}(A; \gamma)$ are I-energetic subsets of X. □

Corollary 2. *If $A = (A_T, A_I, A_F)$ is an $(\in, \in)$-neutrosophic ideal of a BCK/BCI-algebra X, then the strong lower neutrosophic $\in_\Phi$-subsets of X are I-energetic subsets of X where $\Phi \in \{T, I, F\}$.*

Proof. Straightforward. □

Theorem 10. *Let $(\alpha, \beta, \gamma) \in \Lambda_T \times \Lambda_I \times \Lambda_F$, where Λ_T, Λ_I, and Λ_F are subsets of $[0, 1]$. If $A = (A_T, A_I, A_F)$ is an $(\in, \in)$-neutrosophic ideal of a BCK-algebra X, then*

(1) *the (strong) upper neutrosophic $\in_\Phi$-subsets of X are right stable where $\Phi \in \{T, I, F\}$;*
(2) *the (strong) lower neutrosophic $\in_\Phi$-subsets of X are right vanished where $\Phi \in \{T, I, F\}$.*

Proof. (1) Let $x \in X$, $a \in U_T^{\in}(A; \alpha)$, $b \in U_I^{\in}(A; \beta)$, and $c \in U_F^{\in}(A; \gamma)$. Then $A_T(a) \geq \alpha$, $A_I(b) \geq \beta$, and $A_F(c) \leq \gamma$. Because $a * x \leq a$, $b * x \leq b$, and $c * x \leq c$, it follows from Lemma 2 that $A_T(a * x) \geq A_T(a) \geq \alpha$, $A_I(b * x) \geq A_I(b) \geq \beta$, and $A_F(c * x) \leq A_F(c) \leq \gamma$; that is, $a * x \in U_T^{\in}(A; \alpha)$,

$b * x \in U_I^{\in}(A;\beta)$, and $c * x \in U_F^{\in}(A;\gamma)$. Hence the upper neutrosophic $\in_\Phi$-subsets of X are right stable where $\Phi \in \{T, I, F\}$. Similarly, the strong upper neutrosophic $\in_\Phi$-subsets of X are right stable where $\Phi \in \{T, I, F\}$.

(2) Assume that $x * y \in L_T^{\in}(A;\alpha)$, $a * b \in L_I^{\in}(A;\beta)$, and $c * d \in L_F^{\in}(A;\gamma)$ for any $x, y, a, b, c, d \in X$. Then $A_T(x * y) \leq \alpha$, $A_I(a * b) \leq \beta$, and $A_F(c * d) \geq \gamma$. Because $x * y \leq x$, $a * b \leq a$, and $c * d \leq c$, it follows from Lemma 2 that $\alpha \geq A_T(x * y) \geq A_T(x)$, $\beta \geq A_I(a * b) \geq A_I(a)$, and $\gamma \leq A_F(c * d) \leq A_F(c)$; that is, $x \in L_T^{\in}(A;\alpha)$, $a \in L_I^{\in}(A;\beta)$, and $c \in L_F^{\in}(A;\gamma)$. Therefore the lower neutrosophic $\in_\Phi$-subsets of X are right vanished where $\Phi \in \{T, I, F\}$. In a similar way, we know that the strong lower neutrosophic $\in_\Phi$-subsets of X are right vanished where $\Phi \in \{T, I, F\}$. □

Definition 5. *Let $A = (A_T, A_I, A_F)$ be a NS in a BCK/BCI-algebra X and $(\alpha, \beta, \gamma) \in \Lambda_T \times \Lambda_I \times \Lambda_F$, where Λ_T, Λ_I, and Λ_F are subsets of $[0, 1]$. Then (α, β, γ) is called a neutrosophic permeable I-value for $A = (A_T, A_I, A_F)$ if the following assertion is valid:*

$$(\forall x, y \in X) \begin{pmatrix} x \in U_T^{\in}(A;\alpha) \Rightarrow A_T(x * y) \vee A_T(y) \geq \alpha, \\ x \in U_I^{\in}(A;\beta) \Rightarrow A_I(x * y) \vee A_I(y) \geq \beta, \\ x \in U_F^{\in}(A;\gamma) \Rightarrow A_F(x * y) \wedge A_F(y) \leq \gamma \end{pmatrix}. \quad (23)$$

Example 4. *(1) In Example 2, (α, β, γ) is a neutrosophic permeable I-value for $A = (A_T, A_I, A_F)$.*

(2) Consider a BCI-algebra $X = \{0, 1, a, b, c\}$ with the binary operation $*$ that is given in Table 7 (see [14]).

Table 7. Cayley table for the binary operation "$*$".

*	0	1	a	b	c
0	0	0	a	b	c
1	1	0	a	b	c
a	a	a	0	c	b
b	b	b	c	0	a
c	c	c	b	a	0

Let $A = (A_T, A_I, A_F)$ be a NS in X that is given in Table 8.

Table 8. Tabulation representation of $A = (A_T, A_I, A_F)$.

X	$A_T(x)$	$A_I(x)$	$A_F(x)$
0	0.33	0.38	0.77
1	0.44	0.48	0.66
a	0.55	0.68	0.44
b	0.66	0.58	0.44
c	0.66	0.68	0.55

It is routine to check that $(\alpha, \beta, \gamma) \in (0.33, 1] \times (0.38, 1] \times [0, 0.77)$ is a neutrosophic permeable I-value for $A = (A_T, A_I, A_F)$.

Lemma 3. *If a NS $A = (A_T, A_I, A_F)$ in a BCK/BCI-algebra X satisfies the condition of Equation (14), then*

$$(\forall x \in X)\,(A_T(0) \leq A_T(x),\ A_I(0) \leq A_I(x),\ A_F(0) \geq A_F(x)). \quad (24)$$

Proof. Straightforward. □

Theorem 11. *If a NS $A = (A_T, A_I, A_F)$ in a BCK-algebra X satisfies the condition of Equation (14), then every neutrosophic permeable I-value for $A = (A_T, A_I, A_F)$ is a neutrosophic permeable S-value for $A = (A_T, A_I, A_F)$.*

Proof. Let (α, β, γ) be a neutrosophic permeable I-value for $A = (A_T, A_I, A_F)$. Let $x, y, a, b, u, v \in X$ be such that $x * y \in U_T^{\Subset}(A; \alpha)$, $a * b \in U_I^{\Subset}(A; \beta)$, and $u * v \in U_F^{\Subset}(A; \gamma)$. It follows from Equations (23), (3), (III), and (V) and Lemma 3 that

$$\begin{aligned}\alpha &\leq A_T((x*y)*x) \vee A_T(x) = A_T((x*x)*y) \vee A_T(x) \\ &= A_T(0*y) \vee A_T(x) = A_T(0) \vee A_T(x) = A_T(x),\end{aligned}$$

$$\begin{aligned}\beta &\leq A_I((a*b)*a) \vee A_I(a) = A_I((a*a)*b) \vee A_I(a) \\ &= A_I(0*b) \vee A_I(a) = A_I(0) \vee A_I(a) = A_I(a),\end{aligned}$$

and

$$\begin{aligned}\gamma &\geq A_F((u*v)*u) \wedge A_F(u) = A_F((u*u)*v) \wedge A_F(u) \\ &= A_F(0*v) \wedge A_F(u) = A_F(0) \wedge A_F(u) = A_F(u).\end{aligned}$$

Hence $A_T(x) \vee A_T(y) \geq A_T(x) \geq \alpha$, $A_I(a) \vee A_I(b) \geq A_I(a) \geq \beta$, and $A_F(u) \wedge A_F(v) \leq A_F(u) \leq \gamma$. Therefore (α, β, γ) is a neutrosophic permeable S-value for $A = (A_T, A_I, A_F)$. □

Given a NS $A = (A_T, A_I, A_F)$ in a BCK/BCI-algebra X, any upper neutrosophic $\Subset_\Phi$-subsets of X may not be I-energetic where $\Phi \in \{T, I, F\}$, as seen in the following example.

Example 5. *Consider a BCK-algebra $X = \{0, 1, 2, 3, 4\}$ with the binary operation $*$ that is given in Table 9 (see [14]).*

Table 9. Cayley table for the binary operation "$*$".

*	0	1	2	3	4
0	0	0	0	0	0
1	1	0	0	0	0
2	2	1	0	1	0
3	3	1	1	0	0
4	4	2	1	2	0

Let $A = (A_T, A_I, A_F)$ be a NS in X that is given in Table 10.

Table 10. Tabulation representation of $A = (A_T, A_I, A_F)$.

X	$A_T(x)$	$A_I(x)$	$A_F(x)$
0	0.75	0.73	0.34
1	0.53	0.45	0.58
2	0.67	0.86	0.34
3	0.53	0.56	0.58
4	0.46	0.56	0.66

Then $U_T^{\Subset}(A; 0.6) = \{0, 2\}$, $U_I^{\Subset}(A; 0.7) = \{0, 2\}$, and $U_F^{\Subset}(A; 0.4) = \{0, 2\}$. Because $2 \in \{0, 2\}$ and $\{1, 2*1\} \cap \{0, 2\} = \emptyset$, we know that $\{0, 2\}$ is not an I-energetic subset of X.

We now provide conditions for the upper neutrosophic $\Subset_\Phi$-subsets to be I-energetic where $\Phi \in \{T, I, F\}$.

Theorem 12. *Let $A = (A_T, A_I, A_F)$ be a NS in a BCK/BCI-algebra X and $(\alpha, \beta, \gamma) \in \Lambda_T \times \Lambda_I \times \Lambda_F$, where Λ_T, Λ_I, and Λ_F are subsets of $[0, 1]$. If (α, β, γ) is a neutrosophic permeable I-value for $A = (A_T, A_I, A_F)$, then the upper neutrosophic $\Subset_\Phi$-subsets of X are I-energetic subsets of X where $\Phi \in \{T, I, F\}$.*

Proof. Let $x, a, u \in X$ and $(\alpha, \beta, \gamma) \in \Lambda_T \times \Lambda_I \times \Lambda_F$, where Λ_T, Λ_I, and Λ_F are subsets of $[0,1]$ such that $x \in U_T^{\in}(A;\alpha)$, $a \in U_I^{\in}(A;\beta)$, and $u \in U_F^{\in}(A;\gamma)$. Because (α, β, γ) is a neutrosophic permeable I-value for $A = (A_T, A_I, A_F)$, it follows from Equation (23) that

$$A_T(x * y) \vee A_T(y) \geq \alpha,\ A_I(a * b) \vee A_I(b) \geq \beta, \text{ and } A_F(u * v) \wedge A_F(v) \leq \gamma$$

for all $y, b, v \in X$. Hence

$$A_T(x * y) \geq \alpha \text{ or } A_T(y) \geq \alpha, \text{ that is, } x * y \in U_T^{\in}(A;\alpha) \text{ or } y \in U_T^{\in}(A;\alpha);$$
$$A_I(a * b) \geq \beta \text{ or } A_I(b) \geq \beta, \text{ that is, } a * b \in U_I^{\in}(A;\beta) \text{ or } b \in U_I^{\in}(A;\beta);$$

and

$$A_F(u * v) \leq \gamma \text{ or } A_F(v) \leq \gamma, \text{ that is, } u * v \in U_F^{\in}(A;\gamma) \text{ or } v \in U_F^{\in}(A;\gamma).$$

Hence $\{y, x * y\} \cap U_T^{\in}(A;\alpha)$, $\{b, a * b\} \cap U_I^{\in}(A;\beta)$, and $\{v, u * v\} \cap U_F^{\in}(A;\gamma)$ are nonempty, and therefore the upper neutrosophic $\in_\Phi$-subsets of X are I-energetic subsets of X where $\Phi \in \{T, I, F\}$. □

Theorem 13. *Let $A = (A_T, A_I, A_F)$ be a NS in a BCK/BCI-algebra X and $(\alpha, \beta, \gamma) \in \Lambda_T \times \Lambda_I \times \Lambda_F$, where Λ_T, Λ_I, and Λ_F are subsets of $[0,1]$. If $A = (A_T, A_I, A_F)$ satisfies the following condition:*

$$(\forall x, y \in X)\left(\begin{array}{l} A_T(x) \leq A_T(x * y) \vee A_T(y) \\ A_I(x) \leq A_I(x * y) \vee A_I(y) \\ A_F(x) \geq A_F(x * y) \wedge A_F(y) \end{array}\right), \tag{25}$$

then (α, β, γ) is a neutrosophic permeable I-value for $A = (A_T, A_I, A_F)$.

Proof. Let $x, a, u \in X$ and $(\alpha, \beta, \gamma) \in \Lambda_T \times \Lambda_I \times \Lambda_F$, where Λ_T, Λ_I, and Λ_F are subsets of $[0,1]$ such that $x \in U_T^{\in}(A;\alpha)$, $a \in U_I^{\in}(A;\beta)$, and $u \in U_F^{\in}(A;\gamma)$. Using Equation (25), we obtain

$$\begin{aligned} &\alpha \leq A_T(x) \leq A_T(x * y) \vee A_T(y), \\ &\beta \leq A_I(a) \leq A_I(a * b) \vee A_I(b), \\ &\gamma \geq A_F(u) \geq A_F(u * v) \wedge A_F(v), \end{aligned}$$

for all $y, b, v \in X$. Therefore (α, β, γ) is a neutrosophic permeable I-value for $A = (A_T, A_I, A_F)$. □

Combining Theorems 12 and 13, we have the following corollary.

Corollary 3. *Let $A = (A_T, A_I, A_F)$ be a NS in a BCK/BCI-algebra X and $(\alpha, \beta, \gamma) \in \Lambda_T \times \Lambda_I \times \Lambda_F$, where Λ_T, Λ_I, and Λ_F are subsets of $[0,1]$. If $A = (A_T, A_I, A_F)$ satisfies the condition of Equation (25), then the upper neutrosophic $\in_\Phi$-subsets of X are I-energetic subsets of X where $\Phi \in \{T, I, F\}$.*

Definition 6. *Let $A = (A_T, A_I, A_F)$ be a NS in a BCK/BCI-algebra X and $(\alpha, \beta, \gamma) \in \Lambda_T \times \Lambda_I \times \Lambda_F$, where Λ_T, Λ_I, and Λ_F are subsets of $[0,1]$. Then (α, β, γ) is called a neutrosophic anti-permeable I-value for $A = (A_T, A_I, A_F)$ if the following assertion is valid:*

$$(\forall x, y \in X)\left(\begin{array}{lcl} x \in L_T^{\in}(A;\alpha) & \Rightarrow & A_T(x * y) \wedge A_T(y) \leq \alpha, \\ x \in L_I^{\in}(A;\beta) & \Rightarrow & A_I(x * y) \wedge A_I(y) \leq \beta, \\ x \in L_F^{\in}(A;\gamma) & \Rightarrow & A_F(x * y) \vee A_F(y) \geq \gamma \end{array}\right). \tag{26}$$

Theorem 14. *Let $A = (A_T, A_I, A_F)$ be a NS in a BCK/BCI-algebra X and $(\alpha, \beta, \gamma) \in \Lambda_T \times \Lambda_I \times \Lambda_F$, where Λ_T, Λ_I, and Λ_F are subsets of $[0,1]$. If $A = (A_T, A_I, A_F)$ satisfies the condition of Equation (19), then (α, β, γ) is a neutrosophic anti-permeable I-value for $A = (A_T, A_I, A_F)$.*

Proof. Let $x, a, u \in X$ be such that $x \in L_T^{\in}(A;\alpha)$, $a \in L_I^{\in}(A;\beta)$, and $u \in L_F^{\in}(A;\gamma)$. Then

$$A_T(x * y) \wedge A_T(y) \leq A_T(x) \leq \alpha,$$
$$A_I(a * b) \wedge A_I(b) \leq A_I(a) \leq \beta,$$
$$A_F(u * v) \vee A_F(v) \geq A_F(u) \geq \gamma,$$

for all $y, b, v \in X$ by Equation (20). Hence (α, β, γ) is a neutrosophic anti-permeable I-value for $A = (A_T, A_I, A_F)$. □

Theorem 15. *Let $A = (A_T, A_I, A_F)$ be a NS in a BCK/BCI-algebra X and $(\alpha, \beta, \gamma) \in \Lambda_T \times \Lambda_I \times \Lambda_F$, where Λ_T, Λ_I, and Λ_F are subsets of $[0,1]$. If (α, β, γ) is a neutrosophic anti-permeable I-value for $A = (A_T, A_I, A_F)$, then the lower neutrosophic $\in_\Phi$-subsets of X are I-energetic where $\Phi \in \{T, I, F\}$.*

Proof. Let $x \in L_T^{\in}(A;\alpha)$, $a \in L_I^{\in}(A;\beta)$, and $u \in L_F^{\in}(A;\gamma)$. Then $A_T(x * y) \wedge A_T(y) \leq \alpha$, $A_I(a * b) \wedge A_I(b) \leq \beta$, and $A_F(u * v) \vee A_F(v) \geq \gamma$ for all $y, b, v \in X$ by Equation (26). It follows that

$$A_T(x * y) \leq \alpha \text{ or } A_T(y) \leq \alpha, \text{ that is, } x * y \in L_T^{\in}(A;\alpha) \text{ or } y \in L_T^{\in}(A;\alpha);$$
$$A_I(a * b) \leq \beta \text{ or } A_I(b) \leq \beta, \text{ that is, } a * b \in L_I^{\in}(A;\beta) \text{ or } b \in L_I^{\in}(A;\beta);$$

and

$$A_F(u * v) \geq \gamma \text{ or } A_F(v) \geq \gamma, \text{ that is, } u * v \in L_F^{\in}(A;\gamma) \text{ or } v \in L_F^{\in}(A;\gamma).$$

Hence $\{y, x * y\} \cap L_T^{\in}(A;\alpha)$, $\{b, a * b\} \cap L_I^{\in}(A;\beta)$ and $\{v, u * v\} \cap L_F^{\in}(A;\gamma)$ are nonempty, and therefore the lower neutrosophic $\in_\Phi$-subsets of X are I-energetic where $\Phi \in \{T, I, F\}$. □

Combining Theorems 14 and 15, we obtain the following corollary.

Corollary 4. *Let $A = (A_T, A_I, A_F)$ be a NS in a BCK/BCI-algebra X and $(\alpha, \beta, \gamma) \in \Lambda_T \times \Lambda_I \times \Lambda_F$, where Λ_T, Λ_I, and Λ_F are subsets of $[0,1]$. If $A = (A_T, A_I, A_F)$ satisfies the condition of Equation (19), then the lower neutrosophic $\in_\Phi$-subsets of X are I-energetic where $\Phi \in \{T, I, F\}$.*

Theorem 16. *If $A = (A_T, A_I, A_F)$ is an $(\in, \in)$-neutrosophic subalgebra of a BCK-algebra X, then every neutrosophic anti-permeable I-value for $A = (A_T, A_I, A_F)$ is a neutrosophic anti-permeable S-value for $A = (A_T, A_I, A_F)$.*

Proof. Let (α, β, γ) be a neutrosophic anti-permeable I-value for $A = (A_T, A_I, A_F)$. Let $x, y, a, b, u, v \in X$ be such that $x * y \in L_T^{\in}(A;\alpha)$, $a * b \in L_I^{\in}(A;\beta)$, and $u * v \in L_F^{\in}(A;\gamma)$. It follows from Equations (26), (3), (III), and (V) and Proposition 1 that

$$\begin{aligned}\alpha &\geq A_T((x * y) * x) \wedge A_T(x) = A_T((x * x) * y) \wedge A_T(x) \\ &= A_T(0 * y) \wedge A_T(x) = A_T(0) \wedge A_T(x) = A_T(x),\end{aligned}$$

$$\begin{aligned}\beta &\geq A_I((a * b) * a) \wedge A_I(a) = A_I((a * a) * b) \wedge A_I(a) \\ &= A_I(0 * b) \wedge A_I(a) = A_I(0) \wedge A_I(a) = A_I(a),\end{aligned}$$

and

$$\begin{aligned}\gamma &\leq A_F((u * v) * u) \vee A_F(u) = A_F((u * u) * v) \vee A_F(u) \\ &= A_F(0 * v) \vee A_F(u) = A_F(0) \vee A_F(u) = A_F(u).\end{aligned}$$

Hence $A_T(x) \wedge A_T(y) \leq A_T(x) \leq \alpha$, $A_I(a) \wedge A_I(b) \leq A_I(a) \leq \beta$, and $A_F(u) \vee A_F(v) \geq A_F(u) \geq \gamma$. Therefore (α, β, γ) is a neutrosophic anti-permeable S-value for $A = (A_T, A_I, A_F)$. □

4. Conclusions

Using the notions of subalgebras and ideals in BCK/BCI-algebras, Jun et al. [8] introduced the notions of energetic subsets, right vanished subsets, right stable subsets, and (anti-)permeable values in BCK/BCI-algebras, as well as investigated relations between these sets. As a more general platform that extends the concepts of classic and fuzzy sets, intuitionistic fuzzy sets, and interval-valued (intuitionistic) fuzzy sets, the notion of NS theory has been developed by Smarandache (see [1,2]) and has been applied to various parts: pattern recognition, medical diagnosis, decision-making problems, and so on (see [3–6]). In this article, we have introduced the notions of neutrosophic permeable S-values, neutrosophic permeable I-values, $(\in, \in)$-neutrosophic ideals, neutrosophic anti-permeable S-values, and neutrosophic anti-permeable I-values, which are motivated by the idea of subalgebras (s-values) and ideals (I-values), and have investigated their properties. We have considered characterizations of $(\in, \in)$-neutrosophic ideals and have discussed conditions for the lower (upper) neutrosophic $\in_\Phi$-subsets to be S- and I-energetic. We have provided conditions for a triple (α, β, γ) of numbers to be a neutrosophic (anti-)permeable S- or I-value, and have considered conditions for the upper (lower) neutrosophic $\in_\Phi$-subsets to be right stable (right vanished) subsets. We have established relations between neutrosophic (anti-)permeable S- and I-values.

Author Contributions: Y.B.J. and S.-Z.S. initiated the main idea of this work and wrote the paper. F.S. and H.B. performed the finding of the examples and checking of the contents. All authors conceived and designed the new definitions and results and read and approved the final manuscript for submission.

Funding: This research received no external funding.

Acknowledgments: The authors wish to thank the anonymous reviewers for their valuable suggestions.

Conflicts of Interest: The authors declare no conflict of interest.

References

1. Smarandache, F. *A Unifying Field in Logics: Neutrosophic Logic. Neutrosophy, Neutrosophic Set, Neutrosophic Probability*; American Reserch Press: Rehoboth, NM, USA, 1999.
2. Smarandache, F. Neutrosophic set-a generalization of the intuitionistic fuzzy set. *Int. J. Pure Appl. Math.* **2005**, *24*, 287–297.
3. Garg, H.; Nancy. Some new biparametric distance measures on single-valued neutrosophic sets with applications to pattern recognition and medical diagnosis. *Information* **2017**, *8*, 126.
4. Garg, H.; Nancy. Non-linear programming method for multi-criteria decision making problems under interval neutrosophic set environment. *Appl. Intell.* **2017**, doi:10.1007/s10489-017-1070-5.
5. Garg, H.; Nancy. Linguistic single-valued neutrosophic prioritized aggregation operators and their applications to multiple-attribute group decision-making. *J. Ambient Intell. Humaniz. Comput.* **2018**, doi:10.1007/s12652-018-0723-5.
6. Nancy; Garg, H. Novel single-valued neutrosophic aggregated operators under Frank norm operation and its application to decision-making process. *Int. J. Uncertain. Quantif.* **2016**, *6*, 361–375.
7. Jun, Y.B. Neutrosophic subalgebras of several types in BCK/BCI-algebras. *Ann. Fuzzy Math. Inform.* **2017**, *14*, 75–86.
8. Jun, Y.B.; Ahn, S.S.; Roh, E.H. Energetic subsets and permeable values with applications in BCK/BCI-algebras. *Appl. Math. Sci.* **2013**, *7*, 4425–4438.
9. Jun, Y.B.; Smarandache, F.; Bordbar, H. Neutrosophic $\mathcal{N}$-structures applied to BCK/BCI-algebras. *Informations* **2017**, *8*, 128.
10. Jun, Y.B.; Smarandache, F.; Song, S.Z.; Khan, M. Neutrosophic positive implicative $\mathcal{N}$-ideals in BCK-algebras. *Axioms* **2018**, *7*, 3.

11. Öztürk, M.A.; Jun, Y.B. Neutrosophic ideals in BCK/BCI-algebras based on neutrosophic points. *J. Int. Math. Virtual Inst.* **2018**, *8*, 1–17.
12. Song, S.Z.; Smarandache, F.; Jun, Y.B. Neutrosophic commutative $\mathcal{N}$-ideals in BCK-algebras. *Information* **2017**, *8*, 130.
13. Huang, Y.S. *BCI-Algebra*; Science Press: Beijing, China, 2006.
14. Meng, J.; Jun, Y.B. *BCK-Algebras*; Kyungmoon Sa Co.: Seoul, Korea, 1994.

mathematics MDPI

Article

A Novel (R, S)-Norm Entropy Measure of Intuitionistic Fuzzy Sets and Its Applications in Multi-Attribute Decision-Making

Harish Garg * and Jaspreet Kaur

School of Mathematics, Thapar Institute of Engineering & Technology, Deemed University, Patiala 147004, Punjab, India; 30mkaur1995@gmail.com

* Correspondence: harishg58iitr@gmail.com; Tel.: +91-86990-31147

Received: 16 May 2018; Accepted: 28 May 2018; Published: 30 May 2018

Abstract: The objective of this manuscript is to present a novel information measure for measuring the degree of fuzziness in intuitionistic fuzzy sets (IFSs). To achieve it, we define an (R, S)-norm-based information measure called the entropy to measure the degree of fuzziness of the set. Then, we prove that the proposed entropy measure is a valid measure and satisfies certain properties. An illustrative example related to a linguistic variable is given to demonstrate it. Then, we utilized it to propose two decision-making approaches to solve the multi-attribute decision-making (MADM) problem in the IFS environment by considering the attribute weights as either partially known or completely unknown. Finally, a practical example is provided to illustrate the decision-making process. The results corresponding to different pairs of (R, S) give different choices to the decision-maker to assess their results.

Keywords: entropy measure; (R, S)-norm; multi attribute decision-making; information measures; attribute weight; intuitionistic fuzzy sets

1. Introduction

Multi-attribute decision-making (MADM) problems are an important part of decision theory in which we choose the best one from the set of finite alternatives based on the collective information. Traditionally, it has been assumed that the information regarding accessing the alternatives is taken in the form of real numbers. However, uncertainty and fuzziness are big issues in real-world problems nowadays and can be found everywhere as in our discussion or the way we process information. To deal with such a situation, the theory of fuzzy sets (FSs) [1] or extended fuzzy sets such as an intuitionistic fuzzy set (IFS) [2] or interval-valued IFS (IVIFS) [3] are the most successful ones, which characterize the attribute values in terms of membership degrees. During the last few decades, researchers has been paying more attention to these theories and successfully applied them to various situations in the decision-making process. The two important aspects of solving the MADM problem are, first, to design an appropriate function that aggregates the different preferences of the decision-makers into collective ones and, second, to design appropriate measures to rank the alternatives. For the former part, an aggregation operator is an important part of the decision-making, which usually takes the form of a mathematical function to aggregate all the individual input data into a single one. Over the last decade, numerable attempts have been made by different researchers in processing the information values using different aggregation operators under IFS and IVIFS environments. For instance, Xu and Yager [4], Xu [5] presented some weighted averaging and geometric aggregation operators to aggregate the different intuitionistic fuzzy numbers (IFNs). Garg [6] and Garg [7] presented some interactive improved aggregation operators for IFNs using Einstein norm operations. Wang and Wang [8] characterized the preference of the decision-makers in terms of interval-numbers, and then, an MADM was presented corresponding to it with completely unknown weight vectors. Wei [9] presented some induced geometric aggregation operators with intuitionistic fuzzy information. Arora and Garg [10] and Arora and Garg [11] presented some aggregation operators by considering the different parameterization factors in the analysis in the intuitionistic fuzzy soft set environment. Zhou and Xu [12] presented some extreme weighted averaging aggregation operators for solving decision-making problems in terms of the optimism and pessimism points of view. Garg [13] presented some improved geometric aggregation operators for IVIFS. A complete overview about the aggregation operators in the IVIFSs was summarized by Xu and Guo in [14]. Jamkhaneh and Garg [15] presented some new operations for the generalized IFSs and applied them to solve decision-making problems. Garg and Singh [16] presented a new triangular interval Type-2 IFS and its corresponding aggregation operators.

With regard to the information measure, the entropy measure is basically known as the measure for information originating from the fundamental paper "The Mathematical theory of communication" in 1948 by C.E.Shannon [17]. Information theory is one of the trusted areas to measure the degree of uncertainty in the data. However, classical information measures deal with information that is precise in nature. In order to overcome this, Deluca and Termini [18] proposed a set of axioms for fuzzy entropy. Later on, Szmidt and Kacprzyk [19] extended the axioms of Deluca and Termini [18] to the IFS environment. Vlachos and Sergiadis [20] extended their measure to the IFS environment. Burillo and Bustince [21] introduced the entropy of IFSs as a tool to measure the degree of intuitionism associated with an IFS. Garg et al. [22] presented a generalized intuitionistic fuzzy entropy measure of order α and degree β to solve decision-making problems. Wei et al. [23] presented an entropy measure based on the trigonometric functions. Garg et al. [24] presented an entropy-based method for solving decision-making problems. Zhang and Jiang [25] presented an intuitionistic fuzzy entropy by generalizing the measure of Deluca and Termini [18]. Verma and Sharma [26] presented an exponential order measure between IFSs.

In contrast to the entropy measures, the distance or similarity measures are also used by researchers to measure the similarity between two IFSs. In that direction, Taneja [27] presented a theory on the generalized information measures in the fuzzy environment. Boekee and Van der Lubbe [28] presented the R-norm information measure. Hung and Yang [29] presented the similarity measures between the two different IFSs based on the Hausdorff distance. Garg [30], Garg and Arora [31] presented a series of distance and similarity measures in the different sets of the environment to solve decision-making problems. Joshi and Kumar [32] presented an (R, S)-norm fuzzy information measures to solve decision-making problems. Garg and Kumar [33,34] presented some similarity and distance measures of IFSs by using the set pair analysis theory. Meanwhile, decision-making methods based on some measures (such as distance, similarity degree, correlation coefficient and entropy) were proposed to deal with fuzzy IF and interval-valued IF MADM problems [35–38].

In [39–43], emphasis was given by the researchers to the attribute weights during ranking of the alternatives. It is quite obvious that the final ranking order of the alternatives highly depends on the attribute weights, because the variation of weight values may result in a different final ranking order of alternatives [39,44–47]. Now, based on the characteristics of the attribute weights, the decision-making problem can be classified into three types: (a) the decision-making situation where the attribute weights are completely known; (b) the decision-making situation where the attribute weights are completely unknown; (c) the decision-making situation where the attribute weights are partially known. Thus, based on these types, the attribute weights in MADM can be classified as subjective and objective attribute weights based on the information acquisition approach. If the decision-maker gives weights to the attributes, then such information is called subjective. The classical approaches to determine the subjective attribute weights are the analytic hierarchy process (AHP) method [48] and the Delphi method [49]. On the other hand, the objective attribute weights are determined by the decision-making matrix, and one of the most important approaches is the Shannon entropy method [17], which expresses the relative intensities of the attributes' importance to signify the average intrinsic information transmitted to the decision-maker. In the literature, several authors [39,44,50–52] have addressed the MADM problem with subjective weight information. However, some researchers formulated a nonlinear programming model to determine the attribute weights. For instance, Chen and Li [44] presented an approach to assess the attribute weights by utilizing IF entropy in the IFS environment. Garg [53] presented a generalized intuitionistic fuzzy entropy measure to determine the completely unknown attribute weight to solve the decision-making problems. Although some researchers put some efforts into determining the unknown attribute weights [45,46,54,55] under different environments, still it remains an open problem.

Therefore, in an attempt to address such problems and motivated by the characteristics of the IFSs to describe the uncertainties in the data, this paper addresses a new entropy measure to quantify the degree of fuzziness of a set in the IFS environment. The aim of this entropy is to determine the attribute weights under the characteristics of the attribute weights that they are either partially known or completely unknown. For this, we propose a novel entropy measure named the (R,S)-norm-based information measure, which makes the decision more flexible and reliable corresponding to different values of the parameters R and S. Some of the desirable properties of the proposed measures are investigated, and some of their correlations are dreived. From the proposed entropy measures, some of the existing measures are considered as a special case. Furthermore, we propose two approaches for solving the MADM approach based on the proposed entropy measures by considering the characteristics of the attribute weights being either partially known or completely unknown. Two illustrative examples are considered to demonstrate the approach and compare the results with some of the existing approaches' results.

The rest of this paper is organized as follows. In Section 2, we present some basic concepts of IFSs and the existing entropy measures. In Section 3, we propose a new (R,S)-norm-based information measure in the IFS environment. Various desirable relations among the approaches are also investigated in detail. Section 4 describes two approaches for solving the MADM problem with the condition that attribute weights are either partially known or completely unknown. The developed approaches have been illustrated with a numerical example. Finally, a concrete conclusion and discussion are presented in Section 5.

2. Preliminaries

Some basic concepts related to IFSs and the aggregation operators are highlighted, over the universal set X, in this section.

Definition 1. [2] *An IFS A defined in X is an ordered pair given by:*

$$A = \{\langle x, \zeta_A(x), \vartheta_A(x)\rangle \mid x \in X\} \tag{1}$$

where $\zeta_A, \vartheta_A : X \longrightarrow [0,1]$ *represent, respectively, the membership and non-membership degrees of the element* x *such that* $\zeta_A, \vartheta_A \in [0,1]$ *and* $\zeta_A + \vartheta_A \leq 1$ *for all* x. *For convenience, this pair is denoted by* $A = \langle \zeta_A, \vartheta_A \rangle$ *and called an intuitionistic fuzzy number (IFN)* [4,5].

Definition 2. [4,5] *Let the family of all intuitionistic fuzzy sets of universal set X be denoted by FS(X). Let A, B* $\in$ *FS(X) be such that then some operations can be defined as follows:*

1. $A \subseteq B$ *if* $\zeta_A(x) \leq \zeta_B(x)$ *and* $\vartheta_A(x) \geq \vartheta_B(x)$, *for all* $x \in X$;
2. $A \supseteq B$ *if* $\zeta_A(x) \geq \zeta_B(x)$ *and* $\vartheta_A(x) \leq \vartheta_B(x)$, *for all* $x \in X$;
3. $A = B$ *iff* $\zeta_A(x) = \zeta_B(x)$ *and* $\vartheta_A(x) = \vartheta_B(x)$, *for all* $x \in X$;
4. $A \cup B = \{\langle x, \max(\zeta_A(x), \zeta_B(x)), \min(\vartheta_A(x), \vartheta_B(x))\rangle : x \in X\}$;
5. $A \cap B = \{\langle x, \min(\zeta_A(x), \zeta_B(x)), \max(\vartheta_A(x), \vartheta_B(x))\rangle : x \in X\}$;
6. $A^c = \{\langle x, \vartheta_A(x), \zeta_A(x)\rangle : x \in X\}$.

Definition 3. *[19] An entropy E: IFS(X) $\longrightarrow$ R^+ on IFS(X) is a real-valued functional satisfying the following four axioms for $A, B \in IFS(X)$*

(P1) $E(A) = 0$ *if and only if A is a crisp set, i.e., either* $\zeta_A(x) = 1, \vartheta_A(x) = 0$ *or* $\zeta_A(x) = 0, \vartheta_A(x) = 1$ *for all* $x \in X$.
(P2) $E(A) = 1$ *if and only if* $\zeta_A(x) = \vartheta_A(x)$ *for all* $x \in X$.
(P3) $E(A) = E(A^c)$.
(P4) *If* $A \subseteq B$*, that is, if* $\zeta_A(x) \leq \zeta_B(x)$ *and* $\vartheta_A(x) \geq \vartheta_B(x)$ *for any* $x \in X$*, then* $E(A) \leq E(B)$.

Vlachos and Sergiadis [20] proposed the measure of intuitionistic fuzzy entropy in the IFS environment as follows:

$$E(A) = -\frac{1}{n \ln 2} \sum_{i=1}^{n} \left[\zeta_A(x_i) \ln \zeta_A(x_i) + \vartheta_A(x_i) \ln \vartheta_A(x_i) - (1 - \pi_A(x_i)) \ln(1 - \pi_A(x_i)) - \pi_A(x_i) \ln 2 \right] \quad (2)$$

Zhang and Jiang [25] presented a measure of intuitionistic fuzzy entropy based on a generalization of measure of Deluca and Termini [18] as:

$$E(A) = -\frac{1}{n} \sum_{i=1}^{n} \left[\begin{array}{c} \left(\frac{\zeta_A(x_i) + 1 - \vartheta_A(x_i)}{2} \right) \log \left(\frac{\zeta_A(x_i) + 1 - \vartheta_A(x_i)}{2} \right) + \\ \left(\frac{\vartheta_A(x_i) + 1 - \zeta_A(x_i)}{2} \right) \log \left(\frac{\vartheta_A(x_i) + 1 - \zeta_A(x_i)}{2} \right) \end{array} \right] \quad (3)$$

Verma and Sharma [26] proposed an exponential order entropy in the IFS environment as:

$$E(A) = \frac{1}{n(\sqrt{e} - 1)} \sum_{i=1}^{n} \left[\begin{array}{l} \frac{\zeta_A(x_i) + 1 - \vartheta_A(x_i)}{2} e^{1 - \frac{\zeta_A(x_i) + 1 - \vartheta_A(x_i)}{2}} \\ + \frac{\vartheta_A(x_i) + 1 - \zeta_A(x_i)}{2} e^{1 - \frac{\vartheta_A(x_i) + 1 - \zeta_A(x_i)}{2}} - 1 \end{array} \right] \quad (4)$$

Garg et al. [22] generalized entropy measure $E_\alpha^\beta(A)$ of order α and degree β as:

$$E_\alpha^\beta(A) = \frac{2 - \beta}{n(2 - \beta - \alpha)} \sum_{i=1}^{n} \log \left[\begin{array}{l} \left(\zeta_A^{\frac{\alpha}{2-\beta}}(x_i) + \vartheta_A^{\frac{\alpha}{2-\beta}}(x_i) \right) (\zeta_A(x_i) + \vartheta_A(x_i))^{1 - \frac{\alpha}{2-\beta}} \\ + 2^{1 - \frac{\alpha}{2-\beta}} (1 - \zeta_A(x_i) - \vartheta_A(x_i)) \end{array} \right] \quad (5)$$

where log is to the base two, $\alpha > 0$, $\beta \in [0, 1]$, $\alpha + \beta \neq 2$.

3. Proposed (*R*, *S*)-Norm Intuitionistic Fuzzy Information Measure

In this section, we define a new (R, S)-norm information measure, denoted by H_R^S, in the IFS environment. For it, let Ω be the collection of all IFSs.

Definition 4. *For a collection of IFSs* $A = \{(x, \zeta_A(x), \vartheta_A(x)) \mid x \in X\}$*, an information measure* $H_R^S : \Omega^n \to \mathbf{R}$*;* $n \geq 2$ *is defined as follows:*

$$H_R^S(A) = \begin{cases} \frac{R \times S}{n(R-S)} \sum_{i=1}^{n} \left[\begin{array}{c} \left(\zeta_A^S(x_i) + \vartheta_A^S(x_i) + \pi_A^S(x_i) \right)^{\frac{1}{S}} \\ - \left(\zeta_A^R(x_i) + \vartheta_A^R(x_i) + \pi_A^R(x_i) \right)^{\frac{1}{R}} \end{array} \right]; & \text{either } R > 1,\ 0 < S < 1 \text{ or } 0 < R < 1,\ S > 1 \\ \frac{R}{n(R-1)} \sum_{i=1}^{n} \left[1 - (\zeta_A^R(x_i) + \vartheta_A^R(x_i) + \pi_A^R(x_i))^{\frac{1}{R}} \right]; & \text{when } S = 1;\ 0 < R < 1 \\ \frac{S}{n(1-S)} \sum_{i=1}^{n} \left[(\zeta_A^S(x_i) + \vartheta_A^S(x_i) + \pi_A^S(x_i))^{\frac{1}{S}} - 1 \right]; & \text{when } R = 1;\ 0 < S < 1 \\ \frac{-1}{n} \sum_{i=1}^{n} \left[\begin{array}{c} \zeta_A(x_i) \log \zeta_A(x_i) + \vartheta_A(x_i) \log \vartheta_A(x_i) \\ + \pi_A(x_i) \log \pi_A(x_i) \end{array} \right]; & R = 1 = S. \end{cases} \quad (6)$$

Theorem 1. *An intuitionistic fuzzy entropy measure $H_R^S(A)$ defined in Equation (6) for IFSs is a valid measure, i.e., it satisfies the following properties.*

(P1) $H_R^S(A) = 0$ *if and only if A is a crisp set, i.e.,* $\zeta_A(x_i) = 1, \vartheta_A(x_i) = 0$ *or* $\zeta_A(x_i) = 0, \vartheta_A(x_i) = 1$ *for all* $x_i \in X$.

(P2) $H_R^S(A) = 1$ *if and only if* $\zeta_A(x_i) = \vartheta_A(x_i)$ *for all* $x_i \in X$.

(P3) $H_R^S(A) \leq H_R^S(B)$ *if A is crisper than B, i.e., if* $\zeta_A(x_i) \leq \zeta_B(x_i)$ & $\vartheta_A(x_i) \leq \vartheta_B(x_i)$, *for* $\max\{\zeta_B(x_i), \vartheta_B(x_i)\} \leq \frac{1}{3}$ *and* $\zeta_A(x_i) \geq \zeta_B(x_i)$ & $\vartheta_A(x_i) \geq \vartheta_B(x_i)$, *for* $\min\{\zeta_B(x_i), \vartheta_B(x_i)\} \leq \frac{1}{3}$ *for all* $x_i \in X$.

(P4) $H_R^S(A) = H_R^S(A^c)$ *for all* $A \in IFS(X)$.

Proof. To prove that the measure defined by Equation (6) is a valid information measure, we will have to prove that it satisfies the four properties defined in the definition of the intuitionistic fuzzy information measure.

1. Sharpness: In order to prove (P1), we need to show that $H_R^S(A) = 0$ if and only if A is a crisp set, i.e., either $\zeta_A(x) = 1, \vartheta_A(x) = 0$ or $\zeta_A(x) = 0, \vartheta_A(x) = 1$ for all $x \in X$.

   Firstly, we assume that $H_R^S(A) = 0$ for $R, S > 0$ and $R \neq S$. Therefore, from Equation (6), we have:

$$\frac{R \times S}{n(R-S)} \sum_{i=1}^{n} \begin{pmatrix} \left(\zeta_A^S(x_i) + \vartheta_A^S(x_i) + \pi_A^S(x_i)\right)^{\frac{1}{S}} \\ - \left(\zeta_A^R(x_i) + \vartheta_A^R(x_i) + \pi_A^R(x_i)\right)^{\frac{1}{R}} \end{pmatrix} = 0$$

$$\Rightarrow \quad \left(\zeta_A^S(x_i) + \vartheta_A^S(x_i) + \pi_A^S(x_i)\right)^{\frac{1}{S}} - \left(\zeta_A^R(x_i) + \vartheta_A^R(x_i) + \pi_A^R(x_i)\right)^{\frac{1}{R}} = 0 \text{ for all } i = 1, 2, \ldots, n.$$

$$i.e., \quad \left(\zeta_A^S(x_i) + \vartheta_A^S(x_i) + \pi_A^S(x_i)\right)^{\frac{1}{S}} = \left(\zeta_A^R(x_i) + \vartheta_A^R(x_i) + \pi_A^R(x_i)\right)^{\frac{1}{R}} \text{ for all } i = 1, 2, \ldots, n.$$

   Since $R, S > 0$ and $R \neq S$, therefore, the above equation is satisfied only if $\zeta_A(x_i) = 0, \vartheta_A(x_i) = 1$ or $\zeta_A(x_i) = 1, \vartheta_A(x_i) = 0$ for all $i = 1, 2, \ldots, n$.

   Conversely, we assume that set $A = (\zeta_A, \vartheta_A)$ is a crisp set i.e., either $\zeta_A(x_i) = 0$ or 1. Now, for $R, S > 0$ and $R \neq S$, we can obtain that:

$$\left(\zeta_A^S(x_i) + \vartheta_A^S(x_i) + \pi_A^S(x_i)\right)^{\frac{1}{S}} - \left(\zeta_A^R(x_i) + \vartheta_A^R(x_i) + \pi_A^R(x_i)\right)^{\frac{1}{R}} = 0$$

   for all $i = 1, 2, \ldots, n$, which gives that $H_R^S(A) = 0$.

   Hence, $H_R^S(A) = 0$ iff A is a crisp set.

2. Maximality: We will find maxima of the function $H_R^S(A)$; for this purpose, we will differentiate Equation (6) with respect to $\zeta_A(x_i)$ and $\vartheta_A(x_i)$. We get,

$$\frac{\partial H_R^S(A)}{\partial \zeta_A(x_i)} = \frac{R \times S}{n(R-S)} \sum_{i=1}^{n} \left\{ \begin{array}{l} \left(\zeta_A^S(x_i) + \vartheta_A^S(x_i) + \pi_A^S(x_i)\right)^{\frac{1-S}{S}} \left(\zeta_A^{S-1}(x_i) - \pi_A^{S-1}(x_i)\right) \\ - \left(\zeta_A^R(x_i) + \vartheta_A^R(x_i) + \pi_A^R(x_i)\right)^{\frac{1-R}{R}} \left(\zeta_A^{R-1}(x_i) - \pi_A^{R-1}(x_i)\right) \end{array} \right\} \tag{7}$$

   and:

$$\frac{\partial H_R^S(A)}{\partial \vartheta_A(x_i)} = \frac{R \times S}{n(R-S)} \sum_{i=1}^{n} \left\{ \begin{array}{l} \left(\zeta_A^S(x_i) + \vartheta_A^S(x_i) + \pi_A^S(x_i)\right)^{\frac{1-S}{S}} \left(\vartheta_A^{S-1}(x_i) - \pi_A^{S-1}(x_i)\right) \\ - \left(\zeta_A^R(x_i) + \vartheta_A^R(x_i) + \pi_A^R(x_i)\right)^{\frac{1-R}{R}} \left(\vartheta_A^{R-1}(x_i) - \pi_A^{R-1}(x_i)\right) \end{array} \right\} \tag{8}$$

In order to check the convexity of the function, we calculate its second order derivatives as follows:

$$\frac{\partial^2 H_R^S(A)}{\partial^2 \zeta_A(x_i)} = \frac{R\times S}{n(R-S)}\sum_{i=1}^{n}\left\{\begin{array}{l}(1-S)\left(\zeta_A^S(x_i)+\vartheta_A^S(x_i)+\pi_A^S(x_i)\right)^{\frac{1-2S}{S}}\left(\zeta_A^{S-1}(x_i)-\pi_A^{S-1}(x_i)\right)^2\\ +(S-1)\left(\zeta_A^S(x_i)+\vartheta_A^S(x_i)+\pi_A^S(x_i)\right)^{\frac{1-S}{S}}\left(\zeta_A^{S-2}(x_i)+\pi_A^{S-2}(x_i)\right)\\ -(1-R)\left(\zeta_A^R(x_i)+\vartheta_A^R(x_i)+\pi_A^R(x_i)\right)^{\frac{1-2R}{R}}\left(\zeta_A^{R-1}(x_i)-\pi_A^{R-1}(x_i)\right)^2\\ -(R-1)\left(\zeta_A^R(x_i)+\vartheta_A^R(x_i)+\pi_A^R(x_i)\right)^{\frac{1-R}{R}}\left(\zeta_A^{R-2}(x_i)+\pi_A^{R-2}(x_i)\right)\end{array}\right\}$$

$$\frac{\partial^2 H_R^S(A)}{\partial^2 \vartheta_A(x_i)} = \frac{R\times S}{n(R-S)}\sum_{i=1}^{n}\left\{\begin{array}{l}(1-S)\left(\zeta_A^S(x_i)+\vartheta_A^S(x_i)+\pi_A^S(x_i)\right)^{\frac{1-2S}{S}}\left(\vartheta_A^{S-1}(x_i)-\pi_A^{S-1}(x_i)\right)^2\\ +(S-1)\left(\zeta_A^S(x_i)+\vartheta_A^S(x_i)+\pi_A^S(x_i)\right)^{\frac{1-S}{S}}\left(\vartheta_A^{S-2}(x_i)+\pi_A^{S-2}(x_i)\right)\\ -(1-R)\left(\zeta_A^R(x_i)+\vartheta_A^R(x_i)+\pi_A^R(x_i)\right)^{\frac{1-2R}{R}}\left(\vartheta_A^{R-1}(x_i)-\pi_A^{R-1}(x_i)\right)^2\\ -(R-1)\left(\zeta_A^R(x_i)+\vartheta_A^R(x_i)+\pi_A^R(x_i)\right)^{\frac{1-R}{R}}\left(\vartheta_A^{R-2}(x_i)+\pi_A^{R-2}(x_i)\right)\end{array}\right\}$$

and

$$\frac{\partial^2 H_R^S(A)}{\partial \vartheta_A(x_i)\partial \zeta_A(x_i)} = \frac{R\times S}{n(R-S)}\sum_{i=1}^{n}\left\{\begin{array}{l}(1-S)\left(\zeta_A^S(x_i)+\vartheta_A^S(x_i)+\pi_A^S(x_i)\right)^{\frac{1-2S}{S}}\times\\ \quad\times\left(\vartheta_A^{S-1}(x_i)-\pi_A^{S-1}(x_i)\right)\left(\zeta_A^{S-1}(x_i)-\pi_A^{S-1}(x_i)\right)\\ -(1-R)\left(\zeta_A^R(x_i)+\vartheta_A^R(x_i)+\pi_A^R(x_i)\right)^{\frac{1-2R}{R}}\times\\ \quad\times\left(\vartheta_A^{R-1}(x_i)-\pi_A^{R-1}(x_i)\right)\left(\zeta_A^{R-1}(x_i)-\pi_A^{R-1}(x_i)\right)\end{array}\right\}$$

To find the maximum/minimum point, we set $\frac{\partial H_R^S(A)}{\partial \zeta_A(x_i)} = 0$ and $\frac{\partial H_R^S(A)}{\partial \vartheta_A(x_i)} = 0$, which gives that $\zeta_A(x_i) = \vartheta_A(x_i) = \pi_A(x_i) = \frac{1}{3}$ for all i and hence called the critical point of the function H_R^S.

(a) When $R < 1, S > 1$, then at the critical point $\zeta_A(x_i) = \vartheta_A(x_i) = \pi_A(x_i) = \frac{1}{3}$, we compute that:

$$\frac{\partial^2 H_R^S(A)}{\partial^2 \zeta_A(x_i)} < 0$$

$$\text{and}\quad \frac{\partial^2 H_R^S(A)}{\partial^2 \zeta_A(x_i)}\cdot\frac{\partial^2 H_R^S(A)}{\partial^2 \vartheta_A(x_i)} - \left(\frac{\partial^2 H_R^S(A)}{\partial \vartheta_A(x_i)\partial \zeta_A(x_i)}\right)^2 > 0$$

Therefore, the Hessian matrix of $H_R^S(A)$ is negative semi-definite, and hence, $H_R^S(A)$ is a concave function. As the critical point of H_R^S is $\zeta_A = \vartheta_A = \frac{1}{3}$ and by the concavity, we get that $H_R^S(A)$ has a relative maximum value at $\zeta_A = \vartheta_A = \frac{1}{3}$.

(b) When $R > 1, S < 1$, then at the critical point, we can again easily obtain that:

$$\frac{\partial^2 H_R^S(A)}{\partial^2 \zeta_A(x_i)} < 0$$

$$\text{and}\quad \frac{\partial^2 H_R^S(A)}{\partial^2 \zeta_A(x_i)}\cdot\frac{\partial^2 H_R^S(A)}{\partial^2 \vartheta_A(x_i)} - \left(\frac{\partial^2 H_R^S(A)}{\partial \vartheta_A(x_i)\zeta_A(x_i)}\right)^2 > 0$$

This proves that $H_R^S(A)$ is a concave function and its global maximum at $\zeta_A(x_i) = \vartheta_A(x_i) = \frac{1}{3}$.

Thus, for all $R, S > 0; R < 1, S < 1$ or $R > 1, S < 1$, the global maximum value of $H_R^S(A)$ attains at the point $\zeta_A(x_i) = \vartheta_A(x_i) = \frac{1}{3}$, i.e., $H_R^S(A)$ is maximum if and only if A is the most fuzzy set.

3. Resolution: In order to prove that our proposed entropy function is monotonically increasing and monotonically decreasing with respect to $\zeta_A(x_i)$ and $\vartheta_A(x_i)$, respectively, for convince, let $\zeta_A(x_i) = x$, $\vartheta_A(x_i) = y$ and $\pi_A(x_i) = 1 - x - y$, then it is sufficient to prove that for $R, S > 0$, $R \neq S$, the entropy function:

$$f(x,y) = \frac{R \times S}{n(R-S)} \left[(x^S + y^S + (1-x-u)^S)^{\frac{1}{S}} - (x^R + y^R + (1-x-y)^R)^{\frac{1}{R}} \right] \tag{9}$$

where $x, y \in [0,1]$ is an increasing function w.r.t. x and decreasing w.r.t. y.

Taking the partial derivative of f with respect to x and y respectively, we get:

$$\frac{\partial f}{\partial x} = \frac{R \times S}{n(R-S)} \left[\begin{array}{l} \left(x^S(x_i) + y^S(x_i) + (1-x-y)^S(x_i)\right)^{\frac{1-S}{S}} \left(x^{S-1}(x_i) - (1-x-y)^{S-1}\right) \\ - \left(x^R(x_i) + y^R(x_i) + (1-x-y)^R(x_i)\right)^{\frac{1-R}{R}} \left(x^{R-1}(x_i) - (1-x-y)^{R-1}\right) \end{array} \right] \tag{10}$$

and:

$$\frac{\partial f}{\partial y} = \frac{R \times S}{n(R-S)} \left[\begin{array}{l} \left(x^S(x_i) + y^S(x_i) + (1-x-y)^S(x_i)\right)^{\frac{1-S}{S}} \left(y^{S-1}(x_i) - (1-x-y)^{S-1}\right) \\ - \left(x^R(x_i) + y^R(x_i) + (1-x-y)^R(x_i)\right)^{\frac{1-R}{R}} \left(y^{R-1}(x_i) - (1-x-y)^{R-1}\right) \end{array} \right] \tag{11}$$

For the extreme point of f, we set $\frac{\partial f}{\partial x} = 0$ and $\frac{\partial f}{\partial y} = 0$ and get $x = y = \frac{1}{3}$.

Furthermore, $\frac{\partial f}{\partial x} \geq 0$, when $x \leq y$ such that $R, S > 0$, $R \neq S$, i.e., $f(x,y)$ is increasing with $x \leq y$, and $\frac{\partial f}{\partial x} \leq 0$ is decreasing with respect to x, when $x \geq y$. On the other hand, $\frac{\partial f}{\partial y} \geq 0$ and $\frac{\partial f}{\partial y} \leq 0$ when $x \geq y$ and $x \leq y$, respectively.

Further, since $H_R^S(A)$ is a concave function on the IFS A, therefore, if $\max\{\zeta_A(x), \vartheta_A(x)\} \leq \frac{1}{3}$, then $\zeta_A(x_i) \leq \zeta_(x_i)$ and $\vartheta_A(x_i) \leq \vartheta_B(x_i)$, which implies that:

$$\zeta_A(x_i) \leq \zeta_B(x_i) \leq \frac{1}{3}; \vartheta_A(x_i) \leq \vartheta_B(x_i) \leq \frac{1}{3}; \pi_A(x_i) \geq \pi_B(x_i) \geq \frac{1}{3}$$

Thus, we observe that $(\zeta_B(x_i), \vartheta_B(x_i), \pi_B(x_i))$ is more around $(\frac{1}{3}, \frac{1}{3}, \frac{1}{3})$ than $(\zeta_A(x_i), \vartheta_A(x_i), \pi_A(x_i))$. Hence, $H_R^S(A) \leq H_R^B(B)$.

Similarly, if $\min\{\zeta_A(x_i), \vartheta_A(x_i)\} \geq \frac{1}{3}$, then we get $H_R^S(A) \leq H_R^B(B)$.

4. Symmetry: By the definition of $H_R^S(A)$, we can easily obtain that $H_R^S(A^c) = H_R^S(A)$.

Hence $H_R^S(A)$ satisfies all the properties of the intuitionistic fuzzy information measure and, therefore, is a valid measure of intuitionistic fuzzy entropy. □

Consider two IFSs A and B defined over $X = \{x_1, x_2, \ldots, x_n\}$. Take the disjoint partition of X as:

$$\begin{aligned} X_1 &= \{x_i \in X \mid A \subseteq B\}, \\ &= \{x_i \in X \mid \zeta_A(x) \leq \zeta_B(x); \vartheta_A(x) \geq \vartheta_B(x)\} \end{aligned}$$

and:

$$\begin{aligned} X_2 &= \{x_i \in X \mid A \supseteq B\} \\ &= \{x_i \in X \mid \zeta_A(x) \geq \zeta_B(x); \vartheta_A(x) \leq \vartheta_B(x)\} \end{aligned}$$

Next, we define the joint and conditional entropies between IFSs A and B as follows:

1. Joint entropy:

$$H_R^S(A \cup B) = \frac{R \times S}{n(R-S)} \sum_{i=1}^{n} \begin{bmatrix} \left(\zeta_{A\cup B}^S(x_i) + \vartheta_{A\cup B}^S(x_i) + (1 - \zeta_{A\cup B}(x_i) - \vartheta_{A\cup B}(x_i))^S\right)^{\frac{1}{S}} \\ - \left(\zeta_{A\cup B}^R(x_i) + \vartheta_{A\cup B}^R(x_i) + (1 - \zeta_{A\cup B}(x_i) - \vartheta_{A\cup B}(x_i))^R\right)^{\frac{1}{R}} \end{bmatrix}$$

$$= \frac{R \times S}{n(R-S)} \sum_{x_i \in X_1} \begin{bmatrix} \left(\zeta_B^S(x_i) + \vartheta_B^S(x_i) + (1 - \zeta_B(x_i) - \vartheta_B(x_i))^S\right)^{\frac{1}{S}} \\ - \left(\zeta_B^R(x_i) + \vartheta_B^R(x_i) + (1 - \zeta_B(x_i) - \vartheta_B(x_i))^R\right)^{\frac{1}{R}} \end{bmatrix}$$

$$+ \frac{R \times S}{n(R-S)} \sum_{x_i \in X_2} \begin{bmatrix} \left(\zeta_A^S(x_i) + \vartheta_A^S(x_i) + (1 - \zeta_A(x_i) - \vartheta_A(x_i))^S\right)^{\frac{1}{S}} \\ - \left(\zeta_A^R(x_i) + \vartheta_A^R(x_i) + (1 - \zeta_A(x_i) - \vartheta_A(x_i))^R\right)^{\frac{1}{R}} \end{bmatrix}$$

2. Conditional entropy:

$$H_R^S(A|B) = \frac{R \times S}{n(R-S)} \sum_{x_i \in X_2} \begin{bmatrix} \left(\zeta_A^S(x_i) + \vartheta_A^S(x_i) + \pi_A^S(x_i)\right)^{\frac{1}{S}} - \left(\zeta_A^R(x_i) + \vartheta_A^R(x_i) + \pi_A^R(x_i)\right)^{\frac{1}{R}} \\ - \left(\zeta_B^S(x_i) + \vartheta_B^S(x_i) + \pi_B^S(x_i)\right)^{\frac{1}{S}} + \left(\zeta_B^R(x_i) + \vartheta_B^R(x_i) + \pi_B^R(x_i)\right)^{\frac{1}{R}} \end{bmatrix}$$

and:

$$H_R^S(B|A) = \frac{R \times S}{n(R-S)} \sum_{x_i \in X_1} \begin{bmatrix} \left(\zeta_B^S(x_i) + \vartheta_B^S(x_i) + \pi_B^S(x_i)\right)^{\frac{1}{S}} - \left(\zeta_B^R(x_i) + \vartheta_B^R(x_i) + \pi_B^R(x_i)\right)^{\frac{1}{R}} \\ - \left(\zeta_A^S(x_i) + \vartheta_A^S(x_i) + \pi_A^S(x_i)\right)^{\frac{1}{S}} + \left(\zeta_A^R(x_i) + \vartheta_A^R(x_i) + \pi_A^R(x_i)\right)^{\frac{1}{R}} \end{bmatrix}$$

Theorem 2. *Let A and B be the two IFSs defined on universal set $X = \{x_1, x_2, \ldots, x_n\}$, where, $A = \{\langle x_i, \zeta_A(x_i), \vartheta_A(x_i)\rangle \mid x_i \in X\}$ and $B = \{\langle x_i, \zeta_B(x_i), \vartheta_B(x_i)\rangle \mid x_i \in X\}$, such that either $A \subseteq B$ or $A \supseteq B \, \forall \, x_i \in X$, then:*

$$H_R^S(A \cup B) + H_R^S(A \cap B) = H_R^S(A) + H_R^S(B)$$

Proof. Let X_1 and X_2 be the two disjoint sets of X, where,

$$X_1 = \{x \in X : A \subseteq B\}, \quad X_2 = \{x \in X : A \supseteq B\}$$

i.e., for $x_i \in X_1$, we have $\zeta_A(x_i) \le \zeta_B(x_i), \vartheta_A(x_i) \ge \vartheta_B(x_i)$ and $x_i \in X_2$, implying that $\zeta_A(x_i) \ge \zeta_B(x_i), \vartheta_A(x_i) \le \vartheta_B(x_i)$. Therefore,

$$H_R^S(A \cup B) + H_R^S(A \cap B) = \frac{R \times S}{n(R-S)} \sum_{i=1}^{n} \begin{bmatrix} \left(\zeta_{A\cup B}^S(x_i) + \vartheta_{A\cup B}^S(x_i) + (1 - \zeta_{A\cup B}(x_i) - \vartheta_{A\cup B}(x_i))^S\right)^{\frac{1}{S}} \\ - \left(\zeta_{A\cup B}^R(x_i) + \vartheta_{A\cup B}^R(x_i) + (1 - \zeta_{A\cup B}(x_i) - \vartheta_{A\cup B}(x_i))^R\right)^{\frac{1}{R}} \end{bmatrix}$$

$$+ \frac{R \times S}{n(R-S)} \sum_{i=1}^{n} \begin{bmatrix} \left(\zeta_{A\cap B}^S(x_i) + \vartheta_{A\cap B}^S(x_i) + (1 - \zeta_{A\cap B}(x_i) - \vartheta_{A\cap B}(x_i))^S\right)^{\frac{1}{S}} \\ - \left(\zeta_{A\cap B}^R(x_i) + \vartheta_{A\cap B}^R(x_i) + (1 - \zeta_{A\cap B}(x_i) - \vartheta_{A\cap B}(x_i))^R\right)^{\frac{1}{R}} \end{bmatrix}$$

$$= \frac{R \times S}{n(R-S)} \sum_{x_i \in X_1} \left[\left(\zeta_B^S(x_i) + \vartheta_B^S(x_i) + \pi_B^S(x_i)\right)^{\frac{1}{S}} - \left(\zeta_B^R(x_i) + \vartheta_B^R(x_i) + \pi_B^R(x_i)\right)^{\frac{1}{R}}\right]$$

$$+ \frac{R \times S}{n(R-S)} \sum_{x_i \in X_2} \left[\left(\zeta_A^S(x_i) + \vartheta_A^S(x_i) + \pi_A^S(x_i)\right)^{\frac{1}{S}} - \left(\zeta_A^R(x_i) + \vartheta_A^R(x_i) + \pi_A^R(x_i)\right)^{\frac{1}{R}}\right]$$

$$+ \frac{R \times S}{n(R-S)} \sum_{x_i \in X_1} \left[\left(\zeta_A^S(x_i) + \vartheta_A^S(x_i) + \pi_A^S(x_i)\right)^{\frac{1}{S}} - \left(\zeta_A^R(x_i) + \vartheta_A^R(x_i) + \pi_A^R(x_i)\right)^{\frac{1}{R}}\right]$$

$$+ \frac{R \times S}{n(R-S)} \sum_{x_i \in X_2} \left[\left(\zeta_B^S(x_i) + \vartheta_B^S(x_i) + \pi_B^S(x_i)\right)^{\frac{1}{S}} - \left(\zeta_B(x_i)^R + \vartheta_B(x_i)^R + \pi_B^R(x_i)\right)^{\frac{1}{R}}\right]$$

$$= H_R^S(A) + H_R^S(B)$$

□

Theorem 3. *The maximum and minimum values of the entropy $H_R^S(A)$ are independent of the parameters R and S.*

Proof. As from the above theorem, we conclude that the entropy is maximum if and only if A is the most IFS and minimum when A is a crisp set. Therefore, it is enough to show that the value of $H_R^S(A)$ in these conditions is independent of R and S. When A is the most IFS, i.e., $\zeta_A(x_i) = \vartheta_A(x_i)$, for all $x_i \in X$, then $H_R^S(A) = 1$, and when A is a crisp set, i.e., either $\zeta_A(x_i) = 0$, $\vartheta_A(x_i) = 1$ or $\zeta_A(x_i) = 1, \vartheta_A(x_i) = 0$ for all $x_i \in X$, then $H_R^S(A) = 0$. Hence, in both cases, $H_R^S(A)$ is independent of the parameters R and S. □

Remark 1. *From the proposed measure, it is observed that some of the existing measures can be obtained from it by assigning particular cases to R and S. For instance,*

1. *When $\pi_A(x_i) = 0$ for all $x_i \in X$, then the proposed measures reduce to the entropy measure of Joshi and Kumar [32].*
2. *When $R = S$ and $S > 0$, then the proposed measures are reduced by the measure of Taneja [27].*
3. *When $R = 1$ and $R \neq S$, then the measure is equivalent to the R-norm entropy presented by Boekee and Van der Lubbe [28].*
4. *When $R = S = 1$, then the proposed measure is the well-known Shannon's entropy.*
5. *When $S = 1$ and $R \neq S$, then the proposed measure becomes the measure of Bajaj et al. [37].*

Theorem 4. *Let A and B be two IFSs defined over the set X such that either $A \subseteq B$ or $B \subseteq A$, then the following statements hold:*

1. $H_R^S(A \cup B) = H_R^S(A) + H_R^S(B|A)$;
2. $H_R^S(A \cup B) = H_R^S(B) + H_R^S(A|B)$;
3. $H_R^S(A \cup B) = H_R^S(A) + H_R^S(B|A) = H_R^S(B) + H_R^S(A|B)$.

Proof. For two IFSs A and B and by using the definitions of joint, conditional and the proposed entropy measures, we get:

1. Consider:

$$\begin{aligned}
& H_R^S(A \cup B) - H_R^S(A) - H_R^S(B|A) \\
= & \frac{R \times S}{n(R-S)} \sum_{i=1}^{n} \left[\begin{array}{l} \left(\zeta_{A\cup B}^S(x_i) + \vartheta_{A\cup B}^S(x_i) + (1 - \zeta_{A\cup B}(x_i) - \vartheta_{A\cup B}(x_i))^S\right)^{\frac{1}{S}} \\ - \left(\zeta_{A\cup B}^R(x_i) + \vartheta_{A\cup B}^R(x_i) + (1 - \zeta_{A\cup B}(x_i) - \vartheta_{A\cup B}(x_i))^R\right)^{\frac{1}{R}} \end{array} \right] \\
- & \frac{R \times S}{n(R-S)} \sum_{i=1}^{n} \left[\left(\zeta_A^S(x_i) + \vartheta_A^S(x_i) + \pi_A^S(x_i)\right)^{\frac{1}{S}} - \left(\zeta_A^R(x_i) + \vartheta_A^R(x_i) + \pi_A^R(x_i)\right)^{\frac{1}{R}} \right] \\
- & \frac{R \times S}{n(R-S)} \sum_{x_i \in X_1} \left[\begin{array}{l} \left(\zeta_B^S(x_i) + \vartheta_B^S(x_i) + \pi_B^S(x_i)\right)^{\frac{1}{S}} - \left(\zeta_B^R(x_i) + \vartheta_B^R(x_i) + \pi_B^R(x_i)\right)^{\frac{1}{R}} \\ - \left(\zeta_A^S(x_i) + \vartheta_A^S(x_i) + \pi_A^S(x_i)\right)^{\frac{1}{S}} + \left(\zeta_A^R(x_i) + \vartheta_A^R(x_i) + \pi_A^R(x_i)\right)^{\frac{1}{R}} \end{array} \right]
\end{aligned}$$

$$
\begin{aligned}
= &\ \frac{R\times S}{n(R-S)}\sum_{x_i\in X_1}\left[\left(\zeta_B^S(x_i)+\vartheta_B^S(x_i)+\pi_B^S(x_i)\right)^{\frac{1}{S}}-\left(\zeta_B^R(x_i)+\vartheta_B^R(x_i)+\pi_B^R(x_i)\right)^{\frac{1}{R}}\right]\\
+ &\ \frac{R\times S}{n(R-S)}\sum_{x_i\in X_2}\left[\left(\zeta_A^S(x_i)+\vartheta_A^S(x_i)+\pi_A^S(x_i)\right)^{\frac{1}{S}}-\left(\zeta_A^R(x_i)+\vartheta_A^R(x_i)+\pi_A^R(x_i)\right)^{\frac{1}{R}}\right]\\
- &\ \frac{R\times S}{n(R-S)}\sum_{x_i\in X_1}\left[\left(\zeta_A^S(x_i)+\vartheta_A^S(x_i)+\pi_A^S(x_i)\right)^{\frac{1}{S}}-\left(\zeta_A^R(x_i)+\vartheta_A^R(x_i)+\pi_A^R(x_i)\right)^{\frac{1}{R}}\right]\\
+ &\ \frac{R\times S}{n(R-S)}\sum_{x_i\in X_2}\left[\left(\zeta_A^S(x_i)+\vartheta_A^S(x_i)+\pi_A^S(x_i)\right)^{\frac{1}{S}}-\left(\zeta_A^R(x_i)+\vartheta_A^R(x_i)+\pi_A^R(x_i)\right)^{\frac{1}{R}}\right]\\
- &\ \frac{R\times S}{n(R-S)}\sum_{x_i\in X_1}\left[\begin{array}{l}\left(\zeta_B^S(x_i)^{+}\vartheta_B^S(x_i)+\pi_B^S(x_i)\right)^{\frac{1}{S}}-\left(\zeta_B^R(x_i)+\vartheta_B^R(x_i)+\pi_B^R(x_i)\right)^{\frac{1}{R}}\\ -\left(\zeta_A^S(x_i)+\vartheta_A^S(x_i)+\pi_A^S(x_i)\right)^{\frac{1}{S}}+\left(\zeta_A^R(x_i)+\vartheta_A^R(x_i)+\pi_A^R(x_i)\right)^{\frac{1}{R}}\end{array}\right]\\
= &\ 0
\end{aligned}
$$

2. Consider:

$$
\begin{aligned}
&H_R^S(A\cup B)-H_R^S(B)-H_R^S(A|B)\\
= &\ \frac{R\times S}{n(R-S)}\sum_{i=1}^{n}\left[\begin{array}{l}\left(\zeta_{A\cup B}^S(x_i)+\vartheta_{A\cup B}^S(x_i)+(1-\zeta_{A\cup B}(x_i)-\vartheta_{A\cup B}(x_i))^S\right)^{\frac{1}{S}}\\ -\left(\zeta_{A\cup B}^R(x_i)+\vartheta_{A\cup B}^R(x_i)+(1-\zeta_{A\cup B}(x_i)-\vartheta_{A\cup B}(x_i))^R\right)^{\frac{1}{R}}\end{array}\right]\\
- &\ \frac{R\times S}{n(R-S)}\sum_{i=1}^{n}\left[\left(\zeta_B^S(x_i)+\vartheta_B^S(x_i)+\pi_B^S(x_i)\right)^{\frac{1}{S}}-\left(\zeta_B^R(x_i)+\vartheta_B^R(x_i)+\pi_B^R(x_i)\right)^{\frac{1}{R}}\right]\\
- &\ \frac{R\times S}{n(R-S)}\sum_{x_i\in X_2}\left[\begin{array}{l}\left(\zeta_A^S(x_i)^{+}\vartheta_A^S(x_i)+\pi_A^S(x_i)\right)^{\frac{1}{S}}-\left(\zeta_A^R(x_i)+\vartheta_A^R(x_i)+\pi_A^R(x_i)\right)^{\frac{1}{R}}\\ -\left(\zeta_B^S(x_i)+\vartheta_B^S(x_i)+\pi_B^S(x_i)\right)^{\frac{1}{S}}+\left(\zeta_B^R(x_i)+\vartheta_B^R(x_i)+\pi_B^R(x_i)\right)^{\frac{1}{R}}\end{array}\right]\\
= &\ \frac{R\times S}{n(R-S)}\sum_{x_i\in X_1}\left[\left(\zeta_B^S(x_i)+\vartheta_B^S(x_i)+\pi_B^S(x_i)\right)^{\frac{1}{S}}-\left(\zeta_B^R(x_i)+\vartheta_B^R(x_i)+\pi_B^R(x_i)\right)^{\frac{1}{R}}\right]\\
+ &\ \frac{R\times S}{n(R-S)}\sum_{x_i\in X_2}\left[\left(\zeta_A^S(x_i)+\vartheta_A^S(x_i)+\pi_A^S(x_i)\right)^{\frac{1}{S}}-\left(\zeta_A^R(x_i)+\vartheta_A^R(x_i)+\pi_A^R(x_i)\right)^{\frac{1}{R}}\right]\\
- &\ \frac{R\times S}{n(R-S)}\sum_{x_i\in X_1}\left[\left(\zeta_B^S(x_i)+\vartheta_B^S(x_i)+\pi_B^S(x_i)\right)^{\frac{1}{S}}-\left(\zeta_B^R(x_i)+\vartheta_B^R(x_i)+\pi_B^R(x_i)\right)^{\frac{1}{R}}\right]\\
+ &\ \frac{R\times S}{n(R-S)}\sum_{x_i\in X_2}\left[\left(\zeta_B^S(x_i)+\vartheta_B^S(x_i)+\pi_B^S(x_i)\right)^{\frac{1}{S}}-\left(\zeta_B^R(x_i)+\vartheta_B^R(x_i)+\pi_B^R(x_i)\right)^{\frac{1}{R}}\right]\\
- &\ \frac{R\times S}{n(R-S)}\sum_{x_i\in X_2}\left[\begin{array}{l}\left(\zeta_A^S(x_i)+\vartheta_A^S(x_i)+\pi_A^S(x_i)\right)^{\frac{1}{S}}-\left(\zeta_A^R(x_i)+\vartheta_A^R(x_i)+\pi_A^R(x_i)\right)^{\frac{1}{R}}\\ \left(\zeta_B^S(x_i)+\vartheta_B^S(x_i)+\pi_B^S(x_i)\right)^{\frac{1}{S}}-\left(\zeta_B^R(x_i)+\vartheta_B^R(x_i)+\pi_B^R(x_i)\right)^{\frac{1}{R}}\end{array}\right]\\
= &\ 0
\end{aligned}
$$

3. This can be deduced from Parts (1) and (2).

□

Before elaborating on the comparison between the proposed entropy function and other entropy functions, we state a definition [56] for an IFS of the form $A=\langle x,\zeta_A(x_i),\vartheta_A(x_i)\mid x\in X\rangle$ defined on universal set X, which is as follows:

$$
A^n=\{\langle x,[\zeta_A(x_i)]^n,1-[1-\vartheta_A(x_i)]^n\rangle\mid x\in X\} \tag{12}
$$

Definition 5. *The concentration of an IFS A of the universe X is denoted by CON(A) and is defined by:*

$$CON(A) = \{\langle x, \zeta_{CON(A)}(x), \vartheta_{CON(A)}(x)\rangle \mid x \in X\}$$

where $\zeta_{(CON(A))}(x) = [\zeta_A(x)]^2$, $\vartheta_{CON(A)}(x)) = 1 - [1 - \vartheta_A(x)]^2$, *i.e., the operation of the concentration of an IFS is defined by* $CON(A) = A^2$.

Definition 6. *The dilation of an IFS A of the universe X is denoted by DIL(A) and is defined by:*

$$DIL(A) = \{\langle x, \zeta_{DIL(A)}(x), \vartheta_{DIL(A)}(x)\rangle \mid x \in X\}$$

where $\zeta_{DIL(A)}(x) = [\zeta_A(x)]^{1/2}$ *and* $\vartheta_{DIL(A)}(x) = 1 - [1 - \vartheta_A(x)]^{1/2}$, *i.e., the operation of the dilation of an IFS is defined by* $DIL(A) = A^{1/2}$

Example 1. *Consider a universe of the discourse* $X = \{x_1, x_2, x_3, x_4, x_5\}$, *and an IFS A "LARGE" of X may be defined by:*

$$LARGE = \{(x_1, 0.1, 0.8), (x_2, 0.3, 0.5), (x_3, 0.5, 0.4), (x_4, 0.9, 0), (x_5, 1, 0)\}$$

Using the operations as defined in Equation (12), we have generated the following IFSs

$$A^{1/2}, A^2, A^3, A^4,$$

which are defined as follows:

$A^{1/2}$ may be treated as "More or less LARGE"
A^2 may be treated as "very LARGE"
A^3 may be treated as "quite very LARGE"
A^4 may be treated as "very very LARGE"

and their corresponding sets are computed as:

$$\begin{aligned}
A^{\frac{1}{2}} &= \{(x_1, 0.3162, 0.5528), (x_2, 0.5477, 0.2929), (x_3, 0.7071, 0.2254), (x_4, 0.9487, 0), (x_5, 1, 0)\} \\
A^2 &= \{(x_1, 0.01, 0.96), (x_2, 0.09, 0.75), (x_3, 0.25, 0.64), (x_4, 0.81, 0), (x_5, 1, 0)\} \\
A^3 &= \{(x_1, 0.001, 0.9920), (x_2, 0.0270, 0.8750), (x_3, 0.1250, 0.7840), (x_4, 0.7290, 0), (x_5, 1, 0)\} \\
A^4 &= \{(x_1, 0.0001, 0.9984), (x_2, 0.0081, 0.9375), (x_3, 0.0625, 0.8704), (x_4, 0.6561, 0), (x_5, 1, 0)\}
\end{aligned}$$

From the viewpoint of mathematical operations, the entropy values of the above defined IFSs, $A^{1/2}$, A, A^2, A^3 and A^4, have the following requirement:

$$E(A^{1/2}) > E(A) > E(A^2) > E(A^3) > E(A^4) \tag{13}$$

Based on the dataset given in the above, we compute the entropy measure for them at different values of R and S. The result corresponding to these different pairs of values is summarized in Table 1 along with the existing approaches' results. From these computed values, it is observed that the ranking order of the linguistic variable by the proposed entropy follows the pattern as described in Equation (13) for some suitable pairs of (R, S), while the performance order pattern corresponding to [19,21,57] and [58] is $E(A) > E(A^{1/2}) > E(A^2) > E(A^3) > E(A^4)$, which does not satisfy the requirement given in Equation (13). Hence, the proposed entropy measure is a good alternative and performs better than the existing measures. Furthermore, for different pairs of (R, S), a decision-maker may have more choices to access the alternatives from the viewpoint of structured linguistic variables.

Table 1. Entropy measures values corresponding to existing approaches, as well as the proposed approach.

Entropy Measure	$A^{\frac{1}{2}}$	A	A^2	A^3	A^4
$E_{\{BB\}}$ [21]	0.0818	0.1000	0.0980	0.0934	0.0934
$E_{\{SK\}}$ [19]	0.3446	0.3740	0.1970	0.1309	0.1094
$E_{\{ZL\}}$ [57]	0.4156	0.4200	0.2380	0.1546	0.1217
$E_{\{HY\}}$ [58]	0.3416	0.3440	0.2610	0.1993	0.1613
$E_{\{ZJ\}}$ [25]	0.2851	0.3050	0.1042	0.0383	0.0161
$E_{0.4}^{0.2}$ [22]	0.5995	0.5981	0.5335	0.4631	0.4039
H_R^S (proposed measure)					
$R = 0.3, S = 2$	2.3615	2.3589	1.8624	1.4312	1.1246
$R = 0.5, S = 2$	0.8723	0.8783	0.6945	0.5392	0.4323
$R = 0.7, S = 2$	0.5721	0.5769	0.4432	0.3390	0.2725
$R = 2.5, S = 0.3$	2.2882	2.2858	1.8028	1.3851	1.0890
$R = 2.5, S = 0.5$	0.8309	0.8368	0.6583	0.5104	0.4103
$R = 2.5, S = 0.7$	0.5369	0.5415	0.4113	0.3138	0.2538

4. MADM Problem Based on the Proposed Entropy Measure

In this section, we present a method for solving the MADM problem based on the proposed entropy measure.

4.1. Approach I: When the Attribute Weight Is Completely Unknown

In this section, we present a decision-making approach for solving the multi-attribute decision-making problem in the intuitionistic fuzzy set environment. For this, consider a set of 'n' different alternatives, denoted by $A_1, A_2, \ldots, A_n$, which are evaluated by a decision-maker under the 'm' different attributes $G_1, G_2, \ldots, G_m$. Assume that a decision-maker has evaluated these alternatives in the intuitionistic fuzzy environment and noted their rating values in the form of the IFNs $\alpha_{ij} = \langle \zeta_{ij}, \vartheta_{ij} \rangle$ where ζ_{ij} denotes that the degree of the alternative A_i satisfies under the attribute G_j, while ϑ_{ij} denotes the dissatisfactory degree of an alternative A_i under G_j such that $\zeta_{ij}, \vartheta_{ij} \in [0,1]$ and $\zeta_{ij} + \vartheta_{ij} \leq 1$ for $i = 1, 2, \ldots, m$ and $j = 1, 2, \ldots, n$. Further assume that the weight vector $\omega_j (j = 1, 2, \ldots, m)$ of each attribute is completely unknown. Hence, based on the decision-maker preferences α_{ij}, the collective values are summarized in the form of the decision matrix D as follows:

$$D = \begin{array}{c} \\ A_1 \\ A_2 \\ \vdots \\ A_n \end{array} \begin{array}{c} \begin{array}{cccc} G_1 & G_2 & \ldots & G_m \end{array} \\ \begin{pmatrix} \langle \zeta_{11}, \vartheta_{11} \rangle & \langle \zeta_{12}, \vartheta_{12} \rangle & \ldots & \langle \zeta_{1m}, \vartheta_{1m} \rangle \\ \langle \zeta_{21}, \vartheta_{21} \rangle & \langle \zeta_{22}, \vartheta_{22} \rangle & \ldots & \langle \zeta_{2m}, \vartheta_{2m} \rangle \\ \vdots & \vdots & \ddots & \vdots \\ \langle \zeta_{n1}, \vartheta_{n1} \rangle & \langle \zeta_{n2}, \vartheta_{n2} \rangle & \ldots & \langle \zeta_{nm}, \vartheta_{nm} \rangle \end{pmatrix} \end{array} \tag{14}$$

Then, the following steps of the proposed approach are summarized to find the best alternative(s).

Step 1: Normalize the rating values of the decision-maker, if required, by converting the rating values corresponding to the cost type attribute into the benefit type. For this, the following normalization formula is used:

$$r_{ij} = \begin{cases} \langle \zeta_{ij}, \vartheta_{ij} \rangle & ; \quad \text{if the benefit type attribute} \\ \langle \vartheta_{ij}, \zeta_{ij} \rangle & ; \quad \text{if the cost type attribute} \end{cases} \tag{15}$$

and hence, we obtain the normalized IF decision matrix $R = (r_{ij})_{n \times m}$.

Step 2: Based on the matrix R, the information entropy of attribute $G_j (j = 1, 2, \ldots, m)$ is computed as:

$$(H_R^S)_j = \frac{R \times S}{n(R-S)} \sum_{i=1}^{n} \left[\left(\zeta_{ij}^S + \vartheta_{ij}^S + \pi_{ij}^S \right)^{\frac{1}{S}} - \left(\zeta_{ij}^R + \vartheta_{ij}^R + \pi_{ij}^R \right)^{\frac{1}{R}} \right] \tag{16}$$

where $R, S > 0$ and $R \neq S$.

Step 3: Based on the entropy matrix, $H_R^S(\alpha_{ij})$ defined in Equation (16), the degree of divergence (d_j) of the average intrinsic information provided by the correspondence on the attribute G_j can be defined as $d_j = 1 - \kappa_j$ where $\kappa_j = \sum_{i=1}^{n} H_R^S(\alpha_{ij}), j = 1, 2, \ldots, m$. Here, the value of d_j represents the inherent contrast intensity of attribute G_j, and hence, based on this, the attributes weight $\omega_j (j = 1, 2, \ldots, n)$ is given as:

$$\omega_j = \frac{d_j}{\sum_{j=1}^{m} d_j} = \frac{1 - \kappa_j}{\sum_{j=1}^{m} (1 - \kappa_j)} = \frac{1 - \kappa_j}{m - \sum_{j=1}^{m} \kappa_j} \tag{17}$$

Step 4: Construct the weighted sum of each alternative by multiplying the score function of each criterion by its assigned weight as:

$$Q(A_i) = \sum_{j=1}^{m} \omega_j (\zeta_{ij} - \vartheta_{ij}); \quad i = 1, 2, \ldots, n \tag{18}$$

Step 5: Rank all the alternatives $A_i (i = 1, 2, \ldots, n)$ according to the highest value of $Q(A_i)$ and, hence, choose the best alternative.

The above-mentioned approach has been illustrated with a practical example of the decision-maker, which can be read as:

Example 2. *Consider a decision-making problem from the field of the recruitment sector. Assume that a pharmaceutical company wants to select a lab technician for a micro-bio laboratory. For this, the company has published a notification in a newspaper and considered the four attributes required for technician selection, namely academic record* (G_1)*, personal interview evaluation* (G_2)*, experience* (G_3) *and technical capability* (G_4)*. On the basis of the notification conditions, only five candidates* A_1, A_2, A_3, A_4 *and* A_5 *as alternatives are interested and selected to be presented to the panel of experts for this post. Then, the main object of the company is to choose the best candidate among them for the task. In order to describe the ambiguity and uncertainties in the data, the preferences related to each alternative are represented in the IFS environment. The preferences of each alternative are represented in the form of IFNs as follows:*

$$D = \begin{array}{c} \\ A_1 \\ A_2 \\ A_3 \\ A_4 \\ A_5 \end{array} \begin{array}{c} \begin{array}{cccc} G_1 & G_2 & G_3 & G_4 \end{array} \\ \begin{pmatrix} \langle 0.7, 0.2 \rangle & \langle 0.5, 0.4 \rangle & \langle 0.6, 0.2 \rangle & \langle 0.6, 0.3 \rangle \\ \langle 0.7, 0.1 \rangle & \langle 0.5, 0.2 \rangle & \langle 0.7, 0.2 \rangle & \langle 0.4, 0.5 \rangle \\ \langle 0.6, 0.3 \rangle & \langle 0.5, 0.1 \rangle & \langle 0.5, 0.3 \rangle & \langle 0.6, 0.2 \rangle \\ \langle 0.8, 0.1 \rangle & \langle 0.6, 0.3 \rangle & \langle 0.3, 0.7 \rangle & \langle 0.6, 0.3 \rangle \\ \langle 0.6, 0.3 \rangle & \langle 0.4, 0.6 \rangle & \langle 0.7, 0.2 \rangle & \langle 0.5, 0.4 \rangle \end{pmatrix} \end{array} \tag{19}$$

Then, the steps of the proposed approach are followed to find the best alternative(s) as below:

Step 1: Since all the attributes are of the same type, so there is no need for the normalization process.

Step 2: Without loss of generality, we take $R = 0.3$ and $S = 2$ and, hence, compute the entropy measurement value for each attribute by using Equation (16). The results corresponding to it are $H_R^S(G_1) = 3.4064$, $H_R^S(G_2) = 3.372$, $H_R^S(G_3) = 3.2491$ and $H_R^S(G_4) = 3.7564$.

Step 3: Based on these entropy values, the weight of each criterion is calculated as $\omega = (0.2459, 0.2425, 0.2298, 0.2817)^T$.

Step 4: The overall weighted score values of the alternative corresponding to $R = 0.3$, $S = 2$ and $\omega = (0.2459, 0.2425, 0.2298, 0.2817)^T$ obtained by using Equation (18) are $Q(A_1) = 0.3237$, $Q(A_2) = 0.3071$, $Q(A_3) = 0.3294$, $Q(A_4) = 0.2375$ and $Q(A_5) = 0.1684$.

Step 5: Since $Q(A_3) > Q(A_1) > Q(A_2) > Q(A_4) > Q(A_5)$, hence the ranking order of the alternatives is $A_3 \succ A_1 \succ A_2 \succ A_4 \succ A_5$. Thus, the best alternative is A_3.

However, in order to analyze the influence of the parameters R and S on the final ranking order of the alternatives, the steps of the proposed approach are executed by varying the values of R from 0.1 to 1.0 and S from 1.0 to 5.0. The overall score values of each alternative along with the ranking order are summarized in Table 2. From this analysis, we conclude that the decision-maker can plan to choose the values of R and S and, hence, their respective alternatives according to his goal. Therefore, the proposed measures give various choices to the decision-maker to reach the target.

Table 2. Effect of R and S on the entropy measure H_R^S by using Approach I.

S	R	$H_R^S(A_1)$	$H_R^S(A_2)$	$H_R^S(A_3)$	$H_R^S(A_4)$	$H_R^S(A_5)$	Ranking Order
	0.1	0.3268	0.3084	0.3291	0.2429	0.1715	$A_3 \succ A_1 \succ A_2 \succ A_4 \succ A_5$
	0.3	0.3241	0.3081	0.3292	0.2374	0.1690	$A_3 \succ A_1 \succ A_2 \succ A_4 \succ A_5$
1.2	0.5	0.3165	0.2894	0.3337	0.2368	0.1570	$A_3 \succ A_1 \succ A_2 \succ A_4 \succ A_5$
	0.7	0.1688	-0.0988	0.4296	0.2506	-0.0879	$A_3 \succ A_4 \succ A_1 \succ A_5 \succ A_2$
	0.9	0.3589	0.3992	0.3065	0.2328	0.2272	$A_2 \succ A_1 \succ A_3 \succ A_4 \succ A_5$
	0.1	0.3268	0.3084	0.3291	0.2429	0.1715	$A_3 \succ A_1 \succ A_2 \succ A_4 \succ A_5$
	0.3	0.3239	0.3076	0.3293	0.2374	0.1688	$A_3 \succ A_1 \succ A_2 \succ A_4 \succ A_5$
1.5	0.5	0.3132	0.2811	0.3359	0.2371	0.1515	$A_3 \succ A_1 \succ A_2 \succ A_4 \succ A_5$
	0.7	0.4139	0.5404	0.2712	0.2272	0.3185	$A_2 \succ A_1 \succ A_5 \succ A_3 \succ A_2$
	0.9	0.3498	0.3741	0.3125	0.2334	0.2121	$A_2 \succ A_1 \succ A_3 \succ A_4 \succ A_5$
	0.1	0.3268	0.3084	0.3291	0.2429	0.1715	$A_3 \succ A_1 \succ A_2 \succ A_4 \succ A_5$
	0.3	0.3237	0.3071	0.3294	0.2375	0.1684	$A_3 \succ A_1 \succ A_2 \succ A_4 \succ A_5$
2.0	0.5	0.3072	0.2666	0.3396	0.2381	0.1415	$A_3 \succ A_1 \succ A_2 \succ A_4 \succ A_5$
	0.7	0.3660	0.4140	0.3022	0.2308	0.2393	$A_2 \succ A_1 \succ A_3 \succ A_5 \succ A_4$
	0.9	0.3461	0.3631	0.3150	0.2331	0.2062	$A_2 \succ A_1 \succ A_3 \succ A_4 \succ A_5$
	0.1	0.3268	0.3084	0.3291	0.2429	0.1715	$A_3 \succ A_1 \succ A_2 \succ A_4 \succ A_5$
	0.3	0.3235	0.3067	0.3295	0.2376	0.1681	$A_3 \succ A_1 \succ A_2 \succ A_4 \succ A_5$
2.5	0.5	0.3010	0.2517	0.3436	0.2396	0.1308	$A_3 \succ A_1 \succ A_2 \succ A_4 \succ A_5$
	0.7	0.3578	0.3920	0.3074	0.2304	0.2261	$A_2 \succ A_1 \succ A_3 \succ A_4 \succ A_5$
	0.9	0.3449	0.3591	0.3158	0.2322	0.2045	$A_2 \succ A_1 \succ A_3 \succ A_4 \succ A_5$
	0.1	0.3268	0.3084	0.3291	0.2429	0.1715	$A_3 \succ A_1 \succ A_2 \succ A_4 \succ A_5$
	0.3	0.3234	0.3064	0.3296	0.2376	0.1678	$A_3 \succ A_1 \succ A_2 \succ A_4 \succ A_5$
3.0	0.5	0.2946	0.2368	0.3476	0.2417	0.1199	$A_3 \succ A_1 \succ A_4 \succ A_2 \succ A_5$
	0.7	0.3545	0.3829	0.3095	0.2298	0.2209	$A_2 \succ A_1 \succ A_3 \succ A_4 \succ A_5$
	0.9	0.3442	0.3570	0.3161	0.2314	0.2037	$A_2 \succ A_1 \succ A_3 \succ A_4 \succ A_5$
	0.1	0.3268	0.3084	0.3291	0.2429	0.1715	$A_3 \succ A_1 \succ A_2 \succ A_4 \succ A_5$
	0.3	0.3231	0.3058	0.3298	0.2379	0.1674	$A_3 \succ A_1 \succ A_2 \succ A_4 \succ A_5$
5.0	0.5	0.2701	0.1778	0.3638	0.2520	0.0767	$A_3 \succ A_1 \succ A_4 \succ A_2 \succ A_5$
	0.7	0.3496	0.3706	0.3123	0.2277	0.2137	$A_2 \succ A_1 \succ A_3 \succ A_4 \succ A_5$
	0.9	0.3428	0.3532	0.3168	0.2293	0.2020	$A_2 \succ A_1 \succ A_3 \succ A_4 \succ A_5$

4.2. Approach II: When the Attribute Weight Is Partially Known

In this section, we present an approach for solving the multi-attribute decision-making problem in the IFS environment where the information about the attribute weight is partially known. The description of the MADM problem is mentioned in Section 4.1.

Since decision-making during a real-life situation is highly complex due to a large number of constraints, human thinking is inherently subjective, and the importance of the attribute weight vector is incompletely known. In order to represent this incomplete information about the weights, the following relationship has been defined for $i \neq j$:

1. A weak ranking: $\omega_i \geq \omega_j$;
2. A strict ranking: $\omega_i - \omega_j \geq \sigma_i$; $(\sigma_i > 0)$.
3. A ranking with multiples: $\omega_i \geq \sigma_i \omega_j$, $(0 \leq \sigma_i \leq 1)$;
4. An interval form: $\lambda_i \leq \omega_i \leq \lambda_i + \delta_i$, $(0 \leq \lambda_i \leq \lambda_i + \delta_i \leq 1)$;
5. A ranking of differences: $\omega_i - \omega_j \geq \omega_k - \omega_l$, $(j \neq k \neq l)$.

The set of this known weight information is denoted by Δ in this paper.

Then, the proposed approach is summarized in the following steps to obtain the most desirable alternative(s).

Step 1: Similar to Approach I.
Step 2: similar to Approach I.
Step 3: The overall entropy of the alternative $A_i (i = 1, 2, \ldots, n)$ for the attribute G_j is given by:

$$\begin{aligned} H(A_i) &= \sum_{j=1}^{m} H_R^S(\alpha_{ij}) \\ &= \frac{R \times S}{n(R-S)} \sum_{j=1}^{m} \left\{ \sum_{i=1}^{n} \left((\zeta_{ij}^S + \vartheta_{ij}^S + \pi_{ij}^S)^{\frac{1}{S}} - (\zeta_{ij}^R + \vartheta_{ij}^R + \pi_{ij}^R)^{\frac{1}{R}} \right) \right\} \end{aligned} \quad (20)$$

where $R, S > 0$ and $R \neq S$.

By considering the importance of each attribute in terms of weight vector $\omega = (\omega_1, \omega_2, \ldots, \omega_m)^T$, we formulate a linear programming model to determine the weight vector as follows:

$$\begin{aligned} \min H &= \sum_{i=1}^{n} H(A_i) = \sum_{i=1}^{n} \left\{ \sum_{j=1}^{m} \omega_j H_R^S(\alpha_{ij}) \right\} \\ &= \frac{R \times S}{n(R-S)} \sum_{j=1}^{m} \omega_j \left\{ \sum_{i=1}^{n} \left((\zeta_{ij}^S + \vartheta_{ij}^S + \pi_{ij}^S)^{\frac{1}{S}} - (\zeta_{ij}^R + \vartheta_{ij}^R + \pi_{ij}^R)^{\frac{1}{R}} \right) \right\} \\ \text{s.t.} \quad & \sum_{j=1}^{m} \omega_j = 1 \\ & \omega_j \geq 0; \omega \in \Delta \end{aligned}$$

After solving this model, we get the optimal weight vector $\omega = (\omega_1, \omega_2, \ldots, \omega_m)^T$.

Step 4: Construct the weighted sum of each alternative by multiplying the score function of each criterion by its assigned weight as:

$$Q(A_i) = \sum_{j=1}^{m} \omega_j (\zeta_{ij} - \vartheta_{ij}); \quad i = 1, 2, \ldots, n \quad (21)$$

Step 5: Rank all the alternative $A_i (i = 1, 2, \ldots, n)$ according to the highest value of $Q(A_i)$ and, hence, choose the best alternative.

To demonstrate the above-mentioned approach, a numerical example has been taken, which is stated as below.

Example 3. *Consider an MADM problem, which was stated and described in Example 2, where the five alternatives $A_1, A_2, \ldots, A_5$ are assessed under the four attributes G_1, G_2, G_3, G_4 in the IFS environment. Here, we assume that the information about the attribute weight is partially known and is given by the decision-maker* *as* $\Delta = \{0.15 \leq \omega_1 \leq 0.45, 0.2 \leq \omega_2 \leq 0.5, 0.1 \leq \omega_3 \leq 0.3, 0.1 \leq \omega_4 \leq 0.2, \omega_1 \geq \omega_4, \sum_{j=1}^{4} \omega_j = 1\}$. *Then, based on the rating values as mentioned in Equation (19), the following steps of the Approach II are executed as below:*

Step 1: All the attributes are te same types, so there is no need for normalization.

Step 2: Without loss of generality, we take $R = 0.3$ *and* $S = 2$ *and, hence, compute the entropy measurement value for each attribute by using Equation (20). The results corresponding to it are* $H_R^S(G_1) = 3.4064$, $H_R^S(G_2) = 3.372$, $H_R^S(G_3) = 3.2491$ *and* $H_R^S(G_4) = 3.7564$.

Step 3: Formulate the optimization model by utilizing the information of rating values and the partial information of the weight vector $\Delta = \{0.15 \leq \omega_1 \leq 0.45, 0.2 \leq \omega_2 \leq 0.5, 0.1 \leq \omega_3 \leq 0.3, 0.1 \leq \omega_4 \leq 0.2, \omega_1 \geq \omega_4, \sum_{j=1}^{4} \omega_j = 1\}$ *as:*

$$\begin{aligned} \min H &= 3.4064\omega_1 + 3.372\omega_2 + 3.2491\omega_3 + 3.7564\omega_4 \\ \textit{subject to} \quad & 0.15 \leq \omega_1 \leq 0.45, \\ & 0.2 \leq \omega_2 \leq 0.5, \\ & 0.1 \leq \omega_3 \leq 0.3, \\ & 0.1 \leq \omega_4 \leq 0.2, \\ & \omega_1 \geq \omega_4, \\ \textit{and} \quad & \omega_1 + \omega_2 + \omega_3 + \omega_4 = 1. \end{aligned}$$

Hence, we solve the model with the help of MATLAB software, and we can obtain the weight vector as $\omega = (0.15, 0.45, 0.30, 0.10)^T$.

Step 4: The overall weighted score values of the alternative corresponding to $R = 0.3$, $S = 2$ *and* $\omega = (0.15, 0.45, 0.30, 0.10)^T$ *obtained by using Equation (21) are* $Q(A_1) = 0.2700$, $Q(A_2) = 0.3650$, $Q(A_3) = 0.3250$ *and* $Q(A_4) = 0.1500$ *and* $Q(A_5) = 0.1150$.

Step 5: Since $Q(A_2) > Q(A_3) > Q(A_1) > Q(A_4) > Q(A_5)$, *hence the ranking order of the alternatives is* $A_2 \succ A_3 \succ A_1 \succ A_4 \succ A_5$. *Thus, the best alternative is* A_2.

5. Conclusions

In this paper, we propose an entropy measure based on the (R, S)-norm in the IFS environment. Since the uncertainties present in the data play a crucial role during the decision-making process, in order to measure the degree of fuzziness of a set and maintaining the advantages of it, in the present paper, we addressed a novel (R, S)-norm-based information measure. Various desirable relations, as well as some of its properties, were investigated in detail. From the proposed measures, it was observed that some of the existing measures were the special cases of the proposed measures. Furthermore, based on the different parametric values of R and S, the decision-maker(s) may have different choices to make a decision according to his/her choice. In addition to these and to explore the structural characteristics and functioning of the proposed measures, two decision-making approaches were presented to solve the MADM problems in the IFS environment under the characteristics that attribute weights are either partially known or completely unknown. The presented approaches were illustrated with numerical examples. The major advantages of the proposed measure are that it gives various choices to select the best alternatives, according to the decision-makers' desired goals, and hence, it makes the decision-makers more flexible and reliable. From the studies, it is concluded that the proposed work provides a new and easy way to handle the uncertainty and vagueness in the data and, hence, provides an alternative way to solve the decision-making problem in the IFS environment. In the future, the result of this paper can be extended to some other uncertain and fuzzy environments [59–62].

Author Contributions: Conceptualization, Methodology, Validation, H.G.; Formal Analysis, Investigation, H.G., J.K.; Writing-Original Draft Preparation, H.G.; Writing-Review & Editing, H.G.; Visualization, H.G.

Conflicts of Interest: The authors declare no conflict of interest.

References

1. Zadeh, L.A. Fuzzy sets. *Inf. Control* **1965**, *8*, 338–353. [CrossRef]
2. Atanassov, K.T. Intuitionistic fuzzy sets. *Fuzzy Sets Syst.* **1986**, *20*, 87–96. [CrossRef]
3. Atanassov, K.; Gargov, G. Interval-valued intuitionistic fuzzy sets. *Fuzzy Sets Syst.* **1989**, *31*, 343–349. [CrossRef]
4. Xu, Z.S.; Yager, R.R. Some geometric aggregation operators based on intuitionistic fuzzy sets. *Int. J. Gen. Syst.* **2006**, *35*, 417–433. [CrossRef]
5. Xu, Z.S. Intuitionistic fuzzy aggregation operators. *IEEE Trans. Fuzzy Syst.* **2007**, *15*, 1179–1187.
6. Garg, H. Generalized intuitionistic fuzzy interactive geometric interaction operators using Einstein t-norm and t-conorm and their application to decision-making. *Comput. Ind. Eng.* **2016**, *101*, 53–69. [CrossRef]
7. Garg, H. Novel intuitionistic fuzzy decision-making method based on an improved operation laws and its application. *Eng. Appl. Artif. Intell.* **2017**, *60*, 164–174. [CrossRef]
8. Wang, W.; Wang, Z. An approach to multi-attribute interval-valued intuitionistic fuzzy decision-making with incomplete weight information. In Proceedings of the 15th IEEE International Conference on Fuzzy Systems and Knowledge Discovery, Jinan, China, 18–20 October 2008; Volume 3, pp. 346–350.
9. Wei, G. Some induced geometric aggregation operators with intuitionistic fuzzy information and their application to group decision-making. *Appl. Soft Comput.* **2010**, *10*, 423–431. [CrossRef]
10. Arora, R.; Garg, H. Robust aggregation operators for multi-criteria decision-making with intuitionistic fuzzy soft set environment. *Sci. Iran. E* **2018**, *25*, 931–942. [CrossRef]
11. Arora, R.; Garg, H. Prioritized averaging/geometric aggregation operators under the intuitionistic fuzzy soft set environment. *Sci. Iran.* **2018**, *25*, 466–482. [CrossRef]
12. Zhou, W.; Xu, Z. Extreme intuitionistic fuzzy weighted aggregation operators and their applications in optimism and pessimism decision-making processes. *J. Intell. Fuzzy Syst.* **2017**, *32*, 1129–1138. [CrossRef]
13. Garg, H. Some robust improved geometric aggregation operators under interval-valued intuitionistic fuzzy environment for multi-criteria decision -making process. *J. Ind. Manag. Optim.* **2018**, *14*, 283–308. [CrossRef]
14. Xu, Z.; Gou, X. An overview of interval-valued intuitionistic fuzzy information aggregations and applications. *Granul. Comput.* **2017**, *2*, 13–39. [CrossRef]
15. Jamkhaneh, E.B.; Garg, H. Some new operations over the generalized intuitionistic fuzzy sets and their application to decision-making process. *Granul. Comput.* **2018**, *3*, 111–122. [CrossRef]
16. Garg, H.; Singh, S. A novel triangular interval type-2 intuitionistic fuzzy sets and their aggregation operators. *Iran. J. Fuzzy Syst.* **2018**. [CrossRef]
17. Shanon, C.E. A mathematical theory of communication. *Bell Syst. Tech. J.* **1948**, *27*, 379–423. [CrossRef]
18. Deluca, A.; Termini, S. A definition of Non-probabilistic entropy in setting of fuzzy set theory. *Inf. Control* **1971**, *20*, 301–312. [CrossRef]
19. Szmidt, E.; Kacprzyk, J. Entropy for intuitionistic fuzzy sets. *Fuzzy Sets Syst.* **2001**, *118*, 467–477. [CrossRef]
20. Vlachos, I.K.; Sergiadis, G.D. Intuitionistic fuzzy information-application to pattern recognition. *Pattern Recognit. Lett.* **2007**, *28*, 197–206. [CrossRef]
21. Burillo, P.; Bustince, H. Entropy on intuitionistic fuzzy sets and on interval-valued fuzzy sets. *Fuzzy Sets Syst.* **1996**, *78*, 305–316. [CrossRef]
22. Garg, H.; Agarwal, N.; Tripathi, A. Generalized Intuitionistic Fuzzy Entropy Measure of Order α and Degree β and its applications to Multi-criteria decision-making problem. *Int. J. Fuzzy Syst. Appl.* **2017**, *6*, 86–107. [CrossRef]
23. Wei, C.P.; Gao, Z.H.; Guo, T.T. An intuitionistic fuzzy entropy measure based on the trigonometric function. *Control Decis.* **2012**, *27*, 571–574.
24. Garg, H.; Agarwal, N.; Tripathi, A. Entropy based multi-criteria decision-making method under Fuzzy Environment and Unknown Attribute Weights. *Glob. J. Technol. Optim.* **2015**, *6*, 13–20.
25. Zhang, Q.S.; Jiang, S.Y. A note on information entropy measure for vague sets. *Inf. Sci.* **2008**, *178*, 4184–4191. [CrossRef]
26. Verma, R.; Sharma, B.D. Exponential entropy on intuitionistic fuzzy sets. *Kybernetika* **2013**, *49*, 114–127.

27. Taneja, I.J. On generalized information measures and their applications. In *Advances in Electronics and Electron Physics*; Elsevier: New York, NY, USA, 1989; Volume 76, pp. 327–413.
28. Boekee, D.E.; Van der Lubbe, J.C. The R-norm information measure. *Inf. Control* **1980**, *45*, 136–155. [CrossRef]
29. Hung, W.L.; Yang, M.S. Similarity measures of intuitionistic fuzzy sets based on Hausdorff distance. *Pattern Recognit. Lett.* **2004**, *25*, 1603–1611. [CrossRef]
30. Garg, H. Distance and similarity measure for intuitionistic multiplicative preference relation and its application. *Int. J. Uncertain. Quantif.* **2017**, *7*, 117–133. [CrossRef]
31. Garg, H.; Arora, R. Distance and similarity measures for Dual hesistant fuzzy soft sets and their applications in multi criteria decision-making problem. *Int. J. Uncertain. Quantif.* **2017**, *7*, 229–248. [CrossRef]
32. Joshi, R.; Kumar, S. An (R, S)-norm fuzzy information measure with its applications in multiple-attribute decision-making. *Comput. Appl. Math.* **2017**, 1–22. [CrossRef]
33. Garg, H.; Kumar, K. An advanced study on the similarity measures of intuitionistic fuzzy sets based on the set pair analysis theory and their application in decision making. *Soft Comput.* **2018**, 1–12. [CrossRef]
34. Garg, H.; Kumar, K. Distance measures for connection number sets based on set pair analysis and its applications to decision-making process. *Appl. Intell.* **2018**, 1–14. [CrossRef]
35. Garg, H.; Nancy. On single-valued neutrosophic entropy of order α. *Neutrosophic Sets Syst.* **2016**, *14*, 21–28.
36. Selvachandran, G.; Garg, H.; Alaroud, M.H.S.; Salleh, A.R. Similarity Measure of Complex Vague Soft Sets and Its Application to Pattern Recognition. *Int. J. Fuzzy Syst.* **2018**, 1–14. [CrossRef]
37. Bajaj, R.K.; Kumar, T.; Gupta, N. R-norm intuitionistic fuzzy information measures and its computational applications. In *Eco-friendly Computing and Communication Systems*; Springer: Berlin, Germany, 2012; pp. 372–380.
38. Garg, H.; Kumar, K. Improved possibility degree method for ranking intuitionistic fuzzy numbers and their application in multiattribute decision-making. *Granul. Comput.* **2018**, 1–11. [CrossRef]
39. Mei, Y.; Ye, J.; Zeng, Z. Entropy-weighted ANP fuzzy comprehensive evaluation of interim product production schemes in one-of-a-kind production. *Comput. Ind. Eng.* **2016**, *100*, 144–152. [CrossRef]
40. Chen, S.M.; Chang, C.H. A novel similarity measure between Atanassov's intuitionistic fuzzy sets based on transformation techniques with applications to pattern recognition. *Inf. Sci.* **2015**, *291*, 96–114. [CrossRef]
41. Garg, H. Hesitant Pythagorean fuzzy sets and their aggregation operators in multiple attribute decision-making. *Int. J. Uncertain. Quantif.* **2018**, *8*, 267–289. [CrossRef]
42. Chen, S.M.; Cheng, S.H.; Chiou, C.H. Fuzzy multiattribute group decision-making based on intuitionistic fuzzy sets and evidential reasoning methodology. *Inf. Fusion* **2016**, *27*, 215–227. [CrossRef]
43. Kaur, G.; Garg, H. Multi-Attribute Decision-Making Based on Bonferroni Mean Operators under Cubic Intuitionistic Fuzzy Set Environment. *Entropy* **2018**, *20*, 65. [CrossRef]
44. Chen, T.Y.; Li, C.H. Determining objective weights with intuitionistic fuzzy entropy measures: A comparative analysis. *Inf. Sci.* **2010**, *180*, 4207–4222. [CrossRef]
45. Li, D.F. TOPSIS- based nonlinear-programming methodology for multiattribute decision-making with interval-valued intuitionistic fuzzy sets. *IEEE Trans. Fuzzy Syst.* **2010**, *18*, 299–311. [CrossRef]
46. Garg, H.; Arora, R. A nonlinear-programming methodology for multi-attribute decision-making problem with interval-valued intuitionistic fuzzy soft sets information. *Appl. Intell.* **2017**, 1–16. [CrossRef]
47. Garg, H.; Nancy. Non-linear programming method for multi-criteria decision-making problems under interval neutrosophic set environment. *Appl. Intell.* **2017**, 1–15. [CrossRef]
48. Saaty, T.L. Axiomatic foundation of the analytic hierarchy process. *Manag. Sci.* **1986**, *32*, 841–845. [CrossRef]
49. Hwang, C.L.; Lin, M.J. *Group Decision Making under Multiple Criteria: Methods and Applications*; Springer: Berlin, Germany, 1987.
50. Arora, R.; Garg, H. A robust correlation coefficient measure of dual hesistant fuzzy soft sets and their application in decision-making. *Eng. Appl. Artif. Intell.* **2018**, *72*, 80–92. [CrossRef]
51. Garg, H.; Kumar, K. Some aggregation operators for linguistic intuitionistic fuzzy set and its application to group decision-making process using the set pair analysis. *Arab. J. Sci. Eng.* **2018**, *43*, 3213–3227. [CrossRef]
52. Abdullah, L.; Najib, L. A new preference scale mcdm method based on interval-valued intuitionistic fuzzy sets and the analytic hierarchy process. *Soft Comput.* **2016**, *20*, 511–523. [CrossRef]
53. Garg, H. Generalized intuitionistic fuzzy entropy-based approach for solving multi-attribute decision-making problems with unknown attribute weights. *Proc. Natl. Acad. Sci. India Sect. A Phys. Sci.* **2017**, 1–11. [CrossRef]

54. Xia, M.; Xu, Z. Entropy/cross entropy-based group decision-making under intuitionistic fuzzy environment. *Inf. Fusion* **2012**, *13*, 31–47. [CrossRef]
55. Garg, H.; Nancy. Linguistic single-valued neutrosophic prioritized aggregation operators and their applications to multiple-attribute group decision-making. *J. Ambient Intell. Humaniz. Comput.* **2018**, 1–23. [CrossRef]
56. De, S.K.; Biswas, R.; Roy, A.R. Some operations on intuitionistic fuzzy sets. *Fuzzy Sets Syst.* **2000**, *117*, 477–484. [CrossRef]
57. Zeng, W.; Li, H. Relationship between similarity measure and entropy of interval-valued fuzzy sets. *Fuzzy Sets Syst.* **2006**, *157*, 1477–1484. [CrossRef]
58. Hung, W.L.; Yang, M.S. Fuzzy Entropy on intuitionistic fuzzy sets. *Int. J. Intell. Syst.* **2006**, *21*, 443–451. [CrossRef]
59. Garg, H. Some methods for strategic decision-making problems with immediate probabilities in Pythagorean fuzzy environment. *Int. J. Intell. Syst.* **2018**, *33*, 687–712. [CrossRef]
60. Garg, H. Linguistic Pythagorean fuzzy sets and its applications in multiattribute decision-making process. *Int. J. Intell. Syst.* **2018**, *33*, 1234–1263. [CrossRef]
61. Garg, H. Generalized interaction aggregation operators in intuitionistic fuzzy multiplicative preference environment and their application to multicriteria decision-making. *Appl. Intell.* **2017**, 1–17. [CrossRef]
62. Garg, H.; Arora, R. Generalized and Group-based Generalized intuitionistic fuzzy soft sets with applications in decision-making. *Appl. Intell.* **2018**, *48*, 343–356. [CrossRef]

mathematics

MDPI

Article

Hesitant Probabilistic Fuzzy Linguistic Sets with Applications in Multi-Criteria Group Decision Making Problems

Dheeraj Kumar Joshi [1], Ismat Beg [2,*] and Sanjay Kumar [1]

[1] Department of Mathematics, Statistics and Computer Science, G. B. Pant University of Agriculture and Technology, Pantnagar, Uttarakhand 263145, India; maths.dj44010@gmail.com (D.K.J.); skruhela@hotmail.com (S.K.)

[2] Centre for Mathematics and Statistical Sciences, Lahore School of Economics, Lahore 53200, Pakistan

* Correspondence: ibeg@lahoreschool.edu.pk

Received: 3 February 2018; Accepted: 23 March 2018; Published: 26 March 2018

Abstract: Uncertainties due to randomness and fuzziness comprehensively exist in control and decision support systems. In the present study, we introduce notion of occurring probability of possible values into hesitant fuzzy linguistic element (HFLE) and define hesitant probabilistic fuzzy linguistic set (HPFLS) for ill structured and complex decision making problem. HPFLS provides a single framework where both stochastic and non-stochastic uncertainties can be efficiently handled along with hesitation. We have also proposed expected mean, variance, score and accuracy function and basic operations for HPFLS. Weighted and ordered weighted aggregation operators for HPFLS are also defined in the present study for its applications in multi-criteria group decision making (MCGDM) problems. We propose a MCGDM method with HPFL information which is illustrated by an example. A real case study is also taken in the present study to rank State Bank of India, InfoTech Enterprises, I.T.C., H.D.F.C. Bank, Tata Steel, Tata Motors and Bajaj Finance using real data. Proposed HPFLS-based MCGDM method is also compared with two HFL-based decision making methods.

Keywords: hesitant fuzzy set; hesitant probabilistic fuzzy linguistic set; score and accuracy function; multi-criteria group decision making; aggregation operator

1. Introduction

Uncertainties in decision making problems are due to either randomness or fuzziness, or by both and can be classified into stochastic and non-stochastic uncertainty [1]. Stochastic uncertainties in every system may be well captured by the probabilistic modeling [2,3]. Although several theories have been proposed in the literature to deal with non-stochastic uncertainties but among them fuzzy set theory [4,5] is extensively researched and successfully applied in decision making [6–10]. An extensive literature is due to Mardani et al. [11] on the various fuzzy aggregation operators proposed in last thirty years. Type-2 fuzzy sets [5], interval-valued fuzzy set (IVFS) [4], intuitionistic fuzzy sets (IFS) [12] and interval-valued intuitionistic fuzzy sets (IVIFS) [13], Pythagorean fuzzy set [14] and neutrosophic sets [15] are few other extensions of fuzzy sets practiced in MCGDM problems to include non-stochastic uncertainty and hesitation.

Often decision makers (DMs) in multi-criteria group decision making (MCGDM) problems are not in favor of the same assessment on decision criteria and provide different assessment information on each criterion. Difficulty of agreeing on a common assessment is not because of margin of error or some possible distribution as in case of IFS and type-2 fuzzy sets. To address this issue in MCGDM problems Torra and Narukawa [16] and Torra [17] introduced hesitant fuzzy set (HFS) and applied in MCGDM problems [18,19]. Various extensions of HFS e.g., triangular hesitant fuzzy set (THFS), generalized hesitant fuzzy set (GHFS), interval valued hesitant fuzzy set (IVHFS), dual hesitant fuzzy set (DHFS), interval valued intuitionistic hesitant fuzzy set (IVIHFS) and hesitant pythagorean fuzzy set were used in decision making problems [20–28] considering decision hesitancy and prioritization among decision criteria and developed a fuzzy group decision making method to evaluate complex emergency response in sustainable development. Recently, Garg and Arora [29] proposed distance and similarity measures-based MCDM method using dual hesitant fuzzy soft set.

Qualitative and quantitative analysis of decision criteria with hesitant and uncertain information has always been an important issue for researchers in MCGDM problems. Limited knowledge of decision makers (DMs), nature of considered alternatives and unpredictability of events are main constraints in getting sufficient and accurate information about the decision preferences and decision criteria. Many criteria which are difficult to be analyzed quantitatively can be analyzed using linguistic variables [5]. Linguistic variables improve consistency and flexibility of traditional decision making methods [30] and hence many researchers [31–45] have proposed use of linguistic variable in decision making problems. Kobina et al. [46] proposed few probabilistic linguistic aggregation operators for decision making problem. Garg and Kumar [47], Liu et al. [48] and Garg [49] proposed various aggregation operators, prioritized aggregation operators for linguistic IFS and linguistic neutrosophic set and applied them to MCGDM problems. Lin et al. [50] integrated linguistic sets with HFS to define hesitant fuzzy linguistic set (HFLS) to include hesitancy and inconsistencies among DMs in assessment of an alternative with respect to a certain criterion. Ren et al. [51] and Joshi & Kumar [52] proposed TOPSIS method MCGDM using hesitant fuzzy linguistic and IVIHFL information. Recently few researchers [53–55] have proposed generalized single-valued neutrosophic hesitant fuzzy prioritized aggregation operators and linguistic distribution-based decision making methods using hesitant fuzzy linguistic assessment for decision making methods.

Probabilistic and fuzzy approach-based MCGDM method process only either stochastic or non stochastic uncertainty. One of their major limitations is not to handle both types of uncertainties simultaneously. Comprehensive concurrence of stochastic and non stochastic uncertainty in real life problems attracted researchers to incorporate probability theory with fuzzy logic. Idea of integrating fuzzy set theory with probabilistic theory was initiated by Liang and Song [56] and Meghdadi and Akbarzadeh [1]. In 2005, Liu and Li [57] defined probabilistic fuzzy set (PFS) to handle both stochastic and non stochastic uncertainties in a single framework. To handle simultaneous occurrence of both stochastic and non stochastic uncertainties with hesitation, Xu and Zhou [58] introduced probabilistic hesitant fuzzy set (PHFS). PHFS permits more than one membership degree of an element with different probabilities. Recently many applications of PHFS are found in MCGDM problems [58–65].

Earlier in all HFL-based decision making methods, probabilities of occurrence of elements are assumed to be equal. Assumption of equal probabilities in HFL is too hard to be followed by DMs in real life problems of decision making due to their hesitation. For example, a decision maker provides hesitant fuzzy linguistic element (HFLE) $\{s_2, < 0.4, 0.5, 0.6 >\}$ to evaluate the safety level of a vehicle. He or she thinks that the safety level associated with 0.6 and 0.4 are the most and least suitable. However, he or she contradicts with own decision by associating equal probability to each 0.4, 0.5, 0.6. Hence, HFLE $\{s_2, < 0.4, 0.5, 0.6 >\}$ with equal probabilities cannot represent DM's accurate assessment of decision criteria. With this limitation in present form of HFLS, we introduce notion of hesitant probabilistic fuzzy linguistic set (HPFLS). This new class of set undertakes both uncertainties caused by randomness and fuzziness in the environment of hesitation in a single framework.

In the present study, we have proposed HPFLS with expected mean, variance, score and accuracy function and a few operations on its elements. We also develop novel hesitant probabilistic fuzzy linguistic weighted averaging (HPFLWA), hesitant probabilistic fuzzy linguistic weighted geometric (HPFLWG), hesitant probabilistic fuzzy linguistic ordered weighted averaging (HPFLOWA) and hesitant probabilistic fuzzy linguistic ordered weighted geometric (HPFLOWG) aggregation operators to aggregate the HPFL information. A MCGDM method with HPFL information is proposed. Methodology of proposed MCGDM method is illustrated by a numerical example and also applied on a real case study to rank the organizations.

2. Preliminaries

In this section, we briefly review fundamental concepts and definitions of hesitant fuzzy set, linguistic variables, hesitant fuzzy linguistic set and hesitant probabilistic fuzzy set.

Definition 1. *([16,17]) Let X be a reference set. An HFS A on X is defined using a function $I_A(X)$ that returns a subset of [0, 1]. Mathematically, it is symbolized using following expression:*

$$A = \{< x, I_A(x) > | x \in X\} \tag{1}$$

where $I_A(X)$ is hesitant fuzzy element (HFE) having a set of different values lies between [0, 1].

Definition 2. *([32]) Let $S = \{s_i | i = 1, 2, \ldots t\}$ be a finite discrete LTS. Here s_i represents a possible value for a linguistic variable and satisfies the following characteristics:*

1. *The set is ordered: $s_i > s_j$ if $i > j$*
2. *$Max\{s_i, s_j\} = s_i$ if $i \geq j$*
3. *$Min\{s_i, s_j\} = s_j$ if $i \geq j$*

Xu [66] extended finite discrete LTS $S = \{s_i | i = 1, 2, \ldots t\}$ to continuous LTS $S = \{s_\theta | s_0 \leq s_\theta \leq s_t, \theta \in [0, t]\}$ to conserve all the provided information. An LTS is original if $s_\theta \in S$, otherwise it is called virtual.

Definition 3. *([50]) Let X be the reference set and $s_\theta \in S$. A hesitant fuzzy linguistic set A in X is a mathematical object of following form:*

$$A = \{< x, s_\theta(x), h_A(x) > | x \in X\} \tag{2}$$

Here $h_A(x)$ is a set of possible finite number of values belonging to [0, 1] and denotes the possible membership degrees that x belongs to $s_\theta(x)$.

Definition 4. *([58]) Let X be the reference set. An HPFS H_F on X is a mathematical object of following form:*

$$H_P = \{< \gamma_i, h(\gamma_i | p_i) > s_\theta \in X\} \tag{3}$$

Here $h(\gamma_i | p_i)$ is set of elements $\gamma_i | p_i$ expressing the hesitant fuzzy information with probabilities to the set H_P, $0 \leq \gamma_i \leq 1$ $(i = 1, 2, \ldots \#h)$ (number of possible elements in $h(\gamma_i | p_i)$, $p_i \in [0, 1]$ are corresponding probabilities with condition $\sum_{i=1}^{\#h} p_i = 1$ s.

3. Hesitant Probabilistic Fuzzy Linguistic Set (HPFLS) and Hesitant Probabilistic Fuzzy Linguistic Element (HPFLE)

Qualitative and quantitative analysis of decision criteria with hesitant is always been an important issue for researchers in MCGDM problems. Earlier classification of fuzzy sets (hesitant fuzzy set [16,17], hesitant fuzzy linguistic set [50] and probabilistic hesitant fuzzy [58]) are not capable to deal with fuzziness, hesitancy and uncertainty both qualitatively and quantitatively. Keeping in mind the limitations of HFLEs and to fully describe precious information provided by DMs; our aim is to propose a new class of set called HPFLS. This set can easily describe stochastic and non-stochastic uncertainties with hesitant information using both qualitative and quantitative terms. In this section, we also develop expected mean, variance, score and accuracy function of HPFLEs, along with a comparison method. Some basic operations of HPFLEs are also defined in this section.

Definition 5. *Let X and S be the reference set and linguistic term set. An HPFLS H_{PL} on X is a mathematical object of following form:*

$$H_{PL} = \{< x,\ h_{PL}(p_x) > | x \in X\} \tag{4}$$

Here $h_{PL}(p_x) = s_\theta(x|p_k),\ h(\gamma_i|p_i)|\gamma_i, p_i, s_\theta(x) \in S$ and $h_{PL}(p_x)$ is set of some elements x denoting the hesitant fuzzy linguistic information with probabilities to the set H_{PL}, $0 \le \gamma_i \le 1$, $i = 1, 2, \ldots, \#\, h_{PL}$. Here # h_{PL} is the number of possible elements in $h_{PL}\ (p_x)$, $p_i \in [0,\ 1]$ is the hesitant probability of γ_i and $\sum_{i=1}^{\#h_{PL}} p_i = 1$. We call $h_{PL}\ (p_x)$ HPFLE and H_{PL} is set of all HPLFEs.

As an illustration of Definition 5, we assume two HPFLEs $h_{PL}(p_x) = [(s_1|1), \{0.2|0.3, 0.4|0.2, 0.5|0.5\}]$, $h_{PL}(p_y) = [(s_3|1), \{0.4|0.2, 0.5|0.4, 0.2|0.4\}]$ on reference set $X = \{x, y\}$.

An object $H_{PL} = [< x, (s_1|1), \{0.2|0.3, 0.4|0.2, 0.5|0.5\} >, < y, (s_3|1), \{0.4|0.2, 0.5|0.4, 0.2|0.4\}]$ represents an HPFLS.

It is important to note that if the probabilities of the possible values in HPFLEs are equal, i.e., $p_1 = p_2 = \ldots\ p_{\#h}$, then HPFLE reduced to HFLE.

3.1. Some Basic Operations on Hesitant Probabilistic Fuzzy Linguistic Element (HPFLEs)

Based on operational rules of hesitant fuzzy linguistic set [50] and hesitant probabilistic set [61], we propose following operational laws for $h^1_{PL}(p_x) = \langle s_\theta((x)|p),\ h(\gamma_j|p_j)|\gamma_j, p_j\rangle$ and $h^2_{PL}(p_y) = \langle s_\theta((y)|p),\ h(\gamma_k|p_k)|\gamma_k, p_k\rangle$ then

(1) $(h^1_{PL})^\lambda = \langle s_{\theta^\lambda}((x)|p),\ h(\gamma_j{}^\lambda|p_j)|\gamma_j, p_j\rangle$ for some $\lambda > 0$

(2) $\lambda(h^1_{PL}) = \left\langle s_{\lambda\theta}((x)|p),\ h\left(1-(1-\gamma_j)^\lambda|p_j\right)|\gamma_j, p_j\right\rangle$ for some $\lambda > 0$

(3) $h^1_{PL} \oplus h^2_{PL} = \left\langle s_{(\theta(x)+\theta(y))}|p),\ h((\gamma_j+\gamma_k-\gamma_j\gamma_k)|p_jp_k)\right\rangle$

(4) $h^1_{PL} \otimes h^2_{PL} = \left\langle s_{(\theta(x)\theta(y))}|p),\ h((\gamma_j\gamma_k)|p_jp_k)\right\rangle$

(5) $h^1_{PL} \cup h^2_{PL} = \left\langle s_{(\theta(x)\vee\theta(y))}|p),\ h((\gamma_j \vee \gamma_k)|([p_j \vee p_k]/\Sigma(Max(p_j, p_k))\right\rangle$

(6) $h^1_{PL} \cap h^2_{PL} = \left\langle s_{(\theta(x)\wedge\theta(y))}|p),\ h((\gamma_j \wedge \gamma_k)|([p_j \wedge p_k]/\Sigma(Min(p_j, p_k))\right\rangle$

Using definition of $\oplus$ and $\otimes$, it can be easily proved that $h^1_{PL} \oplus h^2_{PL}$ and $h^1_{PL} \otimes h^2_{PL}$ are commutative. In order to show that $(h^1_{PL})^\lambda$, $\lambda(h^1_{PL})$, $h^1_{PL} \oplus h^2_{PL}$, $h^1_{PL} \otimes h^2_{PL}$, $h^1_{PL} \cup h^2_{PL}$ and $h^1_{PL} \cap h^2_{PL}$ are again HPFLE, we assume that $h^1_{PL}(p_x) = [(s_2|1), \{0.2|0.3, 0.4|0.2, 0.5|0.5\}]$ and $h^2_{PL}(p_y) = [(s_3|1), \{0.4|0.2,\ 0.5|0.4,\ 0.2|0.4\}]$ are two HPFLEs on reference set $X = \{x, y\}$ and perform the operation laws as follows:

$$(h^1_{PL})^\lambda = [(s_4|1), \{0.04|0.3, 0.16|0.2, 0.25|0.5\}]$$
$$\lambda(h^1_{PL}) = [(s_4|1), \{0.36|0.3, 0.64|0.2, 0.75|0.5\}]$$
$$h^1_{PL} \oplus h^2_{PL} = [(s_5|1), \{0.54|0.06, 0.58|0.12, 0.28|0.12, 0.76|0.04, 0.82|0.08, 0.52|0.08, 0.8|0.1, 0.8|0.2, 0.5|0.2\}]$$
$$h^1_{PL} \otimes h^2_{PL} = [(s_5|1), \{0.08|0.06, 0.1|0.12, 0.04|0.12, 0.16|0.04, 0.2|0.08, 0.08|0.08, 0.2|0.1, 0.25|0.2, 0.1|0.2\}]$$
$$h^1_{PL} \cup h^2_{PL} = [(s_5|1), \{0.4|0.08, 0.5|0.11, 0.2|0.11, 0.4|0.06, 0.5|0.11, 0.4|0.11, 0.5|0.14, 0.5|0.14, 0.5|0.14\}]$$
$$h^1_{PL} \cap h^2_{PL} = [(s_5|1), \{0.2|0.08, 0.2|0.13, 0.2|0.13, 0.4|0.08, 0.4|0.08, 0.2|0.08, 0.4|0.08, 0.5|0.17, 0.2|0.17\}]$$

3.2. Score and Accuracy Function for Hesitant Probabilistic Fuzzy Linguistic Element (HPFLE)

Comparison is an indispensable and is required if we tend to apply HPFLE in decision making and optimization problems. Hence, we define expected mean, variance, score and accuracy function of HPFLE in this sub section as follows:

Definition 6. *Expected mean* $E(h_{PL}(p_x))$ *and variance* $V(h_{PL}(p_x))$ *for a HPFLE* $h_{PL}(p_x) = \langle s_\theta((x)|p_k),\ h(\gamma_i|p_i)|\gamma_i, p_i\rangle$ *are defined as follows:*

$$E(h_{PL}(p_x)) = \frac{\sum\limits_{i=1}^{\#h} (\gamma_i p_i)}{\#h} \tag{5}$$

$$V(h_{PL}(p_x)) = \sum_{i=1}^{\#h} (\gamma_i - E(h_{PL}(p_x)))^2 p_i \tag{6}$$

Definition 7. *Score function* $S(h_{PL}(p_x))$ *and accuracy function* $A(h_{PL}(p_x))$ *fora HPFLE* $h_{PL}(p_x) = \langle s_\theta((x)|p_k),\ h(\gamma_i|p_i)|\gamma_i, p_i\rangle$ *are defined as follows:*

$$S(h_{PL}(p_x)) = E(h_{PL}(p_x))\ (s_\theta(x)(p_k)) \tag{7}$$

$$A(h_{PL}(p_x)) = V(h_{PL}(p_x))(s_\theta(x)(p_k)) \tag{8}$$

Using score and accuracy functions two HPFLEs $h_{PL}(p_x),\ h_{PL}(p_y)$ can be compared as follows:

(1) If $S(h_{PL}(p_x)) > S(h_{PL}(p_y))$, then $h_{PL}(p_x) > h_{PL}(p_y)$
(2) If $S(h_{PL}(p_x)) < S(h_{PL}(p_y))$, then $h_{PL}(p_x) < h_{PL}(p_y)$
(3) If $S(h_{PL}(p_x)) = S(h_{PL}(p_y))$,

(a) If $A(h_{PL}(p_x)) > A(h_{PL}(p_y))$, then $h_{PL}(p_x) > h_{PL}(p_y)$
(b) If $A(h_{PL}(p_x)) < A(h_{PL}(p_y))$, then $h_{PL}(p_x) < h_{PL}(p_y)$
(c) If $A(h_{PL}(p_x)) = A(h_{PL}(p_y))$, then $h_{PL}(p_x) = h_{PL}(p_y)$

As an illustration of Definitions 6 and 7, we compare two HPFLEs $h_{PL}(p_x) = [(s_2|1), \{0.2|0.3, 0.4|0.2, 0.5|0.5\}]$ and $h_{PL}(p_y) = [(s_3|1), \{0.4|0.2, 0.5|0.4, 0.2|0.4\}]$ using score and accuracy functions as follows:

$$E(h_{PL}(p_x)) = (0.2*0.3 + 0.4*0.2 + 0.5*0.5)/3 = 0.13$$
$$E(h_{PL}(p_y)) = (0.4*0.2 + 0.5*0.4 + 0.2*0.4)/3 = 0.12$$
$$V(h_{PL}(p_x)) = (((0.2-0.13)^2)*0.3 + ((0.4-0.13)^2)*0.2) + ((0.5-0.13)^2)*0.5))/3 = 0.0279$$
$$V(h_{PL}(p_y)) = (((0.4-0.12)^2)*0.2 + ((0.5-0.12)^2)*0.4) + (0.2-0.12)^2*0.4))/3 = 0.0252$$
$$S(h_{PL}(p_x)) = s_{(2*1)((0.2*0.3+0.4*0.2+0.5*0.5)/3)} = s_{0.26}$$
$$S(h_{PL}(p_y)) = s_{(3*1)(0.4*0.2+0.5*0.4+0.2*0.4)/3} = s_{0.36}$$
$$A(h_{PL}(p_x)) = s_{(2*1)((0.2-0.13)^2)*0.3+((0.4-0.13)^2)*0.2)+((0.5-0.13)^2)*0.5)/3} = s_{0.0558}$$
$$A(h_{PL}(p_y)) = s_{(3*1)((0.4-0.12)^2)*0.2+((0.5-0.12)^2)*0.4)+(0.2-0.12)^2*0.4)/3} = s_{0.0756}$$

Since $S(h_{PL}(p_y)) > S(h_{PL}(p_x))$ therefore $h_{PL}(p_y) < h_{PL}(p_x)$.

Different HPFLEs may have different number of PFNs. To make them equal in numbers we extend HPFLEs until they have same number of PFNs. It can be extended according to DMs risk behavior.

4. Aggregation Operators for Hesitant Probabilistic Fuzzy Linguistic Set (HPFLS)

In group decision making problems, an imperative task is to aggregate the assessment information obtained from DMs about alternatives against each criterion. Various aggregation operators for HFLS [18,50] and HPFS [58–63,67] have been developed in the past few decades. As we propose HPFLS for MCGDM problems, we also develop few aggregations operators to aggregate information in the form of HPFLEs. In this section, we define HPFLW and HPFLOW operators.

4.1. Hesitant Probabilistic Linguistic Fuzzy Weighted Aggregation Operators

Let $H_{PL}^{i} = h_{PL}(p_x) = \langle s_{\theta i}((x)|p), h(\gamma_i|p_i)|\gamma_i, p_i, (i = 1,2,\ldots,n)\rangle$ be collection of HPFLEs. Hesitant probabilistic fuzzy linguistic weighted averaging (HPFLWA) operator and hesitant probabilistic fuzzy linguistic weighted geometric (HPFLWG) operator are defined as follows:

Definition 8. HPFLWA is a mapping $H_{PL}^{n} \to H_{PL}$ such that

$$\begin{aligned} \text{HPFLWA}(H_1, H_2, \ldots, H_n) &= \underset{i=1}{\overset{n}{\oplus}} (\omega_i H_i) \\ = \left\langle \left[\sum_{i=1}^{n} \omega_i s_{\theta i}((x)|p)\right], \left[\bigcup_{\gamma_1 \in H_1, \gamma_2 \in H_2, \ldots, \gamma_n \in H_n,} \left\{1 - \prod_{i=1}^{n} (1-\gamma_i)^{\omega_i} | p_1 p_2 \ldots p_n\right\}\right]\right\rangle \end{aligned} \tag{9}$$

Definition 9. *HPFLWG operator is a mapping $H_{PL}^{n} \to H_{PL}$ such that*

$$\begin{aligned} \text{HPFLWG}(H_1, H_2, \ldots, H_n) &= \underset{i=1}{\overset{n}{\otimes}} (H_i)^{\omega_i} \\ = \left\langle \left[\sum_{i=1}^{n} s_{\theta i}{}^{\omega_i}((x)|p)\right], \left[\bigcup_{\gamma_1 \in H_1, \gamma_2 \in H_2, \ldots, \gamma_n \in H_n,} \left\{\prod_{i=1}^{n} (\gamma_i)^{\omega_i} | p_1 p_2 \ldots p_n\right\}\right]\right\rangle \end{aligned} \tag{10}$$

where $\omega = (\omega_1, \omega_2, \ldots, \omega_n)$ is weight vector of $H_i (i = 1,2,\ldots,n)$ with $\omega_i \in [0,1]$ and $\sum_{i=1}^{n} \omega_i = 1$, p_n is the probability of γ_i in the HPFLEs $H_i (i = 1,2,\ldots,n)$. In particular if $\omega = \left(\frac{1}{n}, \frac{1}{n}, \ldots, \frac{1}{n}\right)^T$ then HPFLWA and HPFLWG operator are reduced to following hesitant probabilistic fuzzy linguistic averaging (HPFLA) operator and hesitant probabilistic fuzzy linguistic geometric (HPFLG) operator respectively:

$$\begin{aligned} \text{HPFLA}(H_1, H_2, \ldots, H_n) &= \underset{i=1}{\overset{n}{\oplus}} \left(\tfrac{1}{n} H_i\right) \\ = \left\langle \left[\sum_{i=1}^{n} \tfrac{1}{n} s_{\theta i}((x)|p)\right], \left[\bigcup_{\gamma_1 \in H_1, \gamma_2 \in H_2, \ldots, \gamma_n \in H_n,} \left\{1 - \prod_{i=1}^{n} (1-\gamma_i)^{\frac{1}{n}} | p_1 p_2 \ldots p_n\right\}\right]\right\rangle \end{aligned} \tag{11}$$

$$\begin{aligned} \text{HPFLG}(H_1, H_2, \ldots, H_n) &= \underset{i=1}{\overset{n}{\otimes}} (H_i)^{\frac{1}{n}} \\ = \left\langle \left[\sum_{i=1}^{n} s_{\theta}{}^{\frac{1}{n}}{}_i((x)|p)\right], \left[\bigcup_{\gamma_1 \in H_1, \gamma_2 \in H_2, \ldots, \gamma_n \in H_n,} \left\{\prod_{i=1}^{n} (\gamma_i)^{\frac{1}{n}} | p_1 p_2 \ldots p_n\right\}\right]\right\rangle \end{aligned} \tag{12}$$

Lemma 1. *([17] Let $\alpha_i > 0, \omega_i > 0, i = 1,2,\ldots,n$ and $\sum_{i=1}^{n} \omega_i = 1$, then $\prod_{i=1}^{n} \alpha_i^{\omega_i} \leq \sum_{i=1}^{n} \alpha_i \omega_i$ and equality holds if and only if $\alpha_1 = \alpha_2 = \ldots = \alpha_n$.*

Theorem 1. *Let $H_i = h_{PL}(p_x) = \{s_{\theta i}((x)|p), h(\gamma_i|p_i)|\gamma_i, p_i >\}(i = 1,2,\ldots,n)$ be collection of HPFLEs. Let $\omega = (\omega_1, \omega_2, \ldots, \omega_n)$ be weight vector of $H_i (i = 1,2,\ldots,n)$ with $\omega_i \in [0,1]$ and $\sum_{i=1}^{n} \omega_i = 1$, then*

$$\text{HPFLWG}(H_1, H_2, \ldots, H_n) \leq \text{HPFLWA}(H_1, H_2, \ldots, H_n)$$
$$\text{HPFLG}(H_1, H_2, \ldots, H_n) \leq \text{HPFLA}(H_1, H_2, \ldots, H_n)$$

Proof. Using Lemma l, we have following inequality for any $\gamma_i \in H_i (i = 1,2,\ldots,n)$,

$$\prod_{i=1}^{n} (\gamma_i)^{\omega_i} \leq \sum_{i=1}^{n} \gamma_i \omega_i = 1 - \sum_{i=1}^{n} (1-\gamma_i)\omega_i \leq \prod_{i=1}^{n} (1-\gamma_i)^{\omega_i}$$

Thus, we can obtain the following inequality:

$$\left\langle\left[\bigcup_{\gamma_1\in H_1,\gamma_2\in H_2,\ldots,\gamma_n\in H_n,}\left\{\prod_{i=1}^{n}(\gamma_i)^{\omega_i}|p_1p_2\ldots p_n\right\}\right]\right\rangle \leq \left\langle\left[\bigcup_{\gamma_1\in H_1,\gamma_2\in H_2,\ldots,\gamma_n\in H_n,}\left\{1-\prod_{i=1}^{n}(1-\gamma_i)^{\omega_i}|p_1p_2\ldots p_n\right\}\right]\right\rangle$$

$$\left\langle\left[\sum_{i=1}^{n}s_{\theta i}{}^{\omega_i}((x)|p)\right],\left[\bigcup_{\gamma_1\in H_1,\gamma_2\in H_2,\ldots,\gamma_n\in H_n,}\left\{\prod_{i=1}^{n}(\gamma_i)^{\omega_i}|p_1p_2\ldots p_n\right\}\right]\right\rangle \leq \left\langle\left[\sum_{i=1}^{n}\omega_i s_{\theta i}((x)|p)\right],\left[\bigcup_{\gamma_1\in H_1,\gamma_2\in H_2,\ldots,\gamma_n\in H_n,}\left\{1-\prod_{i=1}^{n}(1-\gamma_i)^{\omega_i}|p_1p_2\ldots p_n\right\}\right]\right\rangle$$

Using definition of score function $S(h_{PL}(p_x)) = \frac{\sum_{i=1}^{\#h}(\gamma_i p_i)}{\#h}(s_\theta(x)(p_k))$, we have HPFLWG $(H_1, H_2, \ldots, H_n) \leq$ HPFLWA$(H_1, H_2, \ldots, H_n)$. Similarly, it can be proved that HPFLG $(H_1, H_2, \ldots, H_n) \leq$ HPFLA $(H_1, H_2, \ldots, H_n)$. □

4.2. Hesitant Probabilistic Fuzzy Linguistic Ordered Weighted Aggregation Operators

Xu and Zhou [58] defined ordered weighted averaging and geometric aggregation operators to aggregate hesitant probabilistic fuzzy information for MCGDM problems. In this sub section we propose hesitant probabilistic fuzzy linguistic ordered weighted averaging (HPFLOWA) operator and hesitant probabilistic fuzzy linguistic ordered weighted geometric (HPFLOWG) operators.

Let $H_{PL}^i = h_{PL}(p_x) = \{s_{\theta i}((x)|p), h(\gamma_i|p_i)|\gamma_i, p_i >\}(i = 1, 2, \ldots, n)$ be collection of HPFLEs, $\omega = (\omega_1, \omega_2, \ldots, \omega_n)$ is weight vector of with $\omega_i \in [0,1]$ and $\sum_{i=1}^{n}\omega_i = 1$. Let p_i is the probability of γ_i in the HPFLEs $H_i(i = 1, 2, \ldots, n)$, $\gamma_{\sigma(i)}$ *i*th be the largest of H_i, $p_{\sigma(i)}$ is the probability of $\gamma_{\sigma(i)}$, and $\omega_{\sigma(i)}$ be the largest of ω. We develop the following two ordered weighted aggregation operators:

Definition 10. *HPFLOWA operator is a mapping* $H_{PL}^n \to H_{PL}$ *such that*

$$\text{HPFLOWA}(H_1, H_2, \ldots, H_n) = \bigoplus_{i=1}^{n}(\omega_i H_{\sigma(i)})$$
$$= \left\langle\left[\sum_{i=1}^{n}\omega_i s_{\theta\sigma(i)}((x)|p)\right],\left[\bigcup_{\gamma_{\sigma(1)1}\in H_{\sigma(1)},\gamma_{\sigma(2)}\in H_{\sigma(2)},\ldots,\gamma_{\sigma(n)}\in H_{\sigma(n)},}\left\{1-\prod_{i=1}^{n}(1-\gamma_{\sigma(i)})^{\omega_i}|p_{\sigma(1)}p_{\sigma(2)}\cdots p_{\sigma(n)}\right\}\right]\right\rangle \quad (13)$$

Definition 11. *HPFLOWG operator is a mapping* $H_{PL}^n \to H_{PL}$ *such that*

$$\text{HPFLOWG}(H_1, H_2, \ldots, H_n) = \bigoplus_{i=1}^{n}(H_{\sigma(i)})^{\omega_i}$$
$$= \left\langle\left[\sum_{i=1}^{n}s_{\theta\sigma(i)}{}^{\omega_i}((x)|p)\right],\left[\bigcup_{\gamma_{\sigma(1)1}\in H_{\sigma(1)},\gamma_{\sigma(2)}\in H_{\sigma(2)},\ldots,\gamma_{\sigma(n)}\in H_{\sigma(n)},}\left\{\prod_{i=1}^{n}(\gamma_{\sigma(i)})^{\omega_i}|p_{\sigma(1)}p_{\sigma(2)}\cdots p_{\sigma(n)}\right\}\right]\right\rangle \quad (14)$$

Similar to Theorem 1, the above ordered weighted operators have the relationship below:

$$\text{HPFLOWG}(H_1, H_2, \ldots, H_n) \leq \text{HPFLOWA}(H_1, H_2, \ldots, H_n)$$

4.3. Properties of Proposed Weighted and Ordered Weighted Aggregation Operators

Following are few properties of proposed weighted and ordered weighted aggregation operators that immediately follow from their definitions.

Property 1. *(Monotonicity). Let* $(H_1, H_2, \ldots, H_n)$ *and* $(H_1', H_2', \ldots, H_n')$ *be two collections of HPFLNs, if* $H_i' \leq H_i$ *for all I = 1,2, ... n, then*

$$\text{HPFLWA}(H_1', H_2', \ldots, H_n') \leq \text{HPFLWA}(H_1, H_2, \ldots, H_n)$$
$$\text{HPFLWG}(H_1', H_2', \ldots, H_n') \leq \text{HPFLWG}(H_1, H_2, \ldots, H_n)$$
$$\text{HPFLOWA}(H_1', H_2', \ldots, H_n') \leq \text{HPFLOWA}(H_1, H_2, \ldots, H_n)$$
$$\text{HPFLOWG}(H_1', H_2', \ldots, H_n') \leq \text{HPFLOWG}(H_1, H_2, \ldots, H_n)$$

Property 2. *(Idempotency). Let* $H_i = H_i', (i = 1,2,\ldots,n)$, *then*

$$\text{HPFLWA}(H_1, H_2, \ldots, H_n) = \text{HPFLWG}(H_1, H_2, \ldots, H_n) = \text{HPFLOWA}(H_1, H_2, \ldots, H_n) = \text{HPFLOWG}(H_1, H_2, \ldots, H_n) = H$$

Property 3. *(Boundedness). All aggregation operators lie between the max and min operators:*

$$\begin{array}{c} \min(H_1, H_2, \ldots, H_n) \leq \text{HPFLWA}(H_1, H_2, \ldots, H_n) \leq \max(H_1, H_2, \ldots, H_n) \\ \min(H_1, H_2, \ldots, H_n) \leq \text{HPFLWG}(H_1, H_2, \ldots, H_n) \leq \max(H_1, H_2, \ldots, H_n) \\ \min(H_1, H_2, \ldots, H_n) \leq \text{HPFLOWA}(H_1, H_2, \ldots, H_n) \leq \max(H_1, H_2, \ldots, H_n) \\ \min(H_1, H_2, \ldots, H_n) \leq \text{HPFLOWG}(H_1, H_2, \ldots, H_n) \leq \max(H_1, H_2, \ldots, H_n) \end{array}$$

5. Application of Hesitant Probabilistic Fuzzy Linguistic Set to Multi-Criteria Group Decision Making (MCGDM)

In this section, we propose a MCGDM method with hesitant probabilistic fuzzy linguistic information. Let $\{A_1, A_2, \ldots, A_m\}$ be set of alternatives to be ranked by a group of DMs $\{D_1, D_2, \ldots, D_k\}$ against criteria $\{C_1, C_2, \ldots, C_n\}$. $w = (w_1, w_2, \ldots, w_n)^T$ is the weight vector of criteria with the condition $0 \leq w_j \leq 1$ and $\sum_{j=1}^{n} w_j = 1$. $\widetilde{H^k} = (H_{ij}{}^k)_{m \times n}$ is HPFL decision matrix where $H_{ij}^k = \{s_\theta((x)|p),\ h(\gamma_t|p_t)|\gamma_t, p_t\}(t = 1,2,\ldots,T)$ denotes HPFLE when alternative A_i is evaluated by $k$thDM under the criteria C_j. If two or more decision makers provide the same value, then the value comes only once in decision matrix. Algorithm of proposed HPFLS-based MCGDM method includes following steps:

Step 1: Construct HPFL decision matrices $\widetilde{H^k} = (H_{ij}{}^k)_{m \times n}$ (i = 1,2,…,m; j = 1,2,…n), according to the preferences information provided by the DMs about the alternative A_i under the criteria C_j denoted by HPFLE $H = h_{PL}(p_x) = \{s_\theta((x)|p),\ h(\gamma_t|p_t)|\gamma_t, p_t >\}(t = 1,2,\ldots,T)$.

Step 2: Use the proposed aggregation operators (HPFLWA and HPFLWG) given in Section 3, to aggregate individual hesitant probabilistic fuzzy linguistic decision matrix information provided by each decision maker into a single HPFL decision matrix $\widetilde{H} = (H_{ij})_{m \times n}$ (i = 1,2,…,m; j = 1,2,…n).

Step 3: Calculate the overall criteria value for each alternative $A_i (i = 1,2,\ldots m)$ by applying the HPFLWA and HPFLWG aggregation operator as follows:

$$H_i(i = 1,2,\ldots,m) = \left\langle \begin{array}{l} \text{HPFLWA}(C_{i1}, C_{i2}, \ldots, C_{in}) \\ = \text{HPFLWA}((C_{11}, C_{12}, \ldots, C_{1n}), (C_{21}, C_{22}, \ldots, C_{2n}), \ldots, (C_{m1}, C_{m2}, \ldots, C_{mn})) \end{array} \right\rangle$$

$$H_i(i = 1,2,\ldots,m) = \left\langle \begin{array}{l} \text{HPFLWG}(C_{i1}, C_{i2}, \ldots, C_{in}) \\ = \text{HPFLWG}((C_{11}, C_{12}, \ldots, C_{1n}), (C_{21}, C_{22}, \ldots, C_{2n}), \ldots, (C_{m1}, C_{m2}, \ldots, C_{mn})) \end{array} \right\rangle$$

Step 4: Use score or accuracy functions to calculate the score values $S(h_{PL}(p_x))$ and accuracy values $A(h_{PL}(p_x))$ of the aggregated hesitant probabilistic fuzzy linguistic preference values $H_i(i = 1,2,\ldots m)$.

Step 5: Rank all the alternatives $A_i (i = 1,2,\ldots m)$ in accordance with $S(h^i{}_{PL}(p_x))$ or $A(h^i{}_{PL}(p_x)), (i = 1,2,\ldots m)$.

6. Illustrative Example

An example is undertaken in this section to understand the implementation methodology of proposed MCGDM method with HPFL information. Further, a real case study is done to rank organizations using proposed MCGDM method. We also compare proposed method with existing HPFL-based MCGDM methods proposed by Lin et al. [50] and Zhou et al. [68].

Example. *Suppose that a group of three decision makers (D_1, D_2, D_3)intend to rank four alternatives (A_1, A_2, A_3, A_4)on the basis of three criteria (C_1, C_2, C_3). All DMs are considered equally important and equal weights are assigned to them. Each DM provides evaluation information of each alternative under each criterion in form of HPFLEs with following LTS:*

$$S = \{s_0 = \text{extremely poor}, s_1 = \text{very poor}, s_2 = \text{poor}, s_3 = \text{fair}, s_4 = \text{good}, s_5 = \text{very good}, s_6 = \text{extremly good}\}$$

Step 1: HPFL decision matrices are constructed according to preferences information provided by DMsD_1, D_2 and D_3 about the alternative A_i (i = 1, 2, 3, 4) under the criteria C_j (i = 1, 2, 3).

Tables 1–3 represent HPFL evaluation matrices provided by DMsD_1, D_2 and D_3.

Table 1. Hesitant probabilistic fuzzy linguistic (HPFL) decision matrix $\widetilde{H^1}$ provided by D_1.

	C_1	C_2	C_3
A_1	$\{(s_1, 0.4 \mid 0.3, 0.5 \mid 0.7)\}$	$\{(s_2, 0.5 \mid 0.2, 0.6 \mid 0.8)\}$	$\{(s_1, 0.4 \mid 1.0)\}$
A_2	$\{(s_3, 0.3 \mid 0.5, 0.4 \mid 0.5,)\}$	$\{(s_4, 0.4 \mid 0.4, 0.5 \mid 0.6)\}$	$\{(s_2, 0.2 \mid 0.6, 0.5 \mid 0.4)\}$
A_3	$\{(s_5, 0.1 \mid 0.5, 0.5 \mid 0.5)\}$	$\{(s_4, 0.3 \mid 0.3, 0.5 \mid 0.7)\}$	$\{(s_1, 0.1 \mid 0.4, 0.2 \mid 0.6)\}$
A_4	$\{(s_2, 0.4 \mid 0.6, 0.5 \mid 0.4)\}$	$\{(s_1, 0.2 \mid 1.0)\}$	$\{(s_3, 0.2 \mid 0.6, 0.5 \mid 0.4)\}$

Table 2. HPFL decision matrix $\widetilde{H^2}$ provided by D_2.

	C_1	C_2	C_3
A_1	$\{(s_1, 0.4 \mid 1.0)\}$	$\{(s_2, 0.2 \mid 0.4, 0.4 \mid 0.6)\}$	$\{(s_1, 0.8 \mid 1.0)\}$
A_2	$\{(s_4, 0.2 \mid 0.5, 0.4 \mid 0.5)\}$	$\{(s_5, 0.4 \mid 0.4, 0.5 \mid 0.6)\}$	$\{(s_2, 0.1 \mid 0.4, 0.5 \mid 0.6)\}$
A_3	$\{(s_2, 0.2 \mid 0.5, 0.5 \mid 0.5)\}$	$\{(s_4, 0.3 \mid 0.6, 0.4 \mid 0.4)\}$	$\{(s_2, 0.1 \mid 1.0)\}$
A_4	$\{(s_3, 0.3 \mid 0.5, 0.5 \mid 0.5)\}$	$\{(s_4, 0.5 \mid 1.0)\}$	$\{(s_3, 0.2 \mid 0.6, 0.5 \mid 0.4)\}$

Table 3. HPFL decision matrix $\widetilde{H^3}$ provided by D_3.

	C_1	C_2	C_3
A_1	$\{(s_2, 0.2 \mid 0.4, 0.4 \mid 0.6)\}$	$\{(s_4, 0.2 \mid 1.0)\}$	$\{(s_4, 0.5 \mid 1.0)\}$
A_2	$\{(s_4, 0.3 \mid 1.0)\}$	$\{(s_5, 0.3 \mid 0.4, 0.4 \mid 0.6)\}$	$\{(s_3, 0.3 \mid 0.6, 0.5 \mid 0.4)\}$
A_3	$\{(s_2, 0.3 \mid 0.5, 0.4 \mid 0.5)\}$	$\{(s_2, 0.2 \mid 0.5, 0.4 \mid 0.5)\}$	$\{(s_4, 0.5 \mid 1.0)\}$
A_4	$\{(s_2, 0.4 \mid 0.6, 0.5 \mid 0.4)\}$	$\{(s_4, 0.5 \mid 1.0)\}$	$\{(s_4, 0.3 \mid 0.5, 0.5 \mid 0.5)\}$

Step 2: Aggregate $(\widetilde{H^1})$, $(\widetilde{H^2})$ and $(\widetilde{H^3})$ into a single HPFL decision matrix $\widetilde{H} = (H_{ij})_{4\times 3}$ (i = 1,2,...,4; j = 1,2,...3) using HPFLWA and HPFLWG operators.

Following is the sample computation process of aggregation of HPFLEs $h_{11}^1, h_{11}^2, h_{11}^3$ into a single H_{11} using proposed HPFLWA and HPFLWG operators.

$$H_{11} = HPFLWA(h_{11}^1, h_{11}^2, h_{11}^3) = [\{(s_1, 0.4|0.3, 0.5|0.7)\}, \{(s_1, 0.4|1.0)\}, \{(s_2, 0.2|0.4, 0.4|0.6)\}]$$

$$= \left[\begin{array}{ll} \{(s_{(1+1+2)/3}, & \{[1-(((1-0.4)^{(1/3)}) * ((1-0.4)^{(1/3)}) * ((1-0.2)^{(1/3)}))] \,|(0.3*1*0.4)\}, \\ & \{[1-(((1-0.4)^{(1/3)}) * ((1-0.4)^{(1/3)}) * ((1-0.2)^{(1/3)}))] \,|(0.3*1*0.4)\}, \\ & \{[1-(((1-0.4)^{(1/3)}) * ((1-0.4)^{(1/3)}) * ((1-0.4)^{(1/3)}))] \,|(0.3*1*0.4)\}, \\ & \{[1-(((1-0.4)^{(1/3)}) * ((1-0.4)^{(1/3)}) * ((1-0.4)^{(1/3)}))] \,|(0.3*1*0.6)\}, \\ & \{[1-(((1-0.5)^{(1/3)}) * ((1-0.4)^{(1/3)}) * ((1-0.2)^{(1/3)}))] \,|(0.7*1*0.4)\}, \\ & \{[1-(((1-0.5)^{(1/3)}) * ((1-0.4)^{(1/3)}) * ((1-0.4)^{(1/3)}))] \,|(0.7*1*0.6)\}, \\ & \{[1-(((1-0.5)^{(1/3)}) * ((1-0.4)^{(1/3)}) * ((1-0.2)^{(1/3)}))] \,|(0.7*1*0.4)\}, \\ & \{[1-(((1-0.5)^{(1/3)}) * ((1-0.4)^{(1/3)}) * ((1-0.4)^{(1/3)}))] \,|(0.7*1*0.6)\} \end{array}\right]$$

$$= [\{(s_{1.3}, 0.34|0.12, 0.34|0.12, 0.4|0.18, 0.4|0.18, 0.38|0.28, 0.44|0.42, 0.38|0.28, 0.44|0.42\}]$$

$$H_{11} = [\{(s_{1.3}, 0.34|0.12, 0.4|0.18, 0.38|0.28, 0.44|0.42\}]$$

$$H_{11} = HPFLWG(h_{11}^1, h_{11}^2, h_{11}^3) = [\{(s_1, 0.4|0.3, 0.5|0.7)\}, \{(s_1, 0.4|1.0)\}, \{(s_2, 0.2|0.4, 0.4|0.6)\}]$$

$$= \left[\begin{array}{ll} \{(s_{(1^{(1/3)}+1^{(1/3)}+2^{(1/3)})}, & \{[((0.4)^{(1/3)}) * ((0.4)^{(1/3)}) * ((0.2)^{(1/3)}))] \,|(0.3*1*0.4)\}, \\ & \{[(((0.4)^{(1/3)}) * ((0.4)^{(1/3)}) * ((0.2)^{(1/3)}))] \,|(0.3*1*0.4)\}, \\ & \{[(((0.4)^{(1/3)}) * ((0.4)^{(1/3)}) * ((0.4)^{(1/3)}))] \,|(0.3*1*0.4)\}, \\ & \{[(((0.4)^{(1/3)}) * ((0.4)^{(1/3)}) * ((0.4)^{(1/3)}))] \,|(0.3*1*0.6)\}, \\ & \{[(((0.5)^{(1/3)}) * ((0.4)^{(1/3)}) * ((0.2)^{(1/3)}))] \,|(0.7*1*0.4)\}, \\ & \{[(((0.5)^{(1/3)}) * ((0.4)^{(1/3)}) * ((0.4)^{(1/3)}))] \,|(0.7*1*0.6)\}, \\ & \{[(((0.5)^{(1/3)}) * ((0.4)^{(1/3)}) * ((0.2)^{(1/3)}))] \,|(0.7*1*0.4)\}, \\ & \{[(((0.5)^{(1/3)}) * ((0.4)^{(1/3)}) * ((0.4)^{(1/3)}))] \,|(0.7*1*0.6)\} \end{array}\right]$$

$$= [\{(s_{3.3}, 0.32|0.12, 0.32|0.12, 0.4|0.2, 0.4|0.2, 0.34|0.3, 0.43|0.38, 0.34|0.3, 0.43|0.38\}]$$

$$H_{11} = [\{(s_{3.3}, 0.32|0.12, 0.4|0.2, 0.3|0.3, 0.4|0.38\}]$$

Similarly other HPFLEs of HPFL decision matrices (Tables 1–3) are aggregated into the single HPFL decision matrix using HPFLWA and HPFLWG operators, and shown in Tables 4 and 5.

Table 4. Aggregated hesitant probabilistic fuzzy linguistic element (HPFLE) group decision matrix using hesitant probabilistic fuzzy linguistic weighted averaging (HPFLWA) operator.

<table>
<tr><th></th><th>C_1</th><th>C_2</th><th>C_3</th></tr>
<tr><td>A_1</td><td>{($s_{1.3}$, 0.34|0.12, 0.4|0.18, 0.38|0.28, 0.44|0.42)}</td><td>{($s_{2.7}$, 0.32|0.08, 0.38|0.12, 0.37|0.32, 0.42|0.48)}</td><td>{(s_2, 0.61|1.0)}</td></tr>
<tr><td>A_2</td><td>{($s_{3.7}$, 0.27|0.25, 0.34|0.25, 0.3|0.25, 0.37|0.25)}</td><td>{($s_{4.7}$, 0.37|0.08, 0.41|0.12, 0.4|0.08, 0.44|0.18, 0.41|0.096 0.44|0.144, 0.47|0.216)}</td><td>{($s_{2.3}$, 0.2|0.144, 0.35|0.22, 0.29|0.144, 0.42|0.144, 0.32|0.096 0.39|0.064, 0.44|0.144, 0.5|0.096)}</td></tr>
<tr><td>A_3</td><td>{(s_3, 0.2|0.125, 0.32|0.13, 0.24|0.125, 0.35|0.125, 0.38|0.125 0.44|0.125, 0.47|0.125)}</td><td>{($s_{3.3}$, 0.27|0.09, 0.3|0.06, 0.34|0.09, 0.37|0.06, 0.35|0.21 0.41|0.21, 0.38|0.14, 0.44|0.14)}</td><td>{($s_{2.3}$, 0.26|0.4, 0.29|0.6)}</td></tr>
<tr><td>A_4</td><td>{($s_{2.3}$, 0.37|0.18, 0.44|0.18, 0.41|0.18, 0.47|0.12, 0.41|0.12 0.44|0.08, 0.47|0.12, 0.5|0.08)}</td><td>{(s_3, 0.42|1.0)}</td><td>{($s_{3.3}$, 0.23|0.18, 0.35|0.12, 0.32|0.18, 0.42|0.12, 0.35|0.12 0.42|0.12, 0.44|0.08, 0.5|0.08)}</td></tr>
</table>

Table 5. Aggregated HPFLE group decision matrix using hesitant probabilistic fuzzy linguistic weighted geometric (HPFLWG) operator.

<table>
<tr><th></th><th>C_1</th><th>C_2</th><th>C_3</th></tr>
<tr><td>A_1</td><td>{($s_{3.3}$, 0.32|0.12, 0.4|0.2, 0.3|0.3, 0.4|0.38)}</td><td>{($s_{4.1}$, 0.27|0.08, 0.3|0.1, 0.3|0.3, 0.36|0.5)}</td><td>{($s_{3.6}$, 0.54|1.0)}</td></tr>
<tr><td>A_2</td><td>{($s_{4.6}$, 0.26|0.25, 0.3|0.25, 0.3|0.3, 0.4|0.2)}</td><td>{(s_5, 0.36|0.08, 0.4|0.12, 0.4|0.22, 0.4|0.14, 0.42|0.1 0.5|0.22)}</td><td>{(s_4, 0.18|0.14, 0.3|0.22, 0.2|0.14, 0.4|0.1, 0.2|0.1 0.3|0.06)}</td></tr>
<tr><td>A_3</td><td>{($s_{4.2}$, 0.18|0.13, 0.2|0.13, 0.3|0.15, 0.3|0.13 0.42|0.2, 0.5|0.23)}</td><td>{($s_{4.4}$, 0.26|0.09, 0.3|0.06, 0.34|0.09, 0.37|0.06, 0.35|0.21 0.41|0.21, 0.38|0.14, 0.44|0.14)}</td><td>{($s_{3.8}$, 0.17|0.4, 0.2|0.6)}</td></tr>
<tr><td>A_4</td><td>{(s_4, 0.36|0.18, 0.43|0.24, 0.39|0.18, 0.46|0.12, 0.42|0.08 0.46|0.12, 0.5|0.08)}</td><td>{($s_{4.2}$, 0.37|1.0)}</td><td>{($s_{4.5}$, 0.23|0.18, 0.31|0.16, 0.27|0.28, 0.42|0.18, 0.42|0.12, 0.5|0.08)}</td></tr>
</table>

Step 3: Aggregate assessment of each alternative $A_i(i = 1,2,3,4)$ against each criteria is calculated using the HPFLWA and HPFLWG aggregation operators with criteria weights $w_1 = 0.4, w_2 = 0.3, w_3 = 0.4$ as follows:

$$H_1 = \text{HPFLWA}\ (C_{11}, C_{12}, C_{13})$$
$$H_1 = [(s_{1.3},\ 0.34|0.12,\ 0.4|0.18,\ 0.38|0.28,\ 0.44|0.42), (s_{2.7},\ 0.32|0.08,\ 0.38|0.12,\ 0.37|0.32,\ 0.42|0.48), (s_2,\ 0.61|1.0)]$$
$$H_1 = \begin{bmatrix} (s_{1.93},\ 0.43|0.01,\ 0.44|0.014,\ 0.44|0.038,\ 0.45|0.058,\ 0.45|0.014,\ 0.46|0.022,\ 0.46|0.058,\ 0.47|0.086, \\ 0.44|0.022,\ 0.45|0.034,\ 0.45|0.09,\ 0.47|0.134,\ 0.46|0.034,\ 0.47|0.05,\ 0.47|0.134,\ 0.49|0.202) \end{bmatrix}$$
$$H_1 = \text{HPFLWG}\ (C_{11}, C_{12}, C_{13})$$
$$H_1 = [(s_{3.3},\ 0.32|0.12,\ 0.4|0.2,\ 0.3|0.3,\ 0.4|0.38), (s_{4.1},\ 0.27|0.08,\ 0.3|0.1,\ 0.3|0.3,\ 0.36|0.5), (s_{3.6},\ 0.54|1.0)]$$
$$H_1 = \begin{bmatrix} (s_{1.93},\ 0.4|0.01,\ 0.42|0.014,\ 0.41|0.038,\ 0.43|0.058,\ 0.42|0.014,\ 0.45|0.022,\ 0.44|0.058,\ 0.46|0.086, \\ 0.41|0.022,\ 0.44|0.034,\ 0.43|0.09,\ 0.45|0.134,\ 0.44|0.034,\ 0.46|0.05,\ 0.46|0.134,\ 0.48|0.202) \end{bmatrix}$$

Similarly, other elements of HPFL decision matrices (Tables 4 and 5) are aggregated into the overall HPFL decision matrix using HPFLWA and HPFLWG operators and shown in Tables 6 and 7.

Table 6. Collective HPFLE group decision matrix using HPFLWA operator.

A_1	{($s_{1.93}$, 0.43 I 0.01, 0.44 I 0.014, 0.44 I 0.038, 0.45 I 0.058, 0.45 I 0.014, 0.46 I 0.022, 0.46 I 0.058, 0.47 I 0.086, 0.44 I 0.022, 0.45 I 0.034, 0.45 I 0.09, 0.47 I 0.134, 0.46 I 0.034, 0.47 I 0.05, 0.47 I 0.134, 0.49 I 0.202
A_2	{($s_{3.57}$, 0.282 I 0.003, 0.321 I 0.004, 0.307 I 0.003, 0.345 I 0.006, 0.295 I 0.003, 0.332 I 0.005, 0.332 I 0.008, 0.323 I 0.004, 0.36 I 0.006, 0.347 I 0.004, 0.383 I 0.01, 0.335 I 0.005, 0.37 I 0.008, 0.37 I 0.012, 0.306 I 0.003, 0.344 I 0.004, 0.33 I 0.003, 0.367 I 0.006, 0.318 I 0.003, 0.354 I 0.005, 0.354 I 0.008, 0.345 I 0.003, 0.381 I 0.004, 0.368 I 0.003, 0.403 I 0.006, 0.357 I 0.003, 0.391 I 0.005, 0.391 I 0.008, 0.315 I 0.002, 0.353 I 0.003, 0.339 I 0.002, 0.375 I 0.004, 0.327 I 0.002, 0.363 I 0.003, 0.363 I 0.005, 0.338 I 0.001, 0.374 I 0.002, 0.361 I 0.001, 0.396 I 0.003, 0.35 I 0.002, 0.384 I 0.002, 0.384 I 0.003, 0.354 I 0.003, 0.39 I 0.004, 0.377 I 0.003, 0.411 I 0.006, 0.366 I 0.003, 0.399 I 0.005, 0.399 I 0.008, 0.375 I 0.002, 0.41 I 0.003, 0.397 I 0.002, 0.431 I 0.004, 0.387 I 0.002, 0.419 I 0.003, 0.419 I 0.005,
A_3	{($s_{2.9}$, 0.241 I 0.005, 0.287 I 0.005, 0.256 I 0.005, 0.301 I 0.005, 0.312 I 0.005, 0.341 I 0.005, 0.354 I 0.005, 0.252 I 0.003, 0.298 I 0.003, 0.268 I 0.003, 0.312 I 0.003, 0.323 I 0.003, 0.351 I 0.003, 0.364 I 0.003, 0.262 I 0.005, 0.307 I 0.005, 0.277 I 0.005, 0.321 I 0.005, 0.332 I 0.005, 0.359 I 0.005, 0.372 I 0.005, 0.274 I 0.003, 0.318 I 0.003, 0.288 I 0.003, 0.332 I 0.003, 0.342 I 0.003, 0.369 I 0.003, 0.382 I 0.003, 0.266 I 0.011,0.31 I 0.011, 0.281 I 0.011, 0.324 I 0.011, 0.335 I 0.011, 0.362 I 0.011, 0.375 I 0.011, 0.287 I 0.011, 0.33 I 0.011, 0.301 I 0.011, 0.344 I 0.011, 0.354 I 0.011, 0.381 I 0.011, 0.393 I 0.011, 0.277 I 0.007, 0.321 I 0.007, 0.292 I 0.007, 0.335 I 0.007, 0.345 I 0.007, 0.372 I 0.007, 0.385 I 0.007, 0.298 I 0.007, 0.34 I 0.007, 0.312 I 0.007, 0.354 I 0.007, 0.364 I 0.007, 0.39 I 0.007, 0.402 I 0.007, 0.241 I 0.007, 0.287 I 0.007,0.256 I 0.007, 0.301 I 0.007, 0.312 I 0.007, 0.341 I 0.007, 0.354 I 0.007, 0.252 I 0.005, 0.298 I 0.005, 0.268 I 0.005, 0.312 I 0.005, 0.323 I 0.005, 0.351 I 0.005, 0.364 I 0.005, 0.262 I 0.007, 0.307 I 0.007, 0.277 I 0.007, 0.321 I 0.007, 0.332 I 0.007, 0.359 I 0.007, 0.372 I 0.007, 0.274 I 0.005, 0.318 I 0.005, 0.288 I 0.005, 0.332 I 0.005, 0.342 I 0.005, 0.369 I 0.005, 0.382 I 0.005, 0.266 I 0.016, 0.31 I 0.016, 0.281 I 0.016, 0.324 I 0.016, 0.335 I 0.016, 0.362 I 0.016, 0.375 I 0.016, 0.287 I 0.016, 0.33 I 0.016, 0.301 I 0.016, 0.344 I 0.016, 0.354 I 0.016, 0.381 I 0.016, 0.393 I 0.016, 0.277 I 0.011, 0.321 I 0.011, 0.292 I 0.011, 0.335 I 0.011, 0.345 I 0.011, 0.372 I 0.011, 0.385 I 0.011, 0.298 I 0.011, 0.34 I 0.011, 0.312 I 0.011, 0.354 I 0.011, 0.364 I 0.011, 0.39 I 0.011, 0.402 I 0.011)}
A_4	{($s_{2.83}$, 0.346 I 0.032, 0.375 I 0.032, 0.362 I 0.032, 0.39 I 0.022, 0.362 I 0.022, 0.377 I 0.014, 0.39 I 0.022, 0.405 I 0.014, 0.376 I 0.022, 0.404 I 0.022, 0.391 I 0.022, 0.418 I 0.014, 0.391 I 0.014, 0.406 I 0.01, 0.418 I 0.014, 0.432 I 0.01, 0.368 I 0.032, 0.396 I 0.032, 0.383 I 0.032, 0.41 I 0.022, 0.383 I 0.022, 0.398 I 0.014, 0.41 I 0.022, 0.424 I 0.014, 0.397 I 0.022, 0.423 I 0.022, 0.411 I 0.022, 0.437 I 0.014, 0.411 I 0.014, 0.426 I 0.01, 0.437 I 0.014, 0.451 I 0.01, 0.376 I 0.022, 0.404 I 0.022, 0.391 I 0.022, 0.418 I 0.014, 0.391 I 0.014, 0.406 I 0.01, 0.418 I 0.014, 0.432 I 0.01, 0.397 I 0.022,0.423 I 0.022, 0.411 I 0.022, 0.437 I 0.014, 0.411 I 0.014, 0.426 I 0.01, 0.437 I 0.014, 0.451 I 0.01,0.405 I 0.014, 0.431 I 0.014, 0.419 I 0.014, 0.445 I 0.01, 0.419 I 0.01, 0.433 I 0.006, 0.445 I 0.01, 0.458 I 0.006, 0.425 I 0.014, 0.45 I 0.014, 0.438 I 0.014, 0.463 I 0.01, 0.438 I 0.01, 0.452 I 0.006, 0.463 I 0.01, 0.476 I 0.006,)}

Table 7. Collective HPFLE group decision matrix using HPFLWG operator.

<table>
<tr><td>A_1</td><td>{($s_{1.93}$, 0.4 | 0.01, 0.42 | 0.014, 0.41 | 0.038, 0.43 | 0.058, 0.42 | 0.014, 0.45 | 0.022, 0.44 | 0.058, 0.46 | 0.086, 0.41 | 0.022,
0.44 | 0.034, 0.43 | 0.09, 0.45 | 0.134, 0.44 | 0.034, 0.46 | 0.05, 0.46 | 0.134, 0.48 | 0.202</td></tr>
<tr><td>A_2</td><td>{($s_{3.57}$, 0.27 | 0.003, 0.31 | 0.004, 0.29 | 0.003, 0.32 | 0.006, 0.28 | 0.003, 0.31 | 0.005, 0.31 | 0.008, 0.32 | 0.004, 0.36 | 0.006,
0.34 | 0.004, 0.38 | 0.01, 0.33 | 0.005, 0.37 | 0.008, 0.36 | 0.012, 0.3 | 0.003, 0.34 | 0.004, 0.33 | 0.003, 0.36 | 0.006, 0.31 | 0.003,
0.35 | 0.005, 0.34 | 0.008, 0.34 | 0.003, 0.38 | 0.004, 0.36 | 0.003, 0.4 | 0.006, 0.35 | 0.003, 0.39 | 0.005, 0.38 | 0.008, 0.31 | 0.002,
0.35 | 0.003, 0.34 | 0.002, 0.37 | 0.004, 0.32 | 0.002, 0.36 | 0.003, 0.35 | 0.005, 0.33 | 0.001, 0.37 | 0.002, 0.36 | 0.001, 0.39 | 0.003,
0.34 | 0.002, 0.38 | 0.002, 0.37 | 0.003, 0.34 | 0.003, 0.39 | 0.004, 0.37 | 0.003, 0.41 | 0.006, 0.35 | 0.003, 0.39 | 0.005, 0.39 | 0.008,
0.36 | 0.002, 0.4 | 0.003, 0.38 | 0.002, 0.42 | 0.004, 0.37 | 0.002, 0.41 | 0.003, 0.4 | 0.005,</td></tr>
<tr><td>A_3</td><td>{($s_{2.9}$, 0.24 | 0.005, 0.29 | 0.005, 0.26 | 0.005, 0.3 | 0.005, 0.31 | 0.005, 0.32 | 0.005, 0.33 | 0.005, 0.25 | 0.003, 0.3 | 0.003,
0.27 | 0.003, 0.31 | 0.003, 0.32 | 0.003, 0.34 | 0.003, 0.35 | 0.003, 0.25 | 0.005, 0.3 | 0.005, 0.27 | 0.005, 0.32 | 0.005, 0.33 | 0.005,
0.35 | 0.005, 0.36 | 0.005, 0.26 | 0.003, 0.31 | 0.003, 0.28 | 0.003, 0.33 | 0.003, 0.34 | 0.003, 0.36 | 0.003, 0.37 | 0.003,
0.26 | 0.011,0.31 | 0.011, 0.28 | 0.011, 0.32 | 0.011, 0.33 | 0.011, 0.35 | 0.011, 0.36 | 0.011, 0.27 | 0.011, 0.32 | 0.011, 0.29 | 0.011,
0.34 | 0.011, 0.35 | 0.011, 0.37 | 0.011, 0.38 | 0.011, 0.26 | 0.007, 0.32 | 0.007, 0.28 | 0.007, 0.33 | 0.007, 0.34 | 0.007, 0.36 | 0.007,
0.37 | 0.007, 0.28 | 0.007, 0.33 | 0.007, 0.3 | 0.007, 0.34 | 0.007, 0.35 | 0.007, 0.37 | 0.007, 0.38 | 0.007, 0.24 | 0.007,
0.29 | 0.007,0.26 | 0.007, 0.3 | 0.007, 0.31 | 0.007, 0.32 | 0.007, 0.33 | 0.007, 0.25 | 0.005, 0.3 | 0.005, 0.27 | 0.005, 0.31 | 0.005,
0.32 | 0.005, 0.34 | 0.005, 0.35 | 0.005, 0.25 | 0.007, 0.3 | 0.007, 0.27 | 0.007, 0.32 | 0.007, 0.33 | 0.007, 0.35 | 0.007, 0.36 | 0.007,
0.26 | 0.005, 0.31 | 0.005, 0.28 | 0.005, 0.33 | 0.005, 0.34 | 0.005, 0.36 | 0.005, 0.37 | 0.005, 0.26 | 0.016, 0.31 | 0.016, 0.28 | 0.016,
0.32 | 0.016, 0.33 | 0.016, 0.35 | 0.016, 0.36 | 0.016, 0.27 | 0.016, 0.32 | 0.016, 0.29 | 0.016, 0.34 | 0.016, 0.35 | 0.016, 0.37 | 0.016,
0.38 | 0.016, 0.26 | 0.011, 0.32 | 0.011, 0.28 | 0.011, 0.33 | 0.011, 0.34 | 0.011, 0.36 | 0.011, 0.37 | 0.011, 0.28 | 0.011, 0.33 | 0.011,
0.3 | 0.011, 0.34 | 0.011, 0.35 | 0.011, 0.37 | 0.011, 0.38 | 0.011)}</td></tr>
<tr><td>A_4</td><td>{($s_{2.83}$, 0.33 | 0.032, 0.36 | 0.032, 0.35 | 0.032, 0.37 | 0.022, 0.35 | 0.022, 0.36 | 0.014, 0.37 | 0.022, 0.38 | 0.014, 0.37 | 0.022,
0.4 | 0.022, 0.39 | 0.022, 0.41 | 0.014, 0.39 | 0.014, 0.4 | 0.01, 0.41 | 0.014, 0.42 | 0.01, 0.36 | 0.032, 0.39 | 0.032, 0.38 | 0.032,
0.4 | 0.022, 0.38 | 0.022, 0.39 | 0.014, 0.4 | 0.022, 0.41 | 0.014, 0.4 | 0.022, 0.42 | 0.022, 0.41 | 0.022, 0.44 | 0.014, 0.41 | 0.014,
0.43 | 0.01, 0.44 | 0.014, 0.45 | 0.01, 0.37 | 0.022, 0.4 | 0.022, 0.39 | 0.022, 0.41 | 0.014, 0.39 | 0.014, 0.4 | 0.01, 0.41 | 0.014,
0.42 | 0.01, 0.4 | 0.022,0.42 | 0.022, 0.41 | 0.022, 0.44 | 0.014, 0.41 | 0.014, 0.43 | 0.01, 0.44 | 0.014, 0.45 | 0.01, 0.4 | 0.014,
0.43 | 0.014, 0.42 | 0.014, 0.44 | 0.01, 0.42 | 0.01, 0.43 | 0.006, 0.44 | 0.01, 0.46 | 0.006, 0.42 | 0.014, 0.45 | 0.014,0.43 | 0.014,
0.46 | 0.01, 0.43 | 0.01, 0.45 | 0.006, 0.46 | 0.01, 0.47 | 0.006,)}</td></tr>
</table>

Step 4: The score values $S(h^i{}_{PL}(p_x))$ $(i = 1,2,3,4)$ of the alternatives $A_i(i = 1,2,3,4)$ are calculated and shown as follows (Table 8):

Table 8. Score values for the alternatives using HPFLWA and HPFLWG operators.

Score	HPFLWA	HPFLWG
$S(h^1{}_{PL}(p_x))$	$s_{0.05652}$	$s_{0.05409}$
$S(h^2{}_{PL}(p_x))$	$s_{0.00559}$	$s_{0.00545}$
$S(h^3{}_{PL}(p_x))$	$s_{0.00745}$	$s_{0.00723}$
$S(h^4{}_{PL}(p_x))$	$s_{0.01901}$	$s_{0.01874}$

Step 5. Finally, alternatives A_i $(i = 1, 2, 3, 4)$ are ranked in accordance with score values $S(h^i{}_{PL}(p_x))$ and shown in Table 9.

Table 9. Ranking of alternatives using proposed HPFLWA and HPFLWG operators.

Method	Ranking	Best/Worst
Using HPFLWA operator	$A_1 > A_4 > A_3 > A_2$	A_1/A_2
Using HPFLWG operator	$A_1 > A_4 > A_3 > A_2$	A_1/A_2

Table 9 confirms that using both proposed HPFLWA and HPFLWG operators best and worst alternatives are A_1 and A_2 respectively.

6.1. A Real Case Study

A real case study is undertaken to rank seven organizations; State Bank of India (A_1), InfoTech Enterprises (A_2), ITC (A_3), H.D.F.C. Bank (A_4), Tata Steel (A_5), Tata Motors (A_6) and Bajaj Finance (A_7) on the basis of their performance against following four criteria.

1. Earnings per share (EPS) of company (C_1)
2. Face value (C_2)
3. Book value (C_3)
4. P/C ratio (Put-Call Ratio) of company (C_4)

In this real case study, C_1, C_2, and C_3 are benefit criteria while C_4 is cost criterion. Real data for each alternative against each criterion are retrieved from http://www.moneycontrol.com from date 20.7.2017 to 27.7.2017. Table 10 shows their average values.

Table 10. Average of actual numerical value of criteria.

	C_1	C_2	C_3	C_4
A_1	13.15	1.00	196.53	19.27
A_2	61.18	5.00	296.12	14.98
A_3	8.54	1.00	37.31	30.52
A_4	59.07	2.00	347.59	28.50
A_5	22.25	2.00	237.82	5.98
A_6	35.47	1.12	511.31	7.95
A_7	36.64	2.00	174.60	45.39

To construct hesitant fuzzy decision matrix (Table 11), we use the method proposed by Bisht and Kumar [69] and fuzzify Table 10 using triangular and Gaussian membership functions.

Table 11. Hesitant fuzzy decision matrix

	C_1	C_2	C_3	C_4
A_1	0.3784, 0.3029	0.6065, 0.50	0.7545, 0.6247	0.997, 0.9614
A_2	0.9676, 0.8718	0.6065, 0.50	0.8964, 0.7662	0.696, 0.5743
A_3	0.1534, 0.0318	0.6065, 0.50	0.1368, 0.0027	0.778, 0.6457
A_4	0.9997, 0.9959	0.6065, 0.50	0.8122, 0.6775	0.7748, 0.6429
A_5	0.949, 0.8382	0.6065, 0.50	0.8655, 0.7312	0.2278, 0.14
A_6	0.7445, 0.6159	0.7491, 0.62	0.9843, 0.9111	0.512, 0.4214
A_7	0.8197, 0.6847	0.6065, 0.50	0.933, 0.8138	0.3055, 0.23

Probabilities are associated with elements of hesitant fuzzy decision matrix (Table 11) to convert it into probabilistic hesitant fuzzy decision matrix $I = [I_{Pij} = (\mu_{ij}|p_{ij})])_{m\times n}$. Probabilities which are associated with first row of hesitant fuzzy decision matrix (Table 11) are as follows:

$$\begin{aligned}
\mu\left(p_{11}^1\right) &= \tfrac{0.3784}{0.3784+0.3029} = 0.5554, \mu\left(p_{11}^2\right) = \tfrac{0.3029}{0.3784+0.3029} = 0.4446\\
\mu\left(p_{12}^1\right) &= \tfrac{0.6065}{0.6065+0.5} = 0.5481, \mu\left(p_{12}^2\right) = \tfrac{0.5}{0.6065+0.5} = 0.4519\\
\mu\left(p_{14}^1\right) &= \tfrac{0.9614}{0.997+0.9614} = 0.9285, \mu\left(p_{14}^2\right) = \tfrac{0.997}{0.9614+0.997} = 0.0715
\end{aligned}$$

Similarly all elements of hesitant fuzzy decision matrix are associated with probabilities and probabilistic hesitant fuzzy decision matrix (Table 12) is obtained.

Following table (Table 13) shows hesitant probabilistic fuzzy linguistic decision matrix.

Table 12. Probabilistic Hesitant fuzzy decision matrix.

	C_1	C_2	C_3	C_4
A_1	{(0.3784 ǀ 0.5554), (0.3029 ǀ 0.4446)}	{(0.6065 ǀ 0.5481), (0.5 ǀ 0.4519)}	{(0.7545 ǀ 0.5471), (0.6247 ǀ 0.4529)}	{(0.997 ǀ 0.0715), (0.9614 ǀ 0.9285)}
A_2	{(0.9676 ǀ 0.5261), (0.8718 ǀ 0.4739)}	{(0.6065 ǀ 0.5481), (0.5 ǀ 0.4519)}	{(0.8964 ǀ 0.5392), (0.7662 ǀ 0.4608)}	{(0.696 ǀ 0.4166), (0.5743 ǀ 0.5834)}
A_3	{(0.1534 ǀ 0.8283), (0.0318 ǀ 0.1717)}	{(0.6065 ǀ 0.5481), (0.5 ǀ 0.4519)}	{(0.1368 ǀ 0.9806), (0.0027 ǀ 0.0194)}	{(0.778 ǀ 0.3852), (0.6457 ǀ 0.6148)}
A_4	{(0.9997 ǀ 0.501), (0.9959 ǀ 0.499)}	{(0.6065 ǀ 0.5481), (0.5 ǀ 0.4519)}	{(0.8122 ǀ 0.5452), (0.6775 ǀ 0.4548)}	{(0.7748 ǀ 0.3867), (0.6429 ǀ 0.6133)}
A_5	{(0.949 ǀ 0.531), (0.8382 ǀ 0.469)}	{(0.6065 ǀ 0.5481), (0.5 ǀ 0.4519)}	{(0.8655 ǀ 0.542), (0.7312 ǀ 0.458)}	{(0.2278 ǀ 0.4731), (0.14 ǀ 0.5269)}
A_6	{(0.7445 ǀ 0.5473), (0.6159 ǀ 0.4527)}	{(0.7492 ǀ 0.5472), (0.62 ǀ 0.4528)}	{(0.9843 ǀ 0.5193), (0.9111 ǀ 0.4807)}	{(0.512 ǀ 0.4575), (0.4214 ǀ 0.5425)}
A_7	{(0.8197 ǀ 0.5449), (0.6847 ǀ 0.4551)}	{(0.6065 ǀ 0.5481), (0.5 ǀ 0.4519)}	{(0.933 ǀ 0.5341), (0.8138 ǀ 0.4659)}	{(0.3055 ǀ 0.4742), (0.23 ǀ 0.5258)}

Table 13. Hesitant probabilistic fuzzy linguistic decision matrix.

	C_1	C_2	C_3	C_4
A_1	{s_1, (0.3784 ǀ 0.5554), (0.3029 ǀ 0.4446)}	{s_1, (0.6065 ǀ 0.5481), (0.5 ǀ 0.4519)}	{s_2, (0.7545 ǀ 0.5471), (0.6247 ǀ 0.4529)}	{s_3, (0.997 ǀ 0.0715), (0.9614 ǀ 0.9285)}
A_2	{s_3, (0.9676 ǀ 0.5261), (0.8718 ǀ 0.4739)}	{s_1, (0.6065 ǀ 0.5481), (0.5 ǀ 0.4519)}	{s_2, (0.8964 ǀ 0.5392), (0.7662 ǀ 0.4608)}	{s_2, (0.696 ǀ 0.4166), (0.5743 ǀ 0.5834)}
A_3	{s_0, (0.1534 ǀ 0.8283), (0.0318 ǀ 0.1717)}	{s_1, (0.6065 ǀ 0.5481), (0.5 ǀ 0.4519)}	{s_0, (0.1368 ǀ 0.9806), (0.0027 ǀ 0.0194)}	{s_2, (0.778 ǀ 0.3852), (0.6457 ǀ 0.6148)}
A_4	{s_3, (0.9997 ǀ 0.501), (0.9959 ǀ 0.499)}	{s_1, (0.6065 ǀ 0.5481), (0.5 ǀ 0.4519)}	{s_2, (0.8122 ǀ 0.5452), (0.6775 ǀ 0.4548)}	{s_2, (0.7748 ǀ 0.3867), (0.6429 ǀ 0.6133)}
A_5	{s_3, (0.949 ǀ 0.531), (0.8382 ǀ 0.469)}	{s_1, (0.6065 ǀ 0.5481), (0.5 ǀ 0.4519)}	{s_3, (0.8655 ǀ 0.542), (0.7312 ǀ 0.458)}	{s_0, (0.2278 ǀ 0.4731), (0.14 ǀ 0.5269)}
A_6	{s_2, (0.7445 ǀ 0.5473), (0.6159 ǀ 0.4527)}	{s_2, (0.7492 ǀ 0.5472), (0.62 ǀ 0.4528)}	{s_3, (0.9843 ǀ 0.5193), (0.9111 ǀ 0.4807)}	{s_1, (0.512 ǀ 0.4575), (0.4214 ǀ 0.5425)}
A_7	{s_2, (0.8197 ǀ 0.5449), (0.6847 ǀ 0.4551)}	{s_1, (0.6065 ǀ 0.5481), (0.5 ǀ 0.4519)}	{s_3, (0.933 ǀ 0.5341), (0.8138 ǀ 0.4659)}	{s_0, (0.3055 ǀ 0.4742), (0.23 ǀ 0.5258)}

Step 2: Assessment of each alternative A_i (i = 1, 2, 3, 4, 5, 6, 7) against each criteria C_j (i = 1, 2, 3, 4) is aggregated using HPFLWA aggregation operator (Equation (9)) as follows:

$$H_1 = \text{HPFLWA}\ (C_{11}, C_{12}, C_{13}, C_{14})$$

$$H_1 = \begin{bmatrix} \{s_1,\ (0.3784|0.5554), (0.3029|0.4446)\}, \{s_1,\ (0.6065|0.5481), (0.5|0.4519), \\ \{s_2,\ (0.7545|0.5471)\},\ (0.6247|0.4529),\ \{s_3,\ (0.997|0.0715), (0.9614|0.9285)\} \end{bmatrix}$$

$$H_1 = \begin{bmatrix} \{s_{1.75}, (0.884|0.012),\ (0.871|0.01),\ (0.877|0.1055),\ (0.8811|0.0095),\ (0.8678|0.0079), \\ (0.3029|0.4446), (0.3784|0.5554), (0.3029|0.4446), (0.3784|0.5554), (0.3029|0.4446), \\ (0.3784|0.5554), (0.3029|0.4446),\ (0.3784|0.5554),\ (0.3029|0.4446),\ (0.3784|0.5554), \\ (0.3029|0.4446)\} \end{bmatrix}$$

In the aggregation of assessment of the alternatives, all criteria are considered of equal weight of 0.25. Similarly other elements of HPFL decision matrix (Table 13) are aggregated and following collective HPFL decision matrix (Table 14) is obtained.

Table 14. Collective hesitant probabilistic fuzzy linguistic decision matrix

A_1	{$s_{1.75}$, (0.884 | 0.012), (0.871 | 0.01), (0.877 | 0.1055), (0.8811 | 0.0095), (0.8678 | 0.0079), (0.3029 | 0.4446), (0.3784 | 0.5554), (0.3029 | 0.4446), (0.3784 | 0.5554), (0.3029 | 0.4446), (0.3784 | 0.5554), (0.3029 | 0.4446), (0.3784 | 0.5554), (0.3029 | 0.4446), (0.3784 | 0.5554), (0.3029 | 0.4446)}
A_2	{s_2, (0.3784 | 0.5554), (0.3029 | 0.4446), (0.3784 | 0.5554), (0.3029 | 0.4446), (0.3784 | 0.5554), (0.3029 | 0.4446), (0.3784 | 0.5554), (0.3029 | 0.4446), (0.3784 | 0.5554), (0.3029 | 0.4446), (0.3784 | 0.5554), (0.3029 | 0.4446), (0.3784 | 0.5554), (0.3029 | 0.4446), (0.3784 | 0.5554), (0.3029 | 0.4446)}
A_3	{$s_{0.75}$, (0.3784 | 0.5554), (0.3029 | 0.4446), (0.3784 | 0.5554), (0.3029 | 0.4446), (0.3784 | 0.5554), (0.3029 | 0.4446), (0.3784 | 0.5554), (0.3029 | 0.4446), (0.3784 | 0.5554), (0.3029 | 0.4446), (0.3784 | 0.5554), (0.3029 | 0.4446), (0.3784 | 0.5554), (0.3029 | 0.4446), (0.3784 | 0.5554), (0.3029 | 0.4446)}
A_4	{s_2, (0.3784 | 0.5554), (0.3029 | 0.4446), (0.3784 | 0.5554), (0.3029 | 0.4446), (0.3784 | 0.5554), (0.3029 | 0.4446), (0.3784 | 0.5554), (0.3029 | 0.4446), (0.3784 | 0.5554), (0.3029 | 0.4446), (0.3784 | 0.5554), (0.3029 | 0.4446), (0.3784 | 0.5554), (0.3029 | 0.4446), (0.3784 | 0.5554), (0.3029 | 0.4446)}
A_5	{$s_{1.75}$, (0.3784 | 0.5554), (0.3029 | 0.4446), (0.3784 | 0.5554), (0.3029 | 0.4446), (0.3784 | 0.5554), (0.3029 | 0.4446), (0.3784 | 0.5554), (0.3029 | 0.4446), (0.3784 | 0.5554), (0.3029 | 0.4446), (0.3784 | 0.5554), (0.3029 | 0.4446), (0.3784 | 0.5554), (0.3029 | 0.4446), (0.3784 | 0.5554), (0.3029 | 0.4446)}
A_6	{s_2, (0.3784 | 0.5554), (0.3029 | 0.4446), (0.3784 | 0.5554), (0.3029 | 0.4446), (0.3784 | 0.5554), (0.3029 | 0.4446), (0.3784 | 0.5554), (0.3029 | 0.4446), (0.3784 | 0.5554), (0.3029 | 0.4446),(0.3784 | 0.5554), (0.3029 | 0.4446), (0.3784 | 0.5554), (0.3029 | 0.4446), (0.3784 | 0.5554), (0.3029 | 0.4446)}
A_7	{$s_{1.5}$, (0.3784 | 0.5554), (0.3029 | 0.4446), (0.3784 | 0.5554), (0.3029 | 0.4446), (0.3784 | 0.5554), (0.3029 | 0.4446), (0.3784 | 0.5554), (0.3029 | 0.4446), (0.3784 | 0.5554), (0.3029 | 0.4446), (0.3784 | 0.5554), (0.3029 | 0.4446), (0.3784 | 0.5554), (0.3029 | 0.4446), (0.3784 | 0.5554), (0.3029 | 0.4446)}

Step 3: The score values $S(h^i{}_{PL}(p_x))$ $(i = 1,2,3,4,5,6,7)$ of the alternatives $A_i(i = 1,2,3,4,5,6,7)$ are calculated using Equation (7) and are shown as follows:

$$\begin{aligned} S(h^1{}_{PL}(p_x)) &= s_{0.0902},\ S(h^2{}_{PL}(p_x)) = s_{0.092},\ S(h^3{}_{PL}(p_x)) = s_{0.019} \\ S(h^4{}_{PL}(p_x)) &= s_{0.101},\ S(h^5{}_{PL}(p_x)) = s_{0.076},\ S(h^6{}_{PL}(p_x)) = s_{0.094} \\ S(h^7{}_{PL}(p_x)) &= s_{0.063} \end{aligned}$$

Step 4: Finally, alternatives $A_i(i = 1,2,3,4,5,6,7)$ are ranked as $A_4 > A_6 > A_2 > A_1 > A_5 > A_7 > A_3$ in accordance with score values $S(h^i{}_{PL}(p_x))$.

6.2. Comparative Analysis

In this section, we compare proposed HPFL-based MCGDM methods with existing HFL-based methods. We apply the proposed method on two different problems which are adapted from Zhou et al. (2016) and Lin et al. (2014) and compare the ranking results. In order to apply the proposed HPFL-based MCGDM on the examples taken by both Lin et al. (2014) and Zhou et al. (2016), we have considered probability of each element of HFL decision matrices as unity.

6.2.1. Comparison 1

In comparison 1, methodology of proposed HPFL-based MCGDM method is applied on the following HFL decision matrix (Table 15) of the problem taken by Lin et al. [50].

Table 15. Hesitant fuzzy linguistic decision matrix ([50]).

	G_1	G_2	G_3	G_4
A_1	<s_5, (0.3, 0.5)>	<s_3, (0.6, 0.7, 0.8)>	<s_2, (0.7, 0.8)>	<s_4, (0.8, 0.9)>
A_2	<s_2, (0.3, 0.4, 0.5)>	<s_5, (0.6, 0.9)>	<s_3, (0.6, 0.7)>	<s_5, (0.4, 0.5)>
A_3	<s_6, (0.4, 0.6)>	<s_2, (0.7, 0.8)>	<s_5, (0.3, 0.5, 0.7)>	<s_3, (0.6, 0.7)>
A_4	<s_5, (0.7, 0.9)>	<s_1, (0.3, 0.4)>	<s_7, (0.5, 0.7)>	<s_2, (0.3, 0.5)>
A_5	<s_4, (0.2, 0.3)>	<s_2, (0.6, 0.7)>	<s_4, (0.5, 0.6)>	<s_2, (0.7, 0.8, 0.9)>

Following table (Table 16) shows the ranking results of the alternatives which are obtained using proposed HPFL and existing HFL-based MCDM method of Lin et al. [50].

Table 16. Comparison of ranking of alternatives.

Method	Ranking	Best Alternative/Worst Alternative
Proposed	$A_4 > A_1 > A_3 > A_2 > A_5$	A_4/A_5
Lin et al. [50]	$A_4 > A_3 > A_1 > A_2 > A_5$	A_4/A_5

On applying the proposed MCGDM method on ranking problem which is adapted from Lin et al. (2014), A_4 and A_5 are ranked again as the best and the worst alternatives respectively.

6.2.2. Comparison 2

In comparison 2, the methodology of proposed HPFL-based MCGDM method is applied on the following HFL decision matrix (Table 17) of the problem taken by Zhou et al. [67].

Table 17. The special linguistic hesitant fuzzy decision matrix ([67]).

	C_1	C_2	C_3	C_4
X_1	$<s_5, (0.3, 0.4)>$	$<s_6, (0.2, 0.4)>$	$<s_5, (0.5, 0.7)>$	$<s_4, (0.4)>$
X_2	$<s_3, (0.5, 0.6)>$	$<s_5, (0.3, 0.5)>$	$<s_4, (0.7)>$	$<s_5, (0.4, 0.6)>$
X_3	$<s_4, (0.4, 0.6)>$	$<s_5, (0.4)>$	$<s_7, (0.7, 0.8)>$	$<s_3, (0.6)>$
X_4	$<s_3, (0.3, 0.4, 0.5)>$	$<s_4, (0.6)>$	$<s_3, (0.4, 0.7)>$	$<s_3, (0.8)>$
X_5	$<s_6, (0.5, 0.7)>$	$<s_6, (0.5, 0.6)>$	$<s_4, (0.6, 0.8)>$	$<s_5, (0.7)>$

Following table (Table 18) shows the ranking results of the alternatives which are obtained using proposed HPFL and existing HFL-based MCDM method of Zhou et al. [67].

Table 18. Comparison of ranking of alternatives.

Method	Ranking	Best Alternative/Worst Alternative
Proposed	$X_5 > X_3 > X_1 > X_2 > X_4$	X_5/X_4
Zhou et al. [67]	$X_5 > X_3 > X_1 > X_2 > X_4$	X_5/X_4

On applying the proposed MCGDM method on ranking problem adapted from Zhou et al. [67], X_5 and X_4 are ranked again as the best and the worst alternatives respectively.

As there is no change found in the ranking results of the alternatives in both the comparisons, it confirms that the proposed HPFL-based MCGDM method is also suitable with HFL information.

7. Conclusions

Uncertainties due to randomness and fuzziness both occur in the system simultaneously. In certain decision making problem, DMs prefer to analyze the alternatives against decision criteria qualitatively using linguistic terms. In this paper, we have proposed hesitant probabilistic fuzzy linguistic set (HPFLS) to integrate hesitant fuzzy linguistic information with probability theory. Prominent characteristic of HPFLS is to associate occurring probabilities to HFLEs which makes it more effective than HFLS. We have investigated the expected mean, variance, score and accuracy function, and basic operations for HPFLEs. We have also defined HPFLWA, HPFLWG, HPFLOWA and HPFLOWG aggregation operators to aggregate hesitant probabilistic fuzzy linguistic information. A novel MCGDM method using HPFLWA, HPFLWG, HPFLOWA and HPFLOWG is also proposed in the present study. Advantage of proposed HPFLS-based MCGDM method is that it associates probabilities to HFLE which makes it competent enough to handle both stochastic and non-stochastic uncertainties with hesitant information using both qualitative and quantitative terms. Another advantage of proposed MCGDM method is that it allows DMs to use their intuitive ability to judge alternatives against criteria using probabilities. This is also important to note that the proposed method can also be used with HFL information if DMs associate equal probabilities to HFLE. Methodology of proposed HPFL-based MCGDM method is illustrated by an example. A real case study to rank the organizations is also undertaken in the present work.

Even though, proposed HPFL-based MCGDM method includes both stochastic and non-stochastic uncertainties along with hesitation, but to determine probabilities of membership grades in linguistic fuzzy set is very difficult in real life problem of decision making. Proposed HPFL-based MCGDM method will be effective when either DMs are expert of their field or they have pre-defined probability distribution function so that the appropriate probabilities could be assigned. Applications of proposed HPFLS with Pythagorean membership grades can also be seen as the scope of future research in decision making problems as an enhancement of the methods proposed by Garg [49].

Author Contributions: Dheeraj Kumar Joshi and Sanjay Kumar defined HPFLS and studied its properties. They together developed MCGDM method using HPFL information. Ismat Beg contributed in verifying the proof of Theorem 1 and the properties of aggregation operators. All authors equally contributed in the research paper.

Conflicts of Interest: Authors declare no conflicts of interest.

References

1. Meghdadi, A.H.; Akbarzadeh-T, M.R. Probabilistic fuzzy logic and probabilistic fuzzy systems. In Proceedings of the 10th IEEE International Conference on Fuzzy Systems, Melbourne, Australia, 2–5 December 2001; Volume 3, pp. 1127–1130.
2. Valavanis, K.P.; Saridis, G.N. Probabilistic modeling of intelligent robotic systems. *IEEE Trans. Robot. Autom.* **1991**, *7*, 164–171. [CrossRef]
3. Pidre, J.C.; Carrillo, C.J.; Lorenzo, A.E.F. Probabilistic model for mechanical power fluctuations in asynchronous wind parks. *IEEE Trans. Power Syst.* **2003**, *18*, 761–768. [CrossRef]
4. Zadeh, L.A. Fuzzy sets. *Inf. Control* **1965**, *8*, 338–353. [CrossRef]
5. Zadeh, L.A. Fuzzy logic and approximate reasoning. *Synthese* **1975**, *30*, 407–428. [CrossRef]
6. Lee, L.W.; Chen, S.M. Fuzzy decision making and fuzzy group decision making based on likelihood-based comparison relations of hesitant fuzzy linguistic term sets. *J. Intell. Fuzzy Syst.* **2015**, *29*, 1119–1137. [CrossRef]
7. Wang, H.; Xu, Z. Admissible orders of typical hesitant fuzzy elements and their application in ordered information fusion in multi-criteria decision making. *Inf. Fusion* **2016**, *29*, 98–104. [CrossRef]
8. Liu, J.; Chen, H.; Zhou, L.; Tao, Z. Generalized linguistic ordered weighted hybrid logarithm averaging operators and applications to group decision making. *Int. J. Uncertain. Fuzz. Knowl.-Based Syst.* **2015**, *23*, 421–442. [CrossRef]
9. Liu, J.; Chen, H.; Xu, Q.; Zhou, L.; Tao, Z. Generalized ordered modular averaging operator and its application to group decision making. *Fuzzy Sets Syst.* **2016**, *299*, 1–25. [CrossRef]
10. Yoon, K.P.; Hwang, C.L. *Multiple Attribute Decision Making: An Introduction*; Sage Publications: New York, NY, USA, 1995; Volume 104.
11. Mardani, A.; Nilachi, M.; Zavadskas, E.K.; Awang, S.R.; Zare, H.; Jamal, N.M. Decision making methods based on fuzzy aggregation operators: Three decades of review from 1986 to 2018. *Int. J. Inf. Technol. Decis. Mak.* **2017**. [CrossRef]
12. Atanassov, K.T. Intuitionistic fuzzy sets. *Fuzzy Sets Syst.* **1986**, *20*, 87–96. [CrossRef]
13. Atanassov, K.T.; Gargov, G. Interval-valued intuitionistic fuzzy sets. *Fuzzy Sets Syst.* **1989**, *31*, 343–349. [CrossRef]
14. Yager, R.R. Pythagorean membership grades in multicriteria decision making. *IEEE Trans. Fuzzy Syst.* **2014**, *22*, 958–965. [CrossRef]
15. Majumdar, P. Neutrosophic Sets and Its Applications to Decision Making. In *Computational Intelligence for Big Data Analysis. Adaptation, Learning, and Optimization*; Acharjya, D., Dehuri, S., Sanyal, S., Eds.; Springer: Cham, Switzerland, 2015; Volume 19.

16. Torra, V.; Narukawa, Y. On hesitant fuzzy sets and decision. In Proceedings of the 18th IEEE International Conference on Fuzzy Systems, Jeju Island, Korea, 20–24 August 2009; pp. 1378–1382.
17. Torra, V. Hesitant fuzzy sets. *Int. J. Intell. Syst.* **2010**, *25*, 529–539. [CrossRef]
18. Xia, M.; Xu, Z. Hesitant fuzzy information aggregation in decision making. *Int. J. Approx. Reason.* **2011**, *52*, 395–407. [CrossRef]
19. Farhadinia, B.; Xu, Z. Distance and aggregation-based methodologies for hesitant fuzzy decision making. *Cogn. Comput.* **2017**, *9*, 81–94. [CrossRef]
20. Qian, G.; Wang, H.; Feng, X. Generalized hesitant fuzzy sets and their application in decision support system. *Knowl.-Based Syst.* **2013**, *37*, 357–365. [CrossRef]
21. Peng, J.J.; Wang, J.Q.; Wang, J.; Yang, L.J.; Chen, X.H. An extension of ELECTRE to multi-criteria decision-making problems with multi-hesitant fuzzy sets. *Inf. Sci.* **2015**, *307*, 113–126. [CrossRef]
22. Chen, S.W.; Cai, L.N. Interval-valued hesitant fuzzy sets. *Fuzzy Syst. Math.* **2013**, *6*, 38–44.
23. Yu, D. Triangular hesitant fuzzy set and its application to teaching quality evaluation. *J. Inf. Comput. Sci.* **2013**, *10*, 1925–1934. [CrossRef]
24. Zhu, B.; Xu, Z.; Xia, M. Dual hesitant fuzzy sets. *J. Appl. Math.* **2012**, *2012*, 879629. [CrossRef]
25. Zhang, Z. Interval-valued intuitionistic hesitant fuzzy aggregation operators and their application in group decision-making. *J. Appl. Math.* **2013**, *2013*, 670285. [CrossRef]
26. Joshi, D.; Kumar, S. Interval-valued intuitionistic hesitant fuzzy Choquet integral based TOPSIS method for multi-criteria group decision making. *Eur. J. Oper. Res.* **2016**, *248*, 183–191. [CrossRef]
27. Garg, H. Hesitant Pythagorean fuzzy sets and their aggregation operators in multiple attribute decision making. *Int. J. Uncertain. Quantif.* **2018**. [CrossRef]
28. Qi, X.-W.; Zhang, J.-L.; Zhao, S.-P.; Liang, C.-Y. Tackling complex emergency response solutions evaluation problems in sustainable development by fuzzy group decision making approaches with considering decision hesitancy and prioritization among assessing criteria. *Int. J. Environ. Res. Public Health* **2017**, *14*, 1165. [CrossRef] [PubMed]
29. Garg, H.; Arora, R. Distance and similarity measures for dual hesitant fuzzy soft sets and their applications in multicriteria decision making problem. *Int. J. Uncertain. Quantif.* **2017**, *7*, 229–248. [CrossRef]
30. Martínez, L.; Ruan, D.; Herrera, F.; Wang, P.P. Linguistic decision making: Tools and applications. *Inf. Sci.* **2009**, *179*, 2297–2298. [CrossRef]
31. Xu, Z. An interactive procedure for linguistic multiple attribute decision making with incomplete weight information. *Fuzzy Optim. Decis. Mak.* **2007**, *6*, 17–27. [CrossRef]
32. Herrera, F.; Martínez, L. A 2-tuple fuzzy linguistic representation model for computing with words. *IEEE Trans. Fuzzy Syst.* **2000**, *8*, 746–752.
33. Herrera, F.; Herrera-Viedma, E.; Alonso, S.; Chiclana, F. Computing with words and decision making. *Fuzzy Optim. Decis. Mak.* **2009**, *8*, 323–324. [CrossRef]
34. Lan, J.; Sun, Q.; Chen, Q.; Wang, Q. Group decision making based on induced uncertain linguistic OWA operators. *Decis. Support Syst.* **2013**, *55*, 296–303. [CrossRef]
35. Beg, I.; Rashid, T. TOPSIS for hesitant fuzzy linguistic term sets. *Int. J. Intell. Syst.* **2013**, *28*, 1162–1171. [CrossRef]
36. Rodríguez, R.M.; Martnez, L.; Herrera, F. A group decision making model dealing with comparative linguistic expressions based on hesitant fuzzy linguistic term sets. *Inf. Sci.* **2013**, *241*, 28–42.
37. Yuen, K.K.F. Combining compound linguistic ordinal scale and cognitive pairwise comparison in the rectified fuzzy TOPSIS method for group decision making. *Fuzzy Optim. Decis. Mak.* **2014**, *13*, 105–130. [CrossRef]
38. Zhang, Z.; Wu, C. Hesitant fuzzy linguistic aggregation operators and their applications to multiple attribute group decision making. *J. Intell. Fuzzy Syst.* **2014**, *26*, 2185–2202.
39. Beg, I.; Rashid, T. Group decision making using comparative linguistic expression based on hesitant intuitionistic fuzzy sets. *Appl. Appl. Math. Int. J.* **2015**, *10*, 1082–1092.
40. Wang, J.Q.; Wang, D.D.; Zhang, H.Y.; Chen, X.H. Multi-criteria group decision making method based on interval 2-tuple linguistic information and Choquet integral aggregation operators. *Soft Comput.* **2015**, *19*, 389–405. [CrossRef]
41. Merigó, J.M.; Palacios-Marqués, D.; Zeng, S. Subjective and objective information in linguistic multi-criteria group decision making. *Eur. J. Oper. Res.* **2016**, *248*, 522–531. [CrossRef]

42. Beg, I.; Rashid, T. Hesitant 2-tuple linguistic information in multiple attributes group decision making. *J. Intell. Fuzzy Syst.* **2016**, *30*, 109–116. [CrossRef]
43. Zhou, W.; Xu, Z. Generalized asymmetric linguistic term set and its application to qualitative decision making involving risk appetites. *Eur. J. Oper. Res.* **2016**, *254*, 610–621. [CrossRef]
44. De Maio, C.; Fenza, G.; Loia, V.; Orciuoli, F. Linguistic fuzzy consensus model for collaborative development of fuzzy cognitive maps: A case study in software development risks. *Fuzzy Optim. Decis. Mak.* **2017**, in press. [CrossRef]
45. Gao, J.; Xu, Z.; Liao, H. A dynamic reference point method for emergency response under hesitant probabilistic fuzzy environment. *Int. J. Fuzzy Syst.* **2017**, *19*, 1261–1278. [CrossRef]
46. Kobina, A.; Liang, D.; He, X. Probabilistic linguistic power aggregation operators for multi-criteria group decision making. *Symmetry* **2017**, *9*, 320. [CrossRef]
47. Garg, H.; Kumar, K. Some aggregation operators for linguistic intuitionistic fuzzy set and its application to group decision-making process using the set pair analysis. *Arbian J. Sci. Eng.* **2017**. [CrossRef]
48. Liu, P.; Mahmood, T.; Khan, Q. Multi-attribute decision-making based on prioritized aggregation operator under hesitant intuitionistic fuzzy linguistic environment. *Symmetry* **2017**, *9*, 270. [CrossRef]
49. Garg, H. Linguistic Pythagorean fuzzy sets and its applications in multi attribute decision making process. *Int. J. Intell. Syst.* **2018**. [CrossRef]
50. Lin, R.; Zhao, X.; Wei, G. Models for selecting an ERP system with hesitant fuzzy linguistic information. *J. Intell. Fuzzy Syst.* **2014**, *26*, 2155–2165.
51. Ren, F.; Kong, M.; Pei, Z. A new hesitant fuzzy linguistic topsis method for group multi-criteria linguistic decision making. *Symmetry* **2017**, *9*, 289. [CrossRef]
52. Joshi, D.; Kumar, S. Trapezium cloud TOPSIS method with interval-valued intuitionistic hesitant fuzzy linguistic information. *Granul. Comput.* **2017**. [CrossRef]
53. Wu, Y.; Li, C.-C.; Chen, X.; Dong, Y. Group decision making based on linguistic distribution and hesitant assessment: Maximizing the support degree with an accuracy constraint. *Inf. Fusion* **2018**, *41*, 151–160. [CrossRef]
54. Wang, R.; Li, Y. Generalized single-valued neutrosophic hesitant fuzzy prioritized aggregation operators and their applications to multiple criteria decision-making. *Information* **2018**, *9*, 10. [CrossRef]
55. Garg, H.; Nancy. Linguistic single-valued neutrosophic prioritized aggregation operators and their applications to multiple-attribute group decision-making. *J. Ambient Intell. Humaniz. Comput.* **2018**. [CrossRef]
56. Liang, P.; Song, F. What does a probabilistic interpretation of fuzzy sets mean? *IEEE Trans. Fuzzy Syst.* **1996**, *4*, 200–205. [CrossRef]
57. Liu, Z.; Li, H.X. A probabilistic fuzzy logic system for modeling and control. *IEEE Trans. Fuzzy Syst.* **2005**, *13*, 848–859.
58. Xu, Z.; Zhou, W. Consensus building with a group of decision makers under the hesitant probabilistic fuzzy environment. *Fuzzy Optim. Decis. Mak.* **2017**, *16*, 481–503. [CrossRef]
59. Hao, Z.; Xu, Z.; Zhao, H.; Su, Z. Probabilistic dual hesitant fuzzy set and its application in risk evaluation. *Knowl.-Based Syst.* **2017**, *127*, 16–28. [CrossRef]
60. Zhou, W.; Xu, Z. Group consistency and group decision making under uncertain probabilistic hesitant fuzzy preference environment. *Inf. Sci.* **2017**. [CrossRef]
61. Zhou, W.; Xu, Z. Expected hesitant VaR for tail decision making under probabilistic hesitant fuzzy environment. *Appl. Soft Comput.* **2017**, *60*, 297–311. [CrossRef]
62. Ding, J.; Xu, Z.; Zhao, N. An interactive approach to probabilistic hesitant fuzzy multi-attribute group decision making with incomplete weight information. *J. Intell. Fuzzy Syst.* **2017**, *32*, 2523–2536. [CrossRef]
63. Li, J.; Wang, J.Q. Multi-criteria outranking methods with hesitant probabilistic fuzzy sets. *Cogn. Comput.* **2017**, *9*, 611–625. [CrossRef]
64. Zhang, S.; Xu, Z.; He, Y. Operations and integrations of probabilistic hesitant fuzzy information in decision making. *Inf. Fusion* **2017**, *38*, 1–11. [CrossRef]
65. Wang, Z.-X.; Li, J. Correlation coefficients of probabilistic hesitant fuzzy elements and their applications to evaluation of the alternatives. *Symmetry* **2017**, *9*, 259. [CrossRef]
66. Xu, Z. A method based on linguistic aggregation operators for group decision making with inguistic preference relations. *Inf. Sci.* **2004**, *166*, 19–30. [CrossRef]

67. Gou, X.; Liao, H.; Xu, Z.; Herrera, F. Double hierarchy hesitant fuzzy linguistic term set and MULTIMOORA method: A case of study to evaluate the implementation status of haze controlling measures. *Inf. Fusion* **2017**, *38*, 22–34. [CrossRef]
68. Zhou, H.; Wang, J.Q.; Zhang, H.Y.; Chen, X.H. Linguistic hesitant fuzzy multi-criteria decision-making method based on evidential reasoning. *Int. J. Syst. Sci.* **2016**, *47*, 314–327. [CrossRef]
69. Bisht, K.; Kumar, S. Fuzzy time series forecasting method based on hesitant fuzzy sets. *Expert Syst. Appl.* **2016**, *64*, 557–568. [CrossRef]

Σ mathematics

MDPI

Article

The Effect of Prudence on the Optimal Allocation in Possibilistic and Mixed Models

Irina Georgescu

Academy of Economic Studies, Department of Economic Cybernetics, Piata Romana No 6 R 70167, Oficiul Postal 22, 010374 Bucharest, Romania; irina.georgescu@csie.ase.ro

Received: 30 May 2018; Accepted: 24 July 2018; Published: 2 August 2018

Abstract: In this paper, several portfolio choice models are studied: a purely possibilistic model in which the return of the risky is a fuzzy number, and four models in which the background risk appears in addition to the investment risk. In these four models, risk is a bidimensional vector whose components are random variables or fuzzy numbers. Approximate formulas of the optimal allocation are obtained for all models, expressed in terms of some probabilistic or possibilistic moments, depending on the indicators of the investor preferences (risk aversion, prudence).

Keywords: prudence; optimal allocation; possibilistic moments

1. Introduction

The standard portfolio choice problem [1–3] considers the determination of the optimal proportion of the wealth an agent invests in a risk-free asset and in a risky asset. The study of this probabilistic model is usually done in the classical expected utility theory. The optimal allocation of a risky asset appears as the solution of a maximization problem. By Taylor approximations, several forms of the solution have been found, depending on different moments of the return of the risky asset, as well as some on indicators of the investors's risk preferences. In the form of the solution from [4] Chapter 2 or [5] Chapter 5, the mean value, the variance, and the Arrow–Pratt index of the investor's utility function appear. The approach from [6–8] led to forms of the approximate solution which depend on the first three moments, the Arrow–Pratt index r_u, and the prudence index P_u [9]. The solution found in [10] is expressed according to the first four moments and the indicators of risk aversion, prudence, and temperance of the utility function. Another form of the solution in which the first four moments appear can be found in [11].

All the above models are probabilistic, the risk is represented by random variables, and the attitude of the agent towards risk is expressed by notions and properties which use the probabilistic indicators (expected value, variance, covariance, moments, etc.). The probabilistic modeling does not cover all uncertainty situations in which risk appears (e.g., when the information is not extracted from a sufficiently large volume of data). Possibility theory, initiated by Zadeh in [12] can model different situations: *"while probability theory offers a quantitative model for randomness and indecisiveness, possibility theory offers a qualitative model of incomplete knowledge"* ([13], p. 277).

In possibility theory, risk is modeled by the notion of possibilistic distribution [14–16]. Fuzzy numbers are the most important class of possibilistic distribution [17]. They generalize real numbers, and by Zadeh's extension principle [12], the operations with real numbers can be extended to operations with fuzzy numbers. So, the set of fuzzy numbers is endowed with a rich algebraic structure, very close to the set of real numbers, and their possibilistic indicators (possibilistic expected value, possibilistic variance, possibilistic moments, etc.) have important mathematical properties [14–19]. Fuzzy numbers are also capable of modelling a large scope of risk situations ([14–16,20–25]). For this, most studies on possibilistic risk have been done in the framework offered by fuzzy numbers, although there exist approaches on possibilistic risk in contexts larger than that offered by fuzzy numbers. For example, in [26] there is a treatment of risk aversion in an abstract framework including fuzzy numbers, random fuzzy numbers, type-2 fuzzy sets, random type-2 fuzzy sets, etc.)

In this paper, several portfolio choice models are studied: a purely possibilistic model, in which the return of the risky asset is represented by a fuzzy number [14,15], and four more models, in which a probabilistic or possibilistic background risk appears. In the formulation of the maximization problem for the first model, the possibilistic expected utility from [16], definition 4.2.7, is used. In the case of the other four models, the notion of bidimensional possibilistic expected utility ([16], p. 60) or the bidimensional mixed expected utility ([16], p. 79) is used. The approximate solutions of these two models are expressed by the possibilistic moments associated with a random variable, a fuzzy number ([14,15,24,25]), and by the indicators on the investor risk preferences.

In the first part of Section 2 the definitions of possibilistic expected utility (cf. [16]) and possibilistic indicators of a fuzzy number (expected value, variance, moments) are presented. The second part of the section contains the definition of a mixed expected utility associated with a mixed vector, a bidimensional utility function, and a weighting function ([16]).

Section 3 is concerned with the possibilistic standard portfolio-choice model, whose construction is inspired by the probabilistic model of [10]. The return of the risky asset is here a fuzzy number, while in [10] it is a random variable. The total utility function of the model is written as a possibilistic expected value. The maximization problem of the model and the first-order conditions are formulated, from which its optimal solution is determined.

Section 4 is dedicated to the optimal asset allocation in the framework of the possibilistic portfolio model defined in the previous section. Using a second-order Taylor approximation, a formula for the approximate calculation of the maximization problem solution is found. In the component of the formula appear the first three possibilistic moments, the Arrow–Pratt index, and the prudence indices of the investor's utility function. The general formula is particularized for triangular fuzzy numbers and HARA (hyperbolic absolute riks aversion) and CRRA (constant relative risk aversion) utility functions.

In Section 5 four moments are defined in which the background risk appears in addition to the investment risk. In these models, risk is represented by a bidimensional vector whose components are either random variables or fuzzy numbers. The agent will have a unidimensional utility function, but the total utility function will be:

- a bidimensional probabilistic expected utility, when both components are random variables;
- a bidimensional possibilistic expected utility ([16], p. 60), when both components are fuzzy numbers;
- a mixed expected utility ([16], p. 79), when a component is a random variable, and the other is a fuzzy number.

Section 6 is dedicated to the determination of an approximate calculation formula for the solution of the optimization problems of the four models with background risk from the previous section. We will study in detail only the model in which the investment risk is a fuzzy number and the background risk is a random variable. For the other three cases, only the approximate calculation formulas of the solutions are presented. The proofs are presented in an Appendix A .

2. Preliminaries

In this section we recall some notions and results on the possibilistic expected utility, mixed expected utility (cf. [16]), and some possibilistic indicators associated with fuzzy numbers (cf. [14,18,19,24,25,27]). For the definition and arithmetical properties of the fuzzy numbers, we refer to [14–16].

2.1. Possibilistic Expected Utility

The classic risk theory is usually developed in the framework of expected utility (EU). The main concept of EU theory is the probabilistic expected utility $E(u(X))$ associated with a utility function u (representing the agent) and a random variable X (representing the risk).

In case of a possibilistic risk EU theory, the agent will be represented by a utility function u, and the risk by a fuzzy number A. Besides these, we will consider a weighting function f. The level-sets $[A]^\gamma$, $\gamma \in [0,1]$ mean a gradualism of risk. By the appearance of f in the definition of possibilistic expected utility and the possibilistic indicators, a weighting of this gradualism is done (by [14], p. 27, *"different weighting functions can give different importances to level-sets of possibility distributions")*.

Thus, we fix a mathematical context consisting of:

- a utility function u of class C^2,
- a fuzzy number A whose level sets are $[A]^\gamma = [a_1(\gamma), a_2(\gamma)]$, $\gamma \in [0,1]$,
- a weighting function $f : [0,1] \to \mathbf{R}$. (f is a non-negative and increasing function that satisfies $\int_0^1 f(\gamma)d\gamma = 1$).

The possibilistic expected utility associated with the triple (u, A, f) is

$$E_f(u(A)) = \frac{1}{2}\int_0^1 [u(a_1(\gamma)) + u(a_2(\gamma))]f(\gamma)d\gamma. \tag{1}$$

In the interpretation from [14], p. 27, the possibilistic expected utility can be viewed as the result of the following process: on each γ-level set $[A]^\gamma = [a_1(\gamma), a_2(\gamma)]$, one considers the uniform distribution. Then, $E_f(u(A))$ is defined as the f-weighted average of the probabilistic expected values of these uniform distributions.

The following possibilistic indicators associated with a fuzzy number A and a weighting function f are particular cases of (1) .

- Possibilistic expected value [18,19]:

$$E_f(A) = \frac{1}{2}\int_0^1 [a_1(\gamma) + a_2(\gamma)]f(\gamma)d\gamma, \tag{2}$$

  (u is the identity function of $\mathbf{R}$).
- Possibilistic variance [18,27]:

$$Var_f(A) = \frac{1}{2}\int_0^1 [(u(a_1(\gamma)) - E_f(A))^2 + (u(a_2(\gamma)) - E_f(A))^2]f(\gamma)d\gamma, \tag{3}$$

  (for $u(x) = (x - E_f(x))^2, x \in \mathbf{R}$).
- The n-th order possibilistic moment [24,25]:

$$M(A^n) = \frac{1}{2}\int_0^1 [u^n(a_1(\gamma)) + u^n(a_2(\gamma))]f(\gamma)d\gamma. \tag{4}$$

Proposition 1. *Let $g : \mathbf{R} \to \mathbf{R}$, $h : \mathbf{R} \to \mathbf{R}$ be two utility functions, $a, b \in \mathbf{R}$ and $u = ag + bh$. Then, $E_f(u(A)) = aE_f(g(A)) + bE_f(h(A))$.*

Corollary 1. $E_f(a + bh(A)) = a + bE_f(h(A))$.

2.2. Mixed Expected Utility

In the financial-economic world, as in the social world, there may be complex situations of uncertainty with multiple risk parameters. In the papers on probabilistic risk, such phenomena are conceptualized by the notion of a random vector (all risk parameters are random variables). However, there can be situations considered as "hybrid", in which some parameters are random variables and others are fuzzy numbers. This is the notion of the mixed vector, which together with a multidimensional and a weighting function, are the basic entities of the mixed EU theory.

In order to treat a risk problem within a mixed EU theory, it is necessary to have a concept of expected utility.

Since two risk parameters appear in the portfolio choice model with background risk from the paper, we will present the definition of mixed expected utility in the bidimensional case.

A bidimensional mixed vector has the form (A, X), where A is a fuzzy number and X is a random variable. We will denote by $M(X)$ the expected value of X. If $g : \mathbf{R} \to \mathbf{R}$ is a continuous function, then $M(g(X))$ is the probabilistic expected utility of X with respect to g.

Let $u : \mathbf{R}^2 \to \mathbf{R}$ be a bidimensional utility function of class $\mathcal{C}^2$, (A, X) a mixed vector, and $f : [0,1] \to \mathbf{R}$ a weighting function. Assume that the level sets of the fuzzy number A are $[A]^\gamma = [a_1(\gamma), a_2(\gamma)], \gamma \in [0,1]$. For any $\gamma \in [0,1]$, we consider the probabilistic expected values $M(u(a_i(\gamma), X)), i = 1,2$.

The mixed expected utility associated with the triple $(u, (A, X), f)$ is:

$$E_f(u(A,X)) = \frac{1}{2}\int_0^1 [M(u(a_1(\gamma), X)) + M(u(a_2(\gamma), X))]f(\gamma)d\gamma. \quad (5)$$

In the definition of $E_f(u(A, X)$, we distinguish the following steps:

- In the first step, the possibilistic risk is parametrized by the decomposition of A in its level sets $[a_1(\gamma), a_2(\gamma)], \gamma \in [0,1]$.
- In the second step, for each level γ one considers the parametrized probabilistic utilities $M(u(a_1(\gamma), X))$ and $M(u(a_2(\gamma), X))$.
- In the third step, the mixed expected utility $E_f(u(A, X))$ is obtained as the f-weighted average of the family of means

$$(\frac{1}{2}[M(u(a_1(\gamma), X)) + M(u(a_2(\gamma))])_{\gamma\in[0,1]}.$$

Remark 1. *If $a \in \mathbf{R}$ then $E_f(u(a, X)) = M(u(a, X))$.*

Proposition 2. *Let g, h be two bidimensional utility functions, $a, b \in \mathbf{R}$ and $u = ag + bh$. Then, $E_f(u(A, X)) = aE_f(g(A, X)) + bE_f(h(A, X))$.*

Propositions 1 and 2 express the linearity of possibilistic expected value and mixed expected utility with respect to the utility functions which appear in the definitions of these two operators.

Corollary 2. *If A is a fuzzy number and Z is a random variable, then $E_f(AZ) = M(Z)E_f(A)$ and $E_f(A^2Z) = M(Z)E_f(A^2)$.*

3. Possibilistic Standard Model

In this section we will present a possibilistic portfolio choice model in which the return of the risky asset is a fuzzy number. Investing an initial wealth between a risk-free asset (bonds) and a risky asset (stocks), an agent seeks to determine that money allocation in the risky asset such that their winnings are maximum.

In defining the total utility of the model, we will use the possibilistic expected utility introduced in the previous section.

We consider an agent (characterized by a utility function u of class $\mathcal{C}^2$, increasing and concave) which invests a wealth w_0 in a risk-free asset and in a risky asset. The agent invests the amount α in a risky asset and $w_0 - \alpha$ in a risk-free asset. Let r be the return of the risk-free asset and x a value of the return of the risky asset. We denote by $w = w_0(1+r)$ the future wealth of the risk-free strategy. The portfolio value $(w_0 - \alpha, \alpha)$ will be (according to [4], pp. 65–66):

$$(w_0 - \alpha)(1+r) + \alpha(1+x) = w + \alpha(x-r). \tag{6}$$

The probabilistic investment model from [4] Chapter 4 or [5] Chapter 5 starts from the hypothesis that the return of the risky asset is a random variable X_0. Then, x is a value of X_0 and (6) leads to the following maximization problem:

$$\max_{\alpha} M[u(w + \alpha(X_0 - r))]. \tag{7}$$

By denoting $X = X_0 - r$ the excess return, the model (7) becomes:

$$\max_{\alpha} M[u(w + \alpha X)]. \tag{8}$$

If we make the assumption that the return of the risky asset is a fuzzy number B_0, then x will be a value of B_0. To describe the possibilistic model resulting from such a hypothesis, we fix a weighting function $f : [0,1] \to \mathbf{R}$. The expression (6) suggests to us the following optimization problem:

$$\max_{\alpha} E_f[u(w + \alpha(B_0 - r))]. \tag{9}$$

By denoting with $B = B_0 - r$ the excess return, the problem (8) becomes:

$$\max_{\alpha} E_f[u(w + \alpha B)]. \tag{10}$$

There is a similarity between the optimization problem (8) and the optimization problem (10). Between the two optimization problems, there are two fundamental differences:

- In (8) there is a probabilistic risk X, and in (10) there is a possibilistic risk A.
- Problem (8) is formulated in terms of a probabilistic expected utility operator $M(u(.))$, while (10) is formulated using the possibilistic expected utility operator $E_f(u(.))$.

Assume that the level sets of the fuzzy number B are $[B]^\gamma = [b_1(\gamma), b_2(\gamma)], \gamma \in [0,1]$. According to (1), the total utility function of the model (10) will have the following form:

$$V(\alpha) = E_f[u(w + \alpha B)] = \frac{1}{2}\int_0^1 [u(w + \alpha b_1(\gamma)) + u(w + \alpha b_2(\gamma))] f(\gamma) d\gamma.$$

Deriving twice, one obtains:

$$V''(\alpha) = \frac{1}{2}\int_0^1 [b_1^2(\gamma) u''(w + \alpha b_1(\gamma)) + b_2^2(\gamma) u''(w + \alpha b_2(\gamma))] f(\gamma) d\gamma.$$

Since $u'' \leq 0$, it follows $V''(\alpha) \leq 0$, thus V is concave.

We assume everywhere in this paper that the portfolio risk is small, thus analogously with [5] (Section 5.2), we can take the possibilistic excess return B as $B = k\mu + A$, where $\mu > 0$ and A is a fuzzy number with $E_f(A) = 0$. Of course $E_f(B) = k\mu$ in that case. The total utility $V(\alpha)$ will be written:

$$V(\alpha) = E_f[u(w + \alpha(k\mu + A))]. \tag{11}$$

Assuming that the level sets of A are $[A]^\gamma = [a_1(\gamma), a_2(\gamma)], \gamma \in [0,1]$, the expression (11) becomes:

$$V(\alpha) = \frac{1}{2}\int_0^1 [u(w + \alpha(k\mu + a_1(\gamma))) + u(w + \alpha(k\mu + a_2(\gamma)))] f(\gamma) d\gamma.$$

By deriving, one obtains:

$$V'(\alpha) = \frac{1}{2}\int_0^1 [(k\mu + a_1(\gamma))u'(w + \alpha(k\mu + a_1(\gamma))) +$$

$$(k\mu + a_2(\gamma))u'(w + \alpha(k\mu + a_2(\gamma)))]f(\gamma)d\gamma,$$

which can be written

$$V'(\alpha) = E_f[(k\mu + A)u'(w + \alpha(k\mu + A))]. \tag{12}$$

Let $\alpha(k)$ be the solution of the maximization problem $\max\limits_{\alpha} V(\alpha)$, with $V(\alpha)$ being written under the form (12). Then, the first order condition $V'(\alpha(k)) = 0$ will be written:

$$E_f[(k\mu + A)u'(w + \alpha(k)(k\mu + A))] = 0. \tag{13}$$

As in [5] (Section 5.2), we assume that $\alpha(0) = 0$.

Everywhere in this paper, we will keep the notations and hypotheses from above.

4. The Effect of Prudence on the Optimal Allocation

The main result of this section is a formula for the approximate calculation of the solution $\alpha(k)$ of Equation (13). In the formula will appear the indicators of absolute risk aversion and prudence, marking how these influence the optimal investment level $\alpha(k)$ in the risky asset.

We will consider the second-order Taylor approximation of $\alpha(k)$ around $k = 0$:

$$\alpha(k) \approx \alpha(0) + k\alpha'(0) + \frac{1}{2}k^2\alpha''(0) = k\alpha'(0) + \frac{1}{2}k^2\alpha''(0). \tag{14}$$

For the approximate calculation of $\alpha(k)$, we will determine the approximate values of $\alpha'(k)$ and $\alpha''(k)$. Note that the calculation of the approximate values of $\alpha'(0)$ and $\alpha''(0)$ follows an analogous line to the one used in [10] in the analysis of the probabilistic model. In the proof of the approximate calculation formulas of $\alpha'(0)$ and $\alpha''(0)$, we will use the properties of the possibilistic expected utility from Section 2.1. Before this, we will recall the Arrow–Pratt index $r_u(w)$ and prudence index $P_u(w)$ associated with the utility function u:

$$r_u(w) = -\frac{u''(w)}{u'(w)}; P_u(w) = -\frac{u'''(w)}{u''(w)}. \tag{15}$$

Proposition 3. $\alpha'(0) \approx \frac{\mu}{E_f(A^2)}\frac{1}{r_u(w)}$.

Proposition 4. $\alpha''(0) \approx \frac{P_u(w)}{(r_u(w))^2}\frac{E_f(A^3)}{(E_f(A^2))^3}\mu^2$.

We recall from Section 3 that $A = B - E_f(B)$. The following result gives us an approximate expression of $\alpha(k)$:

Theorem 1. $\alpha(k) \approx \frac{1}{r_u(w)}\frac{E_f(B)}{Var_f(B)} + \frac{1}{2}\frac{P_u(w)}{((r_u(w))^2}\frac{E_f[(B-E_f(B))^3]}{(Var_f(B))^3}(E_f(B))^2$.

Remark 2. *The previous theorem gives us an approximate solution of the maximization problem* $\max\limits_{\alpha} V(\alpha)$ *with respect to the indices of absolute risk aversion and prudence* $r_u(w)$, $P_u(w)$, *and the first three possibilistic moments* $E_f(B)$, $Var_f(B)$, *and* $E_f[(B - E_f(B))^3]$.

This result can be seen as a possibilistic version of the formula (A.6) of [10], which gives us the optimal allocation of investment in the context of a probabilistic portfolio choice model.

Example 1. We consider the triangular fuzzy number $B = (b, \alpha, \beta)$ defined by:

$$B(t) = \begin{cases} 1 - \frac{b-x}{\alpha} & \text{if} \quad b - \alpha \leq x \leq b, \\ 1 - \frac{x-b}{\beta} & \text{if} \quad b \leq x \leq b + \beta, \\ 0 & \text{otherwise.} \end{cases}$$

The level sets of B are $[B]^{\gamma} = [b_1(\gamma), b_2(\gamma)]$, where $b_1(\gamma) = b - (1-\gamma)\alpha$ and $b_2(\gamma) = b + (1-\gamma)\beta$, for $\gamma \in [0,1]$. We assume that the weighting function f has the form $f(\gamma) = 2\gamma$, for $\gamma \in [0,1]$. Then, by [25], Lemma 2.1:

$$E_f(B) = b + \frac{\beta - \alpha}{6}; Var_f(B) = \frac{\alpha^2 + \beta^2 + \alpha\beta}{18},$$

$$E_f[(B - E_f(B))^2] = \int_0^1 \gamma[(b_1(\gamma) - E_f(B))^2 + (b_2(\gamma) - E_f(B))^2] d\gamma$$

$$= \frac{19(\beta^3 - \alpha^3)}{1080} + \frac{\alpha\beta(\beta - \alpha)}{72}.$$

By replacing these indicators in the formula of Theorem 1, we obtain

$$\alpha(k) \approx \frac{1}{r_u(w)} \frac{b + \frac{\beta-\alpha}{6}}{\frac{\alpha^2+\beta^2+\alpha\beta}{18}} + \frac{1}{2} \frac{P_u(w)}{((r_u(w))^2} \frac{\frac{19(\beta^3-\alpha^3)}{1080} + \frac{\alpha\beta(\beta-\alpha)}{72}}{(\frac{\alpha^2+\beta^2+\alpha\beta}{18})^3} (b + \frac{\beta - \alpha}{6})^2.$$

Assume that the utility function u is HARA-type (see [5], Section 3.6):

$$u(w) = \zeta(\eta + \frac{w}{\gamma})^{1-\gamma}, \text{for } \eta + \frac{w}{\gamma} > 0.$$

Then, according to [5] (Section 3.6):

$$r_u(w) = (\eta + \frac{w}{\gamma})^{-1}; P_u(w) = \frac{\gamma + 1}{\gamma}(\eta + \frac{w}{\gamma})^{-1},$$

$$\frac{1}{r_u(w)} = \eta + \frac{w}{\gamma} \text{ and } \frac{P_u(w)}{((r_u(w))^2} = \frac{\frac{\gamma+1}{\gamma}(\eta + \frac{w}{\gamma})^{-1}}{(\eta + \frac{w}{\gamma})^{-2}} = \frac{\gamma + 1}{\gamma}(\eta + \frac{w}{\gamma}).$$

Replacing in the approximation calculation formula of $\alpha(k)$, it follows:

$$\alpha(k) \approx (\eta + \frac{w}{\gamma}) \frac{b + \frac{\beta-\alpha}{6}}{\frac{\alpha^2+\beta^2+\alpha\beta}{18}} + \frac{1}{2} \frac{\gamma + 1}{\gamma} (\eta + \frac{w}{\gamma}) \frac{\frac{19(\beta^3-\alpha^3)}{1080} + \frac{\alpha\beta(\beta-\alpha)}{72}}{(\frac{\alpha^2+\beta^2+\alpha\beta}{18})^3} (b + \frac{\beta - \alpha}{6})^2.$$

If $B = (b, \alpha)$ is a symmetric triangular fuzzy number ($\alpha = \beta$), then the approximate solution $\alpha(k)$ gets a very simple form:

$$\alpha(k) \approx 18 \frac{b}{\alpha^2}(\eta + \frac{w}{\gamma}).$$

Following [5] (Section 3.6), we consider the CRRA-type utility function:

$$u(w) = \begin{cases} \frac{w^{1-\gamma}}{1-\gamma} & \text{if} \quad \gamma \neq 1, \\ ln(w) & \text{if} \quad \gamma = 1. \end{cases}$$

For $\gamma \neq 1$, we have $r_u(w) = \frac{\gamma}{w}$ and $P_u(w) = \frac{\gamma+1}{w}$. A simple calculation leads to the following form of the solution:

$$\alpha(k) \approx \frac{w}{\gamma} \frac{b + \frac{\beta-\alpha}{6}}{\frac{\alpha^2+\beta^2+\alpha\beta}{18}} + \frac{1}{2} \frac{w(\gamma+1)}{\gamma^2} \frac{\frac{19(\beta^3-\alpha^3)}{1080} + \frac{\alpha\beta(\beta-\alpha)}{72}}{(\frac{\alpha^2+\beta^2+\alpha\beta}{18})^3} (b + \frac{\beta-\alpha}{6})^2.$$

5. Models with Background Risk

In the two standard portfolio choice problems (8) and (10), a single risk parameter appears: in (8) the risk is represented by the random variable X, and in (10) by the fuzzy number B. In both cases, we will call it investment risk. More complex situations may exist in which other risk parameters may appear in addition to the investment risk. This supplementary risk is called background risk (see [4,5]).

For simplicity, in this paper we will study investment models with a single background risk parameter. In the interpretation from [4], this background risk is associated with labor income. Therefore, the considered portfolio choice problems will have two types of risk: investment risk and background risk. Each can be random variables or fuzzy numbers, according to the following table.

The models corresponding to the four cases in Table 1 are obtained by adding in (7) and (9) the background risk as a random variable of a fuzzy number. For each problem we will have an approximate solution expressed in terms of indicators, Arrow–Pratt index, and prudence.

Table 1. Models with background risk.

	Investment Risk	Background Risk
1	probabilistic	probabilistic
2	possibilistic	possibilistic
3	possibilistic	probabilistic
4	probabilistic	possibilistic

Case 1. Besides the return of the risky asset X_0, we will have a probabilistic background risk represented by a random variable Z. Starting from the standard model (7), the following optimization problem is obtained by adding the background risk Z:

$$\max_{\alpha} M[u(w + \alpha(X_0 - r) + Z)]. \tag{16}$$

Case 2. Besides the return of the possibilistic risky-asset B_0, a possibilistic background risk represented by a fuzzy number C appears. In the standard model (8) the fuzzy number C is added and the following optimization problem is obtained:

$$\max_{\alpha} E_f[u(w + \alpha(B_0 - r) + C)]. \tag{17}$$

Case 3. Besides the investment risk B_0 a probabilistic background risk represented by a random variable Z appears. The optimization problem is obtained adding the random variable Z in (9):

$$\max_{\alpha} E_f[u(w + \alpha(B_0 - r) + Z)]. \tag{18}$$

Case 4. Besides the investment risk X_0 of (7) the possibilistic background risk represented by a fuzzy number C appears:

$$\max_{\alpha} E_f[u(w + \alpha(X_0 - r) + C)]. \tag{19}$$

Problem (17) is formulated in terms of a bidimensional possibilistic expected utility (see [16], p. 60), and (18) and (19) use the mixed expected utility defined in Section 2.

By denoting with $X = X_0$ and $B = B_0 - r$ the probabilistic excess return and the possibilistic excess return, respectively, the optimization problems (16)–(19) become

$$\max_{\alpha} M[u(w + \alpha X + Z)], \tag{20}$$

$$\max_{\alpha} E_f[u(w + \alpha B + C)], \tag{21}$$

$$\max_{\alpha} E_f[u(w + \alpha B + Z], \tag{22}$$

$$\max_{\alpha} E_f[u(w + \alpha X + C]. \tag{23}$$

In the following section we will study model 3 in detail, proving an approximate calculation formula of the solution of the optimization problem (18). The proof of the approximate solutions of the other three optimization problems is done similarly.

6. Approximate Solutions of Portfolio Choice Model with Background Risk

In this section we will prove the approximate calculation formulas for the solutions of the optimization problems (20)–(23). These formulas will emphasize how risk aversion and the agent's prudence influence the optimal proportions invested in the risky asset in the case of the four portfolio choice models with background risk. We will study in detail only the mixed model (22), in which, besides this possibilistic risk, a probabilistic background risk may appear, modeled by a random variable Z. This mixed model comes from the possibilistic standard model by adding Z in the composition of the total utility function. More precisely, the total utility function $W(\alpha)$ will be:

$$W(\alpha) = E_f[u(w + \alpha(k\mu + A) + Z)], \tag{24}$$

where the other components of the model have the same meaning as in Section 3.

Assume that the level sets of A are $[A]^{\alpha} = [a_1(\gamma), a_2(\gamma)]$, $\gamma \in [0, 1]$. By definition (5) of the mixed expected utility, formula (24) can be written as:

$$W(\alpha) = \frac{1}{2}\int_0^1 [M(u(w + \alpha(k\mu + a_1(\gamma)) + Z)) + M(u(w + \alpha(k\mu + a_2(\gamma)) + Z))]f(\gamma)d\gamma.$$

One computes the first derivative of $W(\alpha)$:

$$W'(\alpha) = \frac{1}{2}\int_0^1 (k\mu + a_1(\gamma))M(u'(w + \alpha(k\mu + a_1(\gamma)) + Z))f(\gamma)d\gamma +$$

$$+\frac{1}{2}\int_0^1 k\mu + a_2(\gamma))M(u'(w + \alpha(k\mu + a_2(\gamma)) + Z))f(\gamma)d\gamma.$$

$W'(\alpha)$ can be written as:

$$W'(\alpha) = E_f[(k\mu + A)u'(w + \alpha(k\mu + A) + Z)]. \tag{25}$$

By deriving one more time, we obtain:

$$W''(\alpha) = E_f[(k\mu + A)^2 u''(w + \alpha(k\mu + A) + Z)].$$

Since $u'' \leq 0$, it follows that $W''(\alpha) \leq 0$, and thus W is concave. Then, the solution $\beta(k)$ of the optimization problem $\max_{\alpha} W(\alpha)$ will be given by $W'(\beta(k)) = 0$. By (25),

$$E_f[(k\mu + A)u'(w + \beta(k)(k\mu + A) + Z)] = 0. \tag{26}$$

In this case we will also make the natural hypothesis $\beta(0) = 0$.

To compute an approximate value of $\beta(k)$ we will write the second-order Taylor approximation of $\beta(k)$ around $k = 0$:

$$\beta(k) \approx \beta(0) + k\beta'(0) + \frac{1}{2}k^2\beta''(0) = k\beta'(0) + \frac{1}{2}k^2\beta''(0). \quad (27)$$

We propose to find some approximate values of $\beta'(0)$ and $\beta''(0)$.

Proposition 5. $\beta'(0) \approx \frac{\mu}{E_f(A^2)}(\frac{1}{r_u(w)} - M(Z))$.

Proposition 6. $\beta''(0) \approx \frac{P_u(w)(\beta'(0))^2}{Var_f(B)} \frac{E_f[(B-E_f(B))^3]}{1-M(Z)P_u(w)}$.

Theorem 2.

$$\beta(k) \approx \frac{E_f(B)}{Var_f(B)}[\frac{1}{r_u(w)} - M(Z)] + \frac{1}{2}P_u(w)[\frac{1}{r_u(w)} - M(Z)]^2 \frac{E_f^2(B)E_f[(B-E_f(B))^3]}{Var_f^3(B)[1-M(Z)P_u(w)]}.$$

Remark 3. *In the approximate expression of $\beta(k)$ from the previous theorem appear the Arrow index and the prudence index of the utility function u, the possibilistic indicators $E_f(B)$, $Var_f(B)$, and the possibilistic expected value $M(Z)$.*

Example 2. *We consider that the investment risk is represented by a fuzzy number $B = (b, \alpha, \beta)$ and the background risk by the random variable Z with the normal distribution $N(m, \sigma^2)$. We will consider a HARA-type utility:*

$$u(w) = \zeta(\eta + \frac{w}{\gamma})^{1-\gamma}, \text{ for } \eta + \frac{w}{\gamma} > 0.$$

Using the computations from Example 1 and taking into account that $M(Z) = m$, one reaches the following form of the approximate solution:

$$\beta(k) \approx (\eta + \frac{w}{\gamma} - m)\frac{b + \frac{\beta-\alpha}{6}}{\frac{\alpha^2+\beta^2+\alpha\beta}{18}} +$$

$$+\frac{\gamma+1}{2\gamma}\frac{(\eta + \frac{w}{\gamma} - m)^2}{\eta + \frac{w}{\gamma}}\frac{\frac{19(\beta^3-\alpha^3)}{1080} + \frac{\alpha\beta(\beta-\alpha)}{72}}{(\frac{\alpha^2+\beta^2+\alpha\beta}{18})^3}\frac{1}{1 - m\frac{\gamma+1}{\gamma}(\eta + \frac{w}{\gamma})}.$$

Let us assume that the utility function u is of CRRA-type: $u(w) = \frac{w^{1-\gamma}}{1-\gamma}$ if $\gamma \neq 1$ and $u(w) = ln(w)$, if $\gamma = 1$.

For $\gamma \neq 1$ we have $r_u(w) = \frac{\gamma}{w}$, $P_u(w) = \frac{\gamma+1}{w}$, from where it follows:

$$\beta(k) \approx (\frac{w}{\gamma} - m)\frac{b + \frac{\beta-\alpha}{6}}{\frac{\alpha^2+\beta^2+\alpha\beta}{18}} +$$

$$+\frac{(\gamma+1)(\frac{w}{\gamma} - m)^2}{2w}\frac{\frac{19(\beta^3-\alpha^3)}{1080} + \frac{\alpha\beta(\beta-\alpha)}{72}}{(\frac{\alpha^2+\beta^2+\alpha\beta}{18})^3}\frac{1}{1 - \frac{mw(\gamma+1)}{\gamma^2}}.$$

For $\gamma = 1$:

$$\beta(k) \approx (w - m)\frac{b + \frac{\beta-\alpha}{6}}{\frac{\alpha^2+\beta^2+\alpha\beta}{18}} +$$

$$+\frac{(w-m)^2}{w}\frac{\frac{19(\beta^3-\alpha^3)}{1080}+\frac{\alpha\beta(\beta-\alpha)}{72}}{(\frac{\alpha^2+\beta^2+\alpha\beta}{18})^3}\frac{1}{1-2mv}.$$

We will state without proof some results on approximate solutions of the other three models with background risk. For the optimization problems (20) and (23), we will assume that $X = k\mu + Y$, with $\mu > 0$ and $E(Y) = 0$ (according to the model of [5], Section 5.2), and for (21), we will take $B = k\mu + A$, with $B = k\mu + A$, with $\mu > 0$ and $E_f(A) = 0$.

Theorem 3. *An approximate solution $\beta_1(k)$ for the optimization problem (20) is*

$$\beta_1(k) \approx \frac{M(X)}{Var(X)}[\frac{1}{r_u(w)} - M(Z)]+$$

$$+\frac{1}{2}P_u(w)[\frac{1}{r_u(w)} - M(Z)]^2\frac{M^2(Z)M[(X-M(X))^3]}{Var^3(X)[1-M(Z)P_u(w)]}.$$

Example 3. *The formula from Theorem 3 may take different forms, depending on the distributions of the random variables X and Z. If X is the normal distribution $N(m,\sigma)$ then $M(X) = m$, $Var(X) = \sigma^2$ and $M[(X-M(X))^3] = 0$, thus*

$$\beta_1(k) \approx \frac{m}{\sigma^2}[\frac{1}{r_u(w)} - M(Z)].$$

Assume that the utility function u is of HARA-type:

$$u(w) = \zeta(\eta + \frac{w}{\gamma})^{1-\gamma} \text{ for } \eta + \frac{w}{\gamma} > 0,$$

and Z is the distribution $N(0,1)$, we obtain:

$$\beta_1(k) \approx \frac{m}{\sigma^2}\frac{1}{r_u(w)} = \frac{m}{\sigma^2}(\eta + \frac{w}{\gamma}).$$

The form of $\beta_1(k)$ from the previous section extends the approximate calculation formula of the solution of the probabilistic model (8) (see [6,7]). Its proof follows some steps similar to the ones in the formula of $\beta(k)$ from Theorem 2, but uses the probabilistic techniques from [6,7].

Theorem 4. *An approximate solution $\beta_2(k)$ of the optimization problem (21) is*

$$\beta_2(k) \approx \frac{E_f(B)}{Var_f(B)}[\frac{1}{r_u(w)} - E_f(C)]+$$

$$+\frac{1}{2}P_u(w)[\frac{1}{r_u(w)} - E_f(C)]^2\frac{E_f^2(B)E_f[(B-E_f(B))^3]}{Var_f^3(B)[1-E_f(C)P_u(w)]}.$$

Example 4. *We assume that:*

- *B is a triangular fuzzy number $B = (b, \alpha, \beta)$ and C is a symmetric triangular fuzzy number $C = (c, \delta)$,*
- *the utility function u is of HARA-type: $u(w) = \zeta(\eta + \frac{w}{\gamma})^{-1}$ for $\eta + \frac{w}{\gamma} > 0$,*
- *the weighting function f has the form $f(t) = 2t$ for $t \in [0,1]$.*

By taking into account the calculations from Examples 1, 2, and the fact that $E_f(C) = c$, the approximate solution $\beta_2(k)$ becomes:

$$\beta_2(k) \approx (\eta + \frac{w}{\gamma} - c)\frac{b + \frac{\beta-\alpha}{6}}{\frac{\alpha^2+\beta^2+\alpha\beta}{18}}+$$

$$+\frac{\gamma+1}{2\gamma}\frac{(\eta+\frac{w}{\gamma}-c)^2}{\eta+\frac{w}{\gamma}}\frac{\frac{19(\beta^3-\alpha^3)}{1080}+\frac{\alpha\beta(\beta-\alpha)}{72}}{(\frac{\alpha^2+\beta^2+\alpha\beta}{18})^3}\frac{1}{1-c\frac{\gamma+1}{\gamma}(\eta+\frac{w}{\gamma})}.$$

Theorem 5. *An approximate solution $\beta_3(k)$ of the optimization problem (23) is:*

$$\beta_2(k) \approx \frac{M(X)}{Var(X)}[\frac{1}{r_u(w)} - E_f(C)]+$$

$$+\frac{1}{2}P_u(w)[\frac{1}{r_u(w)} - E_f(C)]^2\frac{M^2(X)M[(X-M(X))^3]}{Var^3(X)[1-E_f(C)P_u(w)]}.$$

Example 5. *We consider the following hypotheses:*

- *X has the normal distribution $N(m,\sigma)$ and C is the triangular fuzzy numbers $C=(c,\delta,\epsilon)$,*
- *the utility function u is of HARA-type: $u(w)=\zeta(\eta+\frac{w}{\gamma})^{-1}$ for $\eta+\frac{w}{\gamma}>0$,*
- *the weighting function f has the form: $f(t)=2$ for $t\in[0,1]$.*

Then, $M(X)=m$, $Var(X)=\sigma^2$, $M[(X-M(X))^3]=0$, and $E_f(c)=c+\frac{\epsilon-\delta}{6}$.
It follows the following form of $\beta_3(k)$:

$$\beta_3(k) \approx \frac{m}{\sigma^2}[\frac{1}{r_u(w)} - E_f(C)] = \frac{m}{\sigma^2}[\eta+\frac{w}{\gamma}-c-\frac{\epsilon-\delta}{6}].$$

Author Contributions: The contribution belongs entirely to the author.

Funding: This research received no external funding.

Conflicts of Interest: The author declares no conflict of interest.

Appendix A

Proof of Corollary 2. We take $u(x,z)=xz$, and applying (5), we have

$$E_f(AZ) = \frac{1}{2}\int_0^1[M(a_1(\gamma)Z)+M(a_2(\gamma)Z)]f(\gamma)d\gamma$$

$$=\frac{1}{2}\int_0^1[a_1(\gamma)M(Z)+a_2(\gamma)M(Z)]f(\gamma)d\gamma = M(Z)E_f(A).$$

Taking $u(x,z)=x^2z$, we obtain

$$E_f(A^2Z) = \frac{1}{2}\int_0^1[M(a_1^2(\gamma)Z)+M(a_2^2(\gamma)Z)]f(\gamma)d\gamma$$

$$=\frac{1}{2}\int_0^1[a_1^2(\gamma)M(Z)+a_2^2(\gamma)M(Z)]f(\gamma)d\gamma = M(Z)E_f(A^2).$$

□

Proof of Proposition 3. We consider the Taylor approximation:

$$u'(w+\alpha(k\mu+x)) \approx u'(w)+\alpha(k\mu+x)u''(w).$$

Then, by (11) and Proposition 1

$$V'(\alpha) \approx E_f[(k\mu+A)(u'(w)+u''(w)\alpha(k\mu+A)]$$

$$=u'(w)(k\mu+E_f(A))+\alpha u''(w)E_f[(k\mu+A)^2].$$

The equation $V'(\alpha(k)) = 0$, becomes

$$u'(w)(k\mu + E_f(A)) + \alpha(k)u''(w)E_f[(k\mu + A)]^2 \approx 0.$$

We derive it with respect to k:

$$u'(w)\mu + u''(w)(\alpha'(k)E_f[(k\mu + A)^2] + 2\alpha(k)\mu E_f(k\mu + A)) \approx 0.$$

In this equality we make $k = 0$. Taking into account that $\alpha(0) = 0$, it follows

$$u'(w)\mu + u''(w)\alpha'(0)E_f(A^2) \approx 0,$$

from where we determine $\alpha'(0)$:

$$\alpha'(0) \approx -\frac{\mu}{E_f(A^2)}\frac{u'(w)}{u''(w)} = \frac{\mu}{E_f(A^2)}\frac{1}{r_u(w)}.$$

□

Proof of Proposition 4. To determine the approximate value of $\alpha''(0)$ we start with the following Taylor approximation:

$$u'(w + \alpha(k\mu + x)) \approx u'(w) + \alpha(k\mu + x)u''(w) + \frac{\alpha^2}{2}(k\mu + x)^2 u'''(w),$$

from which it follows:

$$(k\mu + x)u'(w + \alpha(k\mu + x)) \approx u'(w)(k\mu + x) + u''(w)\alpha(k\mu + x)^2 + \frac{u'''(w)}{2}\alpha^2(k\mu + x).$$

Then, by (11) and the linearity of the operator $E_f(.)$

$$V'(\alpha) = E_f[(k\mu + A)u'(w + \alpha(k\mu + A))]$$

$$\approx u'(w)E_f(k\mu + A) + u''(w)\alpha E_f[(k\mu + A)^2] + \frac{u'''(w)}{2}\alpha^2 E_f[(k\mu + A)^3].$$

Using this approximation for $\alpha = \alpha(k)$, the equation $V'(\alpha(k)) = 0$, becomes

$$u'(w)(k\mu + E_f(A)) + u''(w)\alpha(k)E_f[(k\mu + A)^2] + \frac{u'''(w)}{2}(\alpha(k))^2 E_f[(k\mu + A)^3] \approx 0.$$

Deriving with respect to k one obtains:

$$\mu u'(w) + u''(w)[\alpha'(k)E_f((k\mu + A)^2) + 2\mu\alpha(k)E_f(k\mu + A)] +$$

$$+ \frac{u'''(w)}{2}[2\alpha(k)\alpha'(k)E_f((k\mu + A)^3) + 3(\alpha(k))^2\mu E_f((k\mu + A)^2)] \approx 0.$$

We derive one more time with respect to k:

$$u''(w)[\alpha''(k)E_f((k\mu + A)^2) + 2\mu\alpha'(k)E_f(k\mu + A) + 2\mu\alpha'(k)E_f(k\mu + A) +$$

$$2\mu^2\alpha(k)] + \frac{u'''(w)}{2}[2(\alpha'(k))^2 E_f((k\mu + A)^3) + 2\alpha(k)\alpha''(k)E_f((k\mu + A)^3) +$$

$$+ 6\alpha(k)\alpha'(k)E_f((k\mu + A)^2) + 6\mu\alpha(k)\alpha'(k)E_f((k\mu + A)^2) + 6\mu^2(\alpha(k)^2)E_f(k\mu + A)] \approx 0.$$

In the previous relation, we take $k = 0$.

$$u''(w)[\alpha''(0)E_f(A^2) + 2\mu\alpha'(0)E_f(A) + 2\mu\alpha'(0)E_f(A) + 2\mu^2\alpha(0)] +$$

$$+\frac{u'''(w)}{2}[2(\alpha'(0))^2 E_f(A^3) + 2\alpha(0)\alpha''(0)E_f(A^3) + 6\alpha(0)\alpha'(0)E_f(A^2) +$$

$$6\mu\alpha(0)E_f(A^2) + 6\mu^2(\alpha(0))^2 E_f(A)] \approx 0.$$

Taking into account that $\alpha(0) = 0$ and $E_f(A) = 0$, one obtains

$$u''(w)\alpha''(0)E_f(A^2) + u'''(w)(\alpha'(0))^2 E_f(A^3) \approx 0,$$

from where we get $\alpha''(0)$:

$$\alpha''(0) \approx -\frac{u'''(w)}{u''(w)}\frac{E_f(A^3)}{E_f(A^2)}(\alpha'(0))^2.$$

By replacing $\alpha'(0)$ with the expression from Proposition 3 and taking into account (15), it follows:

$$\alpha''(0) = \frac{P_u(w)}{((r_u(w))^2}\frac{E_f(A^3)}{(E_f(A^2))^3}\mu^2.$$

□

Proof of Theorem 1. By replacing in (14) the approximate values of $\alpha'(0)$ and $\alpha''(0)$ given by Propositions 3 and 4 and taking into account that $E_f(B) = k\mu$, one obtains:

$$\alpha(k) \approx k\alpha'(0) + \frac{1}{2}k^2\alpha''(0)$$

$$= \frac{k\mu}{E_f(A^2)}\frac{1}{r_u(w)} + \frac{1}{2}(k\mu)^2\frac{P_u(w)}{(r_u(w))^2}\frac{E_f(A^3)}{(E_f(A^2))^3}$$

$$= \frac{E_f(B)}{E_f(A^2)}\frac{1}{r_u(w)} + \frac{1}{2}(E_f(B))^2\frac{P_u(w)}{(r_u(w))^2}\frac{E_f(A^3)}{(E_f(A^2))^3}.$$

However, $E_f(A^2) = E_f[(B - E_f(B))^2] = Var_f(B)$. Then,

$$\alpha(k) \approx \frac{1}{r_u(w)}\frac{E_f(B)}{Var_f(B)} + \frac{1}{2}\frac{P_u(w)}{(r_u(w))^2}\frac{E_f[(B - E_f(B))^3]}{(Var_f(B))^3}(E_f(B))^2.$$

□

Proof of Proposition 5. We consider the Taylor approximation:

$$u'(w + \alpha(k\mu + x) + z) \approx u'(w) + (\alpha(k\mu + x) + z)u''(w).$$

Then,

$$(k\mu + x)u'(w + \alpha(k\mu + x) + z) \approx u'(w)(k\mu + x) + u''(w)\alpha(k\mu + x)^2 + u''(w)z(k\mu + x).$$

From this relation, from (25) and the linearity of mixed expected utility, it follows:

$$W'(\alpha) \approx u'(w)(k\mu + E_f(A)) + u''(w)\alpha E_f[(k\mu + A)^2] + u''(w)E_f[(k\mu + A)Z].$$

Then, the equation $W'(\beta(k)) = 0$, will be written

$$u'(w)(k\mu + E_f(A)) + u''(w)\beta(k)E_f[(k\mu + A)^2] + u''(w)E_f[(k\mu + A)Z] \approx 0.$$

By deriving with respect to k one obtains:

$$u'(w)\mu + u''(w)(\beta'(k)E_f[(k\mu + A)^2] + 2\beta(k)\mu E_f(k\mu + A)) + u''(w)\mu M(Z) \approx 0.$$

For $k = 0$, it follows

$$u'(w)\mu + u''(w)\mu M(Z) + u''(w)\beta'(0)E_f(A^2) \approx 0,$$

from where $\beta'(0)$ is obtained:

$$\beta'(0) \approx -\frac{(u'(w) + u''(w)M(Z))\mu}{u''(w)E_f(A^2)} = \frac{\mu}{E_f(A^2)}(\frac{1}{r_u(w)} - M(Z)).$$

□

Proof of Proposition 6. We consider the Taylor approximation

$$u'(w + \alpha(k\mu + x) + z) \approx u'(w) + u''(w)[\alpha(k\mu + x) + z] + \frac{1}{2}u'''(w)[\alpha(k\mu + x) + z]^2,$$

from where it follows

$$(k\mu + x)u'(w + \alpha(k\mu + x) + z) \approx u'(w)(k\mu + x) + u''(w)(k\mu + x)[\alpha(k\mu + x) + z]$$
$$+\frac{1}{2}u'''(w)(k\mu + x)[\alpha(k\mu + x) + z]^2.$$

By (25), the previous relation and the linearity of mixed expected utility, we will have

$$W'(\alpha) \approx u'(w)(k\mu + E_f(A)) + u''(w)E_f[(k\mu + A)(\alpha(k\mu + A) + Z)] +$$
$$+\frac{1}{2}u'''(w)E_f[(k\mu + A)(\alpha(k\mu + A) + Z)^2].$$

Then, from $W'(\beta(k)) = 0$, we will deduce:

$$u'(w)(k\mu + E_f(A)) + u''(w)E_f[(k\mu + A)(\beta(k)(k\mu + A) + Z)] +$$
$$+\frac{1}{2}u'''(w)E_f[(k\mu + A)(\beta(k)(k\mu + A) + Z)^2] \approx 0.$$

If we denote

$$g(k) = E_f[(k\mu + A)(\beta(k)(k\mu + A) + Z)], \quad \text{(A1)}$$
$$h(k) = E_f[(k\mu + A)(\beta(k)(k\mu + A) + Z)^2], \quad \text{(A2)}$$

then the previous relation can be written

$$u'(w)(k\mu + E_f(A)) + u''(w)g(k) + \frac{1}{2}u'''(w)h(k) \approx 0.$$

Deriving twice with respect to k, we obtain:

$$u''(w)g''(k) + \frac{1}{2}u'''(w)h''(k) \approx 0. \quad \text{(A3)}$$

We set $k = 0$ in (A2):

$$u''(w)g''(0) + \frac{1}{2}u'''(w)h''(0) \approx 0. \tag{A4}$$

The computation of $g''(0)$. We notice that

$$g(k) = \beta(k)E_f[(k\mu + A)^2] + E_f[(k\mu + A)Z].$$

By denoting $g_1(k) = \beta(k)E_f[(k\mu + A)^2]$ and $g_2(k) = E_f[(k\mu + A)Z]$, we will have $g(k) = g_1(k) + g_2(k)$. One easily sees that $g_2''(k) = 0$, thus $g''(k) = g_1''(k) + g_2''(k) = g_1''(k)$. We derive $g_1(k)$:

$$g_1'(k) = \beta'(k)E_f[(k\mu + A)^2] + 2\mu\beta(k)E_f(k\mu + A)$$
$$= \beta'(k)E_f[(k\mu + A)^2] + 2\mu^2 k\beta(k),$$

Since $E_f(k\mu + A) = k\mu + E_f(A) = k\mu$. We derive one more time

$$g_1''(k) = \beta''(k)E_f[(k\mu + A)^2] + 2\mu\beta'(k)E_f(k\mu + A) + 2\mu^2[\beta(k) + k\beta'(k)].$$

Setting $k = 0$ in the previous relation and taking into account that $\beta(0) = E_f(A) = 0$, it follows

$$g''(0) = \beta''(0)E_f(A^2). \tag{A5}$$

The computation of $h''(0)$. We write $h(k)$ as

$$h(k) = \beta^2(k)E_f[(k\mu + A)^3] + 2\beta(k)E_f[(k\mu + A)^2 Z] + E_f[(k\mu + A)Z^2].$$

We denote

$$h_1(k) = \beta^2(k)E_f[(k\mu + A)^3],$$
$$h_2(k) = \beta(k)E_f[(k\mu + A)^2 Z],$$
$$h_3(k) = E_f[(k\mu + A)Z^2].$$

Then, $h(k) = h_1(k) + h_2(k) + h_3(k)$. One notices that $h_3''(0) = 0$, thus

$$h''(0) = h_1''(0) + 2h_2''(0). \tag{A6}$$

We first compute $h_2''(0)$. One can easily notice that

$$h_2''(k) = \beta''(k)E_f[(k\mu + A)^2 Z] + 2\beta'(k)\frac{d}{dk}E_f[(k\mu + A)^2 Z] + \beta(k)\frac{d^2}{dk^2}E_f[(k\mu + A)^2 Z].$$

Taking into account that

$$\frac{d}{dk}E_f[(k\mu + A)^2 Z] = 2\mu E_f[(k\mu + A)Z],$$

and $\beta(0) = 0$,
We deduce

$$h_2''(0) = \beta''(0)E_f(A^2 Z) + 4\mu\beta'(0)E_f(AZ). \tag{A7}$$

We will compute $h_1''(0)$. We derive twice $h_1(k)$:

$$h_1''(k) = \frac{d^2}{dk^2}(\beta^2(k))E_f[(k\mu + A)^3] + 2\frac{d}{dk}(\beta^2(k))\frac{d}{dk}E_f[(k\mu + A)^3] +$$
$$+\beta^2(k)\frac{d^2}{dk^2}E_f[(k\mu + A)^3].$$

We compute the following derivatives from the last sum:

$$\frac{d}{dk}(\beta^2(k)) = 2\beta(k)\beta'(k),$$

$$\frac{d^2}{dk^2}(\beta^2(k)) = 2[\beta''(k)\beta(k) + (\beta'(k))^2],$$

$$\frac{d}{dk}E_f[(k\mu + A)^3] = 3\mu E_f[(k\mu + A)^2].$$

Then, taking into account $\beta(0) = 0$:

$$\begin{aligned} h_1''(0) &= 2[\beta''(0)\beta(0) + (\beta'(0))^2]E_f(A^3) + 2\beta(0)\beta'(0)3\mu E_f[(k\mu + A)^2] + \beta^2(0) = \tfrac{d}{dk}E_f[(k\mu + A)^3] \\ &= 2(\beta'(0))^2 E_f(A^3). \end{aligned} \tag{A8}$$

By (A6)–(A8):

$$h''(0) = h_1''(0) + 2h_2''(0) = \;= 2(\beta'(0))^2E_f(A^3) + 2(\beta''(0)E_f(A^2Z) + 4\mu\beta'(0)E_f(AZ)). \tag{A9}$$

Replacing in (A3) the values of $g''(0)$ and $h''(0)$ given by (A5) and (A9):

$$u''(w)\beta''(0)E_f(A^2) + \frac{1}{2}u'''(w)[2(\beta'(0))^2E_f(A^3) + 2(\beta''(0)E_f(A^2Z) + 4\mu\beta'(0)E_f(AZ))] \approx 0,$$

from where

$$\beta''(0)[u''(w)E_f(A^2) + u'''(w)E_f(A^2Z)] \approx$$
$$\approx -\beta'(0)u'''(w)[\beta'(0)E_f(A^3) + 4\mu E_f(AZ)].$$

The approximate value of $\beta''(0)$ follows:

$$\beta''(0) \approx -u'''(w)\beta'(0)\frac{\beta'(0)E_f(A^3) + 4\mu E_f(AZ)}{u''(w)E_f(A^2) + u'''(w)E_f(A^2Z)}.$$

According to Corollary 2, the expression above which approximates $\beta''(0)$ can be written:

$$\beta''(0) \approx -u'''(w)\beta'(0)\frac{\beta'(0)E_f(A^3) + 4\mu M(Z)E_f(A)}{u''(w)E_f(A^2) + M(Z)E_f(A^2)}$$

$$= -\frac{u'''(w)\beta'(0)}{E_f(A^2)}\frac{\beta'(0)E_f(A^3)}{u''(w) + M(Z)u'''(w)},$$

Since $E_f(A) = 0$. If we replace A with $B - E_f(B)$, one obtains

$$\beta''(0) \approx -\frac{(\beta'(0))^2u'''(w)}{Var_f(B)}\frac{E_f[(B - E_f(B))^3]}{u''(w) + M(Z)u'''(w)}$$

$$= \frac{P_u(w)(\beta'(0))^2}{Var_f(B)}\frac{E_f[(B - E_f(B))^3]}{1 - M(Z)P_u(w)}.$$

□

Proof of Theorem 2. The approximation formula of $\beta'(0)$ from Proposition 5 can be written:

$$\beta'(0) \approx \frac{\mu}{Var_f(B)}[\frac{1}{r_u(w)} - M(Z)]. \tag{A10}$$

According to (27), (A9), and Proposition 6,

$$\beta(k) \approx k\beta'(0) + \frac{1}{2}k^2\beta''(0)$$

$$= \frac{\mu k}{Var_f(B)}[\frac{1}{r_u(w)} - M(Z)] +$$

$$+\frac{1}{2}P_u(w)\frac{(k\mu)^2 E_f[(B - E_f(B))^3]}{Var_f^3(B)[1 - M(Z)P_u(w)]}[\frac{1}{r_u(w)} - M(Z)]^2.$$

Since $\mu k = E_f(B)$, it follows

$$\beta(k) \approx \frac{E_f(B)}{Var_f(B)}[\frac{1}{r_u(w)} - M(Z)] +$$

$$\frac{1}{2}P_u(w)[\frac{1}{r_u(w)} - M(Z)]^2 \frac{E_f^2(B)E_f[(B - E_f(B))^3]}{Var_f^3(B)[1 - M(Z)P_u(w)]}.$$

□

References

1. Arrow, K.J. *Essays in the Theory of Risk Bearing*; North-Holland Publishing Company: Amsterdam, The Netherlands, 1970.
2. Brandt, M. Portfolio choice problems. In *Handbook of Financial Econometrics: Tools and Techniques*; Ait-Sahalia, Y., Hansen, L.P., Eds.; North-Holland Publishing Company: Amsterdam, The Netherlands, 2009; Volume 1.
3. Pratt, J.W. Risk Aversion in the Small and in the Large. *Econometrica* **1964**, *32*, 122–136. [CrossRef]
4. Eeckhoudt, L.; Gollier, C.; Schlesinger, H. *Economic and Financial Decisions under Risk*; Princeton University Press: Princeton, NJ, USA, 2005.
5. Gollier, C. *The Economics of Risk and Time*; MIT Press: Cambridge, MA, USA, 2004.
6. Athayde, G.; Flores, R. Finding a Maximum Skewness Portfolio A General Solution to Three-Moments Portfolio Choice. *J. Econ. Dyn. Control* **2004**, *28*, 1335–1352. [CrossRef]
7. Garlappi, L.; Skoulakis, G. Taylor Series Approximations to Expected Utility and Optimal Portfolio Choice. *Math. Financ. Econ.* **2011**, *5*, 121–156. [CrossRef]
8. Zakamulin, V.; Koekebakker, S. Portfolio Performance Evaluation with Generalized Sharpe Ratios: Beyond the Mean and Variance. *J. Bank. Financ.* **2009**, *33*, 1242–1254. [CrossRef]
9. Kimball, M.S. Precautionary saving in the small and in the large. *Econometrica* **1990**, *58*, 53–73. [CrossRef]
10. Ñiguez, T.M.; Paya, I.; Peel, D. Pure Higher-Order Effects in the Portfolio Choice Model. *Financ. Res. Lett.* **2016**, *19*, 255–260. [CrossRef]
11. Le Courtois, O. On Prudence, Temperance, and Monoperiodic Portfolio Optimization. In Proceedings of the Risk and Choice: A Conference in Honor of Louis Eeckhoudt, Toulouse, France, 12–13 July 2012.
12. Zadeh, L.A. Fuzzy sets as a basis for a theory of possibility. *Fuzzy Sets Syst.* **1978**, *1*, 3–28. [CrossRef]
13. Dubois, D.; Foulloy, L.; Mauris, G.; Prade, H. Probability–possibility transformations, triangular fuzzy sets and probabilistic inequalities. *Reliab. Comput.* **2004**, *10*, 273–297. [CrossRef]
14. Carlsson, C.; Fullér, R. *Possibility for Decision*; Springer: Berlin, Germany, 2011.
15. Dubois, D.; Prade, H. *Possibility Theory*; Plenum Press: New York, NY, USA, 1988.
16. Georgescu, I. *Possibility Theory and the Risk*; Springer: Berlin, Germany, 2012.
17. Dubois, D.; Prade, H. *Fuzzy Sets and Systems: Theory and Applications*; Academic Press: New York, NY, USA, 1980.
18. Carlsson, C.; Fullér, R. On possibilistic mean value and variance of fuzzy numbers. *Fuzzy Sets Syst.* **2001**, *122*, 315–326. [CrossRef]
19. Fullér, R.; Majlender, P. On weighted possibilistic mean and variance of fuzzy numbers. *Fuzzy Sets Syst.* **2003**, *136*, 363–374.

20. Lucia-Casademunt, A.M.; Georgescu, I. Optimal saving and prudence in a possibilistic framework. In *Distributed Computing and Artificial Intelligence*; Springer: Cham, Switzerland, 2013; Volume 217, pp. 61–68.
21. Collan, M.; Fedrizzi, M.; Luukka, P. Possibilistic risk aversion in group decisions: Theory with application in the insurance of giga-investments valued through the fuzzy pay-off method. *Soft Comput.* **2017**, *21*, 4375–4386. [CrossRef]
22. Majlender, P. A Normative Approach to Possibility Theory and Decision Support. Ph.D. Thesis, Turku Centre for Computer Science, Turku, Finland, 2004.
23. Mezei, J. A Quantitative View on Fuzzy Numbers. Ph.D. Thesis, Turku Centre for Computer Science, Turku, Finland, 2011.
24. Thavaneswaran, A.; Thiagarajahb, K.; Appadoo, S.S. Fuzzy coefficient volatility (FCV) models with applications. *Math. Comput. Model.* **2007**, *45*, 777–786. [CrossRef]
25. Thavaneswaran, A.; Appadoo, S.S.; Paseka, A. Weighted possibilistic moments of fuzzy numbers with applications to GARCH modeling and option pricing. *Math. Comput. Model.* **2009**, *49*, 352–368. [CrossRef]
26. Kaluszka, M.; Kreszowiec, M. On risk aversion under fuzzy random data. *Fuzzy Sets Syst.* **2017**, *328*, 35–53. [CrossRef]
27. Zhang, W.G.; Wang, Y.L. A Comparative Analysis of Possibilistic Variances and Covariances of Fuzzy Numbers. *Fundam. Inform.* **2008**, *79*, 257–263.

Σ mathematics MDPI

Article

The Emergence of Fuzzy Sets in the Decade of the Perceptron—Lotfi A. Zadeh's and Frank Rosenblatt's Research Work on Pattern Classification

Rudolf Seising

The Research Institute for the History of Science and Technology, Deutsches Museum, 80538 Munich, Germany; seising@deutsches-museum.de; Tel.: +49-(0)-89-2179-298

Received: 25 May 2018; Accepted: 19 June 2018; Published: 26 June 2018

Abstract: In the 1950s, the mathematically oriented electrical engineer, Lotfi A. Zadeh, investigated system theory, and in the mid-1960s, he established the theory of Fuzzy sets and systems based on the mathematical theorem of linear separability and the pattern classification problem. Contemporaneously, the psychologist, Frank Rosenblatt, developed the theory of the perceptron as a pattern recognition machine based on the starting research in so-called artificial intelligence, and especially in research on artificial neural networks, until the book of Marvin L. Minsky and Seymour Papert disrupted this research program. In the 1980s, the Parallel Distributed Processing research group requickened the artificial neural network technology. In this paper, we present the interwoven historical developments of the two mathematical theories which opened up into fuzzy pattern classification and fuzzy clustering.

Keywords: pattern classification; fuzzy sets; perceptron; artificial neural networks; Lotfi A. Zadeh; Frank Rosenblatt

---

1. Introduction

> "Man's pattern recognition process—that is, his ability to select, classify, and abstract significant information from the sea of sensory information in which he is immersed—is a vital part of his intelligent behavior."
>
> Charles Rosen [1] (p. 38)

In the 1960s, capabilities for classification, discrimination, and recognition of patterns were demands concerning systems deserving of the label "intelligent". Back then, and from a mathematical point of view, patterns were sets of points in a mathematical space; however, by and by, they received the meaning of datasets from the computer science perspective.

Under the concept of a pattern, objects of reality are usually represented by pixels; frequency patterns that represent a linguistic sign or a sound can also be characterized as patterns. "At the lowest level, general pattern recognition reduces to pattern classification, which consists of techniques to separate groups of objects, sounds, odors, events, or properties into classes, based on measurements made on the entities being classified". This said artificial intelligence (AI) pioneer, Charles Rosen, in the introduction of an article in Science in 1967, he claimed in the summary: "This function, pattern recognition, has become a major focus of research by scientists working in the field of artificial intelligence" [1] (p. 38, 43).

The first AI product that was supposed to solve the classification of patterns, such as handwritten characters, was an artificial neuronal network simulation system named perceptron. Its designer was Frank Rosenblatt, a research psychologist at the Cornell Aeronautical Laboratory in Buffalo, New York.

The historical link between pattern discrimination or classification and fuzzy sets documents a RAND report entitled "Abstraction and Pattern Classification", written in 1964 by Lotfi A. Zadeh, a Berkeley professor of electrical engineering. In this report, he introduced the concept of fuzzy sets for the first time [2]. (The text was written by Zadeh. However, he was not employed at RAND Corporation; Richard Bellman and Robert Kalaba worked at RAND, and therefore, the report appeared under the authorship and order: Bellman, Kalaba, Zadeh; later, the text appeared in the Journal of Mathematical Analysis and Application [3].)

"Pattern recognition, together with learning" was an essential feature of computers going "Steps Toward Artificial Intelligence" in the 1960s as Marvin Lee Minsky postulated already at the beginning of this decade [4] (p. 8). On the 23rd of June of the same year, after four years of simulation experiments, Rosenblatt and his team of engineers and psychologists at the Cornell Aeronautical Laboratory demonstrated to the public their experimental pattern recognition machine, the "Mark I perceptron".

Another historical link connects pattern recognition or classification with the concept of linear separability when Minsky and Seymour Papert showed in their book, "Perceptrons: an introduction to computational geometry" published in 1969, that Rosenblatt's perceptron was only capable of learning linearly separable patterns. Turned to logics, this means that a single-layer perceptron cannot learn the logical connective XOR of the propositional logic.

In addition, a historical link combines Zadeh's research work on optimal systems and the mathematical concept of linear separability, which is important to understand the development from system theory to fuzzy system theory.

We refer to the years from 1957 to 1969 as the decade of the perceptron. It was amidst these years, and it was owing to the research on pattern recognition during the decade of the perceptron, that fuzzy sets appeared as a new "mathematics of fuzzy or cloudy quantities" [5] (p. 857).

This survey documents the history of Zadeh's mathematical research work in electrical engineering and computer science in the 1960s. It shows the intertwined system of research in various areas, among them, mathematics, engineering and psychology. Zadeh's mathematically oriented thinking brought him to fundamental research in logics and statistics, and the wide spectrum of his interests in engineering sciences acquainted him with research on artificial neural networks and natural brains as well.

2. Pattern Separation

Today, algorithms in machine learning and statistics solve the problem of pattern classification, i.e., of separating points in a set. More specifically, and in the case of Euclidean geometry, they determine sets of points to be linearly separable. In the case of only two dimensions in the plane, linear separability of two sets A and B means that there exists at least one line in the plane with all elements of A on one side of the line and all elements of B on the other side.

For n-dimensional Euclidean spaces, this generalizes if the word "line" is replaced by "hyperplane": A and B are linearly separable if there exists at least one hyperplane with all elements of A on one side of the hyperplane and all elements of B on the other side.

Let us consider the case $n = 2$ (see Figure 1): Two subsets $A \subseteq 2^n$, $B \subseteq 2^n$ are linearly separable if there exist $n + 1 = 3$ real numbers w_1, w_2, and for all $\boldsymbol{a} = (a_1, a_2) \in \mathrm{A}$, $\boldsymbol{b} = (b_1, b_2) \in B$ it holds

$$w_1\, a_1 + w_2\, a_2 \leq w_3 \leq w_1\, b_1 + w_2\, b_2.$$

The points $\boldsymbol{x} = (x_1, x_2)$ with $w_1\, x_1 + w_2\, x_2 = w_3$ build the separating line.

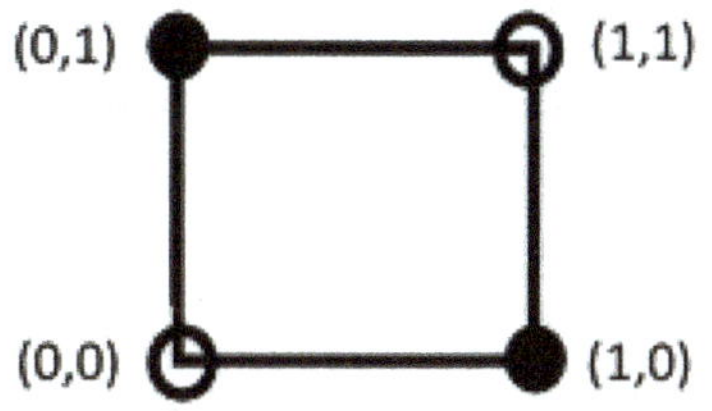

Figure 1. The points (0,1) and (1,0) are not linearly separable.

In 1936, the Polish mathematician, Meier (Maks) Eidelheit (1910–1943), published an article where he proved the later so-called Eidelheit separation theorem concerning the possibility of separating convex sets in normed vector spaces (or local-convex spaces) by linear functionals [6].

One of the researchers who checked the separation theorem for applications in electrical engineering was Lotfi Aliasker Zadeh. He was born in Baku, Azerbaidjan; he studied electrical engineering at the University of Tehran, Iran, and he graduated with a BSc degree in 1942. The following year, he emigrated to the United States (US) via Cairo, Egypt. He landed in Philadelphia, and then worked for the International Electronic Laboratories in New York. In 1944, he went to Boston to continue his studies at the Massachusetts Institute for Technology (MIT). In 1946, Zadeh was awarded a Master's of Science degree at MIT, and then he changed to Columbia University in New York, where he earned his Doctor of Philosophy (PhD) degree in 1950 for his thesis in the area of continuous analog systems [7]. After being appointed assistant professor, he was searching for new research topics. Both information theory and digital technology interested him, and he turned his attention to digital systems. Zadeh, in an interview with the author on 8 September 1999, in Zittau, at the margin of the 7th Zittau Fuzzy Colloquium at the University Zittau/Görlitz said he "was very much influenced by Shannon's talk that he gave in New York in 1946 in which he described his information theory." Zadeh began delivering lectures on automata theory, and in 1949, he organized and moderated a discussion meeting on digital computers at Columbia University, in which Claude E. Shannon, Edmund Berkeley, and Francis J. Murray took part. It was probably the first public debate on this subject ever, as suggested by Zadeh in an interview with the author on 15 June 2001, University of California, Berkeley.)

In the second half of the 1950s, Zadeh (Figure 2) became one of the pioneers of system theory, and among his interests, was the problem of evaluating the performance of systems like electrical circuits and networks with respect to their input and their output. His question was whether such systems could be "identified" by experimental means. His thoughts "On the Identification Problem" appeared in the December 1956 edition of "IRE Transactions on Circuit Theory" of the Institute of Radio Engineers [8]. For Zadeh, a system should be identified given (1) a system as a black box B whose input–output relationship is not known a priori, (2) the input space of B, which is the set of all time functions on which the operations with B are defined, and (3) a black box class A that contains B, which is known a priori. Based on the observed response behavior of B for various inputs, an element of A should be determined that is equivalent to B inasmuch as its responses to all time functions in the input space of B are identical to those of B. In a certain sense, one can claim to have "identified" B by means of this known element of A.

Of course, this "system identification" can turn out to be arbitrarily difficult to achieve. Only insofar as information about black box B is available can black box set A be determined. If B has a "normal" initial state in which it returns to the same value after every input, such as the resting state of a linear system, then the problem is not complicated. If this condition is not fulfilled, however, then B's response behavior depends on a "not normal" initial state, and the attempt to solve the problem gets out of hand very quickly.

All different approaches to solving the problem that was proposed up to that point were of theoretical interest, but they were not very helpful in practice and, on top of that, many of the suggested solutions did not even work when the "black box set" of possible solutions was very limited. In the course of the article, Zadeh only looks at very specific nonlinear systems, which are relatively easy to identify by observation as sinus waves with different amplitudes. The identification problem remained unsolved for Zadeh.

In 1956, Zadeh took a half-year sabbatical at the Institute for Advanced Study (IAS) in Princeton, as disclosed by Zadeh in an interview with the author on 16 June 2001, University of California, Berkeley, that was, for him, the "Mecca for mathematicians". It inspired him very quickly, and he took back to New York many very positive and lasting impressions. As a "mathematical oriented engineer"—he characterized himself that way in one of my interviews on 26 July 2000, University of California, Berkeley—he now started analyzing concepts in system theory from a mathematical point of view, and one of these concepts was optimality.

In his editorial to the March 1958 issue of the "IRE Transactions on Information Theory", Zadeh wrote, "Today we tend, perhaps, to make a fetish of optimality. If a system is not 'best' in one sense or another, we do not feel satisfied. Indeed, we are not apt to place too much confidence in a system that is, in effect, optimal by definition". In this editorial, he criticized scalar-valued performance criteria of systems because "when we choose a criterion of performance, we generally disregard a number of important factors. Moreover, we oversimplify the problem by employing a scalar loss function" [9]. Hence, he suggested that vector-valued loss functions might be more suitable in some cases.

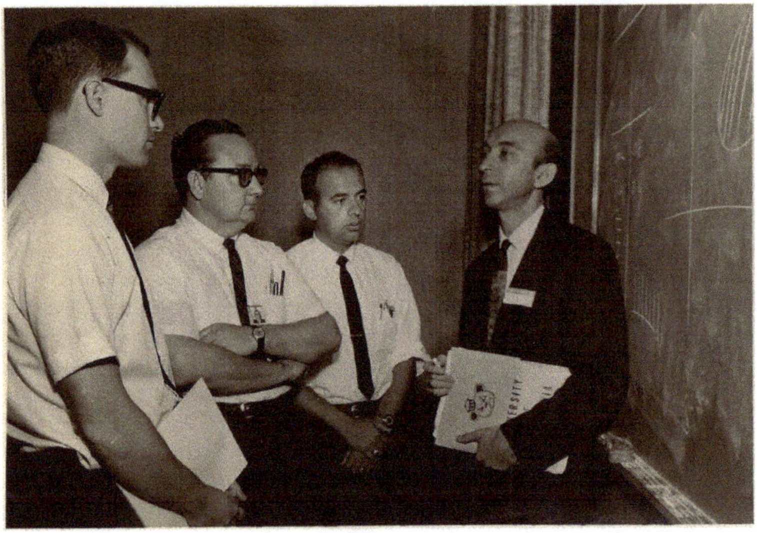

Figure 2. Lotfi A. Zadeh, undated photo, approximately 1950s, photo credit: Fuzzy archive Rudolf Seising.

3. Optimality and Noninferiority

In September 1963, Zadeh continued the mentioned criticism in a correspondence to the "IEEE Transactions on Automatic Control" of the Institute of Electrical and Electronics Engineers [9]. He emphasized, "one of the most serious weaknesses of the current theories of optimal control is that they are predicated on the assumption that the performance of a system can be measured by a single number". Therefore, he sketched the usual reasoning with scalar-valued performance criteria of systems as follows: If $\sum$ is a set of systems and if $P(S)$ is the real-valued performance index of a system S, then a system S_0 is called optimal in the $\sum$ if $P(S_0) \geq P(S)$ for all $S \in \sum$. Thereafter, he criticized that method: "The trouble with this concept of optimality is that, in general, there is more than one consideration that enters into the assessment of performance of S, and in most cases, these considerations cannot be subsumed under a single scalar-valued criterion. In such cases, a system S may be superior to a system S' in some respects and inferior to S' in others, and the class of systems $\sum$ is not completely ordered" [9] (p. 59).

For that reason, Zadeh demanded the distinction between the concepts of "optimality" and "noninferiority". To define what these concepts mean, he considered the "constraint set" $C \subseteq \sum$ that is defined by the constraints imposed on system S, and a partial ordering $\geq$ on $\sum$ by associating with each system S in $\sum$ the following three disjoint subsets of $\sum$:

(1) $\sum > (S)$, the subset of all systems which are "superior" to S.
(2) $\sum \leq (S)$, the subset of all systems, which are inferior or equal ("inferior") to S.
(3) $\sum \sim(S)$, the subset of all systems, which are not comparable with S.

That followed Zadeh's definition of the system's property of "noninferiority":

Definition 1. *A system S_0 in C is noninferior in C if the intersection of C and $\sum_{>}$ (S_0) is empty: $C \cap \sum_{>} (S_0) = \emptyset$.*

Therefore, there is no system in C, which is better than S_0.

The system's property of optimality he defined, as follows:

Definition 2. *A system S_0 in C is optimal in C if C is contained in $\sum \leq (S_0)$: $C \subseteq \sum \leq (S)$.*

Therefore, every system in C is inferior to S_0 or equal to S_0.

These definitions show that an optimal system S_0 is necessarily "noninferior", but not all noninferior systems are optimal.

Zadeh considered the partial ordering of the set of systems, $\sum$, by a vector-valued performance criterion. Let system S be characterized by the vector $x = (x_1, \ldots, x_n)$, whose real-valued components represent, say, the values of n adjustable parameters of S, and let C be a subset of n-dimensional Euclidean space R^n. Furthermore, let the performance of S be measured by an m vector $p(\mathbf{x}) = [p_1(\mathbf{x}), \ldots, p_m(\mathbf{x})]$, where $p_i(\mathbf{x})$, $i = 1, \ldots, m$, is a given real-valued function of x. Then $S \geq S'$ if and only if $p(\mathbf{x}) \geq p(\mathbf{x}')$. That is, $p_i(\mathbf{x}) \geq p_i(\mathbf{x}')$, $i = 1, \ldots, m$.

Figure 3 illustrates "the case where $\sum_{>}$ (S) or, equivalently, $\sum_{>}$ $(\mathbf{x})$ is a fixed cone with a vertex at $\mathbf{x}$, and the constraint set C is a closed bounded subset of R^n" [9] (p. 59).

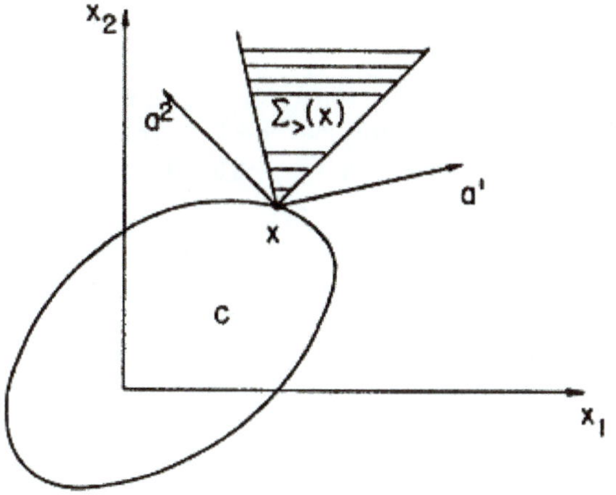

Figure 3. Illustration of the significance of C and $\sum_{>}$ (x) [9].

If $p_i(\mathbf{x}) = a_i^i x_1 + \ldots + a_n x_n^i$, where $a^i = (a_i^i, \ldots, a_n^i)$ is the gradient of $p_i(\mathbf{x})$, a^i = grad $p_i(\mathbf{x})$ (a constant vector), then, $\sum_>(x)$ is the polar cone of the cone spanned by a^i. By definition, noninferior points cannot occur in the interior of the set C. If C is a convex set, then the set of all noninferior points on the boundary of C is the set Γ of all points $\mathbf{x_0}$, through which hyperplanes separating the set C and the set $\sum_>(\mathbf{x_0})$ can be passed. Figure 4 shows the set Γ heavy-lined on the boundary of C.

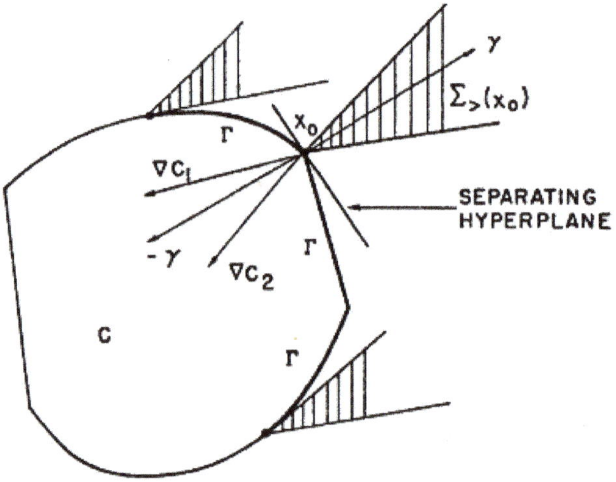

Figure 4. The set of noninferior points on the boundary of C [9].

In this example, $\sum_>(\mathbf{x_0})$ and C are convex sets, and for convex sets, the separation theorem says that there exists a hyperplane, which separates them.

4. Rosenblatt's Perceptron

Among other researchers who studied the separability of data points was Frank Rosenblatt, a research psychologist at the Cornell Aeronautical Laboratory in Buffalo, New York. Rosenblatt was born in New Rochelle, New York on 11 July 1928. In 1957, at the Fifteenth International Congress of Psychology held in Brussels, he suggested a "theory of statistical separability" to interpret receiving and recognizing patterns in natural and artificial systems.

In 1943, Warren McCulloch and Walter Pitts published the first model of neurons that was later called "artificial" or "McCulloch–Pitts neuron". In their article, "A logical calculus of the ideas immanent in nervous activity", they "realized" the entire logical calculus of propositions by "neuron nets", and they arrived at the following assumptions [10] (p. 116):

1. The activity of the neurons is an "all-or-none" process.
2. A certain fixed number of synapses must be excited within the period of latent addition in order to excite a neuron at any time, and this number is independent of previous activity and position on the neuron.
3. The only significant delay within the nervous system is synaptic delay.
4. The activity of any inhibitory synapse prevents without exception the excitation of the neuron at that time.

The structure of the net does not change with time.

In 1949, based on neurophysiological experiments, the Canadian psychologist, Donald Olding Hebb, proposed the later so-called "Hebb learning rule", i.e., a time-dependent principle of behavior of nerve cells: "When an axon of cell A is near enough to excite cell B, and repeatedly or persistently takes part in firing it, some growth process or metabolic change takes place in one or both cells so that A's efficiency, as one of the cells firing B, is increased" [11] (p. 62).

In the same year, the Austrian economist, Friedrich August von Hayek, published "The Sensory Order" [12], in which he outlined general principles of psychology. Especially, he proposed to apply probability theory instead of symbolic logic to model the behavior of neural networks which achieve reliable performance even when they are imperfect by nature as opposed to deterministic machines.

Rosenblatt's theory was in the tradition of Hebb's and Hayek's thoughts. The approach of statistical separability distinguishes his model from former brain models. Rosenblatt "was particularly struck by the fact that all of the mathematically precise, logical models which had been proposed to date were systems in which the phenomenon of distributed memory, or 'equipotentiality' which seemed so characteristic of all biological systems, was either totally absent, or present only as a nonessential artefact, due to postulated 'repetitions' of an otherwise self-contained functional network, which by itself, would be logically sufficient to perform the functions of memory and recall" [13] (p. iii). Therefore, Rosenblatt chose a "model in terms of probability theory rather than symbolic logic" [14] (p. 388).

In his "Probabilistic Model for Visual Perception", as his talk was entitled [15], he characterized perception as a classification process, and in his first project report that appeared in the following year, he wrote, "Elements of stimulation which occur most commonly together are assigned to the same classes in a 'sensory order'. The organization of sensory classes (colors, sounds, textures, etc.) thus comes to reflect the organization of the physical environment from which the sensations originate" ([13], p. 8). To verify his theory, Rosenblatt promised the audience a working electronic model in the near future, and for a start, he presented simulations running on the Weather Bureau's IBM 704. He fed the computer with "two cards, one with squares marked on the left side and the other with squares on the right side". The program differentiated between left and right after "reading" through about 50 punched cards. "It then started registering a 'Q' for the left squares and 'O' for the right squares" [13] (p. 8).

Rosenblatt illustrated the organization of the perceptron via such comparisons with a biological brain, as shown in Figure 5. These illustrations compare the natural brain's connections from the retina to the visual area with a perceptron that connects each sensory point of the "retina" to one or more randomly selected "*A*-units" in the association system. The *A*-units transduce the stimuli, and they increase in value when activated (represented by the red points in Figure 5).

Their responses arrive at "*R*-units", which are binary devices (i.e., "on" or "off", and "neutral" in the absence of any signal because the system will not deliver any output), as Figure 6 shows for a very simple perceptron. The association system has two parts, the upper source set tends to activate the response $R = 1$, and the lower one tends to activate the response $R = 0$. From the responses, a feedback to the source set is generated, and these signals multiply the activity rate of the *A*-unit that receives them. Thus, the activity of the *R*-units shows the response to stimuli as a square or circle, as presented in the environment. "At the outset, when a perceptron is first exposed to stimuli, the responses which occur will be random, and no meaning can be assigned to them. As time goes on, however, changes occurring in the association systems cause individual responses to become more and more specific to such particular, well-differentiated classes of forms as squares, triangles, clouds, trees, or people" [16] (p. 3).

Rosenblatt attached importance to the following "fundamental feature of the perceptron": "When an *A*-unit of the perceptron has been active, there is a persistent after-effect which serves the function of a 'memory trace'. The assumed characteristic of this memory trace is a simple one: whenever a cell is active, it gains in 'strength' so that its output signals (in response to a fixed stimulus) become stronger, or gain in frequency or probability" [16] (p. 3).

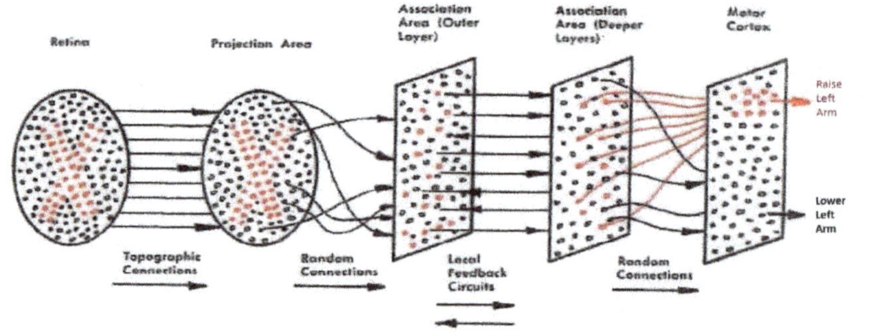

FIG. 1 — Organization of a biological brain. (Red areas indicate active cells, responding to the letter X.)

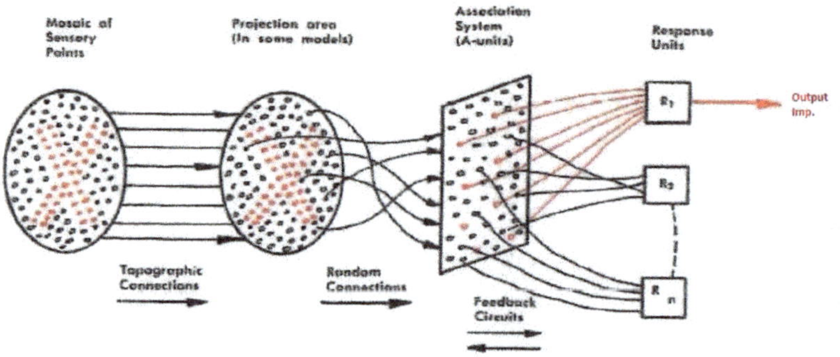

FIG. 2 — Organization of a perceptron.

Figure 5. Organization of a biological brain and a perceptron [16] (p. 2), the picture was modified for better readability).

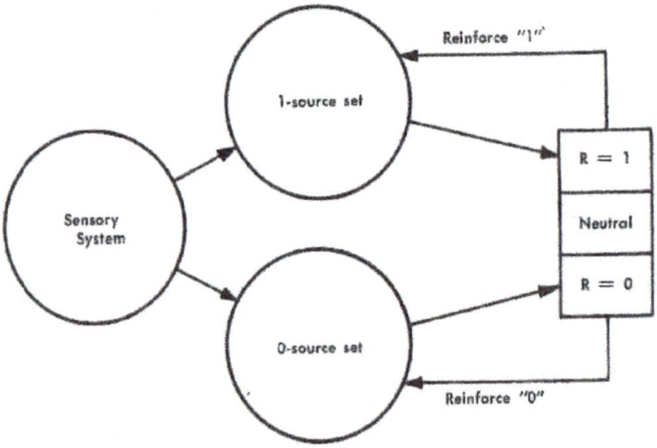

FIG. 3 — Detailed organization of a single perceptron.

Figure 6. Detailed organization of a single perceptron [16] (p. 3).

Rosenblatt presented the results of experiments in which a perceptron had to learn to discriminate between a circle and a square with 100, 200, and 500 *A*-units in each source set of the association system. In Figure 7, the broken curves indicate the probability that the correct response is given when identical stimuli of a test figure were shown during the training period. Rosenblatt called this "the perceptron's capacity to recollect". The solid curves show the probability that the appropriate response for any member of the stimulus class picked at random will be given. Rosenblatt called this "the perceptron's capacity to generalize" [16] (p. 10). Figure 7 shows that both probabilities (capacities) converge in the end to the same limit. "Thus", concluded Rosenblatt, "in the limit it makes no difference whether the perceptron has seen the particular stimulus before or not; it does equally well in either case" [16] (p. 4).

Clearly, probability theory is necessary to interpret the experimental results gathered with his perceptron simulation system. "As the number of association units in the perceptron is increased, the probabilities of correct performance approach unity", Rosenblatt claimed, and with reference to Figure 7, he continued, "it is clear that with an amazingly small number of units—in contrast with the human brain's 10^{10} nerve cells—the perceptron is capable of highly sophisticated activity [16] (p. 4).

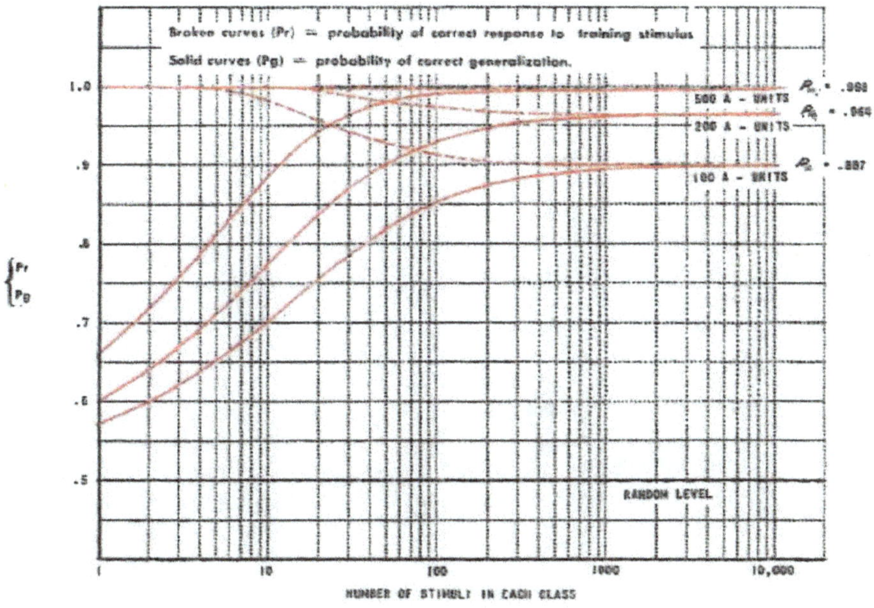

FIG. 5 — Learning curves for three typical perceptrons.

Figure 7. Learning curves for three typical perceptrons [16] (p. 6).

In 1960, on the 23rd of June, a "Perceptron Demonstration", sponsored by the ONR (Office of Naval Research) and the Directorate of Intelligence and Electronic Warfare, Rome Air Development Center, took place at the Cornell Aeronautical Laboratory. After a period of successful simulation experiments, Rosenblatt and his staff had created the experimental machine Mark I perceptron.

Four hundred photocells formed an imaginary retina in this perceptron—a simulation of the retinal tissue of a biological eye—and over 500 other neuronlike units were linked with these photocells by the principle of contingency, so they could supply them with impulses that came from stimuli in the imaginary retina. The actual perceptron was formed by a third layer of artificial neurons, the processing or response layer. The units in this layer formed a pattern associator.

In this classic perceptron, cells can be differentiated into three layers. Staying with the analogy of biological vision, the "input layer" with its (photo) cells or "stimulus units" (S cells) corresponds to the retinal tissue, and the middle "association layer" consists of so-called association units (A cells), which are wired with permanent but randomly selected weights to S cells via randomly linked contacts. Each A cell can, therefore, receive a determined input from the S layer. In this way, the input pattern of the S layer is distributed to the A layer. The mapping of the input pattern from the S layer onto a pattern in the A layer is considered "pre-processing". The "output layer", which actually makes up the perceptron, and which is, thus, also called the "perceptron layer", contains the pattern-processing response units (R cells), which are linked to the A cells. R and A cells are McCulloch–Pitts neurons, but their synapses are variable and are adapted appropriately according to the Hebb rule. When the sensors detect a pattern, a group of neurons is activated, which prompts another neuron group to classify the pattern, i.e., to determine the pattern set to which said pattern belongs.

A pattern is a point in the $x = (x_1, x_2, ..., x_n)$ n-dimensional vector space, and so, it has n components. Let us consider again just the case $n = 2$, then a pattern $x = (x_1, x_2)$ such as this also belongs to one of L "pattern classes". This membership occurs in each individual use case. The perceptron "learned" these memberships of individual patterns beforehand on the basis of known classification examples it was provided. After an appropriate training phase, it was then "shown" a new pattern, which it placed in the proper classes based on what it already "learned". For a classification like this, each unit, r, of the perceptron calculated a binary output value, y_r, from the input pattern, x, according to the following equation: $y_r = \theta\ (w_{r1}x_1 + w_{r2}x_2)$.

The weightings, w_{r1} and w_{r2}, were adapted by the unit, r, during the "training phase", in which the perceptron was given classification examples, i.e., pattern vectors with an indication of their respective pattern class, C_s, such that an output value, $y_r = 1$, occurred only if the input pattern, x, originated in its class, C_r. If an element, r, delivered the incorrect output value, y_r, then its coefficients, w_{r1} and w_{r2}, were modified according to the following formulae:

$$\Delta w_{r1} = \varepsilon_r \cdot (\delta_{rs} - y_r) \cdot x_1 \qquad \text{and} \qquad \Delta w_{r2} = \varepsilon_r \cdot (\delta_{rs} - y_r) \cdot x_2.$$

In doing so, the postsynaptic activity, y_r, used in the Hebb rule is replaced by the difference between the correct output value, δ_{rs}, and the actual output value, y_r. These mathematical conditions for the perceptron were not difficult. Patterns are represented as vectors, and the similarity and disparity of these patterns can be represented if the vector space is normalized; the dissimilarity of two patterns, v_1 and v_2, can then be represented as the distance between these vectors, such as in the following definition: $d(v_1 v_2) = | | v_2 - v_1 | |$.

5. Perceptron Convergence

In 1955, the mathematician Henry David Block came to Cornell where he started in the Department of Mathematics; however, in 1957, he changed to the Department of Theoretical and Applied Mechanics. He collaborated with Rosenblatt, and derived mathematical statements analyzing the perceptron's behavior. Concerning the convergence theorem, the mathematician, Jim Bezdek, said in my interview: "To begin, I note that Dave Block proved the first Perceptron convergence theorem, I think with Nilsson at Stanford, and maybe Novikoff, in about 1962, you can look this up" [17] (p. 5).

Block published his proof in "a survey of the work to date" in 1962 [18]. Nils John Nilsson was Stanford's first Kumagai Professor of Engineering in Computer Science, and Albert Boris J. Novikoff earned the PhD from Stanford. In 1958, he became a research mathematician at the Stanford Research Institute (SRI). He presented a convergence proof for perceptrons at the Symposium on Mathematical Theory of Automata at the Polytechnic Institute of Brooklyn (24–26 April 1962) [19]. Other versions of the algorithm were published by the Russian control theorists, Mark Aronovich Aizerman, Emmanuel M. Braverman, and Lev I. Rozonoér, at the Institute of Control Sciences of the Russian Academy of Sciences, Moscow [20–22].

In 1965, Nilsson wrote the book "Learning Machines: Foundations of Trainable Pattern-Classifying Systems", in which he also described in detail the perceptron's error correction and learning procedure. He also proved that for separable sets, A and B, in the n-dimensional Euclidean space, the relaxation (hill-climbing/gradient) algorithm will converge to a solution in finite iterations.

Judah Ben Rosen, an electrical engineer, who was head of the applied mathematics department in the Shell Development Company (1954–1962) came as a visiting professor to Stanford's computer science department (1962–1964). In 1963, he wrote a technical report entitled "Pattern Separation by Convex Programming" [23], which he later published as a journal article [24].

Coming from the already mentioned separation theorem, he showed "that the pattern separation problem can be formulated and solved as a convex programming problem, i.e., the minimization of a convex function subject to linear constraints" [24] (p. 123). For the n-dimensional case, he proceeded as follows: A number l of point sets in an n-dimensional Euclidean space is to be separated by an appropriate number of hyperplanes. The m_i points in the ith set (where $i = 1, ..., l$) are denoted by n-dimensional vectors, $pi_j, j = 1, ..., mi$. Then, the following matrix describes the points in the ith set:

$$P_i = p_{i1}, p_{i2}, ..., p_{imi}.$$

In the simplest case, at which the Rosenblatt perceptron failed, two points, P_1 and P_2, are to be separated. Rosen provides this definition:

Definition 3. *The point sets, P_1 and P_2, are linearly separable if their convex hulls do not intersect. (The convex hull of a set is the set of all convex combinations of its points. In other words, given the points pi from P and given λi from* **R**, *then the following set is the convex hull of P: conv (P) = $\lambda 1 \cdot p1 + \lambda 2 \cdot p2 + ... + \lambda n \cdot pn$.)*

An equivalent statement is the following:

"The point sets, P_1 and P_2, are linearly separable if and only if a hyperplane $H = H(z,\alpha) = \{p \mid p'z = \alpha\}$ exists such that P_1 und P_2 lie on the opposite sides of H. (p' refers to the transpose of p.)

The orientation of the hyperplane H is, thus, specified by the n-dimensional unit vector, z, and its distance from the origin is determined by a scalar, α. The linear separation of P_1 and P_2 was, therefore, equivalent to demonstrating the existence of a solution to the following system of strict inequalities. (Here $|| \; ||$ denotes the Euclidean norm, and e_i is the m_i-dimensional unit vector):

$$\begin{array}{ll} p_{1j}z > \alpha \; j = 1, ..., m_1 & \\ p_{2j}z < \alpha \; j = 1, ..., m_2 & ||z|| = 1, \\ p_{1j}z > \alpha \; e_1 & \\ p_{2j}z < \alpha \; e_2 & ||z|| = 1. \end{array}$$

Rosen came to the conclusion "that the pattern separation problem can be formulated and solved as a convex programming problem, i.e., the minimization of a convex subject to linear constraints". [24] (p. 1) He considered the two linearly separable sets, P_1 and P_2. The Euclidean distance, δ, between these two sets is then indicated by the maximum value of γ, for which z and α exist such that

$$\begin{array}{ll} P'_1 \; z \geq (\alpha + 1/2 \; \gamma)e_1 & \\ P'_2 \; z \leq (\alpha + 1/2 \; \gamma)e_2 & ||z|| = 1. \end{array}$$

The task is, therefore, to determine the value of the distance, δ, between the sets, P_1 and P_2, formulated as the nonlinear programming problem that can find a maximum, γ, for which the above inequalities are true. Rosen was able to reformulate it into a convex quadratic programming problem that has exactly one solution when the points, P_1 and P_2, are linearly separable. To do so, he introduced a vector, x, and a scalar, β, for which the following applies: $y = \frac{2}{\sqrt{\|x\|}}$, $\alpha = \frac{\beta}{\sqrt{\|x\|}}$, and $z = \frac{x}{\sqrt{\|x\|}}$. Maximizing γ is, thus, equivalent to minimizing the convex function, $\| x \|^2$:

$$\sigma = \begin{matrix} min \\ x, \beta \end{matrix} \left\{ \frac{1}{4}\|x\|^2 | \begin{matrix} p_1' x \geq (\beta + 1)e_1 \\ p_2' x \leq (\beta - 1)e_2 \end{matrix} \right\}$$

After introducing the $(n + 1)$-dimensional vectors, $y = \begin{pmatrix} x \\ \beta \end{pmatrix}$, $q_{ij} = \begin{pmatrix} p_{ij} \\ -1 \end{pmatrix}$, and the $(n + 1) \times m_i$–matrices, $Q_i = [q_{i1}, q_{i2}, \ldots, q_{imi}]$, Rosen could use the standard form of convex quadratic programming, and formulate the following theorem of linear separability:

Theorem 1. *The point sets, P_1 and P_2, are linearly separable if and only if the convex quadratic programming problem*

$$\sigma = \begin{matrix} \boldsymbol{min} \\ y \end{matrix} \left\{ \frac{1}{4}\sum_{i=1}^{n} y_i^2 | \begin{matrix} Q_1' y \geq e_1 \\ Q_2' y \leq e_2 \end{matrix} \right\},$$

has a solution. If P_1 and P_2 are linearly separable, then the distance, δ, between them is given by $\delta = \frac{1}{\sqrt{\sigma}}$, and a unique vector, $y_0 = \begin{pmatrix} x_0 \\ \beta_0 \end{pmatrix}$, achieves the minimum, σ. The separating hyperplane is given by $H(x_0, \beta_0) = \left\{ p / p' x_0 \geq \beta_0 \right\}$.

6. Fuzzy Pattern Classification

In the middle of the 1960s, Zadeh also got back to the topics of pattern classification and linear separability of sets. In the summer of 1964, he and Richard E. Bellman, his close friend at the RAND Corporation, planned on doing some research together. Before that, there was the trip to Dayton, Ohio, where he was invited to talk on pattern recognition in the Wright-Patterson Air Force Base. Here, within a short space of time, he developed his little theory of "gradual membership" into an appropriately modified set theory: "Essentially the whole thing, let's walk this way, it didn't take me more than two, three, four weeks, it was not long", Said (Zadeh in an interview with the author on June 19, 2001, UC Berkeley.) When he finally met with Bellman in Santa Monica, he had already worked out the entire theoretical basis for his theory of fuzzy sets: "His immediate reaction was highly encouraging and he has been my strong supporter and a source of inspiration ever since", said (Zadeh in "Autobiographical Note 1"—an undated two-page typewritten manuscript, written after 1978.)

Zadeh introduced the conceptual framework of the mathematical theory of fuzzy sets in four early papers. Most well-known is the journal article "Fuzzy Sets" [25]; however, in the same year, the conference paper "Fuzzy Sets and Systems" appeared in a proceedings volume [26], in 1966, "Shadows of Fuzzy Sets" was published in Russia [27], and the journal article "Abstraction and Pattern Classification" appeared in print [4]. The latter has three official authors, Bellman, Kalaba, and Zadeh, but it was written by Zadeh; moreover, the text of this article is the same as the text of a RAND memorandum of October 1964 [3]. By the way, preprints of "Fuzzy Sets" and "Shadows of Fuzzy Sets" [28] appeared already as "reports" of the Electronic Research Laboratory, University of California, Berkeley, in 1964 and 1965 ([29], Zadeh 1965c).

Fuzzy sets "do not constitute classes or sets in the usual mathematical sense of these terms". They are "imprecisely defined 'classes'", which "play an important role in human thinking, particularly in the domains of pattern recognition, communication of information, and abstraction", Zadeh wrote in his seminal paper [25] (p. 338). A "fuzzy set" is "a class in which there may be a continuous infinity of grades of membership, with the grade of membership of an object x in a fuzzy set A represented by a number $\mu_A(x)$ in the interval [0, 1]" [26] (p. 29). He defined fuzzy sets, empty fuzzy sets, equal fuzzy sets, the complement, and the containment of a fuzzy set. He also defined the union and intersection of fuzzy sets as the fuzzy sets that have membership functions that are the maximum or minimum, respectively, of their membership values. He proved that the distributivity laws and De Morgan's laws are valid for fuzzy sets with these definitions of union and intersection. In addition, he defined other ways of forming combinations of fuzzy sets and relating them to one another, such as, the "algebraic sum", the "absolute difference", and the "convex combination" of fuzzy sets.

Concerning pattern classification, Zadeh wrote that these "two basic operations: abstraction and generalization appear under various guises in most of the schemes employed for classifying patterns into a finite number of categories" [3] (p. 1). He completed his argument as follows: "Although abstraction and generalization can be defined in terms of operations on sets of patterns, a more natural as well as more general framework for dealing with these concepts can be constructed around the notion of a 'fuzzy' set—a notion which extends the concept of membership in a set to situations in which there are many, possibly a continuum of, grades of membership" [3] (p. 1).

After a discussion of two definitions of "convexity" for fuzzy sets and the definition of "bounded" fuzzy sets, he defined "strictly" and "strongly convex" fuzzy sets. Finally, he proved the separation theorem for bounded convex fuzzy sets, which was relevant to the solution of the problem of pattern discrimination and classification that he perhaps presented at the Wright-Patterson Air Force Base (neither a manuscript nor any other sources exist; Zadeh did not want to either confirm or rule out this detail in the interviews with the author). At any rate, in his first text on fuzzy sets, he claimed that the concepts and ideas of fuzzy sets "have a bearing on the problem of pattern classification" [2] or [3] (p. 1). "For example, suppose that we are concerned with devising a test for differentiating between handwritten letters, O and D. One approach to this problem would be to give a set of handwritten letters, and to indicate their grades of membership in the fuzzy sets, O and D. On performing abstraction on these samples, one obtains the estimates, $\widetilde{\mu_0}$ and $\widetilde{\mu_D}$, of μ_0 and μ_D, respectively. Then, given a letter, x, which is not one of the given samples, one can calculate its grades of membership in O and D, and, if O and D have no overlap, classify x in O or D" [26] (p. 30) (see Figure 8).

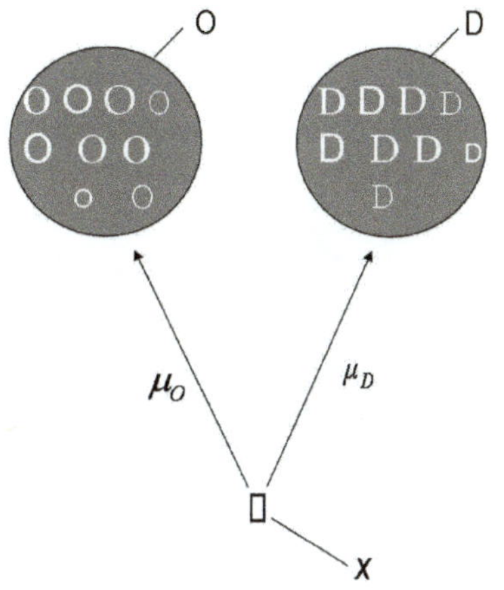

Figure 8. Illustration to Zadeh's view on pattern classification: the sign □ (or x, as Zadeh wrote) belongs with membership value, μ_O(O), to the "class" of Os and with membership value, μ_D(D), to the "class" of Ds.

In his studies about optimality in signal discrimination and pattern classification, he was forced to resort to a heuristic rule to find an estimation of a function $f(x)$ with the only means of judging the "goodness" of the estimate yielded by such a rule lying in experimentation." [3] (p. 3). In the quoted article, Zadeh regarded a pattern as a point in a universe of discourse, Ω, and $f(x)$ as the membership function of a category of patterns that is a (possibly fuzzy) set in Ω.

With reference to Rosen's article, Zadeh stated and proofed an "extension of the separation theorem to convex fuzzy sets" in his seminal paper, of course, without requiring that the convex fuzzy sets A and B be disjoint, "since the condition of disjointness is much too restrictive in the case of fuzzy sets" [25] (p. 351). A hyperplane, H, in an Euclidean space, E^n, is defined by an equation, $h(x) = 0$, then, $h(x) \geq 0$ is true for all points $x \in E^n$ on one side of H, and $h(x) \leq 0$ is true for all points, $x \in E^n$, on the other side of H. If a fuzzy set, A, is on the one side of H, and fuzzy set, B, is on its other side, their membership functions, $f_A(x)$ and $f_B(x)$, and a number, K_H, dependent on H, fulfil the following inequalities:

$$f_A(x) \leq K_H \qquad \text{and} \qquad f_B(x) \geq K_H.$$

Zadeh defined M_H, the infimum of all K_H, and $D_H = 1 - M_H$, the "degree of separation" of A and B by H. To find the highest possible degree of separation, we have to look for a member in the family of all possible hypersurfaces that realizes this highest degree. In the case of hyperplane, H, in E^n Zadeh defined the infimum of all M_H by

$$\overline{M} = Inf_H M_H,$$

and the "degree of separation of A and B" by the relationship,

$$D = 1 - \overline{M}.$$

Thereupon Zadeh presented his extension of the "separation theorem" for convex fuzzy sets:

Theorem 2. *Let A and B be bounded convex fuzzy sets in E^n, with maximal grades, M_A and M_B, respectively, [$M_A = Sup_x f_A(x)$, $M_B = Sup_x f_B(x)$]. Let M be the maximal grade for the intersection, $A \cap B$ ($M = Sup_x$ Min [$f_A(x), f_B(x)$]). Then, $D = 1 - M$ [25] (p. 352) (see Figure 9).*

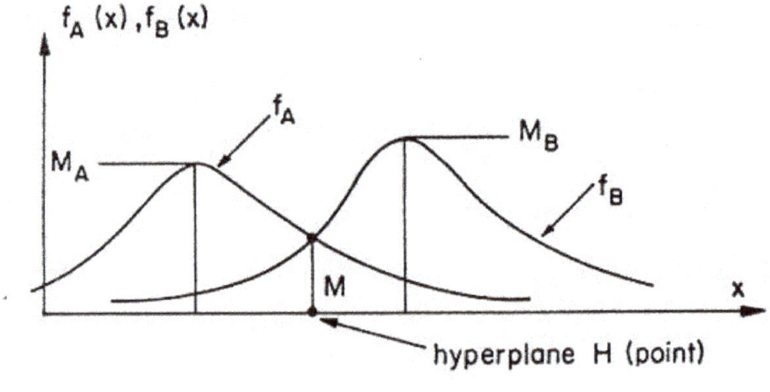

Figure 9. Illustration of the separation theorem for fuzzy sets in E^1 [25].

In 1962, the electrical engineer Chin-Liang Chang came from Taiwan to the US, and in 1964, to UC Berkeley to pursue his PhD under the supervision of Zadeh. In his thesis "Fuzzy Sets and Pattern Recognition" (See: http://www.eecs.berkeley.edu/Pubs/Dissertations/Faculty/zadeh.html), he extended the perception convergence theorem to fuzzy sets, he presented an algorithm for finding a separating hyperplane, and he proved its convergence in finite iterations under a certain condition. A manuscript by Zadeh and Chang entitled "An Application of Fuzzy Sets in Pattern Recognition" with a date of 19 December 1966 (see Figure 10) never appeared published in a journal, but it became part of Chang's PhD thesis.

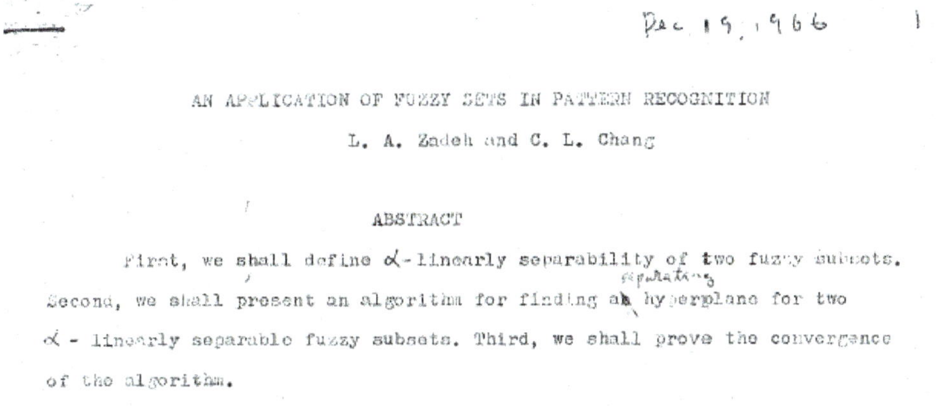
Dec 19, 1966 1

AN APPLICATION OF FUZZY SETS IN PATTERN RECOGNITION

L. A. Zadeh and C. L. Chang

ABSTRACT

First, we shall define α-linearly separability of two fuzzy subsets. Second, we shall present an algorithm for finding a separating hyperplane for two α-linearly separable fuzzy subsets. Third, we shall prove the convergence of the algorithm.

Figure 10. Unpublished manuscript (excerpt) of Zadeh and Chang, 1966. (Fuzzy Archive, Rudolf Seising).

Rosenblatt heralded the perceptron as a universal machine in his publications, e.g., "For the first time, we have a machine which is capable of having original ideas. ... As a concept, it would seem that the perceptron has established, beyond doubt, the feasibility and principle of nonhuman systems which may embody human cognitive functions ... The future of information processing devices which operate on statistical, rather than logical, principles seems to be clearly indicated" [14]. "For the first time we have a machine which is capable of having original ideas", he said in "The New Scientist". "As an analogue of the biological brain the perceptron ... seems to come closer to meeting the requirements of a functional explanation of the nervous system than any system previously proposed" [30] (p. 1392), he continued. To the New York Times, he said, "in principle it would be possible to build brains that could reproduce themselves on an assembly line and which would be conscious of their existence" [31] (p. 25).

The euphoria came to an abrupt halt in 1969, however, when Marvin Minsky and Seymour Papert completed their study of perceptron networks, and published their findings in a book [32]. The results of the mathematical analysis to which they had subjected Rosenblatt's perceptron were devastating: "Artificial neuronal networks like those in Rosenblatt's perceptron are not able to overcome many different problems! For example, it could not discern whether the pattern presented to it represented a single object or a number of intertwined but unrelated objects. The perceptron could not even determine whether the number of pattern components was odd or even. Yet this should have been a simple classification task that was known as a 'parity problem'. What we showed came down to the fact that a Perceptron cannot put things together that are visually nonlocal", Minsky said to Bernstein [33].

Specifically, in their analysis they argued, firstly, that the computation of the XOR had to be done with multiple layers of perceptrons, and, secondly, that the learning algorithm that Rosenblatt proposed did not work for multiple layers.

The so-called "XOR", the either–or operator of propositional logic presents a special case of the parity problem that, thus, cannot be solved by Rosenblatt's perceptron. Therefore, the logical calculus realized by this type of neuronal networks was incomplete.

The truth table (Table 1) of the logical functor, XOR, allocates the truth value "0" to the truth values of the two statements, x_1 and x_2, when their truth values agree, and the truth value "1" when they have different truth values.

Table 1. Truth table of the logical operator XOR.

x_1	x_2	x_1 **XOR** x_2
0	0	0
0	1	1
1	0	1
1	1	0

x_1 and x_2 are components of a vector of the intermediate layer of a perceptron, so they can be interpreted, for example, as the coding of a perception by the retina layer. So, $y = x_1$ *XOR* x_2 is the truth value of the output neuron, which is calculated according to the truth table. The activity of x_1 and x_2 determines this value. It is a special case of the parity problem in this respect. For an even number, i.e., when both neurons are active or both are inactive, the output is 0, while for an odd number, where just one neuron is active, the value is 1.

To illustrate this, the four possible combinations of 0 and 1 are entered into a rectangular coordinate system of x_1 and x_2, and marked with the associated output values. In order to see that, in principle, a perceptron cannot learn to provide the output values demanded by *XOR*, the sum of the weighted input values is calculated by $w_1x_1 + w_2x_2$.

The activity of the output depends on whether this sum is larger or smaller than the threshold value, which results in the plane extending between x_1 and x_2 as follows:

$\Theta = w_1x_1 + w_2x_2$, which results in: $x_2 = -w_1\ w_2\ x_1 + \Theta\ w_2$.

This is the equation of a straight line in which, on one side, the sum of the weighted input values is greater than the threshold value ($w_1x_1 + w_2x_2 > \Theta$) and the neuron is, thus, active (fires); however, on the other side, the sum of the weighted input values is smaller than the threshold value ($w_1x_1 + w_2x_2 < \Theta$), and the neuron is, thus, not active (does not fire).

However, the attempt to find precisely those values for the weights, w_1 and w_2, where the associated line separates the odd number with (0, 1) and (1, 0) from the even number with (0, 0), and (1, 1) must fail (see Figure 10). The proof is very easy to demonstrate by considering all four cases:

$x_1 = 0, x_2 = 1$: y should be $1 \rightarrow w_1 \cdot 0 + w_2 \cdot 1 \geq \Theta \rightarrow$ neuron is active!
$x_1 = 1, x_2 = 0$: y should be $1 \rightarrow w_1 \cdot 1 + w_2 \cdot 0 \geq \Theta \rightarrow$ neuron is active!
$x_1 = 0, x_2 = 0$: y should be $0 \rightarrow w_1 \cdot 0 + w_2 \cdot 0 < \Theta \rightarrow$ neuron is inactive!
$x_1 = 1, x_2 = 1$: y should be $0 \rightarrow w_1 \cdot 1 + w_2 \cdot 1 < \Theta \rightarrow$ neuron is inactive!

Adding the first two equations results in $w_1 + w_2 \geq 2\Theta$.

From the last two equations comes $\Theta > w_1 + w_2 \geq 2\Theta$, and so $\Theta > 2\Theta$.

This applies only where $\Theta < 0$. This is a contradiction of $w10 + w20 < \Theta$. Q.E.D.

The limits of the Rosenblatt perceptron were, thus, demonstrated, and they were very narrow, for it was not even able to classify linearly separable patterns. In their book, Minsky and Papert estimated that more than 100 groups of researchers were working on perceptron networks or similar systems all over the world at that time. In their paper "Adaptive Switching Circuits", Bernard Widrow and Marcian Edward Hoff publicized the linear adaptive neuron model, ADALINE, an adaptive system that was quick and precise thanks to a more advanced learning process which today is known as the "Delta rule" [34]. In his 1958 paper "Die Lernmatrix", German physicist, Karl Steinbuch, introduced a simple technical realization of associative memories, the predecessor of today's neuronal associative memories [35]. In 1959, the paper "Pandemonium" by Oliver Selfridge was published in which dynamic, interactive mechanisms were described that used filtering operations to classify images by means of "significant criteria, e.g., four corners to identify a square". He expected to develop a system that will also recognize "other kinds of features, such as curvature, juxtaposition of singular points, that is, their relative bearings and distances and so forth" [36] (p. 93), [37]. Already since 1955, Wilfred Kenelm Taylor in the Department of Anatomy of London's University College aimed to construct neural analogs to study theories of learning [38].

However, the publication of Minsky and Papert's book disrupted research in artificial neural networks for more than a decade. Because of their fundamental criticism, many of these projects were shelved or at least modified in the years leading up to 1970. In the 15 years that followed, almost no research grants were approved for projects in the area of artificial neuronal networks, especially not by the US Defense Department for DARPA (Defense Advanced Research Projects Agency. The pattern recognition and learning networks faltered on elementary questions of logic in which their competitor, the digital computer, proved itself immensely powerful.

7. Outlook

The disruption of artificial neural networks research later became known as the "AI winter", but artificial neural networks were not killed by Minsky and Papert. In 1988, Seymour Papert did wonder whether this was actually their plan: "Did Minsky and I try to kill connectionism, and how do we feel about its resurrection? Something more complex than a plea is needed. Yes there was some hostility in the energy behind the research reported in Perceptrons, and there is some degree of annoyance at the way the new movement has developed; part of our drive came, as we quite plainly acknowledged in our book, from the fact that funding and research energy were being dissipated on what still appear to me (since the story of new, powerful network mechanisms is seriously exaggerated) to be misleading attempts to use connectionist methods in practical applications. But most of the motivation for Perceptrons came from more fundamental concerns many of which cut cleanly across the division between networkers and programmers" [39] (p. 346).

Independent of artificial neural networks, fuzzy pattern classification became popular in the 1960s. Looking to the emerging field of biomedical engineering, the Argentinian mathematician Enrique Ruspini, a graduate student at the University of Buenos Aires, came "across, however, early literature on numerical taxonomy (then focused on "Biological Systematics", which was mainly concerned with classification of biological species), a field where its seminal paper by Sokal (Robert Reuven Sokal was an Austrian-American biostatistician and entomologist) and Sneath (Peter Henry Andrews Sneath was a British microbiologist. He began working on numerical methods for classifying bacteria in the late 1950s) had been published in 1963" [40,41]. In the interview, he continued, "It is interesting to note that the field was so young that, at the point, there were not even accepted translations to Spanish of words such as 'pattern' or 'clustering'. After trying to understand and formulate the nature of the problem (I am a mathematician after all!), it was clear to me that the stated goal of clustering procedures ('classify similar objects into the same class and different objects into different classes') could not be attained within the framework of classical set theory. By sheer accident I walked one day into the small library of the Department of Mathematics at the School of Science. Perusing through the new-arrivals rack I found the 1965 issue of Information and Control with Lotfi Zadeh's seminal paper [25]. It was clear to me and my colleagues that this was a much better framework to consider and rigorously pose fuzzy clustering problems. Drawing also from results in the field of operations research I was soon able to pose the clustering problem in terms of finding the optimal solution of a continuous variable system with well-defined performance criteria and constraints." [17] (p. 2f)

In the quoted double-interview, Jim Bezdek also looked back: "So, when I got to Cornell in 1969, the same year that the Minsky and Papert book came out, he [Henry David Block] and others (including his best friend, Bernie Widrow, I might add), were in a funk about the apparent death of the ANNs (artificial neural networks). Dave wanted to continue in this field, but funding agencies were reluctant to forge ahead with NNs in the face of the damning damning indictment (which in hindsight was pretty ridiculous) by Minsky and Papert. About 1970, Richard Duda sent Dave a draft of his book with Peter Hart, the now and forever famous 'Duda and Hart' book on Pattern Classification and Scene Analysis, published in 1973 [42,43]. Duda asked Dave to review it. Dave threw it in Joe Dunn's inbox, and from there it made its way to mine. So I read it—cover to cover—trying to find corrections, etc. whilst simultaneously learning the material, and that's how I entered the field of pattern recognition". Bezdek included his best Dave Block story: "In maybe 1971, Dave and I went over to the Cornell

Neurobiology Lab in Triphammer Woods, where we met a young enterprising neuroscientist named Howard Moraff, who later moved to the NSF, where he is (I think still today). Howard was hooking up various people to EEG sensor nodes on their scalps—16 sites at that time—and trying to see if there was any information to be gleaned from the signals. We spent the day watching him, talking to him, etc. Dave was non-committal to Howard about the promise of this enterprise, but as we left the building, Dave turned to me and said 'Maybe there is some information in the signals Jim, but we are about 50 years too early'". Then, he commented this: "I have told this story many times since then (43 years ago now), and I always end it by saying this: 'And if Dave could see the signals today, given our current technology, what do you think he would say now? He would say «Jim, we are about 50 years too soon»'. So, the bottom line for me in 1971 was: don't do NNS, but clustering and classifier design with OTHER paradigms is ok. As it turned out, however, I was out of the frying pan of NNs, and into the fire of Fuzzy Sets, which was in effect a (very) rapid descent into the Maelstrom of probabilistic discontent" [17] (p. 5f). (NFS: National science Foundation, EEG: Electroencephalography)

Since 1981, the psychologists, James L. McClelland and David E. Rumelhart, applied artificial neural networks to explain cognitive phenomena (spoken and visual word recognition). In 1986, this research group published the two volumes of the book "Parallel Distributed Processing: Explorations in the Microstructure of Cognition" [44]. Already in 1982, John J. Hopfield, a biologist and Professor of Physics at Princeton, CalTech, published the paper "Neural networks and physical systems with emergent collective computational abilities" [45] on his invention of an associative neural network (now more commonly known as the "Hopfield Network"), i.e., feedback networks that have only one layer that is both input, as well as output layer, and each of the binary McCulloch–Pitts neurons is linked with every other, except itself. McClelland's research group could show that perceptrons with more than one layer can realize the logical calculus; multilayer perceptrons were the beginning of the new direction in AI: parallel distributed processing.

In the mid-1980s, traditional AI explored their limitations, and with "more powerful hardware" (e.g., parallel architectures) and "new advances made in neural modelling learning methods" (e.g., feedforward neural networks with more than one layer, i.e., multilayer perceptrons), artificial neural modeling has awakened new interest in the fields of science, industry, and governments. In Japan, this resulted in the Sixth Generation Computing Project that started in 1986 [46], in Europe the following year, the interdisciplinary project "Basic Research in Adaptive Intelligence and Neurocomputing" (BRAIN) of the European Economic Community [47], and in the US, the DARPA Neural Network Study (1987–1988) [48].

Today, among other algorithms, e.g., decision trees and random forests, artificial neural networks are enormously successful in data mining, machine learning, and knowledge discovery in databases.

Funding: This research received no external funding.

Acknowledgments: I would like to thank James C. Bezdek, Enrique H. Ruspini, and Chin-Liang Chang for giving interviews in the year 2014 and helpful discussions. I am very thankful to Lotfi A. Zadeh, who sadly passed away in September 2017, for many interviews and discussions in almost 20 years of historical research work. He has encouraged and supported me with constant interest and assistance, and he gave me the opportunity to collect a "digital fuzzy archive" with historical sources.

Conflicts of Interest: The authors declare no conflict of interest.

References

1. Rosen, C.A. Pattern Classification by Adaptive Machines. *Science* **1967**, *156*, 38–44. [CrossRef] [PubMed]
2. Bellman, R.E.; Kalaba, R.; Zadeh, L.A. *Abstraction and Pattern Classification. Memorandum RM-4307-PR*; The RAND Corporation: Santa Monica, CA, USA, 1964.
3. Bellman, R.E.; Kalaba, R.; Zadeh, L.A. Abstraction and Pattern Classification. *J. Math. Anal. Appl.* **1966**, *13*, 1–7. [CrossRef]
4. Minsky, M.L. Steps toward Artificial Intelligence. *Proc. IRE* **1960**, *49*, 8–30. [CrossRef]
5. Zadeh, L.A. From Circuit Theory to System Theory. *Proc. IRE* **1962**, *50*, 856–865. [CrossRef]

6. Eidelheit, M. Zur Theorie der konvexen Mengen in linearen normierten Räumen. *Studia Mathematica* **1936**, *6*, 104–111. [CrossRef]
7. Zadeh, L.A. On the Identification Problem. *IRE Trans. Circuit Theory* **1956**, *3*, 277–281. [CrossRef]
8. Zadeh, L.A. What is optimal? *IRE Trans. Inf. Theory* **1958**, 1.
9. Zadeh, L.A. Optimality and Non-Scalar-Valued Performance Criteria. *IEEE Trans. Autom. Control* **1963**, *8*, 59–60. [CrossRef]
10. McCulloch, W.S.; Pitts, W. A logical calculus of the ideas immanent in nervous activity. *Bull. Math. Biophys.* **1943**, *5*, 115–133. [CrossRef]
11. Hebb, D.O. *The Organization of Behavior: A Neuropsychological Theory*; Wiley and Sons: New York, NY, USA, 1949.
12. Hayek, F.A. *The Sensory Order: An Inquiry into the Foundations of Theoretical Psychology*; University of Chicago Press: Chicago, IL, USA, 1952.
13. Rosenblatt, F. *The Perceptron. A Theory of Statistical Separability in Cognitive Systems, (Project PARA)*; Report No. VG-1196-G-1; Cornell Aeronautical Laboratory: New York, NY, USA, 1958.
14. Rosenblatt, F. The Perceptron. A Probabilistic Model for Information Storage and Organization in the Brain. *Psychol. Rev.* **1958**, *65*, 386–408. [CrossRef] [PubMed]
15. Rosenblatt, F. A Probabilistic Model for Visual Perception. *Acta Psychol.* **1959**, *15*, 296–297. [CrossRef]
16. Rosenblatt, F. The Design of an Intelligent Automaton. *Res. Trends* **1958**, *VI*, 1–7.
17. Seising, R. On the History of Fuzzy Clustering: An Interview with Jim Bezdek and Enrique Ruspini. *IEEE Syst. Man Cybern. Mag.* **2015**, *1*, 20–48. [CrossRef]
18. Block, H.D. The Perceptron: A Model for Brain Functioning I. *Rev. Mod. Phys.* **1962**, *34*, 123–135. [CrossRef]
19. Novikoff, A. On Convergence Proofs for Perceptions. In *Proceedings of the Symposium on Mathematical Theory of Automata*; Polytechnic Institute of Brooklyn: Brooklyn, NY, USA, 1962; Volume XII, pp. 615–622.
20. Aizerman, M.A.; Braverman, E.M.; Rozonoer, L.I. Theoretical Foundations of the Potential Function Method in Pattern Recognition Learning. *Autom. Remote Control* **1964**, *25*, 821–837.
21. Aizerman, M.A.; Braverman, E.M.; Rozonoer, L.I. The Method of Potential Function for the Problem of Restoring the Characteristic of a Function Converter from Randomly Observed Points. *Autom. Remote Control* **1964**, *25*, 1546–1556.
22. Aizerman, M.A.; Braverman, E.M.; Rozonoer, L.I. The Probability Problem of Pattern Recognition Learning and the Method of Potential Functions. *Autom. Remote Control* **1964**, *25*, 1175–1190.
23. Rosen, J.B. *Pattern Separation by Convex Programming*; Technical Report No. 30; Applied Mathematics and Statistics Laboratories, Stanford University: Stanford, CA, USA, 1963.
24. Rosen, J.B. Pattern Separation by Convex Programming. *J. Math. Anal. Appl.* **1965**, *10*, 123–134. [CrossRef]
25. Zadeh, L.A. Fuzzy Sets. *Inf. Control* **1965**, *8*, 338–353. [CrossRef]
26. Zadeh, L.A. Fuzzy Sets and Systems. In *System Theory*; Microwave Research Institute Symposia Series XV; Fox, J., Ed.; Polytechnic Press: Brooklyn, NY, USA, 1965; pp. 29–37.
27. Zadeh, L.A. Shadows of Fuzzy Sets. Problemy peredachi informatsii. In *Akadamija Nauk SSSR Moskva*; Problems of Information Transmission: A Publication of the Academy of Sciences of the USSR; The Faraday Press: New York, NY, USA, 1966; Volume 2.
28. Zadeh, L.A. Shadows of Fuzzy Sets. In *Notes of System Theory*; Report No. 65-14; Electronic Research Laboratory, University of California Berkeley: Berkeley, CA, USA, 1965; Volume VII, pp. 165–170.
29. Zadeh, L.A. *Fuzzy Sets*; ERL Report No. 64-44; University of California at Berkeley: Berkeley, CA, USA, 1964.
30. Rival. *The New Yorker*. 6 December 1958, p. 44. Available online: https://www.newyorker.com/magazine/1958/12/06/rival-2 (accessed on 21 June 2018).
31. New Navy Device learns by doing. Psychologist Shows Embryo of Computer Designed to Read and Crow Wise. *New York Times*, 7 July 1958; 25.
32. Minsky, M.L.; Papert, S. *Perceptrons*; MIT Press: Cambridge, MA, USA, 1969.
33. Bernstein, J.; Profiles, A.I. Marvin Minsky. *The New Yorker*, 14 December 1981; 50–126.
34. Widrow, B.; Hoff, M.E. Adaptive Switching Circu its, IRE Wescon Convention Record. *New York IRE* **1960**, *4*, 96–104.
35. Steinbuch, K. *Die Lernmatrix, Kybernetik*; Springer: Berlin, Germany, 1961; Volume 1.
36. Selfridge, O.G. Pattern Recognition and Modern Computers. In Proceedings of the AFIPS '55 Western Joint Computer Conference, Los Angeles, CA, USA, 1–3 March 1955; ACM: New York, NY, USA, 1955; pp. 91–93.

37. Selfridge, O.G. Pandemonium: A Paradigm for Learning. In *Mechanisation of Thought Processes: Proceedings of a Symposium Held at the National Physical Laboratory on 24th, 25th and 27th November 1958*; National Physical Laboratory, Ed.; Her Majesty's Stationery Office: London, UK, 1959; Volume I, pp. 511–526.
38. Taylor, W.K. Electrical Simulation of Some Nervous System Functional Activities. *Inf. Theory* **1956**, *3*, 314–328.
39. Papert, S.A. One AI or many? In *The Philosophy of Mind*; Beakley, B., Ludlow, P., Eds.; MIT Press: Cambridge, MA, USA, 1992.
40. Sokal, R.R.; Sneath, P.H.A. *Principles of numerical Taxonomy*; Freeman: San Francisco, NC, USA, 1963.
41. Sokal, R.R. *Numerical Taxonomy the Principles and Practice of Numerical Classification*; Freeman: San Francisco, CA, USA, 1973.
42. Duda, R.O.; Hart, P.E. *Pattern Classification and Scene Analysis*, 2nd ed.; Wiley: Hoboken, NJ, USA, 1973.
43. Duda, R.O.; Hart, P.E.; Stork, D.G. *Pattern Classification*; Wiley: Hoboken, NJ, USA, 2000.
44. Rumelhart, D.E.; McClelland, J.L.; The PDP Research Group. *Parallel Distributed Processing. Explorations in the Microstructure of Cognition*; MIT Press: Cambridge, MA, USA, 1986.
45. Hopfield, J.J. Neural networks and physical systems with emergent collective computational abilities. *Proc. Natl. Acad. Sci. USA* **1982**, *79*, 2554–2558. [CrossRef] [PubMed]
46. Gaines, B. Sixth generation computing. A conspectus of the Japanese proposals. *Newsletter* **1986**, *95*, 39–44.
47. Roman, P. The launching of BRAIN in Europe. In *Europea Science Notes*; US Office of Naval Research: London, UK, 1987; Volume 41.
48. DARPA. *Neural Network Study*; AFCEA International Press: Washington, DC, USA, 1988.

mathematics

MDPI

Article

Credibility Measure for Intuitionistic Fuzzy Variables

Mohamadtaghi Rahimi [1,*], Pranesh Kumar [1] and Gholamhossein Yari [2]

[1] Department of Mathematics and Statistics, University of Northern British Columbia, Prince George, BC V2N 4Z9, Canada; Pranesh.Kumar@unbc.ca
[2] Department of Mathematics, Iran University of Science and Technology, 16846-13114 Tehran, Iran; Yari@iust.ac.ir
* Correspondence: mohamadtaghi.rahimi@unbc.ca; Tel.: +1-250-960-6756

Received: 7 March 2018; Accepted: 26 March 2018; Published: 2 April 2018

Abstract: Credibility measures in vague environments are to quantify the approximate chance of occurrence of fuzzy events. This paper presents a novel definition about credibility for intuitionistic fuzzy variables. We axiomatize this credibility measure and to clarify, give some examples. Based on the notion of these concepts, we provide expected values, entropy, and general formulae for the central moments and discuss them through examples.

Keywords: credibility measure; intuitionistic fuzzy variables; expected value; entropy

---

1. Introduction

Fuzzy set theory, proposed by Zadeh [1] and intuitionistic fuzzy set proposed by Atanassov [2,3], have influenced human judgment a great deal and decreased uncertainties in available information. Entropy, as an important tool of measuring the degree of uncertainty, and also the main core of information theory, for the first time was introduced by Shannon [4], and Zadeh [5] was the first who defined entropy for fuzzy sets by introducing weighted Shannon entropy. However, the synthesis of entropy and fuzzy set theory was first defined by Du Luca and Termini [6] using the Shannon function. He replaced the membership degrees of elements with the variable in the Shannon function. Later, this definition was extended by Szmidt and Kacprzyk [7] by introducing the intuitionistic fuzzy sets. The fuzzy and intuitionistic fuzzy entropy are now often employed in various scientific studies. For example, Huang et al. [8] used fuzzy two dimensional entropy to develop a novel approach for the automatic recognition of red Fuji apples in natural scenes, Yari et al. [9,10] employed it in option pricing and portfolio optimization, Song et al. [11] used fuzzy logics in psychology while studying children's emotions, and Farnoosh et al. [12] proposed a method for image processing based on intuitionistic fuzzy entropy. Additionally, many researchers have recently conceptualized fuzzy entropy from different aspects. Some of them can be found in [13–16].

In 2002, using credibility, a new formula was presented by Liu and Liu [17] for expected values of fuzzy variables. By these notations, a new environment has been created in the fuzzy area, both in pure and applied branches. Decision-making, portfolio optimization, pricing models, and supply chain problems are some of the areas which have used these conceptions. Now, in this paper, we define a new concept of credibility measure for intuitionistic fuzzy sets to be used in all of the mentioned areas.

After Liu and Liu [17], new concepts and properties of fuzzy credibility functions were proposed by some researchers. For example, a sufficient and necessary condition was given by Li and Liu [18] for credibility measures and, in 2008, an entropy measure was defined by Li and Liu [19] for discrete and continuous fuzzy variables, based on the credibility distributions. Therefore, in the rest of the paper, several additional concepts of fuzzy credibility functions are presented as a basis of developing the credibility measure.

In this paper, following the introduction in Section 1 and introducing some concepts and knowledge about credibility and fuzzy entropy measures in Section 2, based on the measure defined by Liu and Liu [17], we present a novel definition about credibility for intuitionistic fuzzy variables in Section 3. Section 3 also presents central moments and entropy formulation. All of these definitions are followed by their corresponding examples. We finally discuss and conclude in Section 4.

2. Preliminaries

Suppose that A is a fuzzy subset of the universe of discourse, U. Then the possibility and necessity measures are defined as (Zadeh, [20]):

$$\mathrm{Pos}\{X \text{ is } A\} \triangleq \mathrm{Sup}_{u\in A}\pi_X\{u\} \in [0,1],$$

$$\mathrm{Nec}\{X \text{ is } A\} \triangleq 1 - \mathrm{Sup}_{u\in A^c}\pi_X\{u\},$$

where $\pi_X\{u\}$ is the possibility distribution function of Π_X, a possibility distribution associated with the variable X taking values in U.

For a fuzzy variable, ξ, with membership function μ, the credibility inversion theorem or, in other words, the credibility of $\xi \in \beta \subset \mathbb{R}$ employed by Liu and Liu [17] is:

$$\mathrm{Cr}\{\xi \in \beta\} = \frac{1}{2}(\mathrm{Pos}\{\xi \in \beta\} + \mathrm{Nec}\{\xi \in \beta\}) = \frac{1}{2}(\mathrm{Sup}_{x\in\beta}\,\mu(x) + 1 - \mathrm{Sup}_{x\in\beta^c}\mu(x)) \quad (1)$$

Later, this formula was extended by Mandal et al. [21] to its general form as:

$$\mathrm{Cr}\{\xi \in \beta\} = \rho\mathrm{Pos}\{\xi \in \beta\} + (1-\rho)\mathrm{Nec}\{\xi \in \beta\}, \quad 0 \le \rho \le 1 \quad (2)$$

Liu and Liu [17] also defined the expected value of a fuzzy variable by the credibility function as:

$$E[\xi] = \int_0^{+\infty} \mathrm{Cr}\{\xi \ge r\}dr - \int_{-\infty}^{0} \mathrm{Cr}\{\xi \le r\}dr, \quad (3)$$

Later, Li and Liu [19] formulated a definition of entropy based on the notion of credibility for continuous distributions:

$$H(\xi) = \int_{-\infty}^{+\infty} S(\mathrm{Cr}\{\xi = x\})\, dx, \quad (4)$$

where the integral reduces to sigma for discrete distributions and $S(t) = -t\mathrm{Ln}t - (1-t)\mathrm{Ln}(1-t)$, $0 \le t \le 1$.

Further, De Luca and Termini [6] for the first time debated the fuzzy entropy measure (formulated in the following manner) which was later extended for intuitionistic fuzzy entropy measures by Szmidt and Kacprzyk [7].

Let H be a real-valued function: $F(X) \to [0,1]$. H is an entropy measure of fuzzy set, if it satisfies the four axiomatic requirements:

FS1 : $H(\widetilde{A}) = 0$ iff $\widetilde{A}$ is a crisp set, i.e., $\mu_{\widetilde{A}}(x_i) = 0$ or $1\ \forall x_i \in X$.
FS2 : $H(\widetilde{A}) = 1$ iff $\mu_{\widetilde{A}}(x_i) = 0.5\ \forall x_i \in X$.
FS3 : $H(\widetilde{A}) \le H(\widetilde{B})$ if $\widetilde{A}$ is less fuzzy than $\widetilde{B}\ \forall\ x_i \in X$.
FS4 : $H(\widetilde{A}) = H(\widetilde{A}^c)$, where $\overline{\widetilde{A}}$ is the complement of $\widetilde{A}$.

Here, $H(\widetilde{A}) = \sum_{i=1}^{n} S(\mu_A(x_i))$ or $H(\widetilde{A}) = \int_{-\infty}^{+\infty} S(\mu_A(x))dx$, for discrete and continuous distributions, respectively.

3. Credibility Measures in Intuitionistic Fuzzy Environment

Definition 1. *Determinacy of an intuitionistic fuzzy set*

Let A be an intuitionistic fuzzy subset of the universe of discourse, U; and f: A→B be a function that changes the intuitionistic fuzzy elements $u \in A$ to fuzzy elements $v \in B$. Then, the determinacy measure is defined as follows:

$$det\{X \text{ is } B\} \triangleq Sup_{v\in B}\, \pi_X\{u\} \in [0,1],$$

where $v \in B$ is a fuzzy number with γ and 1-γ as the degrees of membership and non-membership, respectively; γ is the degree of non-membership of the corresponding value of $u \in A$, and $\pi_X\{u\}$ is the possibility distribution function of Π_X.

3.1. Axioms of a Possibility-Determinacy Space

The quadruplet (Θ, P(Θ), Pos, Det) is called a possibility-determinacy space of an intuitionistic fuzzy variable if:

$$\begin{cases} \text{i)}\mathrm{Pos}\{\Theta\} = 1, \\ \text{ii)}\mathrm{Pos}\{\varnothing\} = 0, \\ \text{iii)}0 \le \mathrm{Pos}^+\{A\} = \mathrm{Pos}\{A\} + \mathrm{Det}\{A\} \le 1, \\ \quad \text{and } 0 \le \mathrm{Pos}^-\{A\} = \mathrm{Pos}\{A\} \le 1, & \text{for A in } \mathrm{Pos}\{\Theta\} \\ \text{iv)}\mathrm{Pos}^+\{\mathrm{U_iA_i}\} = \sup_i\{\mathrm{Pos}\{A_i\} + \mathrm{Det}\{A_i\}\}, \\ \text{and } \mathrm{Pos}^-\{\mathrm{U_iA_i}\} = \sup_i\{\mathrm{Pos}\{A_i\}\}, \end{cases} \tag{5}$$

where Θ is a nonempty set, P(Θ) the power set of Θ, Pos a distribution of possibility from 2^U to [0,1] and Det, the determinacy (Definition 1). It is easy to check that the above axioms tend to the possibility space axioms when Det{A}=0; that is, when we have a fuzzy variable.

Possibility and necessity in the intuitionistic fuzzy environment will be denoted by duals (Pos^+, Pos^-) and (Nec^+, Nec^-). These expressions represent the maximum and the minimum of the possibility and necessity, respectively.

3.2. Necessity Measure of an Intuitionistic Fuzzy Set

Following the concepts of the triangular fuzzy numbers by Dubois and Prade [22], let A be an intuitionistic fuzzy variable on a possibility-determinacy space (Θ, P (Θ), Pos, Det). Then, the necessity measure of A is defined as follows:

$$\mathrm{Nec}^+\{A\} = 1 - \mathrm{Pos}^+\{A^c\}, \quad \mathrm{Nec}^-\{A\} = 1 - \mathrm{Pos}^-\{A^c\}, \quad \varphi = \mathrm{Sup}_{x\in\mathbb{R}}\mu\{x\}. \tag{6}$$

Example 1. *Let ξ be a triangular intuitionistic fuzzy number with the following membership and non-membership functions, μ and γ:*

$$\mu(x) = \begin{cases} \dfrac{x-a}{b-a}\varphi, & a \le x < b, \\ \varphi, & x = b, \\ \dfrac{c-x}{c-b}\varphi, & b < x \le c, \\ 0 & \text{otherwise,} \end{cases} \qquad \gamma(x) = \begin{cases} \dfrac{b-x+\omega(x-a)}{b-a}, & a \le x < b, \\ \omega, & x = b, \\ \dfrac{x-b+\omega(c-x)}{c-b}, & b < x \le c, \\ 0 & \text{otherwise,} \end{cases} \qquad 0 \le \omega + \varphi \le 1$$

Then, we have Pos and Det given as:

$$\mathrm{Pos}^{-}\{X \geq x_0\} = \sup_{x \geq x_0} \mu(x) = \begin{cases} \varphi, & a < x_0 \leq b, \\ \dfrac{c - x_0}{c - b}\varphi, & b < x_0 \leq c, \\ 0 & \text{otherwise,} \end{cases}$$

$$\mathrm{Pos}^{+}\{X \geq x_0\} = \sup_{x \geq x_0} \mu(x) = \begin{cases} 1, & a \leq x_0 \leq c, \\ 0 & \text{otherwise,} \end{cases}$$

3.3. Credibility Measure in Intuitionistic Fuzzy Environment

Based on the Pos (Pos^{+} and Pos^{-}) and Nec (Nec^{+} and Nec^{-}) measures, the credibility measure in intuitionistic fuzzy environment is given as:

$$\begin{aligned} &\mathrm{Cr}^{-}\{A\} = \rho\mathrm{Pos}^{-}(A) + (1-\rho)\mathrm{Nec}^{-}(A), \qquad 0 \leq \rho \leq 1 \\ &\Rightarrow \mathrm{Cr}^{-}\{\xi \in B\} = \rho\mathrm{Sup}_{x \in B}\mu\{x\} + (1-\rho)\left(1 - \mathrm{Sup}_{x \in B^c}\mu\{x\}\right). \end{aligned} \tag{7}$$

$$\begin{aligned} &\mathrm{detCr}^{-}\{A\} = \rho(\mathrm{Pos}^{+}\{A\} - \mathrm{Pos}^{-}\{A\}) + (1-\rho)(\mathrm{Nec}^{-}\{A\} - \mathrm{Nec}^{+}\{A\}) \\ &\Rightarrow \mathrm{detCr}^{-}\{\xi \in B\} = \rho\mathrm{Sup}_{x \in B}\gamma\{x\} + (1-\rho)\left(1 - \mathrm{Sup}_{x \in B^c}\gamma\{x\}\right). \end{aligned} \tag{8}$$

Here, the $\mathrm{Cr}^{-}\{A\}$ and $\mathrm{detCr}^{-}\{A\}$ are, respectively, the fixed and the determinacy of the credibility measure, and ξ is an intuitionistic fuzzy variable with membership function μ and non-membership function γ.

We can see that Cr satisfies the following conditions:

(i) $\mathrm{Cr}^{-}\{\varnothing\} = \mathrm{detCr}^{-}\{\varnothing\} = 0,$
(ii) $\mathrm{Cr}^{-}\{\mathbb{R}\} = 1,$
(iii) for $A \subset B$, $\mathrm{Cr}^{-}\{A\} \leq \mathrm{Cr}^{-}\{B\}$ & $\mathrm{detCr}^{-}\{A\} \leq \mathrm{detCr}^{-}\{B\}$, for any $A, B \in 2^{\mathbb{R}}$.

Thus, similar to the credibility measure defined in Liu and Liu [17] and Equation (3.2) in Mandal et al. [10], Cr^{-} is an intuitionistic fuzzy measure on $(\mathbb{R}, 2^{\mathbb{R}})$.

Example 2. *Following Example 1, the credibility for a standard triangular intuitionistic fuzzy number ξ is as follows:*

$$\mathrm{Cr}^{-}\{\xi \geq x_0\} = \begin{cases} \rho\varphi + (1-\rho)\left(1 - \dfrac{x_0 - a}{b - a}\varphi\right)), & a \leq x_0 < b, \\ \rho\left(\dfrac{c - x_0}{c - b}\right)\varphi + (1-\rho)(1-\varphi)), & b < x_0 \leq c, \\ 0, & \text{otherwise.} \end{cases}$$

$$\mathrm{detCr}^{-}\{\xi \geq x_0\} = \begin{cases} \rho(1-\varphi) + (1-\rho)\left(1 - \dfrac{x_0 - a}{b - a}\varphi\right)), & a \leq x_0 < b, \\ \rho\left(1 - \dfrac{c - x_0}{c - b}\right)\varphi + (1-\rho)(1-\varphi)), & b < x_0 \leq c, \\ 0, & \text{otherwise.} \end{cases}$$

Lemma 1. *Let ξ be an intuitionistic fuzzy variable taking values in $\mathbb{R}$. If there exist an interval, B, such that $Cr^{-}\{\xi \in B\} = \varphi$ or $detCr^{-}\{\xi \in B\} = \varphi + \omega$, then for every interval, α (s.t. $\alpha \cap \beta) = \varnothing$, we have $Cr^{-}\{\xi \in \alpha\} = 0$, where φ and ω are, respectively, the supremum values of membership and non-membership functions.*

Proof of Lemma 1. From Equations (7) and (8), the maximum value for Cr^- and $detCr^-$ occur when $Sup_{x\in B^c}\mu(x) = 0$ and $Sup_{x\in B^c}\gamma(x) = 0$, respectively. Then, since $\alpha \subset B^c$, we have $Sup_{x\in\alpha}\mu(x) = Sup_{x\in\alpha}\gamma(x) = 0$. Therefore, $Cr^-\{\xi \in \alpha\} = detCr^-\{\xi \in \alpha\} = 0$. □

Credibility is a value between 0 to φ for possibility function and increases to $\varphi + \omega$ when determinacy is involved. In this lemma, it is shown when we have an interval containing the highest credibility value, the existence of any other disjoint interval containing positive credibility is impossible and, therefore, we can ignore the intervals having no positive possibility and determinacy values in the following definitions, especially, for entropy.

To check this lemma for discrete fuzzy variables, see Li and Liu [19].

Definition 2. *For the expected value and central moments of an intuitionistic fuzzy variable based on Cr^- and $detCr^-$, the general form for the nth moments of real-valued continuous intuitionistic fuzzy variable about a value c are introduced as $E^-\left[(\xi - c)^n\right]$ and $detE^-\left[(\xi - c)^n\right]$, where:*

$$\begin{cases} lE^-[\xi] = \int\limits_0^{+\infty} Cr^-\{\xi \geq r\}dr - \int\limits_{-\infty}^{0} Cr^-\{\xi \leq r\}dr, \\ detE^- = \int\limits_0^{+\infty} Cr^+\{\xi \geq r\}dr - \int\limits_{-\infty}^{0} Cr^+\{\xi \leq r\}dr. \end{cases} \tag{9}$$

Here E^- is the fixed value which is similar to the expected value in fuzzy variables, whereas $detE^-$ measures the determinacy of an expected value. The expected value does not exist and is not defined if the right-hand side of Equation (9) is ∞–∞.

For the central moments such as variance, skewness and kurtosis, similar to Liu and Liu [17], and based on the defined credibility measures in Section 3.3, for each intuitionistic fuzzy variable ξ with finite expected value, we have:

$$CM^-[\xi, n] = E^-\left[\left(\xi - E^-(\xi)\right)^n\right],\ detCM^-[\xi, n] = detE^-\left[\left(\xi - detE^-(\xi)\right)^n\right],$$

where $CM^-[\xi, n]$ and $detCM^-[\xi, n]$ for n = 2, 3, and 4, respectively, represent the variance, skewness, and kurtosis.

Note: In this new Definition 2, the expected value for membership degrees is isolated from the expected value for non-membership degrees wherein both are calculated from the credibility functions. It means that we have a dual $\left[E^-[\xi], detE^-[\xi]\right]$ which denotes the expected values, separately. A linear combination of the elements of this dual can be used as the score function, if one wants to compare some intuitionistic fuzzy variables.

Definition 3. *Entropy for intuitionistic fuzzy variables*

Similar to entropy measure by credibility functions, entropy is formulized for intuitionistic fuzzy variables. In this definition, we have again two measures, fixed and the determinacy entropies:

$$H(\xi) = \int\limits_{\substack{-\infty \\ x \subset \beta}}^{+\infty} S\left(Cr^-\{\xi = x_i\}\right) + S\left(detCr^-\{\xi = x_i\}\right) dx,$$

where according to Lemma 1, β is the smallest interval containing the positive possibilities.

Example 3. *Let ξ be a triangular fuzzy variable with the membership and non-membership functions introduced in Section 3.1. Then, the entropy is:*

$$H(\xi) = \left(\frac{c-a}{2}\right)(2\varphi + \omega).$$

If ξ be a trapezoidal fuzzy variable (a,b,c,d), then:

$$H(\xi) = \left(\left(\frac{d-a}{2}\right) + \left(\ln 2 - \frac{1}{2}\right)(c-b)\right)(2\varphi + \omega).$$

4. Discussion and Conclusions

In this paper, we defined the notion of credibility in intuitionistic fuzzy environment as an extension of credibility for fuzzy values which was not described before and, thus, by this conception, we have created a new environment as we have had for fuzzy in [17]. Based on these conceptions, we presented novel definitions of expected value, entropy and a general formula for central moments of intuitionistic fuzzy variables. In each step, all the definitions and axioms in the paper are provided by illustrative examples.

Acknowledgments: Authors sincerely thank reviewers for their valuable comments and suggestions which have led to the present form of the paper. Mohamadtaghi Rahimi thanks University of Northern British Columbia for awarding the post-doctoral fellowship and acknowledges the partial funding support from the NSERC Discovery Development Grant of Prof. Pranesh Kumar.

Author Contributions: All the authors contributed equally in this work. They read and approved the final manuscript.

Conflicts of Interest: The authors declare no conflict of interest.

References

1. Zadeh, L.A. Fuzzy set. *Inf. Control* **1965**, *8*, 338–353. [CrossRef]
2. Atanassov, K.T. Intuitionistic fuzzy sets. *Fuzzy Sets Syst.* **1986**, *20*, 87–96. [CrossRef]
3. Atanassov, K.T. More on intuitionistic fuzzy sets. *Fuzzy Sets Syst.* **1986**, *33*, 37–46. [CrossRef]
4. Shannon, C.E. A mathematical theory of communication. *Bell Syst. Tech. J.* **1948**, *21*, 379–423. [CrossRef]
5. Zadeh, L.A. Probability measures of fuzzy events. *J. Math. Anal. Appl.* **1968**, *23*, 421–427. [CrossRef]
6. De Luca, A.; Termini, S. A definition of a non-probabilistic entropy in the setting of fuzzy sets theory. *Inf. Control* **1972**, *20*, 301–312. [CrossRef]
7. Szmidt, E.; Kacprzyk, J. Entropy for intuitionistic fuzzy sets. *Fuzzy Sets Syst.* **2001**, *118*, 467–477. [CrossRef]
8. Huang, L.; He, D.; Yang, S.X. Segmentation on Ripe Fuji Apple with Fuzzy 2D Entropy based on 2D histogram and GA Optimization. *Intell. Autom. Soft Comput.* **2013**, *19*, 239–251. [CrossRef]
9. Rahimi, M.; Kumar, P.; Yari, G. Portfolio Selection Using Ant Colony Algorithm And Entropy Optimization. *Pak. J. Stat.* **2017**, *33*, 441–448.
10. Yari, G.; Rahimi, M.; Moomivand, B.; Kumar, P. Credibility Based Fuzzy Entropy Measure. *Aust. J. Math. Anal. Appl.* **2016**, *13*, 1–7.
11. Song, H.S.; Rhee, H.K.; Kim, J.H.; Lee, J.H. Reading Children's Emotions based on the Fuzzy Inference and Theories of Chromotherapy. *Information* **2016**, *19*, 735–742.
12. Farnoosh, R.; Rahimi, M.; Kumar, P. Removing noise in a digital image using a new entropy method based on intuitionistic fuzzy sets. In Proceedings of the 2016 IEEE International Conference on Fuzzy Systems (FUZZ-IEEE), Vancouver, BC, Canada, 24–29 July 2016.
13. Markechová, D. Kullback-Leibler Divergence and Mutual Information of Experiments in the Fuzzy Case. *Axioms* **2017**, *6*, 5. [CrossRef]
14. Markechová, D.; Riečan, B. Logical Entropy of Fuzzy Dynamical Systems. *Entropy* **2016**, *18*, 157. [CrossRef]

15. Markechová, D.; Riečan, B. Entropy of Fuzzy Partitions and Entropy of Fuzzy Dynamical Systems. *Entropy* **2016**, *18*, 19. [CrossRef]
16. Yari, G.; Rahimi, M.; Kumar, P. Multi-period Multi-criteria (MPMC) Valuation of American Options Based on Entropy Optimization Principles. *Iran. J. Sci. Technol. Trans. A Sci.* **2017**, *41*, 81–86. [CrossRef]
17. Liu, B.; Liu, Y.K. Expected value of fuzzy variable and fuzzy expected value models. *IEEE Trans. Fuzzy Syst.* **2002**, *10*, 445–450.
18. Li, X.; Liu, B. A sufficient and necessary condition for credibility measures. *Int. J. Uncertain. Fuzz. Knowl.-Based Syst.* **2006**, *14*, 527–535. [CrossRef]
19. Li, P.; Liu, B. Entropy of Credibility Distributions for Fuzzy Variables. *IEEE Trans. Fuzzy Syst.* **2008**, *16*, 123–129.
20. Zadeh, L.A. Fuzzy Sets as the basis for a theory of possibility. *Fuzzy Sets Syst.* **1978**, *1*, 3–28. [CrossRef]
21. Mandal, S.; Maitya, K.; Mondal, S.; Maiti, M. Optimal production inventory policy for defective items with fuzzy time period. *Appl. Math. Model.* **2010**, *34*, 810–822. [CrossRef]
22. Dubois, D.; Prade, H. *Fuzzy Sets and Systems: Theory and Applications*; Academic Press: New York, NY, USA, 1980.

mathematics

MDPI

Article

Certain Algorithms for Modeling Uncertain Data Using Fuzzy Tensor Product Bézier Surfaces

Musavarah Sarwar and Muhammad Akram *

Department of Mathematics, University of the Punjab, New Campus, Lahore 54590, Pakistan; musavarah656@gmail.com
* Correspondence: makrammath@yahoo.com; Tel.: +92-42-99231241

Received: 31 January 2018; Accepted: 7 March 2018; Published: 9 March 2018

Abstract: Real data and measures are usually uncertain and cannot be satisfactorily described by accurate real numbers. The imprecision and vagueness should be modeled and represented in data using the concept of fuzzy numbers. Fuzzy splines are proposed as an integrated approach to uncertainty in mathematical interpolation models. In the context of surface modeling, fuzzy tensor product Bézier surfaces are suitable for representing and simplifying both crisp and imprecise surface data with fuzzy numbers. The framework of this research paper is concerned with various properties of fuzzy tensor product surface patches by means of fuzzy numbers including fuzzy parametric curves, affine invariance, fuzzy tangents, convex hull and fuzzy iso-parametric curves. The fuzzification and defuzzification processes are applied to obtain the crisp Beziér curves and surfaces from fuzzy data points. The degree elevation and de Casteljau's algorithms for fuzzy Bézier curves and fuzzy tensor product Bézier surfaces are studied in detail with numerical examples.

Keywords: fuzzy tensor product Bézier surface; fuzzy parametric curves; fuzzy iso-parametric curves; degree elevation algorithm; De Casteljau's algorithm

1. Introduction

Data points are usually collected using physical objects to capture their geometric entity and representation in a digital framework, i.e., CAGD and CAD systems. Information is collected by using particular devices such as scanning tools. However, the recorded data do not significantly describe error-free data. This is due to the fact that the errors are produced by limitations of the devices, human errors and environmental factors, etc. Generally, these sorts of data which have uncertain characteristics cannot be used directly to create digitized models. In order to make uncertain data valuable for analysis and modeling, this kind of data have to be characterized in a different approach to handle uncertainties of the measurements.

In curve designing and geometric modeling, control points play a major role in the process of controlling the shape of curves and surfaces. The issue of uncertain shape of surfaces and curves can be handled by using left, crisp, right control points through fuzzy numbers called fuzzy control points [1].

Natural spline, B-spline and Bernstein Bézier functions can be used to produce geometric models with data points [2–4]. The surfaces and curves produced with these functions are the standard approaches to represent a set of given data points. Tensor product Bézier surfaces, also known as Bernstein Bézier surfaces, can be determined by a collection of vertices called control points, which are joined in a sequence to form a closed or open control grid. The shape of the surface changes with the control grid in a smooth fashion. However, there is a major problem in shape designing due to uncertainty, imprecision and vagueness of the real data. The designers and experts are unable to choose an appropriate set of control points due to errors and uncertainties. One of the methods used to handle vagueness and uncertainty issues is the theory of fuzzy sets introduced in [5].

The problem of interpolation was first proposed by Zadeh in [5] stating that if for each $r+1$ distinct real numbers $y_o, y_1, \ldots, y_r$, a fuzzy value is given instead of crisp value, is it possible to construct a smooth curve to fit this fuzzy data of $r+1$ points? To solve Zadeh's proposed problem, Lagrange interpolation polynomial for fuzzy data was first investigated by Lowen [6]. The problem of interpolating fuzzy data using fuzzy splines was also considered by Kaleva [7]. By using spline functions of odd degree, the interpolation of fuzzy data was considered in [8] with complete splines, in [9] with natural splines, and in [10] with fuzzy splines. The concept of a fuzzy tensor product Bézier surface was introduced in [1]. The construction of the fuzzy B-spline model, modeling of uncertain data based on B-spline model curve are discussed in [11–13].

In this research paper, we study various properties of fuzzy tensor product surfaces by means of fuzzy numbers including fuzzy parametric curves, affine invariance, fuzzy tangents, convex hull property and fuzzy iso-parametric curves. We also develop De Casteljau's and degree elevation algorithms for fuzzy Beziér curves and fuzzy tensor product surfaces with numerical examples. We apply the process of fuzzification to obtain the fuzzy interval of fuzzy data points where the crisp solution exists. This is followed by the defuzzification process to construct crisp Beziér curves and surfaces which focus on the defuzzification of fuzzy data points.

We used standard definitions and terminologies in this paper. For other notations, terminologies and applications not mentioned in the paper, the readers are referred to [14–21].

Definition 1 ([5,20])**.** *A fuzzy set λ on a non-empty universe Y is a mapping $\lambda : Y \to [0,1]$. A fuzzy relation on Y is a fuzzy subset v in $Y \times Y$.*

Definition 2 ([21])**.** *A triangular fuzzy number is a fuzzy set on $\mathbb{R}$, denoted by the symbol $A = (\delta, \beta, \gamma)$, $\delta < \beta < \gamma$ $\delta, \beta, \gamma \in \mathbb{R}$, with membership function defined as,*

$$\lambda_A(y) = \begin{cases} \dfrac{y-\delta}{\beta-\delta} & , y \in [\delta, \beta] \\ \dfrac{\gamma - y}{\gamma - \beta} & , y \in [\beta, \gamma] \\ 0 & , \textit{otherwise} \end{cases}$$

The α-cute operation, $0 < \alpha \leq 1$, of triangular fuzzy number is defined as $A_\alpha = [(\beta - \delta)\alpha + \delta, -(\gamma - \beta)\alpha + \gamma]$. For any two triangular fuzzy numbers $A = (\delta_1, \beta_1, \gamma_1)$ and $B = (\delta_2, \beta_2, \gamma_2)$, the sum $A + B = (\delta_1 + \delta_2, \beta_1 + \beta_2, \gamma_1 + \gamma_2)$ is a triangular fuzzy number with membership function defined as, $\mu_{A+B}(z) = \max\limits_{z=x+y} \min\{\mu_A(x), \mu_A(y)\}$. The multiplication of $A = (\delta, \beta, \gamma)$ by a scalar $\omega \neq 0$ is a triangular fuzzy number ωA whose membership function is $\mu_{\omega A}(z) = \max\limits_{\{y:\omega y = z\}} \mu_A(x)$.

Definition 3 ([1])**.** *Let Y be a space and P be a subset of $r+1$ control points in Y. P is said to be a collection of fuzzy control points in Y if there exists $\mu_P : P \to [0,1]$ such that $\mu_P(p_k) = 1$ in which $P = \{(p_k, \mu_P(p_k)) | p_k \in Y\}$. Therefore,*

$$\mu_P(p_k) = \begin{cases} 0 & , \quad p_k \in Y \\ c \in (0,1) & , \quad p_k \widetilde{\in} Y \\ 1 & , \quad p_k \notin Y \end{cases}$$

with $\mu_P(\boldsymbol{p}_k) = (\overleftarrow{\mu}_P(p_k), \mu_P(p_k), \overrightarrow{\mu}_P(p_k))$ where, $\overleftarrow{\mu}_P(p_k)$, $\overrightarrow{\mu}_P(p_k)$ are left-grade and right-grade membership values. $p_k \widetilde{\in} Y$ means that p_k partially belongs to Y. Fuzzy control points can be written as $\boldsymbol{p}_k = (\overleftarrow{p}_k, p_k, \overrightarrow{p}_k)$ where $\overleftarrow{p}_k, p_k, \overrightarrow{p}_k$ are left fuzzy control points, crisp control points and right fuzzy control points, respectively.

Definition 4 ([1]). *Consider a collection of $r+1$ distinct fuzzy control points $\boldsymbol{p}_k^*$, $1 \leq k \leq r$, then a fuzzy Bernstein Beziér (B.B) curve is defined as,*

$$P^*(u) = \sum_{k=0}^{r} B_k^r(u)\boldsymbol{p}_k^*$$

where $B_k^r(u) = \binom{r}{k} u^k(1-u)^{r-k}$ is kth Berstein polynomial of degree r.

Definition 5 ([1]). *Consider a collection of $(r+1) \times (q+1)$ fuzzy control points $\boldsymbol{p}_{k,j}$, $1 \leq k \leq r$ $1 \leq j \leq q$, then a fuzzy Beziér surface is defined as,*

$$P(u,v) = \sum_{k=0}^{r}\sum_{j=0}^{q} B_k^r(u)B_j^q(v)\boldsymbol{p}_{k,j} \qquad u,v \in [0,1].$$

2. Fuzzy Tensor Product Beziér Surfaces

Consider a fuzzy B.B curve,

$$P^*(u) = \sum_{k=0}^{r} B_k^r(u)\boldsymbol{p}_k^* \tag{1}$$

If we define two operators on fuzzy control points, shift operator $E\boldsymbol{p}_k^* = \boldsymbol{p}_{k+1}^*$ and identity operator $I\boldsymbol{p}_k^* = \boldsymbol{p}_k^*$ then, Equation (1) can be written as $P^*(u) = [uE + (1-u)I]^r\boldsymbol{p}_o^*$, $u \in [0,1]$. This is called the symbolic representation of fuzzy B.B curve. For $u \in [0,1]$, a *fuzzy straight line* can be defined as,

$$L(u) = (1-u)\boldsymbol{p}_0 + u\boldsymbol{p}_1$$

where, $\boldsymbol{p}_k = (\overleftarrow{p}_k, p_k, \overrightarrow{p}_k)$ are fuzzy control points. Consider two fuzzy B.B polynomials

$$P^*(u) = \sum_{k=0}^{r} B_k^r(u)\boldsymbol{b}_k^*, \qquad P^*(v) = \sum_{j=0}^{q} B_j^q(v)\boldsymbol{a}_j^*$$

where, $\boldsymbol{b}_k^*$, $0 \leq k \leq r$ and $\boldsymbol{a}_j^*$, $0 \leq j \leq q$ are fuzzy control points. The *fuzzy tensor product surface* or *fuzzy Bernstein Beziér (B.B) surface* can be generated using $P^*(u)$ and $P^*(v)$ as,

$$P(u,v) = \sum_{k=0}^{r}\sum_{j=0}^{q} B_k^r(u)B_j^q(v)\boldsymbol{p}_{k,j} \quad u,v \in [0,1]$$

where, $\boldsymbol{p}_{k,j} = (\overleftarrow{p}_{k,j}, p_{k,j}, \overrightarrow{p}_{k,j})$ are fuzzy control points. For any fuzzy B.B surface, r and q are the degrees of corresponding fuzzy B.B curves. We can say that $P(u,v)$ is a fuzzy B.B surface of degree $r \times q$. If $r = q = 3$, the fuzzy B.B surface is known as *fuzzy cubic by cubic patch.* Likewise, the case $r = q = 2$ is called a fuzzy quadratic by quadratic patch. Also, $(r+1) \times (q+1)$ fuzzy control points are organized into $r+1$ rows and $q+1$ columns. A fuzzy B.B surface of degree 2×2, with fuzzy control points in Table 1, is shown in Figure 1. The fuzzy control points along with dashed lines is called a fuzzy control grid of a fuzzy surface. Each column and row of the fuzzy control points interpret a fuzzy B.B curve. The fuzzy B.B curve defined by the fuzzy control points $\boldsymbol{p}_{k,j}$, $0 \leq j \leq q$ is called kth fuzzy u-curve and the fuzzy B.B curve defined by $\boldsymbol{p}_{k,j}$, $0 \leq k \leq r$ is jth fuzzy v-curve. Consequently, there are $(r+1)$ number of fuzzy u-curves and $(q+1)$ number of fuzzy v-curves. The fuzzy u-curves of Figure 1 are shown in Figure 2. The fuzzy u-curve with 0th row of fuzzy control points is shown with red lines, the 1st row of fuzzy control points is shown in blue and the 2nd row of fuzzy control points is shown in green.

Table 1. Fuzzy control points.

$p_{k,j}$	$\overleftarrow{p}_{k,j}$	$p_{k,j}$	$\overrightarrow{p}_{k,j}$
$p_{0,0}$	$(0.5, 4, -0.5)$	$(1, 4, 0)$	$(1.5, 4, 0.5)$
$p_{0,1}$	$(2.5, -0.5, 0.5)$	$(3, 4, 1)$	$(3.5, 4, 1.5)$
$p_{0,2}$	$(4.5, 0, -0.5)$	$(5, 4, 0)$	$(5.5, 4, 0.5)$
$p_{1,0}$	$(-0.5, 2, -0.5)$	$(0, 2, 0)$	$(0.5, 2, 0.5)$
$p_{1,1}$	$(3, 1.5, 1.5)$	$(3, 2, 1)$	$(3, 2.5, 1.5)$
$p_{1,2}$	$(4.5, 2.5, 0.5)$	$(5, 2, 1)$	$(5.5, 1.5, 1.5)$
$p_{2,0}$	$(0.5, 0, -0.5)$	$(1, 0, 0)$	$(1.5, 0, 0.5)$
$p_{2,1}$	$(2.5, -0.5, 0.5)$	$(3, 0, 1)$	$(3.5, 0.5, 1.5)$
$p_{2,2}$	$(4.5, 0, -0.5)$	$(5, 0, 0)$	$(5.5, 0, 0.5)$

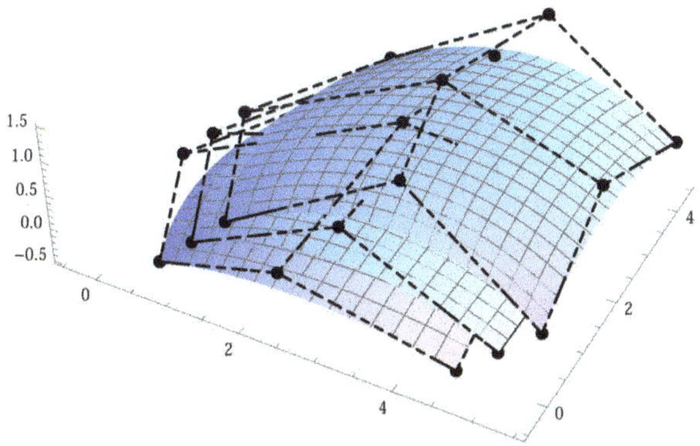

Figure 1. Fuzzy quadratic by quadratic patch.

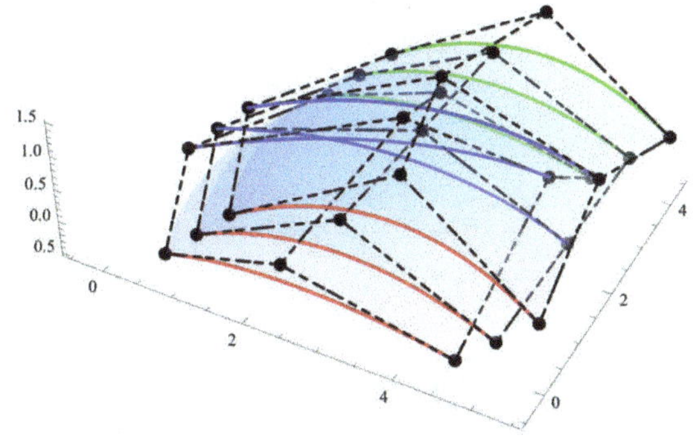

Figure 2. Fuzzy u-curves.

In fuzzy B.B surface $P(u,v)$, $B_k^r(u)$ and $B_j^q(v)$ are basis functions of degree r and q, respectively. There are four fuzzy boundary curves to $P(u,v)$,

$$P(u,0) = \sum_{k=0}^{r} B_k^r(u) p_{k,0} \qquad P(u,1) = \sum_{k=0}^{r} B_k^r(u) p_{k,1}, \qquad u \in [0,1]$$

$$P(0,v) = \sum_{j=0}^{q} B_j^q(v) p_{0,j} \qquad P(1,v) = \sum_{j=0}^{q} B_j^q(v) p_{1,j}, \qquad v \in [0,1].$$

We now present some properties of fuzzy Beziér surfaces.

1. As $P(0,0) = \boldsymbol{p}_{0,0}$, $P(1,0) = \boldsymbol{p}_{1,0}$, $P(0,1) = \boldsymbol{p}_{0,1}$ and $P(1,1) = \boldsymbol{p}_{1,1}$ therefore, $P(u,v)$ interpolates four fuzzy control points.
2. $B_k^r(u)$ and $B_j^q(v)$ are crisp basis functions for all $0 \leq k \leq r$, $0 \leq j \leq q$, $u, v \in [0,1]$ therefore, these are non-negative and $\sum_{k=0}^{r} \sum_{j=0}^{q} B_k^r(u) B_j^q(v) = 1$.
3. Let $f : \widetilde{X} \to \widetilde{Y}$ be an affine transformation where, $\widetilde{X}$ and $\widetilde{Y}$ are sets of triangular fuzzy numbers and $f(y) = By + \hat{\boldsymbol{b}}$ where, the elements of of B are triangular fuzzy numbers and $\hat{\boldsymbol{b}}$ is a 2×1 vector of triangular fuzzy numbers. Fuzzy B.B surface satisfies affine invariance property.

$$\begin{aligned} f(P(u,v)) &= B\left(\sum_{k=0}^{r}\sum_{j=0}^{q} B_k^r(u) B_j^q(v) \boldsymbol{p}_{k,j}\right) + \hat{\boldsymbol{b}} \\ &= \sum_{k=0}^{r}\sum_{j=0}^{q} B_k^r(u) B_j^q(v) B(\boldsymbol{p}_{k,j}) + \sum_{k=0}^{r}\sum_{j=0}^{q} B_k^r(u) B_j^q(v) \hat{\boldsymbol{b}} \\ &= \sum_{k=0}^{r}\sum_{j=0}^{q} B_k^r(u) B_j^q(v) f(\boldsymbol{p}_{k,j}) \end{aligned}$$

4. As $P(u,v)$ is a linear combination of fuzzy control points with non-negative coefficients whose sum is one therefore, fuzzy B.B surface lies in the convex hull defined by the fuzzy control mesh.
5. As $P'(u) = r[uE + (1-u)I]^{r-1}(\boldsymbol{p}_1^* - \boldsymbol{p}_0^*)$ therefore, the fuzzy tangents at the end points of fuzzy Beziér curve can be drawn using a pair of fuzzy control points. For $u = 0$, $P^{*'}(0) = r(\boldsymbol{p}_1^* - \boldsymbol{p}_0^*)$ and for $u = 1$, $P^{*'}(1) = r(\boldsymbol{p}_r^* - \boldsymbol{p}_{r-1}^*)$.
6. At every point of fuzzy Beziér curve, we have two fuzzy tangent directions $\dfrac{\partial P(u,v)}{\partial u}$ and $\dfrac{\partial P(u,v)}{\partial v}$.

   For any fuzzy B.B surface, if we fix one parameter, say $u = a$, then $P(u,v)$ becomes,

$$P(a,v) = \sum_{k=0}^{r}\sum_{j=0}^{q} B_k^r(a) B_j^q(v) \boldsymbol{p}_{k,j} = \sum_{j=0}^{q} B_j^q(v) P_j^*(a)$$

where, $P_j^*(a) = \sum_{k=0}^{r} B_k^r(a) \boldsymbol{p}_{k,j}$. $P(a,v)$ is known as fuzzy u iso-parametric curve. Fuzzy iso-parametric curve on any fuzzy B.B surface can be obtained by fixing one parameter as constant. A fuzzy B.B surface can be considered as a family of fuzzy iso-parametric curves and these fuzzy iso-parametric curves can be studied in terms of fuzzy control curves. Figure 2 represents fuzzy u iso-parametric curves of Figure 1. For any value of a, $P_j^*(a)$ define fuzzy control point positions for fuzzy iso-parametric curve $P(a,v)$.

Clearly, the four fuzzy iso-parametric curves are fuzzy boundary curves $P(u,1)$, $P(u,0)$, $P(1,v)$ and $P(0,v)$. These fuzzy boundary curves are defined by rows and columns of fuzzy control points and are fuzzy control curves. For example, in Figure 1, $P(0,v)$ shown in red color is a fuzzy B.B curve of degree 2 with fuzzy control points $\boldsymbol{p}_{0,0}$, $\boldsymbol{p}_{0,1}$, $\boldsymbol{p}_{0,2}$. Similarly, $P(u,1)$, shown in green, is a fuzzy B.B curve with fuzzy control points $\boldsymbol{p}_{1,0}$, $\boldsymbol{p}_{1,1}$, $\boldsymbol{p}_{1,2}$.

We now describe and design de Casteljau's algorithm to find any point on fuzzy B.B curve.

Algorithm 1. De Casteljau's algorithm for fuzzy B.B curves

Consider fuzzy B.B curve of degree r,

$$P^*(u) = \sum_{k=0}^{r} B_k^r(u)\mathbf{p}_k^*$$

Using symbolic representation, $P^*(u)$ can also be expressed as,

$$\begin{aligned} P^*(u) &= \sum_{k=0}^{r-1} B_k^{r-1}(u)[(1-u)\mathbf{p}_k^* + u\mathbf{p}_{k+1}^*] \\ &= \sum_{k=0}^{r-1} B_k^{r-1}(u)\mathbf{p}_k^{*(1)}(u). \end{aligned} \tag{2}$$

Denote $\mathbf{p}_k^{*(0)}(u) = \mathbf{p}_k^*$ and $\mathbf{p}_k^{*(i)} = (1-u)\mathbf{p}_k^{*(i-1)}(u) + u\mathbf{p}_{k+1}^{*(i-1)}(u)$, $1 \leq i \leq r$ and $0 \leq k \leq r-i$. Equation (2) can also be expressed as, $P^*(u) = \sum_{k=0}^{r-i} B_k^{r-i}(u)\mathbf{p}_k^{*(i)}(u)$ where, $\mathbf{p}_k^{*(i)}(u) = (\overleftarrow{p}_k^{*(i)}(u), p_k^{*(i)}(u), \overrightarrow{p}_k^{*(i)}(u))$ and,

$$\begin{aligned} \overleftarrow{p}_k^{*(i)}(u) &= (1-u)\overleftarrow{p}_k^{*(i-1)}(u) + u\overleftarrow{p}_{k+1}^{*(i-1)}(u) \\ p_k^{*(i)}(u) &= (1-u)p_k^{*(i-1)}(u) + up_{k+1}^{*(i-1)}(u) \\ \overrightarrow{p}_k^{*(i)}(u) &= (1-u)\overrightarrow{p}_k^{*(i-1)}(u) + u\overrightarrow{p}_{k+1}^{*(i-1)}(u). \end{aligned}$$

For $i = r$, $P^*(u) = \mathbf{p}_0^{*(r)}(u) = (\overleftarrow{p}_0^{*(r)}, p_0^{*(r)}, \overrightarrow{p}_0^{*(r)})$.

Example 1. *Consider a fuzzy cubic B.B curve, given in Figure 3, having fuzzy control points as shown in Table 2. We now find* $P^*(\frac{1}{2})$ *using de Casteljau's algorithm. Clearly,* $P^*(\frac{1}{2}) = (\overleftarrow{p}_0^{(3)}(\frac{1}{2}), p_0^{(3)}(\frac{1}{2}), \overrightarrow{p}_0^{(3)}(\frac{1}{2}))$ *where,*

$$\begin{aligned} \overleftarrow{p}_0^{(3)}(\frac{1}{2}) &= \frac{1}{2}\overleftarrow{p}_0^{(2)}(\frac{1}{2}) + \frac{1}{2}\overleftarrow{p}_1^{(2)}(\frac{1}{2}) \\ &= \frac{1}{2}(\frac{1}{2}\overleftarrow{p}_0^{(1)}(\frac{1}{2}) + \frac{1}{2}\overleftarrow{p}_1^{(1)}(\frac{1}{2})) + \frac{1}{2}(\frac{1}{2}\overleftarrow{p}_1^{(1)}(\frac{1}{2}) + \frac{1}{2}\overleftarrow{p}_2^{(1)}(\frac{1}{2})) \\ &= \frac{1}{4}(\frac{1}{2}\overleftarrow{p}_0 + \frac{1}{2}\overleftarrow{p}_1) + \frac{1}{2}(\frac{1}{2}\overleftarrow{p}_1 + \frac{1}{2}\overleftarrow{p}_2) + \frac{1}{4}(\frac{1}{2}\overleftarrow{p}_2 + \frac{1}{2}\overleftarrow{p}_3) \\ &= \frac{1}{8}\overleftarrow{p}_0 + \frac{3}{8}\overleftarrow{p}_1 + \frac{3}{8}\overleftarrow{p}_3 + \frac{1}{8}\overleftarrow{p}_3 = (1.4, 0) \end{aligned}$$

Similarly, $p_0^{(3)}(\frac{1}{2}) = (1.5, 0)$, $\overrightarrow{p}_0^{(3)}(\frac{1}{2}) = (1.6, 0)$ *and therefore,* $P^*(\frac{1}{2}) = ((1.4, 0), (1.5, 0), (1.6, 0))$.

Table 2. Fuzzy control points.

p_k	$\overleftarrow{p}_k$	p_k	$\overrightarrow{p}_k$
p_0	$(-0.1, 0)$	$(0, 0)$	$(0.1, 0)$
p_1	$(0.9, 2)$	$(1, 2)$	$(1.1, 2)$
p_2	$(1.9, -2)$	$(2, -2)$	$(2.1, -2)$
p_3	$(2.9, 0)$	$(3, 0)$	$(3.1, 0)$

Algorithm 2. De Casteljau's algorithm for fuzzy B.B surfaces

Algorithm 1 can be extended to fuzzy B.B surfaces. De Casteljau's can be implemented several times to find $P(u,v)$ for particular (u,v). It is based on fuzzy iso-parametric curves. Consider the equation of fuzzy Beziér surface,

$$P(u,v) = \sum_{k=0}^{r} B_k^r(u)(\sum_{j=0}^{q} B_j^q(v)\boldsymbol{p}_{k,j}) = \sum_{k=0}^{r} B_k^r(u)\boldsymbol{l}_k(v).$$

It clearly shows that $P(u,v)$ can be calculated using $r+1$ fuzzy control points $\boldsymbol{l}_o(v)$, $\boldsymbol{l}_1(v), \ldots, \boldsymbol{l}_r(v)$. The procedure can be illustrated as:

1. As $\boldsymbol{l}_o(v)$ is a fuzzy control point on fuzzy iso-parametric curve defined by the row of fuzzy control points $\boldsymbol{p}_{0,0}, \boldsymbol{p}_{0,1}, \ldots, \boldsymbol{p}_{0,q}$. Therefore, for the fuzzy iso-parametric curve on the 1st row, Algorithm 1 can be applied to compute $\boldsymbol{l}_o(0)$. Repeat this process for all other fuzzy iso-parametric curves.
2. After $r+1$ implementations of de Casteljau's algorithms, we obtain $\boldsymbol{l}_o(v)$, $\boldsymbol{l}_1(v), \ldots, \boldsymbol{l}_r(v)$.
3. At the end, apply de Casteljau's algorithm to $r+1$ fuzzy control points $\boldsymbol{l}_o(v)$, $\boldsymbol{l}_1(v), \ldots, \boldsymbol{l}_r(v)$ with given u to compute $P(u,v)$.

Example 2. *In this example, we illustrate the process of Algorithm 2 for fuzzy quadratic by quadratic surface as shown in Figure 1. We now calculate the value of $P(u,v)$ for $u = v = \frac{1}{2}$.*

Step 1: *For $k = 0$, we compute $\boldsymbol{l}_o(\frac{1}{2}) = (\overleftarrow{l}_o(\frac{1}{2}), l_o(\frac{1}{2}), \overrightarrow{l}_o(\frac{1}{2}))$. The fuzzy control points on the first row are $\boldsymbol{p}_{o,o}$, $\boldsymbol{p}_{o,1}$, $\boldsymbol{p}_{o,2}$. Applying Algorithm 2,*

$$\overleftarrow{l}_o(\frac{1}{2}) = \overleftarrow{p}^{(2)}_{o,o}(\frac{1}{2}) = \frac{1}{2}\overleftarrow{p}^{(1)}_{o,o}(\frac{1}{2}) + \frac{1}{2}\overleftarrow{p}^{(1)}_{o,1}(\frac{1}{2}) = \frac{1}{4}\overleftarrow{p}_{o,o} + \frac{1}{2}\overleftarrow{p}_{o,1} + \frac{1}{4}\overleftarrow{p}_{o,2} = (2.5, 0.75.0)$$

Similarly, $l_o(\frac{1}{2}) = (3,4,0.5)$ and $\overrightarrow{l}_o(\frac{1}{2}) = (3.5, 4.25, 1)$.

Step 2: *Applying Algorithm 1 on all fuzzy iso-parametric curves, we obtain three fuzzy control points as shown in Table 3.*

Step 3: *Applying Algorithm 2 for $u = \frac{1}{2}$, we obtain the following expression,*

$$P(\frac{1}{2},\frac{1}{2}) = \frac{1}{4}l_o(\frac{1}{2}) + \frac{1}{2}l_1(\frac{1}{2}) + \frac{1}{4}l_2(\frac{1}{2}) = ((2.5, 0.8125, 0), (2.875, 2, 0.625), (3.25, 2.1875, 1.125)).$$

Table 3. Fuzzy control points.

Values on Fuzzy Iso-Parametric Curves for $v = \frac{1}{2}$	
$\boldsymbol{l}_o(\frac{1}{2})$	$((2.5, 0.75, 0), (3, 4, 0.5), (3.5, 4.25, 1))$
$\boldsymbol{l}_1(\frac{1}{2})$	$((2.5, 1.375, 0), (2.75, 2, 0.75), (3, 2.125, 1.25))$
$\boldsymbol{l}_2(\frac{1}{2})$	$((2.5, -0.25, 0), (3, 0, 0.5), (3.5, 0.25, 1))$

Upon defining the fuzzy Beziér surface model, the next step is the defuzzification process. This procedure can be applied to obtain the results as a single value. For defining defuzzification, we use the α-cut operation of fuzzy control points on the definition of fuzzy Beziér surface. This is called the *fuzzification process*, and is defined as follows.

Fuzzification process [11]:

If $\{\boldsymbol{p}_{k,j} \mid 0 \le k \le r, 0 \le j \le q\}$ are the set of fuzzy control points then, $\boldsymbol{p}_{k,j_\alpha}$ is the alpa- cut of $\boldsymbol{p}_{k,j}$ and is defined in Equation (3):

$$\begin{aligned} \boldsymbol{p_{k,j}}_{\alpha} &= (\overleftarrow{p_{k,j}}_{\alpha}, p_{k,j_{\alpha}}, \overrightarrow{p_{k,j}}_{\alpha}) \\ &= ([(p_{k,j} - \overleftarrow{p}_{k,j})\alpha + \overleftarrow{p}_{k,j}], p_{k,j}, [(p_{k,j} - \overrightarrow{p}_{k,j})\alpha + \overrightarrow{p}_{k,j}]) \end{aligned} \quad (3)$$

After fuzzification, the next procedure is the defuzzification of fuzzy control points to obtain the crisp solution which is described below.

Defuzzification process [11]:

The defuzzification of fuzzy control point $\boldsymbol{p_{k,j}}_{\alpha}$ is a crisp control point $\overline{p}_{k,j_{\alpha}}$, calculated in Equation (4):

$$\overline{p}_{k,j_{\alpha}} = \frac{1}{3}\{\overleftarrow{p_{k,j}}_{\alpha} + p_{k,j_{\alpha}} + \overrightarrow{p_{k,j}}_{\alpha}\} \quad (4)$$

The fuzzification and defuzzification process is illustrated in Figures 4 and 5. The fuzzification process is applied by means of 0.5-cut operation and a crisp Beziér surface is obtained by applying the defuzzification process.

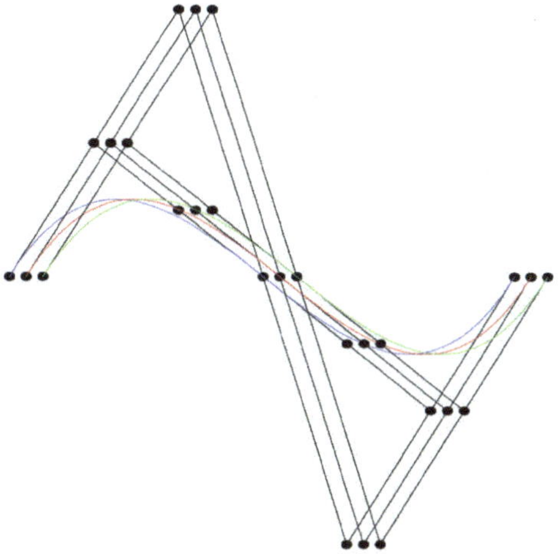

Figure 3. Fuzzy cubic B.B cuve.

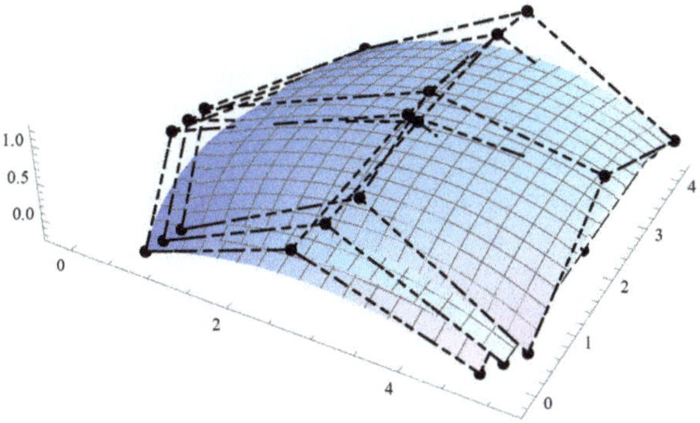

Figure 4. Fuzzification of Figure 1.

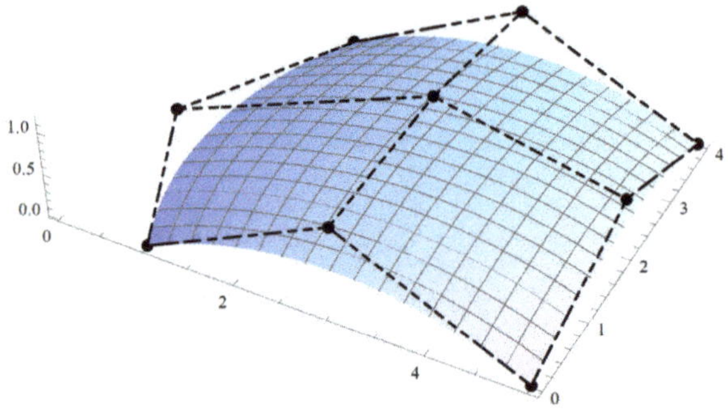

Figure 5. Defuzzification of Figure 4.

Degree Elevation for a Fuzzy B.B Curve

Numerous applications that include more than one fuzzy B.B curve require all the fuzzy curves to have the same degree. Additionally, higher degree fuzzy B.B curves take a longer time to process, but provide more flexibility for designing shapes. The key point is to change the degree of fuzzy B.B curve without changing its shape. This process is called *degree elevation*. We now explain the process of degree raising for a fuzzy B.B curve.

Consider fuzzy B.B curve of degree r having $r+1$ fuzzy control points,

$$P^*(u) = \sum_{k=0}^{r} B_k^r(u) p_k^*. \tag{5}$$

To increase the degree of fuzzy B.B curve to $r+1$, $r+2$ fuzzy control $\mathbf{H}_k^*$, $0 \leq k \leq r+1$, are required. As the fuzzy curve passes through p_o^* and p_r^* therefore, the new set of fuzzy control points must include p_o^* and p_r^*. By replacing u by $\overline{1-u+u}$, Equation (5) can be written as,

$$P^*(u) = \sum_{k=0}^{r+1} B_k^{r+1}(u) H_k^*. \tag{6}$$

where, $H_o^* = p_o^*$, $H_{r+1}^* = p_r^*$ and

$$H_k^* = \frac{k}{r+1} p_{k-1}^* + \left(1 - \frac{k}{r+1}\right) p_k^*, \qquad 1 \leq k \leq r.$$

Each edge of fuzzy control point contains a new fuzzy control point. More precisely, edge $\overrightarrow{p_{k-1}}\overrightarrow{p_k}$ contains $\overrightarrow{H_k}$ in the ratio $\left(1 - \frac{k}{r+1}\right) : \frac{k}{r+1}$. In de Casteljau's algorithm, the fuzzy line segment is divided in the ration $t{:}1-t$. Unlike Algorithm 2, the ratio is not a constant but varies with index k.

Example 3. *Consider a fuzzy quadratic B.B curve having fuzzy control points as shown in Table 4. The fuzzy quadratic B.B curve is shown in Figure 6.*

By applying degree elevation algorithm, the fuzzy cubic B.B curve obtained from Figure 6 is shown in Figure 7.

Table 4. Fuzzy control points.

p_k	$\overleftarrow{p}_k$	p_k	$\overrightarrow{p}_k$
p_0	$(0,0)$	$(0.5,0)$	$(1,0)$
p_1	$(3,5)$	$(3.5,5)$	$(4,5)$
p_2	$(6,0)$	$(6.5,0)$	$(7,0)$

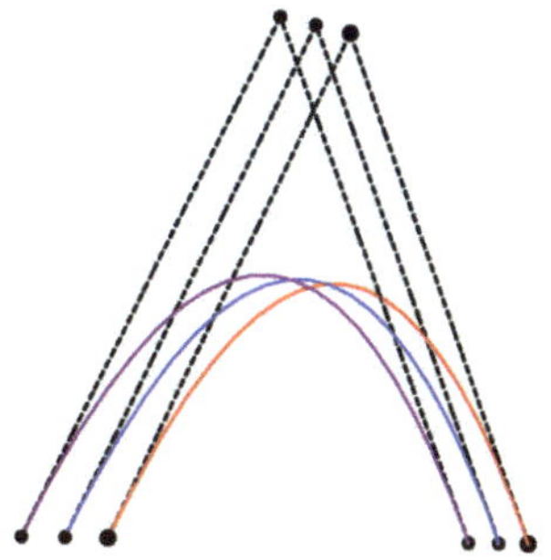

Figure 6. Fuzzy quadratic B.B curve.

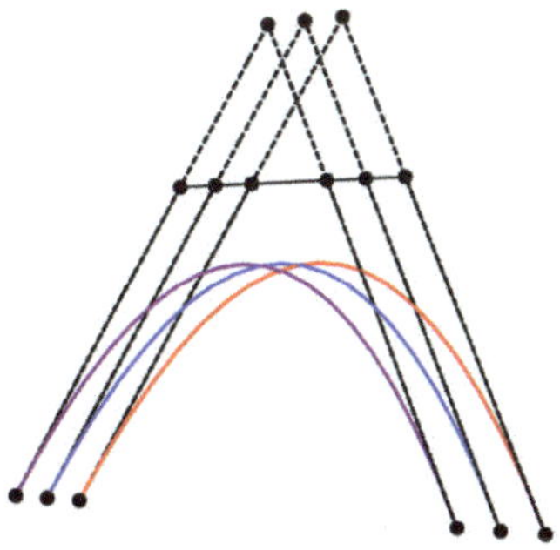

Figure 7. Fuzzy cubic B.B curve.

Fuzzy Rational Beziér Surface Patch

A fuzzy rational Beziér curve (FRB) [13] is defined as,

$$R^*(u) = \frac{\sum_{k=0}^{r} \boldsymbol{w}_k B_k^r(u) \boldsymbol{p}_k^*}{\sum_{k=0}^{r} \boldsymbol{w}_k B_k^r(u)} = \sum_{k=0}^{r} \left[\frac{\boldsymbol{w}_k B_k^r(u)}{\sum_{k=0}^{r} \boldsymbol{w}_k B_k^r(u)} \right] \boldsymbol{p}_k^* = \sum_{k=0}^{r} R_k^r(u) \boldsymbol{p}_k^*$$

where, $\boldsymbol{w}_k = (\overleftarrow{w_k}, w_k, \overrightarrow{w_k})$ are fuzzy weights. Fuzzy rational Beziér curves has several benefits over simple fuzzy Beziér curves. It provides large control to the shape of fuzzy curves. In addition, a 2D FRB curve can represented as a projection of a 3D fuzzy Beziér curve as,

$$R^*(u) = \prod(P^*(u)), \quad P^*(u) = (P_x^*(u), P_y^*(u), P_w^*(u)) = \sum_{k=0}^{r} B_k^r(u) \boldsymbol{P}_k$$

where, $\overleftarrow{P}_k = (\overleftarrow{w_k x_k}, \overleftarrow{w_k y_k}, \overleftarrow{w_k})$, $P_k = (w_k x_k, w_k y_k, w_k)$, $\overrightarrow{P}_k = (\overrightarrow{w_k x_k}, \overrightarrow{w_k y_k}, \overrightarrow{w_k})$ and the operator $\prod$ is defined as $\prod(x, y, w) = (x/w, y/w)$.

The degree elevation and de Casteljau algorithm for fuzzy Beziér curve can be extended to FRB curve. For this, transform the FRB curve into a 3D fuzzy Beziér curve as discussed above. Next, apply the algorithms to 3D fuzzy Beziér curve. Finally, convert the 3D fuzzy Beziér curve to 2D fuzzy curve by applying the projection operator $\prod$. The resulting fuzzy control points turn out to be the fuzzy weights of given FRB curve.

3. Conclusions

Fuzzy splines are the most useful mathematical and graphical tools to reduce uncertainty in curve and surface modeling. In this research paper, various properties of fuzzy tensor product surface patches are studied using fuzzy numbers including fuzzy parametric curves, affine invariance, fuzzy tangents, convex hull and fuzzy iso-parametric curves. The degree elevation and de Casteljau's algorithms for fuzzy Bézier curves, fuzzy tensor product Bézier surfaces and FRB curves are presented. The proposed techniques are useful to visualize uncertain and vague measures via surface modeling. The process of fuzzification is applied to obtain the fuzzy interval of fuzzy data points where the crisp solution exists. It is then followed by the defuzzification process to construct crisp Beziér curves and surfaces which are focused on the defuzzification of fuzzy data points. Finally, to check the effectiveness of Beziér surfaces this process is applied to numerical examples. We aim to extend the theory of fuzzy splines to find its applications in geometric modeling, representing fuzzy data points using fuzzy numbers and fuzzy spline approximation problems.

Author Contributions: Musavarah Sarwar and Muhammad Akram conceived of the presented idea. Musavarah Sarwar developed the theory and performed the computations. Muhammad Akram verified the analytical methods.

Conflicts of Interest: The authors declare no conflict of interest.

References

1. Wahab, A.F.; Ali, J.M.; Majid, A.A. Fuzzy Geometric Modeling. In Proceedings of the IEEE 2009 Sixth International Conference on Computer Graphics, Imaging and Visualization, Tianjin, China, 11–14 August 2009; pp. 276–280.
2. Farin, G. *Curves and Surfaces for CAGD: A Practical Guide*, 5th ed.; Morgan Kaufmann: Burlington, MA, USA, 2002.
3. Rogers, D.F. *An Introduction to NURBS: With Historical Perspective*, 1st ed.; Morgan Kaufmann: Burlington, MA, USA, 2000.
4. Yamaguchi, F. *Curves and Surfaces in Computer Aided Geometric Design*; Springer Science & Business Media: Berlin/Heidelberg, Germany, 1988.
5. Zadeh, L.A. Fuzzy sets. *Inf. Control* **1965**, *8*, 338–353.
6. Lowen, R. A Fuzzy Lagrange interpolation Theorem. *Fuzzy Sets Syst.* **1990**, *34*, 33–38.
7. Kaleva, O. Interpolation of fuzzy data. *Fuzzy Sets Syst.* **1994**, *61*, 63–70.
8. Abbasbandy, S. Interpolation of fuzzy data by complete splines. *J. Appl. Math. Comput.* **2001**, *8*, 587–594.
9. Abbasbandy, S.; Babolian, E. Interpolation of fuzzy data by natural splines. *J. Appl. Math. Comput.* **1998**, *5*, 457–463.
10. Abbasbandy, S.; Ezzati, R.; Behforooz, H. Interpolation of fuzzy data by using fuzzy splines. *Int. J. Uncertain. Fuzziness Knowl. Based Syst.* **2008**, *16*, 107–115.
11. Zakaria, R.; Wahab, A.F.; Gobithaasan, R.U. Fuzzy B-Spline surface modeling. *J. Appl. Math.* **2014**, *2014*, 285045.
12. Zakaria, R.; Wahab, A.B. Fuzzy B-spline modeling of uncertainty data. *Appl. Math. Sci.* **2012**, *6*, 6971–6991.
13. Wahab, A.F.; Zakaria, R.; Ali, J.M. Fuzzy interpolation rational bezier curve. In Proceedings of the IEEE 2010 Seventh International Conference on Computer Graphics, Imaging and Visualization (CGIV), Sydney, NSW, Australia, 7–10 August 2010; pp. 63–67.
14. Anile, A.M.; Falcidieno, B.; Gallo, G.; Spagnuolo, M.; Spinello, S. Modeling uncertain data with fuzzy B-splines. *Fuzzy Sets Syst.* **2000**, *113*, 397–410.
15. Behforooz, H.; Ezzati, R.; Abbasbandy, S. Interpolation of fuzzy data by using E(3) cubic splines. *Int. J. Pure Appl. Math.* **2010**, *60*, 383–392.
16. Dubois, D.; Prade, H. Operations on fuzzy numbers. *Int. J. Syst. Sci.* **1978**, *9*, 613–626.
17. Fortuna, L.; Muscato, G. A roll stabilization system for a monohull ship: modeling, identification, and adaptive control. *IEEE Trans. Control Syst. Technol.* **1996**, *4*, 18–28.
18. Sarwar, M.; Akram, M. An algorithm for computing certain metrics in intuitionistic fuzzy graphs. *J. Intell. Fuzzy Syst.* **2016**, *30*, 2405–2416.

19. Sarwar, M.; Akram, M. Certain algorithms for computing strength of competition in bipolar fuzzy graphs. *Int. J. Uncertain. Fuzziness Knowl. Based Syst.* **2017**, *25*, 877–896.
20. Zadeh, L.A. Similarity relations and fuzzy orderings. *Inf. Sci.* **1971**, *3*, 177–200.
21. Chang, S.S.L.; Zadeh, L.A. On fuzzy mapping and control. *IEEE Trans. Syst. Man Cybern.* **1972**, *2*, 30–34.

Σ mathematics

MDPI

Article

Numerical Methods for Solving Fuzzy Linear Systems

Lubna Inearat and Naji Qatanani *

Department of Mathematics, An–Najah National University, Nablus, P.O. Box 7, Palestine; lubna_inearat@hotmail.com
* Correspondence: nqatanani@najah.edu

Received: 21 November 2017; Accepted: 29 January 2018; Published: 1 February 2018

Abstract: In this article, three numerical iterative schemes, namely: Jacobi, Gauss–Seidel and Successive over-relaxation (SOR) have been proposed to solve a fuzzy system of linear equations (FSLEs). The convergence properties of these iterative schemes have been discussed. To display the validity of these iterative schemes, an illustrative example with known exact solution is considered. Numerical results show that the SOR iterative method with $\omega = 1.3$ provides more efficient results in comparison with other iterative techniques.

Keywords: fuzzy system of linear equations (FSLEs); iterative schemes; strong and weak solutions; Hausdorff

---

1. Introduction

The subject of Fuzzy System of Linear Equations (FSLEs) with a crisp real coefficient matrix and with a vector of fuzzy triangular numbers on the right-hand side arise in many branches of science and technology such as economics, statistics, telecommunications, image processing, physics and even social sciences. In 1965, Zadeh [1] introduced and investigated the concept of fuzzy numbers that can be used to generalize crisp mathematical concept to fuzzy sets.

There is a vast literature on the investigation of solutions for fuzzy linear systems. Early work in the literature deals with linear equation systems whose coefficient matrix is crisp and the right hand vector is fuzzy. That is known as FSLEs and was first proposed by Friedman et al. [2]. For computing a solution, they used the embedding method and replaced the original fuzzy $n \times n$ linear system by a $2n \times 2n$ crisp linear system. Later, several authors studied FSLEs. Allahviranloo [3,4] used the Jacobi, Gauss–Seidel and Successive over-relaxation (SOR) iterative techniques to solve FSLEs. Dehghan and Hashemi [5] investigated the existence of a solution provided that the coefficient matrix is strictly diagonally dominant matrix with positive diagonal entries and then applied several iterative methods for solving FSLEs. Ezzati [6] developed a new method for solving FSLEs by using embedding method and replaced an $n \times n$ FSLEs by two $n \times n$ crisp linear system. Furthermore, Muzziolia et al. [7] discussed FSLEs in the form of $A_1x + b_1 = A_2x + b_2$ with A_1, A_2 being square matrices of fuzzy coefficients and b_1, b_2 fuzzy number vectors. Abbasbandy and Jafarian [8] proposed the steepest descent method for solving FSLEs. Ineirat [9] investigated the numerical handling of the fuzzy linear system of equations (FSLEs) and fully fuzzy linear system of equations (FFSLEs).

Generally, FSLEs is handled under two main headings: square $(n \times n)$ and nonsquare $(m \times n)$ forms. Most of the works in the literature dealwith square form. For example, Asady et al. [10], extended the model of Friedman for $n \times n$ fuzzy linear system to solve general $m \times n$ rectangular fuzzy linear system for $\times n$, where the coefficients matrix is crisp and the right-hand side column is a fuzzy number vector. They replaced the original fuzzy linear system $m \times n$ by a crisp linear system $2m \times 2n$. Moreover, they investigated the conditions for the existence of a fuzzy solution.

Fuzzy elements of this system can be taken as triangular, trapezoidal or generalized fuzzy numbers in general or parametric form. While triangular fuzzy numbers are widely used in earlier works, trapezoidal fuzzy numbers have neglected for along time. Besides, there exist lots of works using the parametric and level cut representation of fuzzy numbers.

The paper is organized as follows: In Section 2, a fuzzy linear system of equations is introduced. In Section 3, we present the Jacobi, Gauss–Seidel and SOR iterative methods for solving FSLEs with convergence theorems. The proposed algorithms are implemented using a numerical example with known exact solutions in Section 4. Conclusions are drawn in Section 5.

2. Fuzzy Linear System

Definition 1. In Reference [11]: *An arbitrary fuzzy number in parametric form is represented by an ordered pair of functions* $(\overline{v(r)}, \underline{v(r)})$, $0 \le r \le 1$, *which satisfy the following requirements:*

(1) $\underline{v(r)}$ *is a bounded left-continuous non-decreasing function over* $[0,\ 1]$.

(2) $\overline{v(r)}$ *is a bounded left-continuous non-increasing function over* $[0,\ 1]$.

(3) $\underline{v(r)} \le \overline{v(r)}; 0 \le r \le 1$.

Definition 2. In Reference [12]: *For arbitrary fuzzy numbers u and vthe quantity*

$$D(u, v) = \sup_{0 \le r \le 1} \{\max\{|\underline{u}^r - \underline{v}^r|, |\overline{u}^r - \overline{v}^r|\}\}$$

is called the Hausdorff distance between u and v.

Definition 3. In Reference [13]: *The* $n \times n$ *linear system*

$$\begin{cases} a_{11}x_1 + a_{12}x_2 + \ \dots\ + \ a_{1n}x_n = b_1 \\ a_{21}x_1 + a_{22}x_2 + \ \dots\ + \ a_{2n}x_n = b_2 \\ \quad \vdots \\ a_{n1}x_1 + a_{n2}x_2 + \ \dots\ + a_{nn}x_n = b_n \end{cases} \tag{1}$$

where the coefficients matrix $A = (a_{ij})$, $1 \le i, j \le n$ *is a crisp* $n \times n$ *matrix and each* $b_i \in E^1$, $1 \le i \le n$, *is fuzzy number, is called FSLEs.*

Definition 4. In Reference [13]: *A fuzzy number vector* $X = (x_1, x_2, \dots, x_n)^t$ *given by* $x_i = (\underline{x_i}(r), \overline{x_i}(r)), 1 \le i \le n$, $0 \le r \le 1$ *is called (in parametric form) a solution of the FSLEs (1) if*

$$\begin{aligned} \underline{\sum_{j=1}^{n} a_{ij}x_j} &= \sum_{j=1}^{n} \underline{a_{ij}x_j} = \underline{b_i}, \\ \overline{\sum_{j=1}^{n} a_{ij}x_j} &= \sum_{j=1}^{n} \overline{a_{ij}x_j} = \overline{b_i}. \end{aligned} \tag{2}$$

Following Friedman [2] we introduce the notations below:

$$x = \left(\underline{x_1}, \underline{x_2}, \dots \underline{x_n}, -\overline{x_1}, -\overline{x_2}, \dots - \overline{x_n}\right)^t$$

$$b = \left(\underline{b_1}, \underline{b_2}, \dots \underline{b_n}, -\overline{b_1}, -\overline{b_2}, \dots - \overline{b_n}\right)^t$$

$S = (s_{ij})$, $1 \le i,\ j \le 2n$, *where* s_{ij} are determined as follows:

$$\begin{aligned} a_{ij} \ge 0 &\Rightarrow s_{ij} = a_{ij},\ s_{i+n,j+n} = a_{ij}, \\ a_{ij} < 0 &\Rightarrow s_{i,j+n} = -a_{ij},\ s_{i+n,j} = -a_{ij}. \end{aligned} \tag{3}$$

and any s_{ij} which is not determined by Equation (3) is zero. Using matrix notation, we have

$$SX = b \tag{4}$$

The structure of S implies that $s_{ij} \geq 0$ and thus

$$S = \begin{pmatrix} B & C \\ C & B \end{pmatrix} \tag{5}$$

where B contains the positive elements of A , C contains the absolute value of the negative elements of A and $A = B - C$. An example in the work of Friedman [2] shows that the matrix S may be singular even if A is nonsingular.

Theorem 1. In Reference [2]: *The matrix S is nonsingular matrixif and only if the matrices $A = B - C$ and $B + C$ are both nonsingular.*

Proof. By subtracting the jth column of S, from its $(n+j)$th column for $1 \leq j \leq n$ we obtain

$$S = \begin{pmatrix} B & C \\ C & B \end{pmatrix} \to \begin{pmatrix} B & C-B \\ C & B-C \end{pmatrix} = S_1.$$

Next, we adding the $(n+i)$ throw of S to its ith row for $1 \leq i \leq n$ then we obtain

$$S_1 = \begin{pmatrix} B & C-B \\ C & B-C \end{pmatrix} \to \begin{pmatrix} B+C & 0 \\ C & B-C \end{pmatrix} = S_2.$$

Clearly, $|S| = |S_1| = |S_2| = |B+C||B-C| = |B+C||A|$.
Therefore
$|S| \neq 0$ if and only if $|A| \neq 0$ and $|B+C| \neq 0$,
These concludes the proof. □

Corollary 1. In Reference [2]: If a crisp linear system does not have a unique solution, the associated fuzzy linear system does not have one either.

Definition 5. In Reference [14]: *If $X = \left(\underline{x_1}, \underline{x_2}, \ldots \underline{x_n}, -\overline{x_1}, -\overline{x_2}, \ldots, -\overline{x_n}\right)^{\mathrm{T}}$ is a solution of system (4) and for each $1 \leq i \leq n$, when the inequalities $\underline{x_i} \leq \overline{x_i}$ hold, then the solution $X = \left(\underline{x_1}, \underline{x_2}, \ldots \underline{x_n}, -\overline{x_1}, -\overline{x_2}, \ldots, -\overline{x_n}\right)^{\mathrm{T}}$ is called a strong solution of the system (4) .*

Definition 6. In Reference [14]: *If $X = \left(\underline{x_1}, \underline{x_2}, \ldots \underline{x_n}, -\overline{x_1}, -\overline{x_2}, \ldots, -\overline{x_n}\right)^{\mathrm{T}}$ is a solution of system (4) and for some $i \in [1,n]$, when the inequality $\underline{x_i} \geq \overline{x_i}$ hold, then the solution $X = \left(\underline{x_1}, \underline{x_2}, \ldots \underline{x_n}, -\overline{x_1}, -\overline{x_2}, \ldots, -\overline{x_n}\right)^{\mathrm{T}}$ is called a weak solution of the system (4).*

Theorem 2. In Reference [14]: *Let $S = \begin{pmatrix} B & C \\ C & B \end{pmatrix}$ be a nonsingular matrix. Then the system (4) has a strong solution if and only if $(B+C)^{-1}\left(\underline{b} - \overline{b}\right) \leq 0$.*

Theorem 3. In Reference [14]: *The FSLEs (1) has a unique strong solution if and only if the following conditions hold:*

(1) *The matrices*

$A = B - C$ and $B + C$ are both invertible matrices.

(2) $(B+C)^{-1}\left(\underline{b} - \overline{b}\right) \leq 0$.

3. Iterative Schemes

In this section we will present the following iterative schemes for solving FSLEs.

3.1. The Jacobi and Gauss–Seidel Iterative Schemes

An iterative technique for solving an $n \times n$ linear system $AX = b$ involves a process of converting the system $AX = b$ into an equivalent system $X = TX + C$. After selecting an initial approximation X^0, a sequence $\{X^k\}$ is generated by computing

$$X^k = TX^{k-1} + C \quad k \geq 1.$$

Definition 7. In Reference [4]: *A square matrix A is called diagonally dominant matrix if* $|a_{ij}| \geq \sum_{i=1,i\neq j}^{n} |a_{ij}|$, $j = 1,2,\ldots,n$. *A is called strictly diagonally dominant if* $|a_{ij}| > \sum_{i=1,i\neq j}^{n} |a_{ij}|$, $j = 1,2,\ldots,n$.

Next, we are going to present the following theorems.

Theorem 4. In Reference [3]: *Let the matrix A in Equation (1) be strictly diagonally dominant then both the Jacobi and the Gauss–Seidel iterative techniques converge to* $A^{-1}Y$ *for any* X^0.

Theorem 5. In Reference [3]: *The matrix A in Equation (1) is strictly diagonally dominant if and only if matrix S is strictly diagonally dominant.*

Proof. For more details see [3].

From [3], without loss of generality, suppose that $s_{ii} > 0$ for all $i = 1,2,\ldots,2n$. Let $S = D + L + U$ where

$$D = \begin{bmatrix} D_1 & 0 \\ 0 & D_1 \end{bmatrix}, \quad L = \begin{bmatrix} L_1 & 0 \\ S_2 & L_1 \end{bmatrix}, \quad U = \begin{bmatrix} U_1 & S_2 \\ 0 & U_1 \end{bmatrix}$$

$(D_1)_{ii} = s_{ii} > 0$, $I = 1,2,\ldots,n$, and assume $S_1 = D_1 + L_1 + U_1$. In the Jacobi method, from the structure of $SX = Y$ we have

$$\begin{bmatrix} D_1 & 0 \\ 0 & D_1 \end{bmatrix}\begin{bmatrix} \underline{X} \\ \overline{X} \end{bmatrix} + \begin{bmatrix} L_1 + U_1 & S_2 \\ S_2 & L_1 + U_1 \end{bmatrix}\begin{bmatrix} \underline{X} \\ \overline{X} \end{bmatrix} = \begin{bmatrix} \underline{Y} \\ \overline{Y} \end{bmatrix}$$

then

$$\begin{aligned} \underline{X} &= D_1^{-1}\underline{Y} - D_1^{-1}(L_1 + U_1)\underline{X} - D_1^{-1}S_2\overline{X}, \\ \overline{X} &= D_1^{-1}\overline{Y} - D_1^{-1}(L_1 + U_1)\overline{X} - D_1^{-1}S_2\underline{X}. \end{aligned} \tag{6}$$

Thus, the Jacobi iterative technique will be

$$\begin{aligned} \underline{X}^{k+1} &= D_1^{-1}\underline{Y} - D_1^{-1}(L_1 + U_1)\underline{X}^k - D_1^{-1}S_2\overline{X}^k, \\ \overline{X}^{k+1} &= D_1^{-1}\overline{Y} - D_1^{-1}(L_1 + U_1)\overline{X}^k - D_1^{-1}S_2\underline{X}^k, \quad k = 0,1,\ldots \end{aligned} \tag{7}$$

The elements of $X^{k+1} = \left(\underline{X}^{k+1}, \overline{X}^{k+1}\right)^t$ are

$$\begin{aligned} \underline{x}_i^{k+1}(r) &= \frac{1}{s_{i,i}}\left[\underline{y}_i(r) - \sum_{j=1,j\neq i}^{n} s_{i,j}\underline{x}_j^k(r) - \sum_{j=1}^{n} s_{i,n+j}\overline{x}_j^k(r)\right], \\ \overline{x}_i^{k+1}(r) &= \frac{1}{s_{i,i}}\left[\overline{y_i}(r) - \sum_{j=1,j\neq i}^{n} s_{i,j}\overline{x}_j^k(r) - \sum_{j=1}^{n} s_{i,n+j}\underline{x}_j^k(r)\right], \\ & k = 0,1,2,\ldots, \quad i = 1,2,\ldots,n. \end{aligned}$$

The result in the matrix form of the Jacobi iterative technique is $X^{k+1} = PX^k + C$ where

$$P = \begin{bmatrix} -D_1^{-1}(L_1 + U_1) & -D_1^{-1}S_2 \\ -D_1^{-1}S_2 & -D_1^{-1}(L_1 + U_1) \end{bmatrix}, \quad C = \begin{bmatrix} D_1^{-1}\underline{Y} \\ D_1^{-1}\overline{Y} \end{bmatrix}, \quad X = \begin{bmatrix} \underline{X} \\ \overline{X} \end{bmatrix}.$$

For the Gauss–Seidel method, we have:

$$\begin{bmatrix} D_1 + L_1 & 0 \\ S_2 & D_1 + L_1 \end{bmatrix} \begin{bmatrix} \underline{X} \\ \overline{X} \end{bmatrix} + \begin{bmatrix} U_1 & S_2 \\ 0 & U_1 \end{bmatrix} \begin{bmatrix} \underline{X} \\ \overline{X} \end{bmatrix} = \begin{bmatrix} \underline{Y} \\ \overline{Y} \end{bmatrix} \tag{8}$$

then

$$\begin{aligned} \underline{X} &= (D_1 + L_1)^{-1}\underline{Y} - (D_1 + L_1)^{-1}U_1\underline{X} - (D_1 + L_1)^{-1}S_2\overline{X}, \\ \overline{X} &= (D_1 + L_1)^{-1}\overline{Y} - (D_1 + L_1)^{-1}U_1\overline{X} - (D_1 + L_1)^{-1}S_2\underline{X}. \end{aligned} \tag{9}$$

Thus, the Gauss–Seidel iterative technique becomes

$$\begin{aligned} \underline{X}^{k+1} &= (D_1 + L_1)^{-1}\underline{Y} - (D_1 + L_1)^{-1}U_1^{-1}\underline{X}^k - (D_1 + L_1)^{-1}S_2\overline{X}^k, \\ \overline{X}^{k+1} &= (D_1 + L_1)^{-1}\overline{Y} - (D_1 + L_1)^{-1}U_1^{-1}\overline{X}^k - (D_1 + L_1)^{-1}S_2\underline{X}^k, \quad k = 0, 1, \ldots \end{aligned} \tag{10}$$

So the elements of $X^{k+1} = \left(\underline{X}^{k+1}, \overline{X}^{k+1}\right)^t$ are

$$\begin{aligned} \underline{x}_i^{k+1}(r) &= \frac{1}{s_{i,i}}\left[\underline{y}_i(r) - \sum_{j=1}^{i-1} s_{i,j}\underline{x}_j^{k+1}(r) - \sum_{j=i+1}^{n} s_{i,j}\underline{x}_j^k(r) - \sum_{j=1}^{n} s_{i,n+j}\overline{x}_j^k(r)\right], \\ \overline{x}_i^{k+1}(r) &= \frac{1}{s_{i,i}}\left[\overline{y}_i(r) - \sum_{j=1}^{i-1} s_{i,j}\overline{x}_j^k(r) - \sum_{j=i+1}^{n} s_{i,j}\overline{x}_j^k(r) - \sum_{j=1}^{n} s_{i,n+j}\underline{x}_j^k(r)\right], \\ & k = 0, 1, 2, \ldots, \quad i == 1, 2, \ldots, n. \end{aligned}$$

This results in the matrix form of the Gauss–Seidel iterative technique as

$$X^{k+1} = PX^k + C$$

$$P = \begin{bmatrix} -(D_1 + L_1)^{-1}U_1 & -(D_1 + L_1)^{-1}S_2 \\ -(D_1 + L_1)^{-1}S_2 & -(D_1 + L_1)^{-1}U_1 \end{bmatrix}, \quad C = \begin{bmatrix} (D_1 + L_1)^{-1}\underline{Y} \\ (D_1 + L_1)^{-1}\overline{Y} \end{bmatrix}, \quad X = \begin{bmatrix} \underline{X} \\ \overline{X} \end{bmatrix}.$$

□

From Theorems 4 and 5, both Jacobi and Gauss–Seidel iterative schemes converge to the unique solution $X = A^{-1}Y$, for any X^0, where $X \in R^{2n}$ and $(\underline{X}, \overline{X}) \in E^n$. For a given tolerance $\epsilon > 0$ the decision to stop is

$$\frac{\left\|\overline{X}^{k+1} - \overline{X}^k\right\|}{\left\|\overline{X}^{k+1}\right\|} < \epsilon, \quad \frac{\left\|\underline{X}^{k+1} - \underline{X}^k\right\|}{\left\|\underline{X}^{k+1}\right\|} < \epsilon, \quad k = 0, 1, \ldots$$

3.2. *Successive over-Relaxation (SOR) Iterative Method*

In this section we turn next to a modification of the Gauss–Seidel iteration which known as SOR iterative method. By multiplying system (8) by D^{-1} gives,

$$\begin{bmatrix} I + D_1^{-1}L_1 & 0 \\ S_2 & I + D_1^{-1}L_1 \end{bmatrix} \begin{bmatrix} \underline{X} \\ \overline{X} \end{bmatrix} + \begin{bmatrix} D_1^{-1}U_1 & S_2 \\ 0 & D_1^{-1}U_1 \end{bmatrix} \begin{bmatrix} \underline{X} \\ \overline{X} \end{bmatrix} = \begin{bmatrix} D_1^{-1}\underline{Y} \\ D_1^{-1}\overline{Y} \end{bmatrix} \tag{11}$$

Let $D_1^{-1}U_1 = U_1, D_1^{-1}L_1 = L_1$ then

$$\begin{bmatrix} I+L_1 & 0 \\ S_2 & I+L_1 \end{bmatrix}\begin{bmatrix} \underline{X} \\ \overline{X} \end{bmatrix}+\begin{bmatrix} U_1 & S_2 \\ 0 & U_1 \end{bmatrix}\begin{bmatrix} \underline{X} \\ \overline{X} \end{bmatrix}=\begin{bmatrix} D_1^{-1}\underline{Y} \\ D_1^{-1}\overline{Y} \end{bmatrix} \tag{12}$$

Hence

$$(I+L_1)\underline{X}=D^{-1}\underline{Y}-U_1\underline{X}-S_2\overline{X},(I+L_1)\overline{X}=D^{-1}\overline{Y}-U_1\overline{X}-S_2\underline{X} \tag{13}$$

for some parameter ω :

$$\begin{aligned} (I+\omega L_1)\underline{X} &= \omega D^{-1}\underline{Y}-[(1-\omega)I+\omega U_1]\underline{X}-\omega S_2\overline{X}, \\ (I+\omega L_1)\overline{X} &= \omega D^{-1}\overline{Y}-[(1-\omega)I+\omega U_1]\overline{X}-\omega S_2\underline{X}. \end{aligned} \tag{14}$$

If $\omega=1$, then clearly X is just the Gauss–Seidel solution (13). Then the SOR iterative method takes the form:

$$\begin{aligned} \underline{X}^{k+1} &= (I+\omega L_1)^{-1}\omega D^{-1}\underline{Y}-(I+\omega L_1)^{-1}[(1-\omega)I+\omega U_1]\underline{X}^k-(I+\omega L_1)^{-1}\omega S_2\overline{X}^k, \\ \overline{X}^{k+1} &= (I+\omega L_1)^{-1}\omega D^{-1}\overline{Y}-(I+\omega L_1)^{-1}[(1-\omega)I+\omega U_1]\overline{X}^k-(I+\omega L_1)^{-1}\omega S_2\underline{X}^k. \end{aligned} \tag{15}$$

Consequently, this results in the matrix form of the SOR iterative method as $X^{K+1}=PX^K+C$ where

$$P=\begin{bmatrix} -(I+\omega L_1)^{-1}[(1-\omega)I+\omega U_1] & -(I+\omega L_1)^{-1}\omega S_2 \\ -(I+\omega L_1)^{-1}\omega S_2 & -(I+\omega L_1)^{-1}[(1-\omega)I+\omega U_1] \end{bmatrix},$$

$$C=\begin{bmatrix} (I+\omega L_1)^{-1}\omega D^{-1} \\ (I+\omega L_1)^{-1}\omega D^{-1} \end{bmatrix}.$$

For $0<\omega<1$ this method is called the successive under-relaxation method that can be used to achieve convergence for systems that are not convergent by the Gauss–Seidel method.

For $\omega>1$ the method is called the SOR method that can be used to accelerate of convergence of linear systems that are already convergent by the Gauss–Seidel method.

Theorem 6. In Reference [4]: *If S is a positive definite matrix and $0<\omega<2$ then the SOR method converges for any choice of initial approximate vector X^0.*

4. Numerical Example and Results

To demonstrate the efficiency and accuracy of the proposed iterative techniques, we consider the following numerical example with known exact solution.

Example 1. *Consider the 6×6 non-symmetric fuzzy linear system*

$$\begin{aligned} 9x_1+2x_2-x_3+x_4+x_5-2x_6 &= (-53+8r,-25-20r) \\ -x_1+10x_2+2x_3+x_4-x_5-x_6 &= (-13+9r,\ 18-22r) \\ x_1+3x_2+9x_3-x_4+x_5+2x_6 &= (18+17r,73-38r) \\ 2x_1-x_2+x_3+10x_4-2x_5+3x_6 &= (31+16r,61-14r) \\ x_1+x_2-x_3+2x_4+7x_5-x_6 &= (34+8r,58-16r) \\ 3x_1+2x_2+x_3+x_4-x_5+10x_6 &= (51+26r,99-22r) \end{aligned} \tag{16}$$

The extended 12×12 matrix is

$$S = \begin{bmatrix} 9 & 2 & 0 & 1 & 1 & 0 & 0 & 0 & -1 & 0 & 0 & -2 \\ 0 & 10 & 2 & 1 & 0 & 0 & -1 & 0 & 0 & 0 & -1 & -1 \\ 1 & 3 & 9 & 0 & 1 & 2 & 0 & 0 & 0 & -1 & 0 & 0 \\ 2 & 0 & 1 & 10 & 0 & 3 & 0 & -1 & 0 & 0 & -2 & 0 \\ 1 & 1 & 0 & 2 & 7 & 0 & 0 & 0 & -1 & 0 & 0 & -1 \\ 3 & 2 & 1 & 1 & 0 & 10 & 0 & 0 & 0 & 0 & -1 & 0 \\ 0 & 0 & -1 & 0 & 0 & -2 & 9 & 2 & 0 & 1 & 1 & 0 \\ -1 & 0 & 0 & 0 & -1 & -1 & 0 & 10 & 2 & 1 & 0 & 0 \\ 0 & 0 & 0 & -1 & 0 & 0 & 1 & 3 & 9 & 0 & 1 & 2 \\ 0 & -1 & 0 & 0 & -2 & 0 & 2 & 0 & 1 & 10 & 0 & 3 \\ 0 & 0 & -1 & 0 & 0 & -1 & 1 & 1 & 0 & 2 & 7 & 0 \\ 0 & 0 & 0 & 0 & -1 & 0 & 3 & 2 & 1 & 1 & 0 & 10 \end{bmatrix}$$

$$X = S^{-1}Y = \begin{bmatrix} 0.1136 & -0.0220 & 0.0041 & -0.0050 & -0.0148 & -0.0020 & -0.0088 & -0.0046 & 0.0100 & 0.0011 & -0.0050 & 0.0167 \\ 0.0007 & 0.1034 & -0.0206 & -0.0117 & 0.0020 & 0.0096 & 0.0073 & -0.0065 & 0.0010 & -0.0057 & 0.0116 & 0.0122 \\ -0.0041 & -0.0267 & 0.1184 & 0.0080 & -0.0130 & -0.0266 & -0.0023 & 0.0044 & -0.0035 & 0.0133 & -0.0045 & -0.0081 \\ -0.0126 & 0.0090 & -0.0075 & 0.1023 & 0.0022 & -0.0258 & -0.0006 & 0.0081 & -0.0023 & -0.0071 & 0.0272 & 0.0012 \\ -0.0130 & -0.0136 & 0.0037 & -0.0253 & 0.1450 & 0.0042 & -0.0049 & -0.0064 & 0.0150 & 0.0028 & -0.0100 & 0.0067 \\ -0.0330 & -0.0130 & -0.0067 & -0.0069 & 0.0041 & 0.1046 & 0.0002 & 0.0001 & -0.0023 & -0.0023 & 0.0114 & -0.0063 \\ -0.0088 & -0.0046 & 0.0100 & 0.0011 & -0.0050 & 0.0167 & 0.1136 & -0.0220 & 0.0041 & -0.0050 & -0.0148 & -0.0020 \\ 0.0073 & -0.0065 & 0.0010 & -0.0057 & 0.0116 & 0.0122 & 0.0007 & 0.1034 & -0.0206 & -0.0117 & 0.0020 & 0.0096 \\ 0.0023 & 0.0044 & -0.0035 & 0.0133 & -0.0045 & -0.0081 & -0.0041 & -0.0267 & 0.1184 & 0.0080 & -0.0130 & -0.0266 \\ -0.0006 & 0.0081 & -0.0023 & -0.0071 & 0.0272 & 0.0012 & -0.0126 & 0.0090 & -0.0075 & 0.1023 & 0.0022 & -0.0258 \\ -0.0049 & -0.0064 & 0.0150 & 0.0028 & -0.0100 & 0.0067 & -0.0130 & -0.0136 & 0.0037 & -0.0253 & 0.1450 & 0.0042 \\ 0.0002 & 0.0001 & -0.0023 & -0.0023 & 0.0114 & -0.0063 & -0.0330 & -0.0130 & -0.0067 & -0.0069 & 0.0041 & 0.1046 \end{bmatrix} \begin{bmatrix} -53+8r \\ -13+9r \\ 18+17r \\ 31+16r \\ 34+8r \\ 51+26r \\ -25-20r \\ 18-22r \\ 73-38r \\ 61-14r \\ 58-16r \\ 99-22r \end{bmatrix}$$

The exact solution is

$$x_1 = \left(\underline{x_1}(r), \overline{x_1}(r)\right) = (-4.12 + 0.12r,\ -2.88 - 1.12r),$$

$$x_2 = \left(\underline{x_2}(r), \overline{x_2}(r)\right) = (-0.25 + 0.25r,\ 1.25 - 1.25r),$$

$$x_3 = \left(\underline{x_3}(r), \overline{x_3}(r)\right) = (0.78 + 1.22r,\ 5.22 - 3.22r),$$

$$x_4 = \left(\underline{x_4}(r), \overline{x_4}(r)\right) = (3.6 + 0.4r,\ 4.4 - 0.4r),$$

$$x_5 = \left(\underline{x_5}(r), \overline{x_5}(r)\right) = (6.66 + 0.34r,\ 8.34 - 1.34r),$$

$$x_6 = \left(\underline{x_6}(r), \overline{x_6}(r)\right) = (6.78 + 2.22r,\ 10.22 - 1.22r).$$

The exact and approximate solution using the Jacobi, Gauss–Seidel and the SOR iterative schemes are shown in Figures 1–3 respectively. The Hausdoeff distance of solutions with $\varepsilon = 10^{-3}$ in the Jacobi method is 0.4091×10^{-3} in the Gauss–Seidel method is 0.4335×10^{-4} and in the SOR method with $\omega = 1.3$ is 5.5611×10^{-4}.

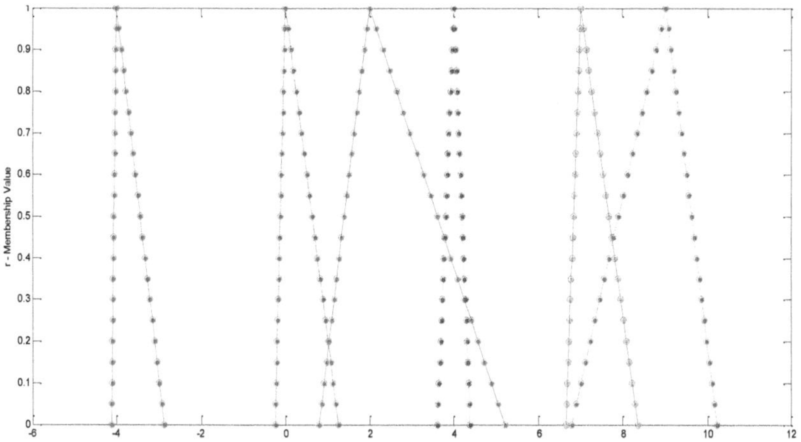

Figure 1. The Hausdorff distance of solutions with $\varepsilon = 10^{-3}$, in the Jacobi method is 0.4091×10^{-3}.

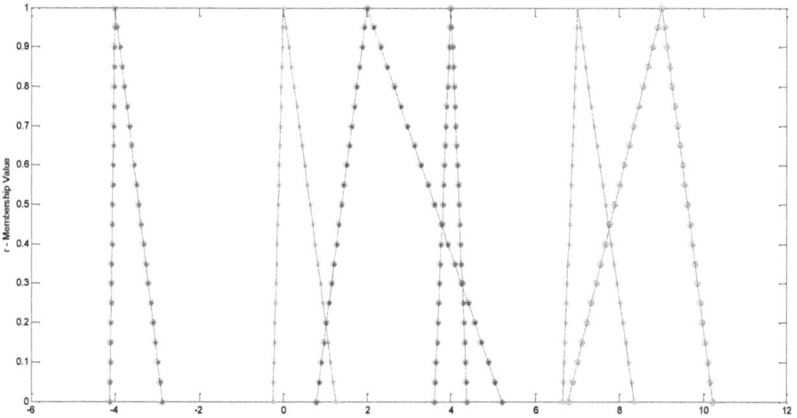

Figure 2. The Hausdorff distance of solutions with $\varepsilon = 10^{-3}$ in the Gauss–Seidel method is 0.4335×10^{-4}.

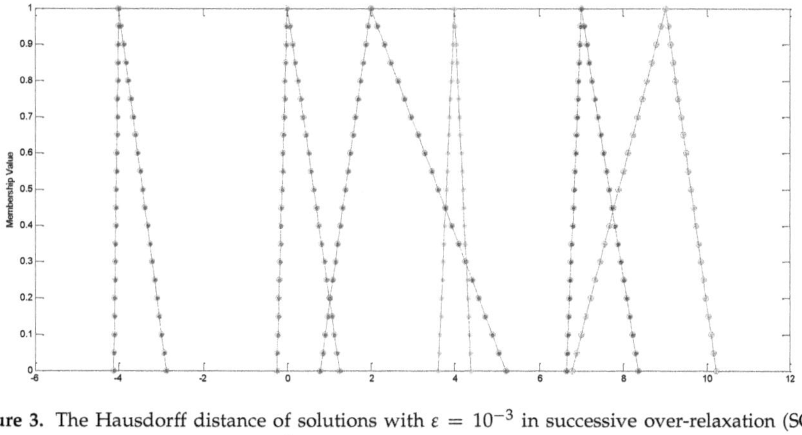

Figure 3. The Hausdorff distance of solutions with $\varepsilon = 10^{-3}$ in successive over-relaxation (SOR) method with $\omega = 1.3$ is 5.5611×10^{-4}.

5. Conclusions

In this article the Jacobi, Gauss–Seidel and SOR iterative methods have been used to solve the FSLEs where the coefficient matrix arrays are crisp numbers, the right-hand side column is an arbitrary fuzzy vector and the unknowns are fuzzy numbers. The numerical results have shown to be in a close agreement with the analytical ones. Moreover, Figures 1–3 containing the Hausdorff distance of solutions show clearly that the SOR iterative method is more efficient in comparison with other iterative techniques.

Author Contributions: Lubna Inearat and Naji Qatanani conceived and designed the experiments, both performed the experiments. Lubna Inearat and Naji Qatanani analyzed the data, both contributed reagents/materials/analysis tools and wrote the paper.

Conflicts of Interest: The authors declare no conflicts of interest.

References

1. Zadeh, L.A. Fuzzy sets. *Inf. Control* **1965**, *8*, 338–353. [CrossRef]
2. Friedman, M.; Ming, M.; Kandel, A. Fuzzy linear systems. *Fuzzy Sets Syst.* **1998**, *96*, 201–209. [CrossRef]
3. Allahviraloo, T. Numerical methods for fuzzy system of linear equations. *Appl. Math. Comput.* **2004**, *155*, 493–502.
4. Allahviraloo, T. Successive over relaxation iterative method for fuzzy system of linear equations. *Appl. Math. Comput.* **2005**, *62*, 189–196.
5. Dehghan, M.; Hashemi, B. Iterative solution of fuzzy linear systems. *Appl. Math. Comput.* **2006**, *175*, 645–674. [CrossRef]
6. Ezzati, R. Solving fuzzy linear systems. *Soft Comput.* **2011**, *15*, 193–197. [CrossRef]
7. Muzzioli, S.; Reynaerts, H. Fuzzy linear systems of the form $A_1x + b_1 = A_2x + b_2$. *Fuzzy Sets Syst.* **2006**, *157*, 939–951.
8. Abbasbandy, S.; Jafarian, A. Steepest descent method for system of fuzzy linear equations. *Appl. Math. Comput.* **2006**, *175*, 823–833. [CrossRef]
9. Ineirat, L. Numerical Methods for Solving Fuzzy System of Linear Equations. Master's Thesis, An-Najah National University, Nablus, Palestine, 2017.
10. Asady, B.; Abbasbandy, S.; Alavi, M. Fuzzy general linear systems. *Appl. Math. Comput.* **2005**, *169*, 34–40. [CrossRef]
11. Senthilkumar, P.; Rajendran, G. An algorithmic approach to solve fuzzy linear systems. *J. Inf. Comput. Sci.* **2011**, *8*, 503–510.
12. Bede, B. Product type operations between fuzzy numbers and their applications in geology. *Acta Polytech. Hung.* **2006**, *3*, 123–139.
13. Abbasbandy, S.; Alavi, M. A method for solving fuzzy linear systems. *Iran. J. Fuzzy Syst.* **2005**, *2*, 37–43.
14. Amrahov, S.; Askerzade, I. Strong solutions of the fuzzy linear systems. *CMES-Comput. Model. Eng. Sci.* **2011**, *76*, 207–216.

MDPI
St. Alban-Anlage 66
4052 Basel
Switzerland
Tel. +41 61 683 77 34
Fax +41 61 302 89 18
www.mdpi.com

Mathematics Editorial Office
E-mail: mathematics@mdpi.com
www.mdpi.com/journal/mathematics

www.ingramcontent.com/pod-product-compliance
Lightning Source LLC
Chambersburg PA
CBHW051720210326
41597CB00032B/5552
* 9 7 8 3 0 3 8 9 7 3 2 2 5 *